Bioremediation and Phytoremediation of Chlorinated and Recalcitrant Compounds

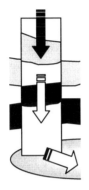

EDITORS
Godage B. Wickramanayake, Arun R. Gavaskar,
Bruce C. Alleman, and Victor S. Magar
Battelle

The Second International Conference
on Remediation of Chlorinated and
Recalcitrant Compounds

Monterey, California, May 22–25, 2000

BATTELLE PRESS
Columbus • Richland

Library of Congress Cataloging-in-Publication Data

International Conference on Remediation of Chlorinated and Recalcitrant Compounds (2nd : 2000 : Monterey, Calif.)
    Bioremediation and phytoremediation of chlorinated and recalcitrant compounds (C2-4) / edited by Godage B. Wickramanayake ... [et al.].
      p. cm.
    "The Second International Conference on Remediation of Chlorinated and Recalcitrant Compounds, Monterey, California, May 22–25, 2000."
    Includes bibliographical references and index.
    ISBN 1-57477-098-5 (alk. paper)
      1. Organochlorine compounds--Biodegradation--Congresses. 2. Hazardous waste site remediation--Congresses. 3. Bioremediation--Congresses. 4. Phytoremediation--Congresses. I. Wickramanayake, Godage B., 1953– II. Title.

TD1066.O73 I58 2000
628.5'2--dc21

                                                                                                                        00-034242

Printed in the United States of America

Copyright © 2000 Battelle Memorial Institute. All rights reserved. This document, or parts thereof, may not be reproduced in any form without the written permission of Battelle Memorial Institute.

Battelle Press
505 King Avenue
Columbus, Ohio 43201-2693, USA
614-424-6393 or 1-800-451-3543
Fax: 1-614-424-3819
E-mail: press@battelle.org
Website: www.battelle.org/bookstore

For information on future environmental remediation meetings, contact:
Remediation Conferences
Battelle
505 King Avenue
Columbus, Ohio 43201-2693, USA
Fax: 614-424-3667
Website: www.battelle.org/conferences

CONTENTS

Foreword   ix

Potential Electron Donors

Vegoil: A Novel Approach for Stimulating Reductive Dechlorination.
K.J. Boulicault, R.E. Hinchee, T.H. Wiedemeier, S.W. Hoxworth, T.P. Swingle, E. Carver, and P.E. Haas   1

Economic Considerations in Enhanced Anaerobic Biodegradation.
M.R. Harkness   9

A Field Demonstration Showing Enhanced Reductive Dechlorination Using Polymer Injection. *T.T. Schuhmacher, W.A. Bow, and J.P. Chitwood*   15

Full-Scale Anaerobic Bioremediation Using Acetate and Lactate Electron Donors. *S.A. Fam, M. Findlay, S. Fogel, T. Pirelli, and T. Sullivan*   23

Results from Two Direct Hydrogen Delivery Field Tests for Enhanced Dechlorination. *C.J. Newell, P.E. Haas, J.B. Hughes, and T. Khan*   31

Time-Release Electron Donor Technology for Accelerated Biological Reductive Dechlorination. *S.S. Koenigsberg, W.A. Farone, and C.A. Sandefur*   39

Insoluble Substrates for Reductive Dehalogenation in Permeable Reactive Barriers. *M.J. Zenker, R.C. Borden, M.A. Barlaz, M.T. Lieberman, and M.D. Lee*   47

PCE Dechlorination with Complex Electron Donors.
T.D. DiStefano and R. Baral   55

Anaerobic Oxidation of (Chlorinated) Hydrocarbons.
J. Dijk, J.M. de Bont, X. Lu, P.M. Becker, T.N.P. Bosma, H.H.M. Rijnaarts, and J. Gerritse   63

In Situ Biowall Containing Organic Mulch Promotes Chlorinated Solvent Bioremediation. *P.E. Haas, P. Cork, C.E. Aziz, and M. Hampton*   71

Laboratory Studies Using Edible Oils to Support Reductive
Dechlorination. *M.D. Lee, R.J. Buchanan, and D.E. Ellis*   77

Bioremediation Microcosm and Laboratory Studies

Behavior of Nutrients in the Biodegradation of Azo Dye.
H. Ozaki and K. Sharma   85

Anaerobic Biostimulation for the In Situ Treatment of Carbon
Tetrachloride. *M.F. DeFlaun and R.J. Steffan*   93

Site Classification for Bioremediation of Chlorinated Compounds
Using Microcosm Studies. *W.A. Farone, S.S. Koenigsberg, T. Palmer,
and D. Brooker*   101

Microbial Degradation of 1,2-Dichloropropane and
2,2-Dichlorodiisopropyl Ether in Continuous Reactors.
R. Hauck and W. Hegemann   107

Leachate Permeation in Soils: Its Implications for Clogging.
N. Singhal and J. Islam   115

Determination of Fluoranthene Biodegradation Metabolites and
Their Toxicity. *E. Sepic, M. Bricelj, and H. Leskovsek*   123

Transformation of Chlorobenzene in a Laboratory Aquifer Column.
C. Vogt, H. Lorbeer, L. Wuensche, and W. Babel   133

Bioremediation of MTBE

Biodegradation of MTBE by Cometabolism in Laboratory-Scale
Fermentations. *P. Piveteau, F. Fayolle, Y. Le Penru, and F. Monot*   141

Cometabolic Degradation of MTBE by Iso-Alkane-Utilizing Bacteria
from Gasoline-Impacted Soils. *M. Hyman, C. Taylor, and K. O'Reilly*   149

In Situ Application of Propane Sparging for MTBE Bioremediation.
*R.J. Steffan, C. Condee, J. Quinnan, M. Walsh, S.H. Abrams,
and J. Flanders*   157

Biotreatment of MTBE with a New Bacterial Isolate. *R.J. Steffan,
S. Vainberg, C. Condee, K. McClay, and P. Hatzinger*   165

Factors Influencing Biological Treatment of MTBE in Fixed-Film Reactors. *W.T. Stringfellow, R.D. Hines, D.K. Cockrum, and S.T. Kilkenny* 175

Cometabolic Degradation of MTBE by a Cyclohexane-Oxidising Bacteria. *D. Corcho, R.J. Watkinson, and D.N. Lerner* 183

Mineralization of MTBE With Various Primary Substrates. *G.J. Wilson, A.P. Richter, M.T. Suidan, and A.D. Venosa* 191

Field Applications of Enhanced Reductive Dechlorination of Chlorinated Solvents

Dechlorination of Tetrachloroethylene by a Membrane-Associated Dehalogenase from *Clostridium bifermentans* DPH-1. *B.C. Okeke, Y.C. Chang, M. Hatsu, and K. Takamizawa* 197

Enhanced Biological Reductive Dechlorination at a Dry-Cleaning Facility. *M. Lodato, D. Graves, and J. Kean* 205

Remedial Action Using HRC™ Under a State Dry Cleaning Program. *D. Anderson, M. Ochsner, C. Sandefur, and S. Koenigsberg* 213

Enhanced Reductive Dechlorination of Ethenes — Large-Scale Pilot Testing. *J.A. Peeples, J.M. Warburton, I.A. Al-Fayyomi, and J. Haff* 221

Enhanced Anaerobic In Situ Bioremediation of Chloroethenes at NAS Point Mugu. *D.P. Leigh, C.D. Johnson, R.S. Skeen, M.G. Butcher, L.A. Bienkowski, and S. Granade* 229

Enhanced Bioremediation Using Hydrogen Release Compound (HRC™) in Clay Soils. *Z.M. Zahiraleslamzadeh and J.C. Bensch* 237

A Reductive Dechlorination Treatability Study of a Shallow Alluvial Aquifer. *V. Murt, T. Huscher, and D. Easley* 245

Enhanced Closure of a TCE Site Using Injectable HRC™. *S.L. Boyle, V.B. Dick, M.N. Ramsdell, and T.M. Caffoe* 255

Enhanced Reductive Dechlorination: Lessons Learned at over Twenty Sites. *M.A. Hansen, J. Burdick, F.C. Lenzo, and S. Suthersan* 263

Comparison of Natural and Enhanced Attenuation Rates Through Substrate Amendments. *J.F. Horst, K.A. Beil, J.S. Burdick, and S.S. Suthersan* ... 271

Optimization Strategy for Enhancing Biodegradation in an Upland-Wetland Plume. *W.A. Jackson, J.H. Pardue, G. Nemeth, T. DeReamer, D. McInnis, and D. Green* ... 279

HRC™-Enhanced Bioremediation of Chlorinated Solvents. *W. Murray, M. Dooley and S. Koenigsberg* ... 287

HRC™-Enhanced Reductive Dechlorination of Source Trichloroethene in an Unconfined Aquifer. *W.D. Harms, K.A. Taylor, and B.S. Taylor* ... 295

Evaluation of In Situ Biological Dehalogenation Activity Using a Reactive Tracer. *S. Vancheeswaran, G. Hickman, R. Pratt, S. McKinley, M. Germon, J. Gross, L. Semprini, and J. Istok* ... 303

Engineered PCE Dechlorination Incorporating Competitive Biokinetics: Optimization and Transport Modeling. *M. Willis and C. Shoemaker* ... 311

Aerobic/Anaerobic Mechanisms

Performance of Field-Scale Sequential Anaerobic/Aerobic In Situ Bioremediation Demonstration. *A. Turpie, C. Lizotte, M.F. DeFlaun, J. Quinnan, and M. Marley* ... 319

Anaerobic/Aerobic Treatment of PCE Using a Single Microbial Consortia. *S. Hoxworth, A.A. Randall, and T.M. McCue* ... 327

Enhanced Biodegradation of Organic Compounds in Groundwater via Nutrient Injection. *E.J. Raes, M.A. Baviello, J. Cook, and S. Radel* ... 337

Treatment of Organic-Contaminated Water in Microbial Mat Bioreactors. *W.L. O'Niell and V.A. Nzengung* ... 347

TCE Degradation in Anaerobic/Aerobic Circulating Column. *A. Narjoux, Y. Comeau, J.-C. Frigon, and S.R. Guiot* ... 353

Enhanced In Situ Degradation of Residual Chemicals in Groundwater Near a Closed Waste Disposal Cell. *S.D. Warner, J.H. Honniball, T.A. Delfino, and C.E.B. Goering* ... 361

Contents vii

Aerobic and Anaerobic Bioremediation of cis-1,2-Dichloroethene and Vinyl Chloride. *T.S. Cornuet, C. Sandefur, W.M. Eliason, S.E. Johnson, and C. Serna*   373

Bioaugmentation and Biomonitoring

Design of In Situ Microbial Filter for the Remediation of Naphthalene. *M. Warith and L. Fernandes*   381

Bioaugmentation Potential at a Carbon Tetrachloride-Contaminated Site. *S.M. Pfiffner, T.J. Phelps, and A.V. Palumbo*   389

The Effect of Soil Heterogeneity on the Vadose Zone Transport of Bacteria for Bioaugmentation. *B.L. Kinsall, G.V. Wilson, and A.V. Palumbo*   395

Anaerobic Bioremediation of Chlorinated VOCs in a Fractured Bedrock Aquifer — Preparations for an In Situ Pilot Study. *R.J. Fiacco, M.H. Daly, J.D. Fitzgerald, G. Demers, M.D. Lee, and D. Wanty*   405

Fungal Treatment for Wastewater Containing Recalcitrant Compounds. *S. Srinivasan and D.V.S. Murthy*   413

Cometabolic Processes

Cometabolism of Poorly Biodegradable Ethers in Engineered Bioreactors. *M.J. Zenker, R.C. Borden, and M.A. Barlaz*   421

Predicting Enhanced Bioremediation Performance at the SRS Sanitary Landfill. *N.D. Durant, P.A. Weeber, P. Andersen, D.G. Jackson, and B.J. Travis*   429

Enhanced Bioremediation of Solvents, Acetone, and Isopropanol in Bedrock Groundwater. *G.L. Carter, T. Dalton, J.C. Vincent, B.B. Lemos, and R. Kryczkowski*   437

A Designing Method for In Situ Groundwater Bioremediation through Precise Evaluation of Transport Characteristics. *M. Nakamura, T. Kawai, and J. Kawabata*   445

Cometabolid Degradation of Chlorinated Solvent Contaminants by Actinomycetes. *S.-B. Lee. S.E. Strand, and H.D. Stensel*   455

Phytoremediation

Phytoremediation of Organic Solvents in Groundwater: Pilot Study at a Superfund Site. *A. Ferro, B. Chard, M. Gefell, B. Thompson, and R. Kjelgren* — 461

Monitoring Site Constraints at NUWC Keyport's Hybrid Poplar Phytoremediation Plantation. *W.L. Rohrer, L. Newman, M. Sharp, P. Heilman, and B.R. Wallis* — 467

The Influence of an Integrated Remedial System on Groundwater Hydrology. *W.H. Schneider, J.G. Wrobel, S.R. Hirsh, H.R. Compton, and D. Haroski* — 477

Uptake of Arsenic by Tamarisk and Eucalyptus Under Saline Conditions. *R.W. Tossell, K. Binard, and M.T. Rafferty* — 485

Phytoremediation of a Creosote-Contaminated Site — A Field Study. *J.T. Novak, M.A. Widdowson, M. Elliott and S. Robinson* — 493

In Situ Bioremediation of #2 Fuel Oil Utilizing Phytoremediation. *E.P. Carman, T.L. Crossman, and K.L. Daleness* — 501

Author Index — 509

Keyword Index — 531

FOREWORD

Biologically based approaches to site remediation and closure seek to take advantage of natural processes for reducing contamination. At first, such approaches were largely limited to sites contaminated with hydrocarbons, which are relatively vulnerable to biological treatment. However, progress continues to be made in applying a variety of biological approaches to sites contaminated with more resistant contaminants, such as MTBE, chlorinated solvents, and metals. *Bioremediation and Phytoremediation of Chlorinated and Recalcitrant Compounds* provides a comprehensive overview of the latest laboratory studies and field applications in chapters that cover potential electron donors, bioremediation microcosm and laboratory studies, bioremediation of MTBE, field applications of enhanced reductive dechlorination of chlorinated solvents, aerobic/anaerobic mechanisms, bioaugmentation and biomonitoring, cometabolic processes, and phytoremediation applications.

This is one of seven volumes resulting from the Second International Conference on Remediation of Chlorinated and Recalcitrant Compounds (May 22–25, 2000, Monterey, California). Like the first meeting in the series, which was held in May 1998, the 2000 conference focused on the more problematic contaminants—chlorinated solvents, pesticides/herbicides, PCBs/dioxins, MTBE, DNAPLs, and explosives residues—in all environmental media and on physical, chemical, biological, thermal, and combined technologies for dealing with these compounds. The conference was attended by approximately 1,450 environmental professionals involved in the application of environmental assessment and remediation technologies at private- and public-sector sites around the world.

A short paper was invited for each presentation accepted for the program. Each paper submitted was reviewed by a volume editor for general technical content. Because of the need to complete publication shortly after the Conference, no in-depth peer review, copy-editing, or detailed typesetting was performed for the majority of the papers in these volumes. Papers for 60% of the presentations given at the conference appear in the proceedings. Each section in this and the other six volumes corresponds to a technical session at the Conference. Most papers are printed as submitted by the authors, with resulting minor variations in word usage, spelling, abbreviations, the manner in which numbers and measurements are presented, and formatting.

We would like to thank the people responsible for the planning and conduct of the Conference and the production of the proceedings. Valuable input to our task of defining the scope of the technical program and delineating sessions was provided by a steering committee made up of several Battelle scientists and engineers – Bruce Alleman, Abraham Chen, James Gibbs, Neeraj Gupta, Mark Kelley, and Victor Magar. The committee members, along with technical reviewers from Battelle and many other organizations, reviewed more than 600 abstracts submitted for the Conference and determined the content of the individual sessions. Karl Nehring provided valuable advice on the development of the program schedule and the organization of the proceedings volumes. Carol Young, with assistance from Gina Melaragno, maintained program data, corresponded with speakers and authors, and compiled the final program and abstract books. Carol and Lori Helsel were responsible for the proceedings production effort, receiving assistance on specific aspects from Loretta Bahn, Tom Wilk, and Mark Hendershot. Lori, in particular, spent many hours examining papers for format and contacting authors as necessary to obtain revisions. Joe Sheldrick, the manager of Battelle Press, provided valuable production-planning advice; he and Gar Dingess designed the volume covers.

Battelle organizes and sponsors the Conference on Remediation of Chlorinated and Recalcitrant Compounds. Several organizations made financial contributions toward the 2000 Conference. The co-sponsors were EnviroMetal Technologies Inc. (ETI); Geomatrix Consultants, Inc.; the Naval Facilities Engineering Command (NAVFAC); Parsons Engineering Science, Inc.; and Regenesis.

As stated above, each article submitted for the proceedings was reviewed by a volume editor for basic technical content. As necessary, authors were asked to provide clarification and additional information. However, it would have been impossible to subject more than 300 papers to a rigorous peer review to verify the accuracy of all data and conclusions. Therefore, neither Battelle nor the Conference co-sponsors or supporting organizations can endorse the content of the materials published in these volumes, and their support for the Conference should not be construed as such an endorsement.

Godage B. Wickramanayake and Arun R. Gavaskar
Conference Chairs

VEGOIL: A NOVEL APPROACH FOR STIMULATING REDUCTIVE DECHLORINATION

Kent J. Boulicault, Robert E. Hinchee, Todd H. Wiedemeier,
Scott W. Hoxworth, and Todd P. Swingle
(Parsons Engineering Science, Inc.)
Ed Carver (45th Space Wing, Cape Canaveral Air Station, Florida)
Patrick E. Haas (Air Force Center for Environmental Excellence)

Abstract: Reductive dechlorination has been shown to be the dominant mechanism in the natural attenuation of chlorinated ethenes. The presence of an electron donor appears to be the limiting factor in the rate of contaminant biodegradation. A variety of supplemental electron donors have been developed to enhance the reductive dechlorination process with some level of success as well as limitations, namely cost. In an effort to identify a cost-effective solution, a pilot study has been conducted evaluating the use of vegetable oil as an electron donor. The study included the installation of six monitoring wells: one injection well, four observation wells downgradient of the injection well and one upgradient well for use as a background location. The wells were each installed in an area where the concentrations of TCE ranged from 88 to 190 mg/L. Soybean oil was injected into the contaminated aquifer with free oil recovered immediately following the injection. Field and laboratory analytical results for six months of post-injection monitoring show a significant decrease in TCE concentrations at the injection point and evidence of enhanced reductive dechlorination at the downgradient monitoring points. An expanded version of the pilot study is planned.

INTRODUCTION

The widespread use of chlorinated solvents, primarily tetrachloroethene (PCE), trichloroethene (TCE), 1,1,1-trichloroethane (TCA), and carbon tetrachloride (CT) have resulted in contaminated soil and groundwater worldwide. Reductive dechlorination is the most important biological mechanism for remediation of these compounds. The addition of supplemental electron donors, such as the injection of lactate, has been shown to stimulate this process. However, experience has shown that an obstacle to successful bioremediation is often the cost of nutrient/carbon addition as well as permitting and regulatory obstacles. For example, with lactate addition, the carbon source may need to be continuously injected, thereby reducing the cost effectiveness of the method. The United States Air Force Center for Environmental Excellence, Technology Transfer Division, in conjunction with Parsons Engineering Science, Inc., the 45th Space Wing Installation Restoration Program at Cape Canaveral Air Station, FL, and several Department of Defense facilities is evaluating the addition of vegetable oil as an electron donor to stimulate reductive dechlorination. The advantages of vegetable oil are that it is an inexpensive, innocuous food-grade carbon source that is not regulated as a contaminant by the EPA. Perhaps most importantly though, vegetable oil is a nonaqueous-phase liquid so it will slowly dissolve into groundwater. Thus, the potential exists that a single, low-cost, injection could provide sufficient carbon to drive reductive dechlorination for many years. In addition, the oil can be injected using a variety of techniques such as: directly injected using conventional wells or Geoprobe® points; application in a trench or by direct trenching; or by placement in an excavation. The objective of this work is to evaluate the effectiveness of vegetable oil as an electron donor and its ability to stimulate the reductive dechlorination process.

Site Background. Hangar K, which is located within the industrial area of Cape Canaveral Air Station (CCAS), Florida, was formerly operated as a missile assembly building. The chlorinated solvent, TCE, was among the chemicals known to have been used and stored at the site. The use and storage of this chemical has resulted in a 160 acre chlorinated ethene groundwater contaminant plume comprised primarily of TCE, dichloroethene (DCE), and vinyl chloride (VC). The site is underlain by an unconfined aquifer. The geology within the source area consists of very fine to course sands with silt content increasing with depth to approximately 33 feet (10 meters [m]) below ground surface (bgs), where a 1 foot (0.31 m) clay unit was identified. Baseline contaminant and general water chemistry information for this area is presented in Table 1.

TABLE 1. Baseline Concentrations and Groundwater Chemistry

Parameter	Concentration	Parameter	Concentration
TCE	88 - 190 mg/L	Total Alkalinity	128 - 270 mg/L
Cis-1,2-DCE	48 - 78 mg/L	Total Organic Carbon	28 - 55 mg/L
Trans-1,2-DCE	0.7 – 1.5 mg/L	Chloride	71 - 92 mg/L
VC	0.3 – 1 mg/L	Nitrate-N	<0.05 - 0.1 mg/L
Total CVOCs	149 - 247 mg/L	Sulfate	19 - 39 mg/L
TPH	<0.1-1.4 mg/L	Methane	0.09 - 0.96 mg/L
pH	7.03 - 7.84	Ethane	<0.009 - 0.40 mg/L
Conductivity	721 - 765 µS/cm	Ethene	0.02 - 0.62 mg/L
Dissolved O_2	0.19 – 0.24 mg/L		

Pilot Study Setup. The pilot study included the installation of six wells in the source area: one injection well, four observation wells downgradient of the injection well, and one upgradient well for use as a background location (also in the source area). The pilot study layout is shown in Figure 1. The downgradient observation wells were located at approximately 3 ft (0.9 m), 6 ft (1.8 m), 9 ft (2.7 m), and 12 ft (3.7 m) from the injection well. The background well was approximately 20 ft (6.1 m) upgradient. Each of the wells was installed using rotosonic drilling techniques to a depth of 33 feet (10 m) bgs and constructed of 2-inch (5.1 cm) diameter PVC casing coupled to a 1-foot (0.3 m) section of well screen.

Oil Injection/Extraction. Soybean oil was selected for injection based on its ready availability, chemical stability, high energy content, low viscosity, and low melting point. The oil, produced by Bakers Chef, was purchased from a local food retailer for $2.40/gallon (approximately $0.30/pound). Over a two day period, approximately 110 gallons (416 liters [L]) of the oil was injected into the contaminated aquifer through a single well (VEG-2) using an air driven diaphragm pump at a measured back pressure ranging from 20 to 32 psi. The average oil injection rate was 0.36 gallons per minute (1.36 Lpm). Immediately following the injection each day, free oil was extracted back out of the aquifer. Periodic free oil recovery continued over the next three weeks with a total volume of recovered oil at 63 gallons (238 L).

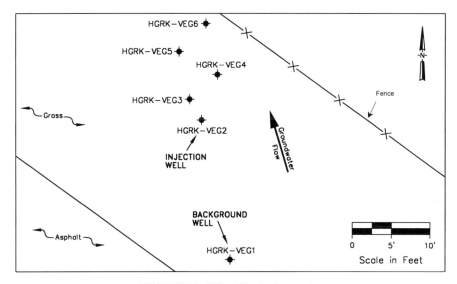

FIGURE 1. Pilot Study Layout

Sampling and Analysis. Groundwater samples were collected from each of the six wells as a baseline prior to oil injection and then again each month for six months following the injection. Field analyses included pH, temperature, conductivity, DO, and ORP using Horiba U-10 and U-22 water analyzers. Ferrous iron (Fe^{+2}) and nitrite-N (NO_2^-) were also conducted in the field using HACH test kits. Fixed-base laboratory analyses included VOCs by EPA Method SW8260, total alkalinity by EPA 310.1, total organic carbon by EPA 415.1, chloride by EPA 325.3, nitrate-N (NO_3^-) by EPA 353.3, sulfate by EPA 375.4, and methane, ethane, and ethene by RSK 175 modified. With the exception of samples for VOC analysis, low-flow sampling techniques using a peristaltic pump and disposable Teflon® tubing were employed for the collection of all samples. Samples for VOC analyses were collected using disposable Teflon® bailers.

RESULTS AND DISCUSSION

Field Monitoring. No significant trends in groundwater geochemistry were observed through the field analyses conducted. Table 2 presents the ranges of values observed over the monitoring period at the observation wells. As shown by Table 2, groundwater geochemistry at the site is suitable for reductive dechlorination. For example, with the exception of one measurement during the second monthly sampling event, all DO concentrations were below 0.5 mg/L and the majority of the measurements were at 0.0 mg/L. ORP at all locations slowly increased from an average of approximately –260 to –180 over the six month period. The increase in ORP values was not sufficient to indicate an impact to the anaerobic environment. Vegetable oil was consistently observed at the injection well and not observed in samples from the other wells.

TABLE 2. Field Analyses – Concentrations Observed Over Six-Months

Parameter	Concentration	Parameter	Concentration
pH	5.88 - 7.84	ORP	-278 to -147 mV
Temperature	25 – 28 °C	Nitrite-N	<0.05 - 0.1 mg/L
Conductivity	721-1,440 µS/cm	Ferrous Iron	2.1 - >5.1 mg/L
Dissolved O_2	0.19 – 0.24 mg/L		

Laboratory Analyses. Figure 2 presents the molar fractions of the chlorinated ethenes in samples from the injection well during the study period. Following oil injection, the relative concentration of cis-1,2-DCE to TCE began to rise indicating the occurrence of reductive dechlorination. The molar fraction of cis-1,2-DCE increased at the injection well from less than 40% before injection to over 70% within 2 months. Molar fraction here refers to the molar concentration of a specific constituent relative to the total concentration of TCE plus DCE plus VC.

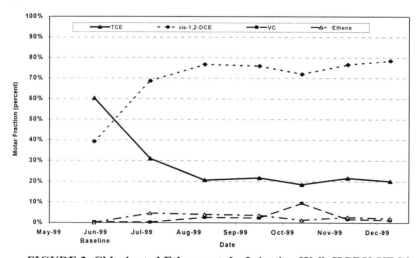

FIGURE 2. Chlorinated Ethenes at the Injection Well, HGRK-VEG2

Potential Electron Donors 5

A significant decrease in TCE concentration was observed at the injection well 30 days after injection of the oil. The TCE concentrations decreased from 100 mg/L to 0.1 mg/L. CVOC analyses of oil samples from the injection well indicate that the CVOCs are partitioning into the oil. This was expected. Over time, TCE and the other CVOCs will partition out of the oil and re-dissolve slowly, at a rate similar to the rate of vegetable oil dissolution. The vegetable oil which codissolves with the CVOCs will serve as a carbon source for degradation of the CVOCs. It is likely that much of the initial decline in CVOC concentrations in the injection well was due to this partitioning rather than biodegradation.

A similar trend in the molar fraction of TCE was observed at the other downgradient wells, although there was more of a delay before the transformations were observed. Over the 6-month period, concentrations of cis-1,2-DCE and VC increased at the downgradient locations. Figure 3 presents the molar fractions of the chlorinated ethenes at observation well, HGRK-VEG3. This trend in the ratio of TCE to cis-1,2-DCE provides good evidence that reductive dechlorination has been enhanced.

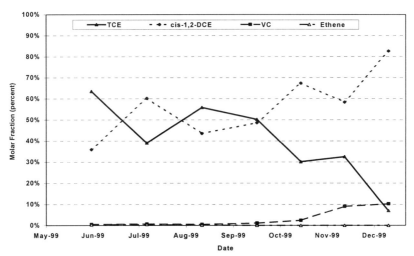

FIGURE 3. Chlorinated Ethenes at Observation Well, HGRK-VEG3

Both total alkalinity and total organic carbon showed significant increases at the injection well and at HGRK-VEG3. Sulfate is present and shows an overall decreasing trend. Definite trends were not observed for the other analytes. It should be noted that this aquifer was anaerobic before oil injection, so little or no dissolved oxygen or nitrate was present at the beginning of the test. Figures 4, 5, and 6 present the results for total alkalinity, total organic carbon, and sulfate, respectively.

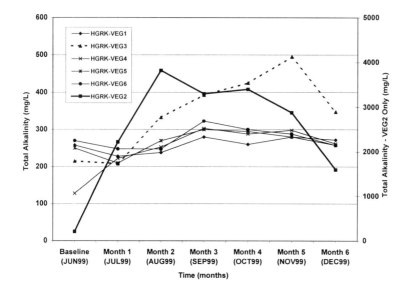

FIGURE 4. Total Alkalinity Concentrations

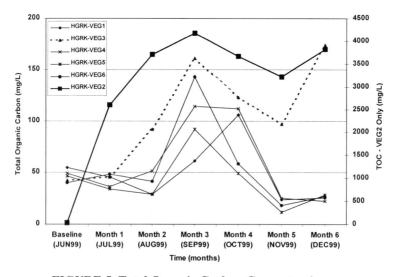

FIGURE 5. Total Organic Carbon Concentrations

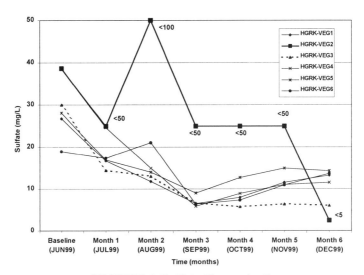

FIGURE 6. Sulfate Concentrations

CONCLUSIONS

During the six months of monitoring after the injection of vegetable oil, it is apparent that the anaerobic dechlorination of CVOCs has been stimulated. The molar concentrations of cis-1,2-DCE have risen relative to TCE, and VC has begun to increase toward the end of the study. The presence of sulfate may indicate competition for the electron donor. However, as time progresses, the sulfate should be reduced and reductive dechlorination further enhanced. It does not appear that the six month period has been adequate to see the anaerobic dechlorination mature and reach steady state, but it is clear that the reaction has been stimulated. Additionally it appears that at least at the injection well, CVOC concentrations have declined substantially. This is due at least in part to partitioning into the oil. Due to the hydrophobic nature of the oil and TCE, this was anticipated and should not present a problem in the application of the technology, but must be considered in evaluating concentration changes. An expanded pilot study is currently in progress and is intended to further evaluate and confirm these results.

ECONOMIC CONSIDERATIONS IN ENHANCED ANAEROBIC BIODEGRADATION

Mark R. Harkness (GE Research & Development, Schenectady, NY)

ABSTRACT: Enhanced anaerobic biodegradation has become a viable commercial remedial process for treating chlorinated solvents in groundwater. Enhanced biodegradation processes can be divided into systems that use the continuous or semi-continuous addition of soluble electron donors to stimulate the reductive dechlorination of chlorinated compounds and those that use slow-release substrates to stimulate such activity. A comparative economic analysis is presented which shows that the major cost for continuous or semi-continuous systems is in the operation, maintenance, and labor costs associated with substrate delivery, whereas the major cost for slow-release systems is the substrate itself. Therefore, designers of cost effective enhanced anaerobic biodegradation remedies should focus on optimizing the delivery system in the first case and on optimizing the cost and efficiency of the electron donor in the second case.

INTRODUCTION

Enhanced anaerobic biodegradation involves the addition of an electron donor and supplemental nutrients to the subsurface in order to stimulate the reductive dechlorination of chlorinated solvents. Soluble electron donors such as lactate, molasses, or methanol typically must be added continuously or semi-continuously in order to sustain activity. Recently, interest has grown in the use of slow-release substrates, which require less frequent donor addition, and therefore can have significant operational advantages over continuous systems. These slow-release donors are usually injected directly into the subsurface using a Geoprobe® or similar means and don't require engineered injection systems. Commercial slow-release electron donor formulations currently exist (Koenigsberg & Farone, 1999). In addition, a variety of natural slow-release substances, including corn oil, cheese whey, manure, and sludge leachate, also support the biodegradation of chlorinated aliphatics (Dybas et al., 1977; Lee et al., 1998).

Development of efficient and economical enhanced anaerobic biodegradation systems will require understanding both the way electron donors are utilized in the subsurface and the key cost drivers inherent in the technology. The analysis presented here compares the distribution of cost elements in continuous and semi-continuous feed systems to those in slow-release systems. It is not meant to be a definitive economic analysis, but is performed to highlight relevant cost issues and associated development needs in the technology.

MATERIALS AND METHODS

The economic analysis presented here is based upon a method developed by DuPont used to compare in situ remediation technologies on a consistent economic basis (Quinton et al., 1997). The methodology uses a template site containing a perchloroethene (PCE) plume and considers variables such as duration of the remediation, design, modeling, and engineering costs, capital equipment costs, operation, maintenance, and labor costs, substrate costs, and monitoring costs in comparing remedial options.

Dupont's original analysis (Quinton et al., 1997) compared a conventional continuous-feed enhanced anaerobic biodegradation system to monitored natural attenuation, a permeable reactive barrier, and pump and treat. The present analysis will compare three configurations of enhanced anaerobic biodegradation systems, with the goal of breaking out the component costs and identifying the significant cost drivers. All economic comparisons will be made on a net present value basis, assuming the total lifetime of the remediation project is 30 years and using a combined discount rate of 8.7% for annual costs.

The template site contains a PCE plume that is 1000 feet long by 400 feet wide by 60 feet deep (300 x 125 x 18 meters) and has an average PCE concentration in the dissolved phase of 1 mg/L. The source of the plume is assumed to be 6000 lbs (2725 kgs) of PCE, present as residual phase in the saturated zone. This amount is equivalent to ten 55 gallon drums of PCE. The depth to groundwater is 20 feet (6 meters) below ground surface (bgs). A clay aquitard underlies the site at 80 feet (25 meters) bgs. The average linear groundwater velocity is 1.0 foot/day (0.3 meters/day) and the soil porosity is 0.25.

In this analysis enhanced anaerobic biodegradation will be used to treat the spill zone, an area defined to be 75 x 100 feet (25 x 30 meters). The approach assumes that PCE in the plume beyond the spill zone will undergo natural attenuation and that the water exiting the site boundary will meet a risk-based cleanup standard.

The monitoring network at the site consists of ten monitoring wells and is the same for all the cases considered. It's assumed that groundwater sampling will be performed by site personnel and occur quarterly for the first year and semi-annually thereafter. Monitoring costs include equipment, analytical, and reporting costs.

Continuous Injection Configuration. The base case analysis considers a recirculating groundwater system operating within the spill (source) zone and is the same system described by Quinton et al. (1997). The system consists of four injection and four extraction wells placed up gradient and down gradient from the spill zone. Sodium benzoate is used as the electron donor and costs $0.77/lb. The sodium benzoate is diluted with water, mixed with low levels of ammonium phosphate, and added continuously or semi-continuously using an automated system consisting of tanks, piping, controllers, and valves. Due to the presence of other biological processes such as methanogenesis that compete for electron donor, it's assumed that 35 lbs (16 kgs) of sodium benzoate are required to sustain degradation of each lb of PCE. This equates to an average injection rate of 115 lbs/day (52 kg/day) for sodium benzoate. A single operator is present at the site for 50/hours each week, performing routine operations and maintenance and sampling, as needed. It's assumed that the PCE will be dechlorinated to ethene and that it will take five years to dissolve and degrade the residual source.

Batch Injection Configuration. The second case analysis considers periodic batch injection of a soluble substrate within the spill zone without groundwater recirculation. In this case, ten injection wells are used to insure that the electron donor is uniformly distributed across the site. It's assumed that the natural groundwater gradient will move the substrate across the spill zone. Sodium benzoate is again used as the electron donor. It's mixed with ammonium phosphate, diluted, and pressure injected into the formation twice a month, such that the total monthly substrate loading is the same as the previous case. It's conservatively assumed that two people need to be present at the site for 40 hours each month to inject the electron donor and carry out periodic operation, maintenance, and sampling activities. As in the previous case, it's assumed that it will take five years to remove the residual source material using this approach.

Slow-release Configuration. The last case considers periodic injection of a slow-release electron donor into the spill zone, once again without groundwater recirculation. It's assumed that the electron donor is injected into the formation at 100 injection points across the spill zone using a Geoprobe® or similar method and that this operation costs $10,000 (excluding electron donor costs) each time it is performed. It's conservatively assumed the electron donor will last six months, so that reinjection will be required twice a year, and that it will take five years to remove the residual source material using this approach. The cost of the slow-release donor is arbitrarily set at $3.00/lb and it's assumed that the total amount of electron donor required is equivalent to the other two systems evaluated. Because there are no injection or extraction wells, it's assumed there is no maintenance associated with this system.

RESULTS AND DISCUSSION

The net present cost breakdown of the various configurations of enhanced anaerobic biodegradation systems appears in Table 1. The costs are divided into capital costs (primarily monitoring wells, injection wells, extractions wells, tankage, and piping), design costs (including modeling and tracer studies), operation, maintenance, and labor costs, monitoring costs, and substrate costs. The total net present cost of each configuration is also shown, along with the percentage of the total contained in each category.

As shown in Table 1, operation, maintenance, and labor costs dominate in both the continuous addition and batch addition systems, making up 57% and 43% of the total costs, respectively. The continuous addition system has a total net present cost of $1300K, compared to $840K for the batch addition system. The difference is primarily in labor costs. Man-hours required on-site are reduced by 68% in the batch system, reflecting reduced need to monitor the delivery process and keep the injection wells open and free from biofouling. The batch system also has lower capital costs, reflecting savings in vessel, piping and control costs. Design, monitoring, and substrate costs are similar for both systems.

In contrast to the continuous and batch systems, substrate costs dominate in the slow-release process, making up 56% of the total net present cost of $900K.

In this case the operation, maintenance, and labor costs are reduced to those associated with the injection of the substrate. Capital costs are also lower because no injection wells are required, whereas design and monitoring costs are similar to the previous cases.

Table 1: Comparison of Cost Breakdown for Enhanced Anaerobic Biodegradation Systems

Cost Component	Net Present Value ($K)	% of Total
Case I : Continuous System		
Capital	$155	12 %
Design	$140	11 %
O&M/Labor	$745	57 %
Monitoring	$130	10 %
Substrate	$130	10 %
Total	$1300	
Case II : Batch System		
Capital	$100	12 %
Design	$120	14 %
O&M/Labor	$360	43 %
Monitoring	$130	15 %
Substrate	$130	15 %
Total	$840	
Case III : Slow-release System		
Capital	$50	6 %
Design	$120	13 %
O&M/Labor	$100	11 %
Monitoring	$130	14 %
Substrate	$500	56 %
Total	$900	

These results suggest that the most effective way to improve the economics of continuous or batch enhanced anaerobic systems is to reduce the operational, maintenance, and labor costs associated with substrate delivery. The continuous addition example shown here probably overstates the labor requirements for a system that would be built today. The need for constant human oversight could be reduced in continuous addition systems with good control systems and remote alarms, and in many cases existing plant personnel could be used to operate the system. Use of batch substrate addition further simplifies the process, particularly if groundwater recirculation is not required. However, the amount of substrate added and interval between batch additions must be determined on a site-specific basis to ensure subsurface electron donor concentrations are sufficient to adequately support dechlorination.

Biofouling around injection wells is a significant issue, particularly in continuous addition systems, and can contribute significantly to operation and maintenance costs when injection well screens must be opened regularly. Batch addition of electron donor appears to help mitigate biofouling, and therefore can help reduce these costs as well.

The results also suggest that efforts to reduce cost in slow-release electron donor systems should focus on reducing the cost and improving the efficiency of the electron donor itself. Table 2 provides some examples of bulk prices for both soluble and commercial and natural slow-release electron donors that have been

Table 2: Example Bulk Prices for Soluble and Slow-release Electron Donors

Electron Donor	$/lb
Soluble Donors	
Methanol	0.05
Ethanol	0.20 - 0.25
Molasses	0.20 - 0.40
Sodium Lactate	2.20
Slow-release Donors	
Edible Oils	0.20 - 0.50
Cellulose	0.40 - 0.80
Chitin	2.25 - 3.00
Methyl Cellulose	4.00 - 5.00
HRC™ (Regenesis commercial material)	12.00

Note : Prices reflect those obtainable for bulk chemicals in 1999. Price per pound is based upon weight of actual electron donor. Where present, weight of water or viscosity reducing agents are factored out of cost calculation.

shown to support the reductive dechlorination of PCE or TCE at GE and elsewhere. The prices shown are based upon weight of actual electron donor. Where present, the weight of water or viscosity reducing agents are factored out of the cost calculation so that fair comparisons can be made. The table indicates that slow-release electron donors are generally more expensive per pound than soluble donors, although there is a significant range in price for slow-release donors. Practitioners are now beginning to explore the utility of some of the less expensive natural materials listed in Table 2.

In addition to price, the life and efficiency of the slow-release are also important. The life of the donor includes the amount of reducing equivalents produced per pound of donor and rate of release of those equivalents in the subsurface. Efficiency relates to the degree to which those equivalents support the reductive dechlorination of chlorinated solvents, as opposed to methanogenesis or other competing biological processes. Increasing the life and efficiency of the donor reduces both the total amount of electron donor required and the frequency with which it needs to be added to the subsurface. These issues are the focus of current research, both in the laboratory and in the field.

It should be noted that the source remediation example given here is only one application of slow-release electron donors. These materials can also used to produce biobarriers to remove solvents in the dissolved phase of the plume. Because the duration of treatment is typically longer and the mass of solvents treated much lower, slow-release systems will have significant advantages over continuous or semi-continuous systems in this application. However, substrate costs will drive the economics here as well.

Finally, the original DuPont economic study calculated the net present cost for monitored natural attenuation for this template site to be $890K. The batch addition and slow-release enhanced biodegradation systems shown in the present analysis have comparable net present costs. This occurs because the source treatment is assumed to reduce the level and duration of monitoring required at the site after source remediation is complete. This suggests that treatment of the source area using enhanced anaerobic biodegradation may be as cost effective as monitored natural attenuation alone in some cases.

ACKNOWLEDGEMENTS

The author would like to thank Ron Buchanan of DuPont for providing detailed information on the DuPont cost model and Michael Lee of Terra Systems and Jeffrey Burdick of ARCADIS Geraghty and Miller for helpful discussions regarding alternative configurations of enhanced anaerobic systems.

REFERENCES

Dybas, M. J., G. M. Tatara, M. E. Witt, and C. S. Criddle. 1997. "Slow-Release Substrates for Transformation of Carbon Tetrachloride by *Pseudomonas* Strain KC." In *In Situ and On-Site Bioremediation: Vol. 3.* Battelle Press, Columbus, OH. p. 59.

Koenigsberg, S.S. and W.A. Farone. 1999. "The Use of Hydrogen Release Compound (HRC™) for CAH Bioremediation." In *Engineered Approaches for In Situ Bioremediation of Chlorinated Solvent Contamination*. Battelle Press, Columbus, OH. pp. 67-72.

Lee, M.D., J.M. Odom, R.J. Buchanan, Jr. 1998. "New Perspectives on Microbial Dehalogenation of Chlorinated Solvents : Insights from the Field." *Annu. Rev. Microbiol.* 52:423-452.

Quinton, G.E, R.J. Buchanan Jr., D.E. Ellis, and S.H. Shoemaker. 1997. "A Method to Compare Groundwater Cleanup Technologies." *Remediation*. Autumn pp. 7-16

A FIELD DEMONSTRATION SHOWING ENHANCED REDUCTIVE DECHLORINATION USING POLYMER INJECTION

Thea T. Schuhmacher (LFR Levine·Fricke, Inc., Costa Mesa, California)
William A. Bow and John P. Chitwood (CADDIS, Inc., St. Charles, Illinois)

ABSTRACT: A large field-scale pilot study was conducted at a manufacturing facility in southern Illinois to evaluate the effectiveness of enhanced natural attenuation for in situ treatment of chlorinated hydrocarbons in groundwater. A biodegradable food-grade polymer, Hydrogen Release Compound (HRC™), was injected into the subsurface to enhance in situ bioremediation of tetrachloroethene (PCE) and its daughter products through processes of reductive dechlorination. Significant biodegradation of PCE and trichloroethene (TCE) occurred in the source area within several weeks following HRC™ injection. Thirty days after the injection, significant reductions in PCE concentrations (from 7 mg/L to 2 mg/L) were observed in observation wells downgradient of the HRC™ barrier with corresponding increases in TCE and dichloroethene (DCE) concentrations. After 172 days, PCE concentrations in the source area were further reduced to 0.5 mg/L. In some areas of the dissolved phase groundwater plume, reductions in PCE and TCE concentrations were much less significant. These areas also contained high sulfate concentrations (from 300 to 500 mg/L), a competing electron acceptor, which may have limited the effectiveness of the HRC™ in those areas. Reinjection in those areas at higher doses of HRC™ resulted in significant reductions of sulfate and PCE within 30 days of the injection.

INTRODUCTION

Historical releases of PCE and TCE from an underground storage tank (UST) at a manufacturing facility in southern Illinois impacted shallow groundwater, creating a plume that extends approximately 800 feet (240 m) in length and 400 feet (120 m) in width (Figure 1). Groundwater sampling since October 1993 indicated that chlorinated hydrocarbons (PCE, TCE, cis-DCE and vinyl chloride (VC)) exceeded Illinois State criteria for Class I groundwater resulting in requirements for groundwater remediation at this site (Illinois Pollution Control Board, 1998). The primary objectives of remediation were to reduce the mass of chlorinated hydrocarbons in groundwater and to prevent off-site migration of the leading edge of the groundwater plume. Evaluation of the groundwater analytical results suggested that anaerobic reductive dechlorination of PCE and TCE was occurring at the site under natural conditions.

During the reductive dechlorination process, the chlorinated hydrocarbon (chlorinated ethene) is used as an electron acceptor, and a chlorine atom is removed and replaced with a hydrogen atom. In general, the transformation of chlorinated ethenes occurs by sequential dechlorination from PCE to TCE to DCE to VC to ethene (Equations 1 through 4; U.S. Environmental Protection Agency, 1998).

$$C_2Cl_4 \text{ (PCE)} + H^+ + 2e^- \rightarrow C_2HCl_3 \text{ (TCE)} + Cl^- \qquad (1)$$
$$C_2HCl_3 \text{ (TCE)} + H^+ + 2e^- \rightarrow C_2H_2Cl_2 \text{ (cis-DCE)} + Cl^- \qquad (2)$$
$$C_2H_2Cl_2 \text{ (cis-DCE)} + H^+ + 2e^- \rightarrow C_2H_3Cl \text{ (VC)} + Cl^- \qquad (3)$$
$$C_2H_3Cl \text{ (VC)} + H^+ + 2e^- \rightarrow C_2H_4 \text{ (ethene)} + Cl^- \qquad (4)$$

The rate of the dechlorination process has been reported to decrease as the degree of chlorination decreases (Vogel and McCarty, 1985). Therefore, PCE and TCE are most susceptible to reductive dechlorination and an accumulation of DCE and VC may be observed as they are the least oxidized of the chlorinated hydrocarbons (Murray and Richardson, 1993). Because chlorinated hydrocarbons are used as electron acceptors during reductive dechlorination, there must be an appropriate electron donor for the process to occur. HRC™ is a proprietary, food-grade, polylactate ester that slowly releases lactic acid upon hydrolysis. Lactic acid is then metabolized by native microbes to hydrogen, which is a suitable electron donor for the reductive dechlorination process (Koenigsberg and Farone, 1999).

Further evaluation of site conditions favored the injection of HRC™. The HRC™ used at this site was formulated and sold by Regenesis Bioremediation Products in San Clemente, California and provided a passive and cost-effective means of reducing chlorinated hydrocarbon concentrations to acceptable regulatory levels and preventing off-site migration of the groundwater plume.

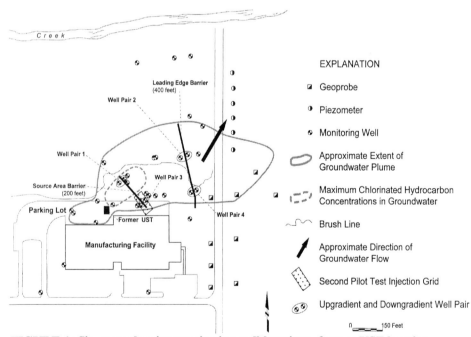

FIGURE 1: Site map showing monitoring well locations, former UST location, groundwater flow direction, extent of chlorinated hydrocarbon contamination and HRC™ injection locations with associated monitoring well pairs

Site Description. The southern portion of the groundwater plume extends under an asphalt parking lot and the northern portion of the plume is beneath a grassy area (Figure 1). The shallow permeable interval containing elevated concentrations of contaminants is confined between an overlying predominantly clay layer and an underlying fine grained glacial till. The permeable interval consists of silty sand to coarse sand and is present from approximately 14 to 20 feet below ground surface (bgs). The depth to groundwater is relatively shallow at the site and generally ranges from 8 to 10 feet bgs. Groundwater flow velocity was estimated to be approximately 0.5 feet per day toward a creek in the northeastern part of the site (Figure 1).

Groundwater sampling results indicated that PCE, TCE, cis-DCE and VC concentrations exceeded State of Illinois cleanup objectives for Class I groundwater for human ingestion. The highest chlorinated hydrocarbon concentrations were detected directly downgradient of the former UST (Figure 1) The maximum detected concentrations of the chlorinated hydrocarbons downgradient of the former UST are presented in Table 1.

TABLE 1. Maximum chlorinated hydrocarbon concentrations detected in groundwater between 1993 and 1998

Compound	Illinois Class I Groundwater Cleanup Objective (mg/L)	Maximum Groundwater Concentration (mg/L)
PCE	0.005	9.58
TCE	0.005	3.43
cis-DCE	0.07	14.1
VC	0.002	2.27

APPROACH

Prior to field scale implementation, a bench-scale experiment was conducted to determine whether naturally occurring microbes in site soils were suitable for enhancing reductive dechlorination processes in the presence of HRC™. The bench-scale treatability test utilized test tubes containing approximately 10 grams of soil with increasing amounts of TCE and HRC™. The test tubes were analyzed for TCE and lactic acid over a 35 day period and bacterial counts were determined at the end of the experiment. Volatile acids such as lactic acid were analyzed in water samples from the test tubes to indicate the presence of HRC™, as lactic acid is the hydrolysis product of the HRC™ polymer (Koeningsberg and Farone, 1999). In the presence of HRC™, an 89 to 99 percent reduction in TCE concentrations was observed in the test tube experiments. The bench test results were therefore very encouraging and field scale application followed.

The initial field-scale pilot test was conducted in December 1998. The direct-push drilling method was used to deliver the HRC™ polymer into the permeable contaminated interval. The polymer was injected on 5-foot centers along two separate lines, one in the source area downgradient of the UST and another at the leading edge of the groundwater plume. The first injected barrier was a 200-foot (60 m) long barrier of dissolved lactic acid that extends through the center of the source area, the second barrier was 400-foot (120 m) long at the leading edge of the plume. The bottom of the permeable interval was easily identified by the resistance of the direct push rod when it encountered the underlying till. Identification of the bottom of the permeable interval was important to ensure that HRC™ was injected throughout the entire contaminated permeable zone thickness. The volume of polymer injected into the subsurface during this pilot test was approximately 2.8 pounds per vertical foot (4.2 kilograms per meter).

Groundwater quality was subsequently monitored from four sets of well pairs (one upgradient and one well downgradient of the barrier in each pair) to evaluate the success of the remediation technology (Figure 1). Groundwater samples were collected and analyzed from each monitoring well prior to polymer injection to establish a baseline. After polymer injection, groundwater samples were collected from upgradient and downgradient well pairs, 30, 94, 129, and 172 days after injection. Groundwater samples were analyzed for chlorinated hydrocarbons using EPA method 8260, volatile acids (lactic, acetic, propionic, and pyruvic), total organic carbon (TOC), total dissolved iron, sulfate, nitrate, and chloride. Field measurements for ferrous iron, sulfide, pH, temperature, and oxygen-reduction potential were also conducted.

A subsequent pilot study was conducted in August 1999, near well pair 3, where less significant reductions of PCE had been observed (Figure 1). A total of 15 additional injection points in a 10 by 20 foot (3 by 6 meter) area were injected with 8 pounds of HRC™ per vertical foot (12 kilograms per meter). Three additional monitoring wells were installed downgradient from the 15 additional injection points to monitor the performance of the increased HRC™ dosage. In addition to the groundwater analyses performed in the first pilot study, groundwater samples were also analyzed for total dissolved manganese, carbon dioxide (CO_2), methane, and dissolved oxygen.

RESULTS AND DISCUSSION

Sampling results indicated that HRC™ was successful in enhancing biodegradation of PCE and TCE at the site. Thirty days after injection, significant reductions in PCE concentrations (from 7 mg/L to 2 mg/L) were observed in observation wells downgradient of the source area with subsequent increases in TCE and DCE concentrations (Figure 2). After 172 days, PCE concentrations in the source area have been reduced to 0.5 mg/L and TCE concentrations also started to decrease. At 172 days, however, DCE concentrations started to level off and there was not a further significant increase in VC concentrations. This may indicate that the reductive dechlorination process is not degrading DCE to a large extent or that VC concentrations were too low for accurate laboratory analysis.

Fluctuations in PCE concentrations were also observed in upgradient monitoring wells that were not influenced by the dissolved lactic acid plume. These fluctuations may have been due to natural fluctuations in groundwater levels. Therefore, to evaluate reductive dechlorination processes, molar ratios of PCE to TCE and TCE to cis-DCE were compared. Molar ratios between parent compound and daughter product should remain constant if no enhanced biodegradation is occurring. The molar ratios show significant evidence of enhanced reductive dechlorination processes downgradient of the source area just 172 days after injection (Figure 3).

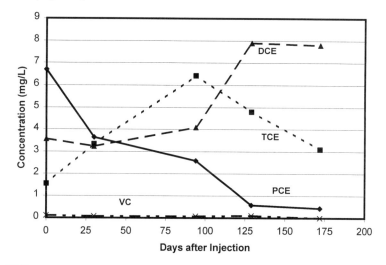

FIGURE 2: Chlorinated hydrocarbon concentrations in groundwater downgradient of source area (well pair 1).

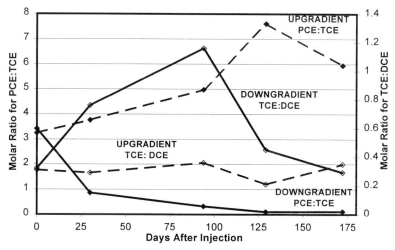

FIGURE 3: Molar ratios of PCE:TCE and TCE:cis-DCE in groundwater from source area monitoring wells (well pair 1)

The rate of reductive dechlorination appeared to decrease over time. This indicates the probable need for additional injection of the HRC™ polymer if sufficient decreases in the chlorinated hydrocarbons concentrations are not observed.

Evidence for enhanced reductive dechlorination was also observed in monitoring wells downgradient from the HRC™ barrier, at the leading edge of the groundwater plume at well pair 2 (Figure 1). The PCE concentrations were reduced from 2.6 to 0.005 mg/L (PCE cleanup objective) in the downgradient monitoring well 129 and 172 days after injection. TCE and cis-DCE concentrations subsequently increased due to dechlorination of PCE but started to decrease during the last sampling event.

Some areas of the dissolved phase groundwater plume treated with HRC™ did not show significant reductions in chlorinated hydrocarbon concentrations (well pair 2 and 4; Figure 1). These areas also contained high sulfate concentrations that may have limited the effectiveness of the HRC™ on the reduction of chlorinated hydrocarbons. The hydrogen that is produced through biotransformation of dissolved lactic acid may be consumed by sulfate instead of the chlorinated hydrocarbon therefore competing for hydrogen (Equation 5)

$$8e^- + 10H^+ + SO_4^{2-} \rightarrow H_2S^0 + 4H_2O \tag{5}$$

To evaluate whether sulfate was limiting the effectiveness of HRC, another pilot study was conducted in the well pair 3 area near the southern end of the 200 foot long barrier (Figure 1). In August 1999, a higher dose of HRC™ (8 pounds per vertical foot) was injected near the source area upgradient of well pair 2 and 121 days after the injection, PCE concentrations in downgradient monitoring wells were reduced from 3.8 mg/L to 0.042 mg/L (Figure 4).

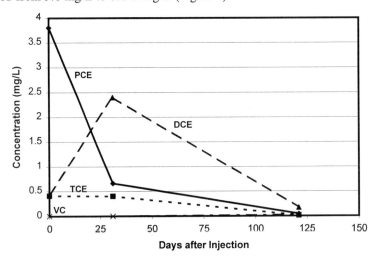

FIGURE 4: Chlorinated hydrocarbon concentrations for downgradient monitoring well (well pair 3)

TCE concentrations also decreased significantly within 31 days, because only an increase in DCE was observed. After 121 days, most of the cis-DCE was also reduced, with a corresponding increase in VC observed from 0.005 mg/L at day 0 to 0.021 mg/L at day 121. Addition of HRC™ also caused a reduction in sulfate concentrations from 370 mg/L to approximately 98 mg/L. Therefore, it appears that sulfate can inhibit the effectiveness of the HRC™ for reducing PCE concentrations, but this may be mitigated by injecting higher doses of HRC™.

Several other trends were also observed in the groundwater chemistry after HRC™ injection (example on Table 2). Similar to sulfate, decreasing concentrations of other electron acceptors such as nitrate and dissolved oxygen were observed. There were no trends observed for chloride concentrations in groundwater, however, addition of HRC™ did increase concentrations of TOC, volatile organic acids (with the exception of lactic acid for the example given below), CO_2, dissolved iron, dissolved manganese, and methane. These trends are indicative of anaerobic biotransformation processes and are therefore useful in validating that enhanced reductive dechlorination may be occurring in groundwater in the presence of HRC™ (U.S. Environmental Protection Agency, 1998). The increase in methane is also an indicator that methanogenesis is occurring. There is evidence that suggests that there is competition between reductive dehalogenators and methanogens in which the methanogens compete for the use of hydrogen in the conversion of CO_2 to methane (Koenigsberg and Farone, 1999). However, during this pilot study the significant decreases observed in PCE concentrations suggest that reductive dechlorination occurs in the presence of methanogenesis.

TABLE 2. Groundwater concentrations of various compounds for the same downgradient monitoring well before and after HRC™ injection

Compound	Pre HRC™ Injection (mg/L)	Post HRC™ Injection (mg/L)
Sulfate	370	98
Nitrate	17	9.8
Dissolved Oxygen	0.79	0.57
Chloride	not available	49
TOC	5.2	530
Lactic Acid	<100	<25
Propionic Acid	4	460
Butyric Acid	<1	110
Acetic Acid	13	370
Carbon Dioxide	81	408
Dissolved Iron	0.15	139
Dissolved Manganese	0.77	6.3
Methane	0.14	2.7

CONCLUSIONS

HRC™ injection was successful in enhancing reductive dechlorination processes of chlorinated hydrocarbons in groundwater at a large manufacturing facility in southern Illinois. Significant reductions in PCE concentrations were observed in monitoring wells downgradient from an HRC™ barrier. Decreases in PCE concentrations also resulted in increases of TCE and DCE concentrations due to reductive dechlorination. In the source area, PCE concentrations were reduced from 7 to 0.5 mg/L after 172 days and reductions in TCE and DCE concentrations were observed in some monitoring wells during the last sampling event. Some areas of the dissolved phase groundwater plume treated with HRC™ showed relatively less significant reductions in PCE and TCE concentrations. These areas also contained high sulfate concentrations (from 300 to 500 mg/L) which may have limited the effectiveness of the HRC™. However, re-injection of HRC™ in those areas at increased HRC™ doses resulted in significant reductions of sulfate and PCE within 30 days of the injection. Other trends in the groundwater chemistry were also observed that were indicative that reductive dechlorination processes were occurring. Therefore, it appears that sulfate or high concentrations of other electron acceptors may inhibit the effectiveness of the HRC™ on reducing PCE concentrations but this may be mitigated by injecting higher doses of HRC™.

REFERENCES

Illinois Pollution Control Board. 1998. *State of Illinois Tiered Approach to Corrective Action Objectives*. 35 Illinois Administrative Code, Part 742.

Koenigsberg, S.S., and W.A. Farone. 1999. "The Use of Hydrogen Release Compound (HRC™) for CAH Bioremediation." *Engineering Approaches for In Situ Bioremediation of Chlorinated Solvent Contamination*, pp. 67-72. Batelle Press, Columbus, Ohio.

Murray, W.D. and Richardson, M. 1993. "Progress Toward the Biological Treatment of C_1 and C_2 Halogenated Hydrocarbons". *Crit. Rev. Environ. Sci. Technol.* 23(3): 195-217.

Vogel, T.M. and McCarty, P.L.. 1985. "Biotransformation of Tetrachloroethylene to Trichloroethylene, Dichloroethylene, Vinyl Chloride, and Carbon Dioxide under Methanogenic Conditions." *Appl. Environ. Microbiol.* 49(5): 1080-1083.

Wiedemeier, T.H., Swanson, M.A., Moutoux, D.E., Gorden, E.K., Wilson, J.T., Wilson, B.H., Kampbell, D.H., Haas, P.E., Miller, R.N., Hansen, J.E., and Chapelle F.H. (Eds.). 1998. *Technical Protocol for Evaluating Natural Attenuation of Chlorinated Solvents in Ground Water*. U.S. Environmental Protection Agency Technical Report, EPA/600/R-98/128, Cincinnati, Ohio.

FULL-SCALE ANAEROBIC BIOREMEDIATION USING ACETATE AND LACTATE ELECTRON DONORS

Sami A. Fam, Ph.D., P.E., L.S.P., (Innovative Engineering Solutions, Inc., Needham, MA), Margaret Findlay, Ph.D., (Bioremediation Consulting, Inc., Newton, MA), Sam Fogel, Ph.D., (Bioremediation Consulting, Inc., Newton, MA) Tony Pirelli, P.G., (Menomonee Falls, WI), Tom Sullivan, P.G., (Bascor Environmental Inc., Madison, WI)

ABSTRACT: This paper describes an ongoing site remedial program, which includes addition of acetate, lactate and ammonium phosphate to the groundwater to enhance anaerobic degradation of chlorinated organics. This is likely one of the largest full-scale anaerobic dechlorination systems in the world. The remedial measures involve extraction of impacted groundwater/injection of amendments from a network of 36 wells (3-acre area). The set of pumping and non-pumping (amendment injection) wells is modified every month. Groundwater and soil vapor are extracted using seven liquid ring pumps due to the site's low permeability. Off gases from vapor extraction and above-ground groundwater aeration are treated by on-site steam regenerable activated carbon. A site perimeter bio-enhancement reactive wall is used to add electron donors to mitigate off site impacts. The remedial system has greatly improved site conditions over its 21 months of operation.

INTRODUCTION

The site, located in the southern U.S., is impacted with halogenated and non-halogenated volatile organic compounds (VOCs) resulting from industrial use over the past thirty years. Highest concentrations of total VOCs were typically in the 500 mg/L range and consist primarily of tetrachloroethene (PCE), trichloroethene, 1,1,1 trichlorethane (1,1,1 TCA), methylene chloride (MC), acetone, and toluene. The impacted area is roughly 3-acres in size as shown in Figure 1. The overburden generally consists of ten to twenty five feet of silty-clay with slightly more permeable fill/silty-sand seams within a historical drainage trough that runs through the site. The 25-foot depth to the underlying limestone bedrock is largest within the trough as shown in Figure 2.

The initial biogeochemical investigation showed that extensive biodegradation is occurring at the site, but that phosphate and electron donor deficiencies may be limiting degradation. A lab-scale treatability test was conducted to confirm that electron donor and phosphate addition would enhance biodegradation of the VOCs. The microcosm study, which were done in accordance with the US Air Force RABITT protocol (Morse et al., 1998), was also intended to provide a basis for initial dosing requirements for the selected amendments during full-scale implementation.

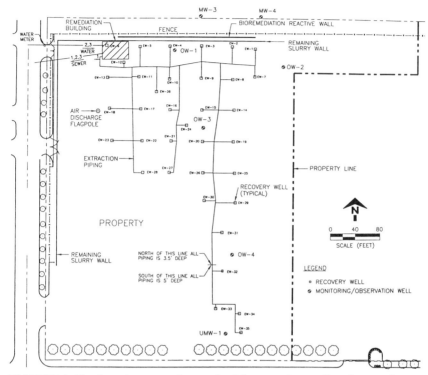

FIGURE 1. Site location map shown the 36 extraction wells, observation wells, property boundary bioremediation reactive wall, and the remediation system building

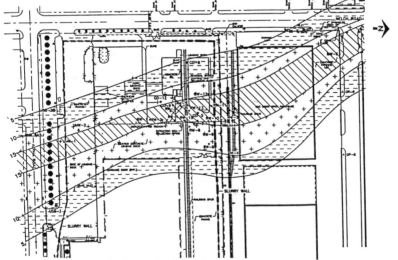

FIGURE 2. Depth to bedrock in feet, showing trough in the middle of the site

Groundwater samples were collected from 2 site wells for treatability testing. The samples from each well were subdivided using anaerobic sterile technique into multiple 160 ml. serum bottles, and each bottle was subjected to varying treatment scenarios as follows: 1) Killed immediately with HCl; 2)Unamended; 3)Amended with Phosphate (P) only; 4) Amended with Acetate (electron donor) and P; 5)Amended with Lactate (electron donor) and P;

The various bottles were analyzed at different times for VOCs, electron donors, and degradation products (ethane, ethene, CO2). In general, the data showed that TCE degradation at the site is enhanced by the addition of electron donors and phosphate. The greatest enhancement appeared to be from Phosphate and Acetate addition. This is to be expected, since acetate and lactate were found in the groundwater (already acclimated microorganisms), and are likely the most naturally available electron donors (acetate is produced from the degradation of many compounds such as methylene chloride, which is abundant at the site). c-DCE concentrations increased initially, concomitantly until TCE loss, consistent with conversion of TCE to DCE.

REMEDIAL SYSTEM DESCRIPTION

The dual extraction portion of the system is expected to remediate the vadose zone and the dewatered soil, and the bioremediation component of the system is expected to remediate the portions of the saturated zone that are not dewatered.

The in-situ remediation system consists of a network of 36 dual extraction wells (all individually piped to the remediation system) covering approximately a three-acre parcel of land impacted by halogenated and mono aromatic volatile organic compounds (VOCs). Seven water-cooled liquid ring pumps (LRPs) extract groundwater and soil vapors. Extracted groundwater is removed from the vapor/liquid stream in a large knock-out tank and is subsequently treated using an aeration tray system manufactured by North East Environmental of Lebanon, New Hampshire. Groundwater yield from the overburden groundwater is generally in the range of 1to10 gallons per minute (from all wells combined). Treated groundwater is discharged to the municipal sewer system. A sequestering agent is added to the groundwater to minimize fouling. Each extraction well is also equipped with an air bleed line, in order to ensure groundwater withdrawal in very low permeability wells.

Extracted air (approximately 300 cubic feet per minute {CFM}) is treated by an on-site steam regenerable granular activated carbon system (GAC), manufactured by Westport, Inc. of Westport, Massachusetts. The dual bed system provides automated regeneration of the spent carbon bed based upon FID readings or timed operational cycles. The GAC system also treats emissions from the tray stripper (an additional air flow of 300 CFM). Only one bed is operational at any one time, the second bed is on stand-by. Air effluent from the primary bed is polished by one 2,400 lb. GAC vessel. This GAC vessel is regenerated off-site.

The steam regeneration system cycle operates 4 times per day. Condensed steam from regeneration is slowly metered back into the main knock out tank.

Recovered non aqueous phase liquid (NAPL) from regeneration is stored in an on-site, 550 gallon tank for eventual disposal.

The remediation program also calls for periodic batch addition of electron donors and minerals to 7 of the 36 on-site extraction wells. Each feeding consists of approximately 300 pounds of acetate, 300 pounds of lactate and 50 pounds of ammonium phosphate. The amendment dose will likely be adjusted in the future based upon on-going testing, and will also likely be increased since residual levels of electron donors in the groundwater continue to be limiting the degradation. Wells to which amendments are added do not operate (i.e. fluids are not extracted) for one month after the addition dose. The electron donor and nutrient addition is intended to enhance anaerobic dechlorination in the portion of the site that is not dewatered. The addition is conducted on a monthly basis by mixing the additives in an on-site 250 gallon tank and turning on the batch addition pump for a period of approximately two to three days. Additives flow to the wells through the air bleed lines. Following electron donor and inorganic nutrient addition, the air bleed lines are flushed using an on-site air compressor. Additives are expected to be distributed across the site by the pumping action of the operating wells. Figure 3 presents a process flow diagram for the installed remediation system.

SYSTEM OPERATIONAL RESULTS

As of December 31, 1999, approximately 1 million gallons of groundwater have been pumped, treated and discharged by the remediation system. Over a 21 month period, with an operational time of 60%, this is an average flow rate of 2.5 gallons per minute (the system was designed for flows up to 20 gallons per minute, which occurs during severe rainstorms). The vapor extraction component of the dual extraction system has generally averaged 300 cubic feet per minute (the system was designed for flows up to 350 cubic feet per minute). Highest air flow rates (350 CFM) are achieved in seasonal low water conditions (fall) and the lowest air flow rates (250 CFM) are achieved in the wet season (winter and spring).

Cumulative VOC removal in the vapor phase by vapor extraction and water phase extraction of VOCs were calculated for the Subject Property based upon the screening measurements (using PID and FID detectors) and the combined air flow to the steam regenerable gas phase activated carbon treatment system. Calculations of VOC removal are graphically presented in Figure 4 (95% from vapor extraction and 5% from groundwater extraction). Mass removal in the vapor phase is calculated as the product of extracted air volume and extracted vapor concentration as measured by the screening instrument. The estimate of 30,000 pounds removed matches fairly well with the estimate of NAPL removed for off site disposal of approximately 2,100 gallons of mixed solvents. Using a specific gravity of 1.2, the 2,100 gallons equates to approximately 21,000 pounds. The mass removal estimate from disposal is likely low since the extracted air is not treated with 100% efficiency and the regeneration process likely results in some evaporative loss. The true estimate of mass removal from the subsurface is

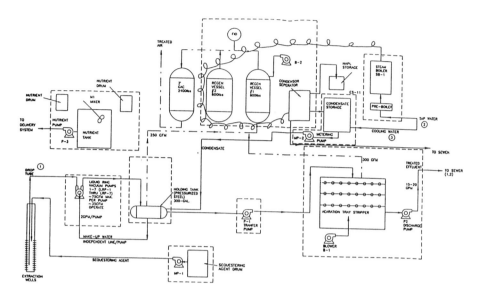

FIGURE 3. System process flow diagram

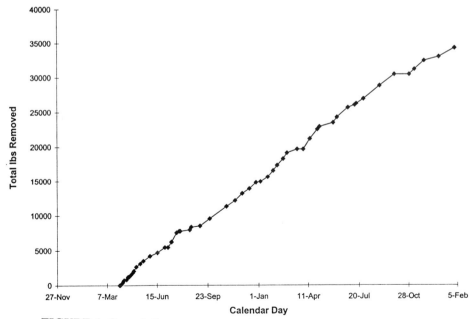

FIGURE 4. Cumulative mass removal in total pounds of VOCs by dual extraction from April 1998 to February 2000

likely between 21,000 and 30,000 pounds of VOCs. This estimate does not include mass that has been biodegraded, which likely exceeds 20,000 pounds based upon our rough calculations.

The full-scale bioenhancement system has been operating for 21 months and has greatly improved site conditions. Total VOC concentrations have dropped from 460 mg/L to approximately 41 mg/L (in September 1999) in the most impacted well (OW-3). Of the 41 mg/l of total VOCs (halogenated and non-halogenated) in OW-3, approximately 25 mg/l are vinyl chloride and chloroethane (91.7% of the total halogenated VOCs). In contrast, prior to initiating active remediation vinyl chloride and chloroethane represented only 7% of the total halogenated VOCs in OW-3. Concentrations of ethane and ethene in the observation wells have increased by approximately 50 fold in the impacted zone since initiation of the bio-enhancement program. Ethene concentrations were measured to be approximately 7 ppm in December 1998. Figures 5 and 6 illustrate the groundwater quality improvement trends in OW-3. Figures 7 and 8 show the groundwater quality trends in OW-1. Total VOCs have been reduced from 4.24 ppm in OW-2 (June 1998) to 2.1 mg/L in September 1999. Total VOCs in OW-4 have remained stable at approximately 0.5 mg/L for the past two years, although the percentage of parent compounds has been reduced from 35% in 1998 to approximately 1% in September 1999.

Electron donor and ammonium phosphate amendment is added to the boundary reactive wall (Figure 1) on a monthly basis effective October 1999. The state regulatory agency required a 1-year on-site operational track record prior to initiating off site amendment addition and a demonstration that bio-enhancement will not degrade groundwater quality. Data is therefore not currently available to evaluate the effectiveness of the off-site enhancement system. The groundwater plume extends approximately 500 feet off site.

REFERENCES

Morse J.J., Alleman, B.C., et al. (1998). *Draft Technical protocol: A Treatability Test for Evaluating the Potential Applicability of the Reductive Anaerobic Biological In Situ Treatment Technology (RABITT) to Remediate Chloroethenes.* NTIS document ADA352416. Battelle Memorial Institute, Columbus Ohio.

Potential Electron Donors 29

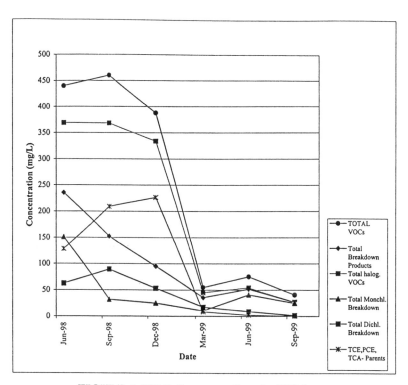

FIGURE 5. VOC Concentrations in OW-3

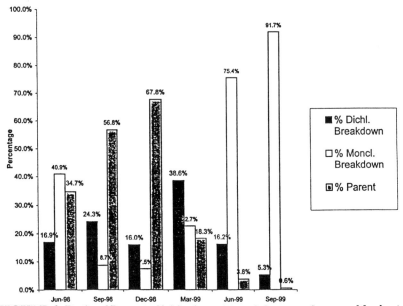

FIGURE 6: Ratio of Parent, dichlorinated breakdown and monochlorinated breakdown products in OW-3

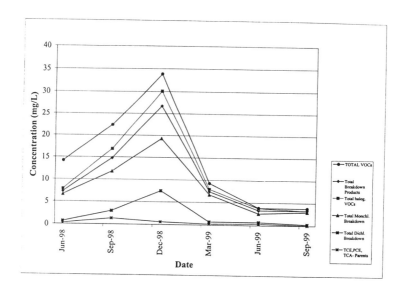

FIGURE 7. VOC Concentrations in OW-1

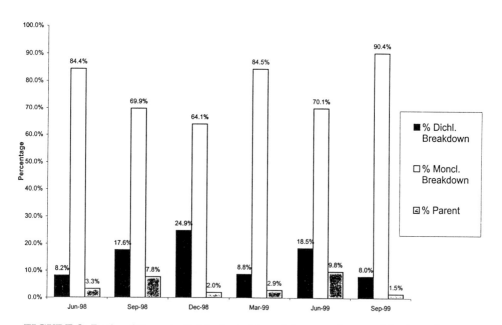

FIGURE 8: Ratio of Parent, dichlorinated breakdown and monochlorinated breakdown products in OW-1

RESULTS FROM TWO DIRECT HYDROGEN DELIVERY FIELD TESTS FOR ENHANCED DECHLORINATION

Charles J. Newell, Ph.D., P.E., Groundwater Services, Inc., Houston, TX,
Patrick E. Haas, Air Force Center for Environmental Excellence, Brooks AFB, TX
Joseph B. Hughes, Ph.D, Rice University Dept. Envi. Sci. and Eng., Houston, TX
Tariq Khan, Groundwater Services, Inc., Houston, TX

ABSTRACT: Direct hydrogen addition, wherein hydrogen is delivered without the use of fermentation substrates or carbon sources is an in-situ bioremediation technology for chlorinated solvent plumes that is currently under development. A multi-site project to field test the applicability and feasibility of direct hydrogen addition has been initiated by the Technology Transfer Division at the Air Force Center for Environmental Excellence (AFCEE).

Results from a *pull-push-pull treatability test* designed by one of the authors (Haas) conducted at Offutt AFB in Nebraska in Nov. 1998 showed > 99% reduction of the cis-DCE in a 900 L test volume over the 41 hour test period (0.43 to < 0.005 mg/L) and > 99% decrease in the amended hydrogen.

Results from the four-month sampling event from a planned 18-month *low-volume pulsed biosparging test* at Cape Canaveral Air Station Florida also showed apparent biological dechlorination in a 48 x 90 ft test zone located 15-20 ft below the water table in a sandy aquifer. Analysis of the gas tracers after one week indicated extensive biological utilization of hydrogen. At the four month interval, data from 14 sampling points in the test area indicated significant reductions in TCE (e.g., from 48 mg/L to < 0.005 mg/L in one well) and cis-DCE concentrations (e.g., 140 mg/L to 4 mg/L in one well) compared to the 2 points in the nitrogen control area. No excessive methane production was observed in the hydrogen delivery zone.

INTRODUCTION

Hydrogen is now widely recognized as a key electron donor required for the biologically-mediated dechlorination of chlorinated compounds (e.g., see Wiedemeier et al. 1999). In this process, hydrogen acts as an *electron donor* and halogenated compounds such as chlorinated solvents act as *electron acceptors,* becoming reduced in the reductive dechlorination process.

While much of the early research has focused on indirect hydrogen addition of via the delivery of fermentation substrates such as methanol, toluene, lactate, benzoate, etc. and the subsequent fermentation of these substrates to form hydrogen, it is also possible to deliver the hydrogen directly without the fermentation step. Direct delivery methods that have been proposed by Hughes, Newell, and Fisher (1997) include circulation of groundwater containing dissolved hydrogen, placement of chemical agents that release dissolved hydrogen, electrolysis of water with subsurface electrodes, use of colloidal gas aphrons (foams), and low-volume pulsed biosparging.

Work performed by Carr and Hughes (1998) showed no microbiological constraints on the direct addition of hydrogen (i.e., no restrictions on the dechlorinating bacteria due to competition effects with other bacteria). Therefore they concluded that selection of the most appropriate method of hydrogen delivery (e.g., addition of fermentation substrates vs. direct delivery) is based on factors such as cost and the relative efficiency of hydrogen distribution within the treatment area (Carr and Hughes, 1998).

Several direct hydrogen delivery field tests, one with introduction of water amended with dissolved hydrogen, and one with low-volume pulsed biosparging of hydrogen have been completed under the AFCEE test program. Two of these projects are described below.

TEST 1: PULL-PUSH-PULL TREATABILITY TEST

Test Procedure. A pull-push-pull treatability test was performed at the Fire Protection Training Area 3 (FPTA 3) site at Offutt AFB, Nebraska. Across the majority of the site, groundwater occurs at 8 to 10 ft below ground surface (BGS) within a sand unit. Starting concentrations of key contaminants at the site were less than 1 mg/L of cis-DCE and less than 1 mg/L of total BTEX. As described in Fisher et al., 1999, the hydrogen pull-push-pull treatability test consists of the following steps:

1. *Initial Groundwater Extraction ("pull"):* Extraction of a known quantity of groundwater (e.g., 750 L) from within the test area through an existing monitoring well.
2. *Amendment Addition:* Addition of known quantities of hydrogen and various volatile and non-volatile tracers (e.g., bromide, helium, sulfur-hexafluoride (SF_6)) to the extracted groundwater, followed by thorough mixing to create a homogeneous test solution.
3. *Initial Sampling:* Collection of a representative test solution sample which is analyzed for chlorinated organic compounds, hydrogen, tracers, and other constituents of interest (e.g., oxygen, nitrate, sulfate, etc.).
4. *Re-Injection of Groundwater Test Solution:* Pulse injection ("push") of amended groundwater into the saturated zone through the same monitoring well used for groundwater extraction.
5. *Final Groundwater Extraction:* Extraction ("pull") of the test solution/ groundwater mixture from the test well following a contact/reaction period (typically 24 to 48 hr). Sampling is conducted during the extraction.
6. *Final Sampling:* Collection of a final representative test solution sample which is again analyzed for chlorinated organic compounds, hydrogen, tracers, and other constituents of interest.

Test Results. cis-1,2-DCE concentrations dropped by more than 99% (0.43 to < 0.005 mg/L) over the course of the 41-hour test, while much lower reductions in Total BTEX was observed (34% reduction). Due to the expected similar volatilization and adsorption behavior of DCE and BTEX (based on similar Henry's Law and organic carbon partitioning coefficients), the observed substantial loss of DCE does not appear to be fully accounted for by these physical mechanisms, and a biological loss is indicated. Tracer recovery results indicate a minimum 39% mass loss of hydrogen to biological consumption over the course of the 41-hour treatability test at FPTA 3 (0.012 mg/L hydrogen consumed). Although a biological reduction of all of the cis-1,2-DCE is indicated, it could not be confirmed by observed changes in vinyl chloride or ethene concentrations.

TEST 1: LOW-VOLUME PULSED BIOSPARGING TEST

Test Procedure. Cape Canaveral Air Station is located on a barrier island along the Atlantic coast of Florida, separated from the Florida mainland by the Banana River. Launch Complex 15 is one of a series of rocket launching facilities located along the easternmost edge of the Base, adjoining the Atlantic Ocean. The near-surface soil/aquifer material at Launch Complex 15 consists of silica sand with some shell, and little clay or organic matter. The sand unit is continuous from the surface to the maximum explored depth of approximately 70 ft below ground surface (BGS), with some silt and clay lenses at depth. Groundwater is typically encountered at 6 to 7 feet below ground surface. The observed potentiometric surface is very flat, with less than 0.1 ft of head difference over the 90 ft.

The hydrogen biosparging pilot test system at Launch Complex 15 utilizes a 4-sparge point, 20-monitoring point well network in a 48 x 90 ft area as shown in Figure 1. Three hydrogen sparge points, spaced 12 ft apart, and 16 monitoring wells located within the treatment zone are being used to determine the rate and extent of chlorinated solvent degradation over the test period. The fourth sparge point and an additional 4 monitoring wells are being used as a nitrogen sparge control for purposes of assessing chlorinated solvent losses occurring through volatilization as a result of sparging activities.

In addition to the four rows of upgradient/downgradient monitoring wells, additional monitoring points in the form of six multi-level saturated zone samplers and three multi-level vadose zone samplers were installed immediately adjacent to the hydrogen sparge points. These samplers allow for the collection of data concerning the distribution of sparged gases within the treatment zone.

Hydrogen biosparging equipment at Launch Complex 15 consists of 5 T-size compressed gas cylinders (256 scf each) linked to a common manifold, an adjustable pressure regulator, an automatic timer-activated solenoid valve, direct reading flow gages, and stainless steel tubing connecting the gas cylinders to the three hydrogen sparge wells. A parallel nitrogen sparging system utilizes a single T-size compressed gas cylinder connected to a single sparge well. System power is provided by a 12-volt gel-cell battery equipped with a solar-powered recharging circuit. The initial gas mixture consisted of 48% hydrogen, 48% helium, and 2% SF_6 so that all compounds would have a theoretical equilibrium concentration with the water of about 0.8 mg/L.

The sparging system operates on a pulse cycle controlled by a timer such that gas is sparged in a burst of a set duration, occurring at a specified frequency.

Injection pressures are approximately 20 psig, sufficient to overcome hydrostatic pressure within the sparge wells (approx. 9 psig). Approximately 130 scf of a 49% hydrogen, 49% helium, and 2% SF_6 gas mixture was pulsed into each of the three sparge points (located on 12 ft centers) on the first day of sampling (2/7/99). Note that no breakthrough of hydrogen gas to the surface was detected during this initial sparging event (vadose zone thickness: 5 ft). After the first day, smaller 1-minute "maintenance" pulses consisting of 15-20 scf of research grade hydrogen gas were added to each sparge point once per day. Recently the system has been reconfigured to delivery one large pulse of hydrogen once a week.

Test Results. Results from the four-month sampling event of the planned 18-month low-volume pulsed biosparging test at Cape Canaveral Air Station showed hydrogen transfer, hydrogen consumption, and apparent biological dechlorination in the test zone located 15-20 ft below the water table in a sandy aquifer.

After 1 week, hydrogen and tracer concentrations were measured in the multi-level samplers located in the sparging row (see Figure 2). The maximum observed hydrogen concentration was 0.47 mg/L, or a little over half of the equilibrium hydrogen concentration of 0.8 mg/L for the initial injection gas (comprised of 48% hydrogen, 48% helium, and 2% SF_6). Overall significant reductions in hydrogen concentration were observed horizontally and vertically away from the gas sparge points. Comparison with the helium and SF_6 showed a larger distribution of these tracers, indicating biological consumption of the injected hydrogen. Overall each sparge point appeared to deliver hydrogen concentrations in about a 5 ft radius away from the sparge point (10 ft diameter).

At the four month interval, data from 14 sampling points in the test area indicated significant reductions in TCE (see Table 1 and Figure 3) and cis-DCE concentrations (see Table 1 and Figure 3) compared to the 2 points in the nitrogen control area (see Table 2). No excessive methane production was observed in the hydrogen delivery zone (see Table 1).

TABLE 1. Results from Cape Canaveral Low-Volume Pulsed Hydrogen Field Test: Hydrogen Sparge Test Section.

Constituent	Geometric Mean of 6 CLOSE Sampling Pts in H_2 Test Zone (3-6 ft horizontally from sparge points)				Geo. Mean of 6 MIDDLE Sampling Pts in H_2 Test Zone (15 ft horizontally from sparge pts)		
	2/7/99 (mg/L)	2/13/99 (mg/L)	6/23/99 (mg/L)	4 Month Change (mg/L)	2/7/99 (mg/L)	6/23/99 (mg/L)	4 Month Change (mg/L)
TCE	14.1	8.1	0.5	**- 13.6**	15.7	7.9	**- 7.8**
cis-DCE	237.	239.	88.	**- 149.**	195.	165.	**- 30.**
Vinyl Chl.	39.	22.	21.	**- 18.**	41.	44.	**+ 3.**
Ethene	2.6	1.8	1.5	**- 1.1**	5.4	5.6	**+ 0.2**
Methane	0.50	0.23	0.07	**- 0.43**	1.8	0.9	**- 0.86**

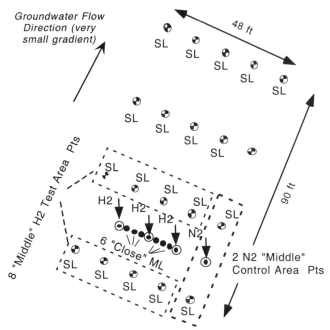

FIGURE 1. Plan view of test plot showing sparging wells (arrows), multi-level wells (ML), and single-level monitoring points (SL). Data for "Close" (3-6 ft from sparging point) and "Middle" (~15 ft from sparging point) monitoring groups are shown on Table 1 and Table 2.

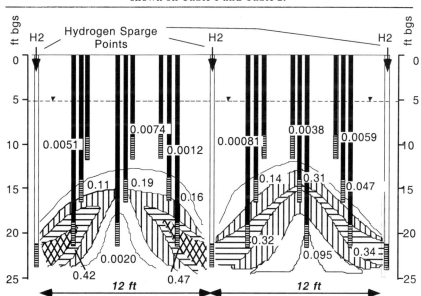

FIGURE 2. Dissolved hydrogen concentration (mg/L) after one week of low-volume pulsed biosparging. Note all single-level monitoring points were screened in same interval as deepest multi-level points (20-25 ft bgs).

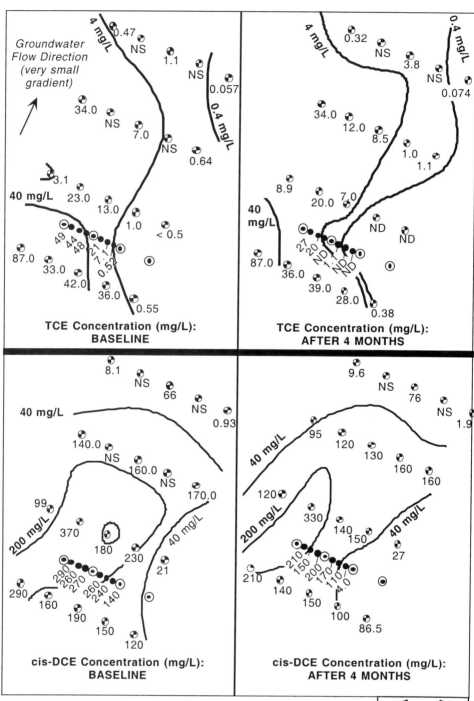

FIGURE 3. Changes in concentration of TCE and cis-DCE after four months of low-volume pulsed biosparging.

TABLE 2. Results from Cape Canaveral Low-Volume Pulsed Hydrogen Field Test: Nitrogen Sparge Control Section.

Constituent	Geo. Mean of 2 MIDDLE Sampling Pts in N_2 Control (15 ft horiz. from sparge pt)		
	2/7/99 (mg/L)	6/23/99 (mg/L)	4 Month Change (mg/L)
TCE	0.55	0.39	**- 0.17**
cis-DCE	50.	45.	**- 5.**
Vinyl Chl.	24.	31.	**+ 7.**
Ethene	3.8	4.9	**+ 1.1**
Methane	2.0	3.2	**+ 1.2**

CONCLUSIONS

Two field tests of direct hydrogen delivery showed reductions in chlorinated solvent concentrations and consumption of hydrogen. In addition, a push-pull-push test showed conservation of benzene compared to consumption of cis-DCE, indicating that biological processes were responsible for the removal of cis-DCE rather than physical processes. Reductions in TCE and cis-DCE in a low-volume pulsed biosparging test were higher in the hydrogen test zone compared to a nitrogen control zone.

Additional pilot tests using injection water amended with dissolved hydrogen are now being planned. The low-volume pulsed biosparging system is still being operated, with sampling events planned for Feb. 2000 (one year of operation) and August 2000 (eighteen months of operation).

REFERENCES

Carr, C.S., and J.B. Hughes. 1998. "Enrichment of High Rate PCE Dechlorination and Comparative Study of Lactate, Methanol, and Hydrogen as Electron Donors to Sustain Activity." *Environmental Science and Technology* 32(12):1817-1824.

Fisher, R.T., C.J. Newell, P.E. Haas, and J.B. Hughes, 1999. "Treatability Studies of Hydrogen-Enhanced Bioremediation of Chlorinated Solvent-Impacted Media." Battelle In Situ and On-Site Bioremediation, San Diego, California, Battelle Press.

Hughes, J.B., C.J. Newell, and R.T. Fisher. 1997. Process for In-Situ Biodegradation of Chlorinated Aliphatic Hydrocarbons by Subsurface Hydrogen Injection. U.S. Patent No. 5,602,296, issued February 11, 1997.

Wiedemeier, T.H., Rifai, H.S., Newell, C.J., and Wilson, J.W. 1999. *Natural Attenuation of Fuels and Chlorinated Solvents,* John Wiley & Sons, New York.

TIME-RELEASE ELECTRON DONOR TECHNOLOGY FOR ACCELERATED BIOLOGICAL REDUCTIVE DECHLORINATION

Stephen S. Koenigsberg (Regenesis, San Clemente, CA)
William A. Farone (Applied Power Concepts, Anaheim, CA)
Craig A. Sandefur (Regenesis, San Clemente, CA)

ABSTRACT: A significant body of literature supports the validity of enhancing biological reductive dechlorination with organic substrates that are fermented to produce hydrogen. Hydrogen then supplies electrons that are used by microorganisms in the benign transformation of a wide range of chlorinated hydrocarbons and selected inorganic compounds such as perchlorate. Implementation of an appropriate time-release electron donor system can eliminate major design, capital and operational costs, is minimally invasive and is invisible during the working phase. Also, some researchers believe that lower hydrogen partial pressures in the aquifer, that can be facilitated by a controlled release technology, may favor reductive dechlorination over competing methanogenic processes.

Hydrogen Release Compound (HRC™) is one option currently available to deliver electrons in a time-release fashion. HRC is a food grade, polylactate ester which is produced as a semi-solid material for bore hole implantation to the saturated zone or, it can be thinned to a moderately flowable liquid and injected. These formulations, depending on their viscosity and degree of esterification, will slowly hydrolyze over a period of months to years and generate readily fermentable lactic acid. This technology is most applicable to the passive, long term and low cost treatment of dissolved phase plumes and associated hydrophobically sorbed contaminants. Results from several field applications, not already being presented by others at this Conference, are reviewed.

INTRODUCTION

Hydrogen Release Compound (HRC) offers a passive, low-cost treatment option for in-situ anaerobic bioremediation of chlorinated hydrocarbons and selected inorganic compounds such as perchlorate. HRC is a proprietary, environmentally safe, food quality, polylactate ester specially formulated for the slow release of lactic acid upon hydration. Further details can be found in the following references which are derived from the Proceedings of the Fifth International Symposium on In-Situ and On-Site Bioremediation (1999 Battelle Conference – San Diego, CA). These papers present: basic HRC chemistry and laboratory performance verification (Koenigsberg and Farone, 1999); a mathematical model for HRC performance (Farone et al., 1999); two single well field tests (Wu, 1999, Kallur and Koenigsberg, 1999); a recirculation well field test (Dooley et al., 1999) and an injection field pilot test (Sheldon et al., 1999).

Several presentations are also being made at this Conference that represent the following areas of interest: a series of microcosm-to-field comparative studies (Farone et al., 2000); a laboratory study with perchlorate (Logan et al., 2000); a long term field implant study (Sheldon and Armstrong, 2000); two pilot barrier applications (Anderson et al., 2000, Dooley and Murray, 2000); a full scale barrier application (Schuhmacher et al., 2000); a pilot source treatment (Zahiraleslamzadeh and Bensch, 2000) and four full scale source treatments (Boyle et al., 2000, Harms, 2000, Lodato et al., 2000, Rhodes et al., 2000). A comprehensive peer reviewed compilation of these studies and subsequent field work can also be found in Koenigsberg and Sandefur (1999).

At this writing HRC has been applied experimentally and commercially at over 50 sites nationwide; 12 of them being advanced enough to warrant presentation. The results at nine of the sites are being presented by the above referenced authors. This presentation will cover three additional projects: a pilot barrier study, a pilot source treatment and a full scale source treatment.

RESULTS

Pilot Barrier Study. At a dry cleaner site in New Jersey, HRC was injected into an aquifer contaminated with PCE at levels of approximately 300 ug/L. The aquifer consisted of fine to medium sand with some silt, and the application thickness was 5 ft (1.5 m). 210 lb (95 kg) of HRC was injected into the aquifer via eight points placed in an arc downgradient of existing monitoring well MW-3 (Figure 1). In order to monitor conditions both upgradient and downgradient of the application zone, a monitoring well was installed 5.5 ft (1.7 m) downgradient of MW-3 shortly after HRC application.

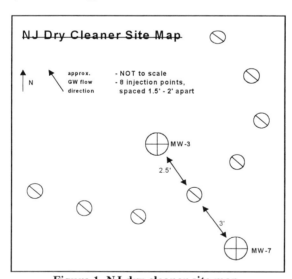

Figure 1. NJ dry cleaner site map.

At 72 days after HRC application, PCE concentrations in the upgradient well rose to 1,500 ug/L while concentrations in the downgradient well remained at 290 ug/L, suggesting the ability of HRC to manage and control contaminant flux.

Similarly, TCE concentrations increased to 50 ug/L and DCE concentrations increased to 420 ug/L in the upgradient well, while TCE and DCE concentrations remained stable in the downgradient well at 5.3 ug/L and 28 ug/L, respectively. The detection of significant levels of lactic acid and by-products indicated the presence of HRC-derived reducing power in the groundwater.

Pilot Source Treatment. PCE contaminated the groundwater at an industrial site in New Jersey with concentrations ranging from 4,000 to 6,000 ug/L. 1,080 lb (490 kg) of HRC was injected via 23 direct-push points, arranged in a grid, covering an area of approximately 1,100 ft^2 (102 m^2). The HRC was applied to the unconsolidated aquifer over a 10 ft (3 m) interval, and monitoring results were taken from three existing wells located within the HRC application area.

Figure 2 shows changes in PCE, TCE, 1,2-DCE and VC concentrations through 124 days. PCE concentrations at the site decreased by an average of 85%, from 5,280 ug/L at day 0 to 775 ug/L at day 124. In conjunction with this decrease, TCE concentrations rose from 22.7 ug/L to 1,841 ug/L, 32 days after HRC application, and have since dropped to 866 ug/L. DCE concentrations at the site increased from 10 ug/L to 3,353 ug/L. The rise and fall in TCE along with the increase in biodegradation daughter product 1,2-DCE, shows the ability of HRC to enhance the anaerobic biodegradation of PCE.

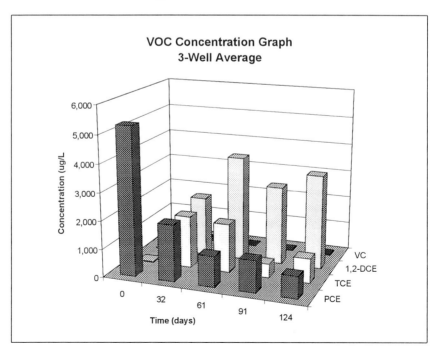

Figure 2. VOC concentration graph for NJ industrial site.

With reference to Table 1, the detection of significant levels of lactic acid and by-products and a concomitant rise in TOC indicates the presence of HRC-derived reducing power in the groundwater. The drop in sulfate and the rise in dissolved iron are geochemical indicators that anaerobic conditions favoring reductive dechlorination were developed.

3-well average	Units	baseline	day 32	day 61	day 91	day 124
Acetic Acid	mg/L	<1	37.7	112.7	359.0	326.7
Butyric Acid	mg/L	<1	10.7	18.7	107.0	160.7
Lactic Acid	mg/L	<1	35.0	42.7	496.7	274.0
Propionic Acid	mg/L	<1	54.3	136.7	373.3	428.0
Pyruvic Acid	mg/L	<0.1	0.2	<0.1	0.6	0.6
Sulfate	mg/L	47.7	42.3	38.3	12.7	NA
Iron, dissolved	mg/L	0.05	<0.2	1.0	25.0	22.5
Total Organic Carbon	mg/L	NA	105.9	154.2	659.5	691.7

Table 1. Metabolic acids and geochemical parameters at NJ industrial site.

Full Scale Source Treatment: A commercial dry cleaning facility released PCE into an aquifer underlying a strip mall in Washington. PCE concentrations within the gravelly sand aquifer reached levels of 500 to 600 ug/L. Approximately 1,000 lb (454 kg) of HRC was injected into a 2,000 ft^2 (186 m^2) source area via 55 direct-push points. Soon after the HRC application, a sewer leak (discovered post-injection) caused PCE concentrations to rise to approximately 67,000 ug/L. Despite the extremely high VOC levels caused by the sewer leak, HRC was able to enhance anaerobic bioremediation within the aquifer. The sewer leak was eliminated 170 days after the HRC application, and an additional 180 lb (81.7 kg) of HRC was added as a backfill amendment.

Overall VOC mass degradation results through 328 days are presented in Figure 3. Since the point of greatest contamination (the result of the sewer leak), PCE concentrations have decreased 99% (from 67,400 ug/L down to 259 ug/L). Concentrations of TCE initially increased (as a result of PCE degradation), but have since decreased 97% (from a high of 11,700 ug/L to 404 ug/L). The degradation of PCE and TCE has resulted in increased concentrations of daughter products cis-1,2-DCE and VC. These end products, while able to further degrade anaerobically, are now less oxidized and can degrade more readily when exposed to aerobic conditions which might occur at the at the edge of the plume. Contour plots depicting concentrations over time are presented in Figure 4.

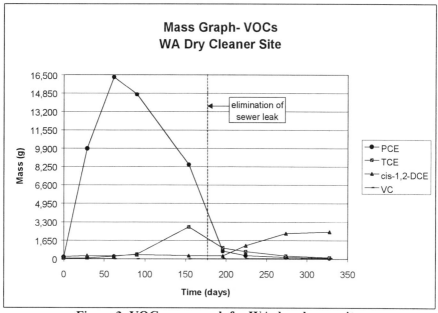

Figure 3. VOC mass graph for WA dry cleaner site.

CONCLUSIONS

In spite of our technological sophistication, we are still humbled by the fact that aquifer remediation is a macroscopic, multi-variable problem. There are exceptions; but, in general, groundwater contamination problems are hard to characterize, understand, monitor, and solve. Brute force is not usually the answer—at least not at an affordable price and monitored natural attenuation has serious limitations. The new paradigm being advanced here, and by others, is that the most sensible and proactive thing to do to close sites is to implement enhanced natural attenuation.

Through this terminology – enhanced natural attenuation - we establish that nature offers a solution while invoking the promise that there is a way to accelerate it. Unassisted natural remediation processes operate on a time scale that does not a satisfy many who are responsible for our environmental health and safety.

As demonstrated here and in other published work, the advent of controlled release electron donors such as HRC offers a passive, low-cost treatment option for in situ anaerobic bioremediation of chlorinated hydrocarbons and certain other inorganic compounds of concern that are subject to remediation by chemical reduction.

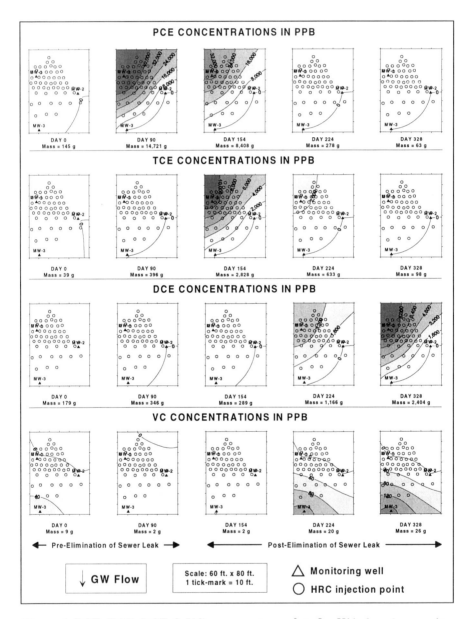

Figure 4. PCE, TCE, DCE & VC mass contour plots for WA dry cleaner site.

REFERENCES

Anderson, D. "Remedial Action Using HRC Under a State Dry Cleaning Program". *2nd International Conference on Remediation of Chlorinated and Recalcitrant Compounds*. May 22-25, 2000, Monterey, California (in press).

Boyle, S.L., V.B. Dick, M.N. Ramsdell and T.M. Caffoe. "Case Study: Chlorinated VOC Site Bioremediation Using Injectable HRC^{TM}". *2nd International Conference on Remediation of Chlorinated and Recalcitrant Compounds*. May 22-25, 2000, Monterey, California (in press).

Dooley, M., W. Murray and S. Koenigsberg. 1999. "Passively Enhanced In Situ Biodegradation of Chlorinated Solvents". In: A. Leeson and B.C. Alleman (Eds.), *Engineered Approaches for In Situ Bioremediation of Chlorinated Solvent Contamination*, pp. 121-127. Battelle Press, Columbus, OH.

Dooley, M. and W. Murray. "HRC-Enhanced Bioremediation of Chlorinated Solvents". *2nd International Conference on Remediation of Chlorinated and Recalcitrant Compounds*. May 22-25, 2000, Monterey, California (in press).

Farone, W.A., S.S. Koenigsberg and J. Hughes. 1999. "A Chemical Dynamics Model for CAH Remediation with Polylactate Esters". In: A. Leeson and B.C. Alleman (Eds.), *Engineered Approaches for In Situ Bioremediation of Chlorinated Solvent Contamination*, pp. 287-292. Battelle Press, Columbus, OH.

Farone, W.A., S. Koenigsberg, T. Palmer and D. Brooker. "Site Classification for Bioremediation of Chlorinated Compounds Using Microcosm Studies". *2nd International Conference on Remediation of Chlorinated and Recalcitrant Compounds*. May 22-25, 2000, Monterey, California (in press).

Harms, W.D. "Enhanced Reductive Dechlorination of Subsurface Chlorinated Solvents". *2nd International Conference on Remediation of Chlorinated and Recalcitrant Compounds*. May 22-25, 2000, Monterey, California (in press).

Kallur, S. and S. Koenigsberg. 1999. "Enhanced Bioremediation of Chlorinated Solvents- A Single Well Pilot Study". In: A. Leeson and B.C. Alleman (Eds.), *Engineered Approaches for In Situ Bioremediation of Chlorinated Solvent Contamination*, pp. 181-184. Battelle Press, Columbus, OH.

Koenigsberg, S.S. and W. Farone. 1999. "The Use of Hydrogen Release Compound (HRC^{TM}) for CAH Bioremediation". In: A. Leeson and B.C. Alleman (Eds.), *Engineered Approaches for In Situ Bioremediation of Chlorinated Solvent Contamination*, pp. 67-72. Battelle Press, Columbus, OH.

Koenigsberg, S.S. and C.A. Sandefur. 1999. "The Use of Hydrogen Release Compound for the Accelerated Bioremediation of Anaerobically Degradable Contaminants: The Advent of Time-Release Electron Donors". *Remediation Journal.* 10(1): 31-53.

Lodato, M., D. Graves and J. Kean. "Enhanced Biological Reductive Dechlorination at a Dry-Cleaning Facility". *2nd International Conference on Remediation of Chlorinated and Recalcitrant Compounds.* May 22-25, 2000, Monterey, California (in press).

Logan, B.E., J. Wu, A. Elkin, K. Becker, R. Unz and S.S. Koenigsberg. "The Potential for In Situ Perchlorate Degradation". *2nd International Conference on Remediation of Chlorinated and Recalcitrant Compounds.* May 22-25, 2000, Monterey, California (in press).

Rhodes, J.A., P.T. Mottola and S. Gupta. "Large-Scale Remediation using HRC^{TM} on Groundwater Contaminated with Chlorinated VOCs". *2nd International Conference on Remediation of Chlorinated and Recalcitrant Compounds.* May 22-25, 2000, Monterey, California (in press).

Schumacher, T., W. Bow and J. Chitwood. "A Field Demonstration Showing Enhanced Reductive Dechlorination Using Polymer Injection". *2nd International Conference on Remediation of Chlorinated and Recalcitrant Compounds.* May 22-25, 2000, Monterey, California (in press).

Sheldon, J.K., S.S. Koenigsberg, K.J. Quinn and C.A. Sandefur. 1999. "Field Application of a Lactic Acid Ester for PCE Bioremediation". In: A. Leeson and B.C. Alleman (Eds.), *Engineered Approaches for In Situ Bioremediation of Chlorinated Solvent Contamination,* pp. 61-66. Battelle Press, Columbus, OH.

Sheldon, J.K. and K.G. Armstrong. "Barrier Implants for the Accelerated Bioattenuation of TCE". *2nd International Conference on Remediation of Chlorinated and Recalcitrant Compounds.* May 22-25, 2000, Monterey, California (in press).

Wu, M. 1999. "A Pilot Study Using HRC^{TM} to Enhance Bioremediation of CAHs". In: A. Leeson and B.C. Alleman (Eds.), *Engineered Approaches for In Situ Bioremediation of Chlorinated Solvent Contamination,* pp. 177-180. Battelle Press, Columbus, OH.

Zahiraleslamzadeh, Z.M. and J.C. Bensch. "Enhanced Bioremediation Using Hydrogen Release Compound (HRC^{TM}) in Clay Soils". *2nd International Conference on Remediation of Chlorinated and Recalcitrant Compounds.* May 22-25, 2000, Monterey, California (in press).

INSOLUBLE SUBSTRATES FOR REDUCTIVE DEHALOGENATION IN PERMEABLE REACTIVE BARRIERS

Matthew J. Zenker, North Carolina State University, Raleigh, NC, USA
Robert C. Borden, North Carolina State University, Raleigh, NC, USA
Morton A. Barlaz, North Carolina State University, Raleigh, NC, USA
M. Tony Lieberman, Solutions Industrial & Environmental Services, Raleigh, NC, USA
Michael D. Lee, Terra Systems, Inc., Wilmington, DE, USA

ABSTRACT: Laboratory screening studies were conducted to evaluate the suitability of several different edible oils for enhancing reductive dehalogenation in a permeable reactive barrier (PRB). Corn oil degraded most rapidly followed by liquid soybean oil, solid soybean oil and semi-solid soybean oil in buffered liquid media with an anaerobic digester sludge inoculum. In microcosm studies using sediment from a chlorinated solvent contaminated aquifer, semi-solid soybean oil supported rapid conversion of trichloroethene (TCE) and *cis*-1,2-dichloroethene (*cis*-DCE) to vinyl chloride (VC) and then to ethene. Molasses and liquid soybean oil also stimulated reductive dehalogenation; however, ethene production was slower than for the semi-solid soybean oil. Extensive field pilot tests are being conducted in a chlorinated solvent contaminated aquifer to evaluate two different approaches for enhancing reductive dehalogenation in permeable reactive barriers.

INTRODUCTION

Enhanced anaerobic bioremediation has been proposed as a method for remediating aquifers contaminated with chlorinated solvents. In this process, an organic substrate is injected into the aquifer to stimulate the growth of anaerobic dechlorinating bacteria by providing an electron donor for energy generation and carbon source for cell growth (Lee et al., 1998; McCarty and Semprini, 1994). For example tetrachloroethene (PCE) and TCE can be treated by the following reaction.

$$PCE \rightarrow TCE \rightarrow cis\text{-}DCE \rightarrow VC \rightarrow ethene$$

cis-DCE and VC are produced as intermediate compounds in this reaction. However, when a suitable microbial population and sufficient substrate are present, *cis*-DCE and VC will be completely degraded to the non-toxic end-product ethene.

A number of investigators have attempted to stimulate anaerobic biodegradation of chlorinated solvents by continuously flushing a soluble, readily biodegradable substrate through the contaminated zone. While this approach can be successful, there are a number of very important limitations.

- There is a significant initial capital cost associated with installation of the required tanks, pumps, mixers, injection and pumping wells, and process controls required to continuously feed a soluble, easily biodegradable substrate.

- Operation and maintenance (O&M) costs can be high because of frequent clogging of injection wells and the labor for extensive monitoring and process control. In a field demonstration by Remediation Technologies Development Forum using lactate (RTDF, 1996), injection wells initially had to be cleaned once per month. By the end of the project they had to be cleaned weekly to prevent excessive head buildup.

- In many aquifers, the cleanup rate is controlled by the rate of contaminant dissolution and transport by the mobile groundwater. When DNAPLs are present or contaminants are present in lower permeability zones, dissolution rates will be slow and a long time will be required for aquifer cleanup. Under these conditions, high O&M costs are a major problem.

Permeable reactive barriers (PRBs) are being considered at many sites because they are expected to have much lower O&M costs than active pumping systems. As solvents or other contaminants migrate through a PRB, the contaminants are removed or degraded. We are developing an alternative barrier system for controlling the migration of chlorinated solvents. An 'insoluble,' slowly degradable substrate will be injected into the aquifer in a barrier configuration to serve as an electron donor and carbon source for reductive dechlorination. This approach is expected to be applicable to many of the same contaminants that can be remediated using metallic iron barriers. However, capital and long-term O&M costs for these systems are expected to be much lower than for competing processes.

Objectives. The overall objective of this work was to develop an effective, low-cost barrier system for controlling the migration of chlorinated solvents by providing a slowly degradable, 'insoluble' substrate to enhance biological reductive dehalogenation. Laboratory studies were first performed to identify organic substrates that would be suitable for use in these barriers. The ideal substrate would have the following characteristics.

- Moderately resistant to biodegradation to sustain long-term dehalogenation in a subsurface environment.

- Sufficiently biodegradable to support complete reductive dehalogenation of chlorinated ethenes to non-toxic end products.

- Easily distributed throughout the barrier zone.

- Inexpensive and safe.

Pilot studies were then conducted at a site contaminated with chlorinated solvents to evaluate different approaches for distributing and immobilizing the organic substrate.

LABORATORY STUDIES

Biodegradability Screening. Preliminary biodegradability screening studies were first conducted to evaluate the following edible oils for their potential use in a biologically active barrier system: 1) corn oil; 2) liquid soybean oil (100); 3) viscous, liquid soybean oil (230); 4) semi-solid soybean oil (560); and 5) solid soybean oil (670). Parallel incubations where conducted using ethanol, corn syrup, and molasses for comparison.

Each carbon source (50 mg) was added to triplicate 160 mL glass serum bottle containing 100 mLs of anaerobic medium and 5 mL of anaerobic digester sludge, and sealed with a thick black butyl rubber stopper and aluminum crimp top. Bottles were incubated at 35 °C for approximately two months. Control bottles containing only digester sludge and media were incubated in parallel. As an indicator of substrate biodegradability, total gas production ($CH_4 + CO_2$) was monitored over time using a wetted syringe. Specific gas production was calculated as:

[(gas production – avg. of gas production in control)/ weight of carbon substrate]

Results from the initial screening assays are shown in Figure 1. As expected, ethanol, corn syrup and molasses all degraded very quickly with rapid gas production (data not shown for ethanol and corn syrup). Gas production was somewhat slower for corn oil, followed by the two liquid soybean oils (100 and 230), and the solid soybean oil (670). Gas production was slowest for the semi-solid soybean oil (560). It is not clear why gas production was more rapid for the solid soybean oil than for the semi-solid; however, this pattern was consistent in all three triplicate semi-solid and solid soybean oil incubations.

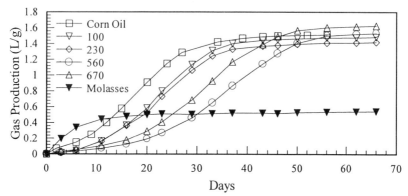

FIGURE 1. Specific gas production versus time in biodegradability screening assays. Results are the average of triplicate incubations.

Chlorinated Solvent Biodegradation. Microcosm experiments were then conducted to evaluate the suitability of liquid soybean oil (100) and the semi-solid hydrogenated soybean oil (560) for enhancing reductive dehalogenation of chlorinated ethenes. Parallel incubations were also conducted using molasses as a positive control. Sediment with previous exposure to tetrachloroethene was obtained from a PCE contaminated site in the North Carolina coastal plain and placed in serum bottles along with native groundwater and a substrate. The native groundwater contained approximately 15 mg/L of TCE and 80 mg/L of cis-1,2-dichloroethene. Each treatment was conducted in triplicate with matching abiotic and live controls. Each bottle was monitored for the dissolved and headspace chlorinated ethenes, dissolved organic carbon, and chloride ion.

Reductive dehalogenation was most rapid in the microcosms amended with semi-solid soybean oil (Figure 2). TCE and DCE were reduced to below detection within two months with concurrent production of vinyl chloride and ethene. After 130 days of incubation, vinyl chloride in the headspace was reduced to near the analytical detection limit with essentially complete conversion of TCE to ethene. Molasses and liquid soybean oil also stimulated reductive dehalogenation (data not shown); however ethene production was slower than for the semi-solid soybean oil. The reason for the slower ethene production using the more rapidly degraded substrates is not understood at this time.

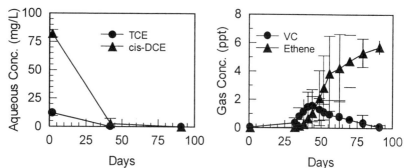

FIGURE 2. TCE and *cis*-DCE biotransformation to ethene in microcosms amended with semi-solid hydrogenated oil. Error bars are the standard deviation of triplicate microcosms.

FIELD PILOT TEST

An extensive pilot test of this process is being conducted in a chlorinated solvent plume at Dover Air Force Base near Dover, DE. Two different barrier configurations are being evaluated: 1) injection of liquid soybean oil in closely spaced wells; and 2) injection of a soybean oil-in-water emulsion in moderately spaced wells (see Figure 3).

Each barrier is constructed with 1-inch dia. continuously screened direct push wells. In Barrier 1, ~20 gallons of liquid soybean oil are being injected into each well followed by ~50 gallons of groundwater resulting in 18 to 24 inch cylindrical plugs of oil spaced 24-inches on center (OC). In Barrier 2, ~200 gallons of a soybean oil-in-water emulsion are being injected into each well followed by ~2,000 gallons of groundwater to distribute the oil. This should result in 6 to 8 ft.-diameter cylindrical columns of treated sediment spaced 5 ft. OC. Groundwater upgradient and downgradient of each barrier is being monitored to evaluate the effectiveness of each approach for distributing the oil and enhancing chlorinated solvent biodegradation.

ACKNOWLEDGEMENTS

Financial support the laboratory screening and microcosm experiments was provided by a grant from S&ME, Inc. to North Carolina State University. Support for the field pilot studies is being provided by the Technology Transfer Division of the Air Force Center for Environmental Excellence.

REFERENCES

Lee, M.D., J.M. Odom, and R.J. Buchanan, Jr. 1998. "New perspectives on biological dehalogenation of chlorinated solvents: Insights from the field." *Ann. Rev. Microbiol. 52*: 423-452.

McCarty, P.L., and L. Semprini. 1994. "Ground-water treatment for chlorinated solvents." In Norris et al. (Eds.) *Handbook of Bioremediation*, pp. 87–116. Lewis Publishers, Boca Raton, FL.

U. S. Environmental Protection Agency Newsletter, EPA 542-F-96-010B. Remediation Technologies Development Forum-Bioremediation of Chlorinated Solvents Consortium. September 1996.

Potential Electron Donors 53

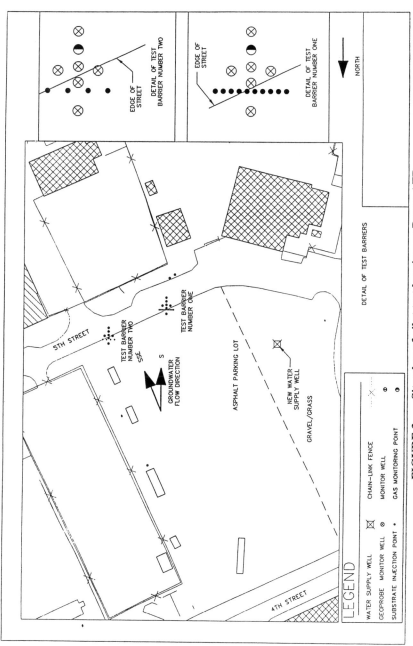

FIGURE 3. Site plan of pilot test barriers at Dover AFB

PCE DECHLORINATION WITH COMPLEX ELECTRON DONORS

Thomas D. DiStefano, Rishi Baral
Bucknell University, Lewisburg, Pennsylvania 17837 USA

ABSTRACT: A laboratory-scale study was completed to compare the ability of sugar, flour, corn steep liquor, molasses, non-fat milk, and whey to support anaerobic dechlorination of tetrachloroethene (PCE). The electron donors were compared based on startup time to achieve dechlorination, the extent of PCE dechlorination facilitated by each donor, the minimum electron donor dose necessary to achieve complete PCE removal, and unit cost. The time required to achieve routine dechlorination for each donor was (in days): corn steep liquor (10), whey (10), milk (10), molasses (14), sugar (26) and flour (30). Regarding the extent of PCE dechlorination, ethene production was achieved by milk- (115) and whey (115) -fed cultures. The other electron donors did not produce ethene. Corn steep liquor-, whey-, and sugar-fed cultures needed 5 times the stoichiometric amount to facilitate complete PCE removal, whereas cultures fed milk, molasses, and flour needed 20 times the stoichiometric amount. Based on minimum dose requirements, costs to achieve stoichiometric conversion of PCE to cis-DCE are estimated to be ($ per lb. PCE) whey (0.04), molasses (0.16), milk (0.18), sugar (0.40), corn steep liquor (0.44), and flour (0.85).

INTRODUCTION

Anaerobic consortia that dechlorinate PCE require the availability of an electron donor to sequentially reduce less chlorinated daughter products. Many electron donors have been shown to support PCE dechlorination, such as glucose, acetate, formate, methanol, hydrogen (Freedman and Gossett, 1989), ethanol, lactate (De Bruin, et al., 1992), butyrate, propionate (Fennell, et al., 1997), benzoate (Scholtz-Muramatsu, et al., 1990), and toluene (Sewell and Gibson, 1991).

For this study, non-fat milk, molasses, flour, corn steep liquor (light steep water - LSW), sugar, and whey were evaluated as potential electron donors for PCE dechlorination. Three major issues were examined during this experiment: (1) the time lag before the onset of PCE dechlorination (i.e. the startup time), (2) the extent of PCE dechlorination that was facilitated by each electron donor, and (3) the minimum dose needed to support dechlorination.

MATERIALS AND METHODS

Chemicals and Stock Solutions. PCE, TCE, and cis-1,2-dichloroethene (DCE) were obtained in neat liquid form (Aldrich Chemical Co.) for use as analytical standards. Spectrophotometric grade PCE (>99% pure, ACROS) and high resolution gas chromatography (GC Resolv) grade methanol (99.9% pure, Fisher Scientific) were utilized as culture substrates. Vinyl chloride (VC), ethene (ETH) and methane (Matheson Gas Products, Inc.) were obtained as gases for development of analytical standards.

Yeast extract (Difco Laboratories) solution was prepared as a 50 g/L stock. Corn steep liquor (light steep water - LSW) was obtained from Staley's (Sagamore, IN), whey was obtained from Sigma Chemical, and dried non-fat milk, molasses, flour, and sugar were purchased locally.

Bacterial Cultures and Procedures. All cultures were maintained in 70-mL serum bottles containing 50 mL of liquid. The bottles were sealed with Teflon™-lined butyl rubber stoppers and aluminum crimp caps. For all studies, a water chiller (Neslab) was employed to maintain culture bottles at 20°C.

Inocula was obtained from anaerobic digesters in the Hamilton Township (NJ) Wastewater Treatment Facility. All anaerobic cultures were maintained in anaerobic basal salts medium (BSM) according to a recipe modified from Zeikus (Zeikus, 1977).

Culture Preparation and Maintenance. Fresh anaerobic sludge that had not been previously exposed to the electron donors was employed. Cultures employed 6 mL anaerobic sludge and 44 mL basal salts media. Duplicate bottles were employed for each electron donor. Methanol-fed cultures and yeast extract controls were maintained for comparison. Each bottle was routinely amended with electron donor (4 mg COD/bottle, or 80 mg/L), PCE (0.26 µmole) and yeast extract (20 mg/L).

Fresh serum bottles were prepared to determine the minimum electron donor dose. The resulting cultures were initially maintained at an electron donor dose of five times the stoichiometric electron equivalent requirement to convert 0.26 µmole PCE to ETH.

All bottles were maintained at a 100-day solids retention time. No additional PCE was added until the previous dose declined significantly.

Analytical Procedures. The total mass of each volatile organic within serum-bottle cultures was determined using a 0.3-mL headspace gas injection into a gas chromatograph (Hewlett-Packard 6890) equipped with a flame ionization detector (FID). Samples were injected into an 8-ft x 1/8-in stainless steel column packed with one percent SP-1000 on 60/80 Carbopack-B™ (Supelco, Inc.). The carrier gas was nitrogen. The output from the FID was sent to a computer equipped with ChemStation™ software (Hewlett-Packard). The oven was temperature-programmed as follows: 60°C for two minutes, temperature ramp of 20°C per minute to 150°C, temperature ramp of 10°C per minute to 200°C, hold at 200°C for 3.2 minutes.

The chemical oxygen demand (COD) of each electron donor was determined so that each donor could be compared on an equivalent COD basis. To experimentally determine COD, stock solutions of each electron donor were prepared and measured in triplicate (Clesceri, et al., 1989).

RESULTS

Average COD for each electron donor was determined to be (mg COD/mg electron donor): molasses (3.19), methanol (1.49), non-fat milk (1.33), sugar (1.12), whey (1.04), flour (0.71), and light steep water (0.25).

Table 1 summarizes results regarding startup time and extent of dechlorination achieved. For this study, routine dechlorination is specifically defined as the time required for microbes to facilitate 0.26 µmole (0.71 mg/L aqueous) PCE dechlorination over a 48-hour incubation.

Note that light steep water, whey, milk, molasses, and methanol bottles exhibited routine PCE dechlorination in 10-14 days whereas sugar and flour required 26-30 days. The yeast extract control bottles achieved routine dechlorination in 45 days. LSW, whey, milk, molasses, and methanol should likely be considered equivalent with respect to startup, whereas sugar flour, and yeast extract performed very similarly. Therefore, sugar or flour as electron donors offer no advantage over yeast extract from a startup perspective.

In Table 1, "PCE to TCE" means that added PCE was converted to mostly TCE, whereas "PCE to DCE" and "PCE to VC" refers to complete dechlorination of PCE to DCE, or VC, respectively.

Electron donors with the least acclimation time also facilitated complete PCE dechlorination to DCE in the least time. From Table 1, LSW achieved complete dechlorination of PCE to DCE in 10 days, and methanol in 39 days. All other donors (including the yeast extract control) performed similarly in that they facilitated PCE to DCE dechlorination in 44-48 days. As shown, methanol, milk, and whey achieved stoichiometric conversion of PCE to VC. Methanol achieved PCE to VC conversion in 72 days, whereas milk and whey, needed 110 to 120 days. Also, methanol, milk, and whey facilitated ETH production; these donors would therefore be worthy of consideration for strictly anaerobic conditions. However, further studies are necessary to determine if stoichiometric production of ETH is possible. Given that LSW-fed cultures were maintained for 70 days, it possible that LSW may facilitate PCE dechlorination to VC and ETH.

TABLE 1. Performance of Electron Donors

Electron Donor	Days to achieve			ETH produced
	PCE to TCE	PCE to DCE	PCE to VC	
Methanol	10	39 on	72	37
Whey	10	46 on	120	115
Non-fat milk	10	48 on	110	115
LSW[†]	—	10 on	none*	none
Molasses	14	48 on	none*	none
Sugar	26	48 on	none	none
Flour	30	48 on	>140?	none
Yeast control	45	44 on	none	none

* some VC detected after 48 days (LSW) and 115 days (molasses).
[†] LSW culture was monitored for 70 days.

Minimum electron donor requirements and costs to achieve complete removal of PCE are summarized in Table 2. For the purpose of this study, the minimum electron donor dose is defined as that dose that facilitates dechlorination of 0.26 µmole PCE over a two-day period. Yeast extract was not employed for this study.

To achieve complete dechlorination of a 0.26 µmole PCE in 48 hours, light steep water, sugar, and whey are the most efficient electron donors in that 5 times the stoichiometric requirement was sufficient. By contrast, milk, molasses, and flour required a minimum dose of 20 times the stoichiometric requirement to achieve PCE dechlorination. Methanol was least efficient in that 50 times the stoichiometric requirement was needed.

Among the candidate electron donors, whey is the cheapest ($0.04/ lb. PCE), followed by molasses ($0.16/lb. PCE), non-fat milk ($0.18/lb. PCE), sugar ($0.40/lb. PCE), light steep water ($0.44/lb. PCE), methanol ($0.64/lb. PCE), and flour ($0.85/lb. PCE). Note that these costs are referenced to the *Engineering News Record* Construction Cost Index (CCI) of 6091 (August 1999). Because whey is often considered a waste product by some food processors, a cheaper price may be negotiated directly with a producer. Note that this comparison considers only the unit cost to dechlorinate PCE to DCE.

TABLE 2. Electron Donor Costs (ENR CCI= 6091 August,1999)

Electron Donor	Min. Dose*	Unit Cost (Reference)	$/lb. Donor	$/lb. COD	lb. PCE / lb.COD++	$/lb. PCE
Whey	5X	$0.02/lb. (1)	0.02	0.02	0.52	0.04
Molasses	20X	$58.15/ton (2)	0.03	0.02	0.13	0.16
Milk	20X	$0.025/lb. (1)	0.03	0.02	0.13	0.18
Sugar	5X	$0.22/lb. (2)	0.22	0.21	0.52	0.40
LSW	5X	$50/ton (4)	0.03	0.23	0.52	0.44
Methanol	50X	$0.30/gallon (3)	0.05	0.03	0.05	0.64
Flour	20X	$11.62/100 lb. (2)	0.12	0.11	0.13	0.85

Reference: 1. Chicago Mercantile Exchange; 2. The CRB Commodity Yearbook 1998, Bridge Commodity Research Bureau, Chicago; 3. Chemical Market Reporter (Nov. 28, 1998); 4. Staley's, Sagamore, IN.
*The minimum cost required for each electron donor was determined to facilitate dechlorination of 0.26 μmol PCE to DCE in 48 hours. No yeast extract was added.
++Mass PCE dechlorinated per mass COD electron donor added

DISCUSSION

The relative acclimation time of these complex donors may be useful in choosing between donors of similar cost and effectiveness. Electron donors that achieve rapid startup and partial dechlorination to, say DCE or VC, may be useful in a sequential anaerobic/aerobic process or for in-situ applications with downgradient aerobic regions. Aerobic biodegradation of DCE and VC has been observed (Hartmans, et al., 1985; Hartmans and de Bont, 1992; Freedman and Herz, 1996). From a startup perspective, LSW achieved complete PCE dechlorination to DCE about 30-40 days faster than the other donors tested. The other donors should likely be considered equivalent regarding the ability to initiate PCE to DCE dechlorination in a 40-50-day time frame. Perhaps some donors facilitate more rapid development of a diverse mixed culture that supports dechlorination with reducing equivalents, and possibly, other nutritional requirements.

Electron donors that indicate the capability to achieve ETH production may be useful for strictly anaerobic conditions. Whey, milk, methanol eventually achieved complete conversion of PCE to VC and some ETH production. Additional study may be useful in determining if these donors can support complete dechlorination to ETH. Furthermore, longer-term studies may demonstrate little difference in performance of all donors in that, given ample time, all donors may eventually achieve ETH production. Such findings from long-term studies have been demonstrated previously (Carr and Hughes, 1998). However, these investigators employed much higher ratios of donor to PCE than used in this study. As recommended by Fennell and Gossett (Fennell and Gossett, 1999), the concentration of PCE and its potential inhibitory effect on methanogens and other non-dechlorinating anaerobes that compete for electron donor may be an important considering in determining the best electron donor for a particular application.

Electron donor requirements obtained in this study are likely less than those recommended by others because dechlorination of PCE was limited to DCE. DCE or VC would require further treatment — by aerobic or anaerobic means — which would increase treatment costs. Furthermore, the costs determined in this study were obtained from laboratory-scale experimentation. Full-scale costs would likely be greater due to competition for donor by indigenous microbes, difficulty in delivering donor to the entire contaminant plume, and the desired extent of dechlorination.

SUMMARY AND CONCLUSIONS

In a laboratory-based study, molasses, whey, dried non-fat milk, sugar, corn steep liquor, flour were examined as potential electron donors to determine their effectiveness in facilitating dechlorination of PCE. The effectiveness of each donor was evaluated with respect to start-up, maturation of the dechlorination process, ultimate extent of dechlorination, and unit cost. Based on 140 days of operation, all donors supported dechlorination, and differences in performance were noted in the time needed to achieve complete PCE removal and the ultimate extent of dechlorination. Economic analysis combined with minimum required dose of donor suggests that cost savings may be realized by prudent choice of these donors; however, the cost difference will likely be significant only for large sites with large quantities of PCE to dechlorinate. Other factors, which were not considered, such as ease of delivery of donor and related biomass growth, may be significant factors that could influence the choice of electron donor.

ACKNOWLEDGMENT

This research was supported by the U.S. Air Force STTR program and was completed by Bucknell University, in conjunction with Envirogen, Inc. The project was managed by the Air Force Research Laboratory, Airbase and Environmental Technology Division, Tyndall Air Force Base, Florida.

REFERENCES

Carr, C.S., and Hughes, J.B. (1998). Enrichment of high-rate PCE dechlorination and comparative study of lactate, methanol, and hydrogen as electron donors to sustain activity. *Environ. Sci. Technol.*, **32**(12), 1817-1824.

Clesceri, L.S., Greenberg, A.E., and Trussel, R.R. (Ed.). (1989). Standard methods for the examination of water and wastewater (17th ed.). American Public Health Association.

De Bruin, W.P., Kotterman, M.J.J., Posthumus, M.A., Schraa, G., and Zehnder, A.J.B. (1992). Complete biological reductive transformation of tetrachloroethene to ethane. *Appl. Environ. Microbiol.*, **58**(6), 1996-2000.

Fennell, D.E., and Gossett, J.M. (1999). Comment on "Enrichment of high-rate PCE dechlorination and comparative study of lactate, methanol, and hydogen as electron donors to sustain activity". *Environ. Sci. Technol.*, **33**(15), 2681-2682.

Fennell, D.E., Gossett, J.M., and Zinder, S.H. (1997). Comparison of butyric acid, ethanol, lactic acid, and propionic acid as hydrogen donors for the reductive dechlorination of tetrachloroethene. *Environ. Sci. Technol.*, **31**(3), 918-926.

Freedman, D.L., and Gossett, J.M. (1989). Biological reductive dechlorination of tetrachloroethylene and trichloroethylene to ethylene under methanogenic conditions. *Appl Environ Microbiol*, **55**(9), 2144-2151.

Freedman, D.L., and Herz, S.D. (1996). Use of ethylene and ethane as primary substrates for aerobic cometabolism of vinyl chloride. *Water Environ. Res.*, **68**(3), 320-328.

Hartmans, S., de Bont, J., Tramper, J., and Luyben, K. (1985). Bacterial degradation of vinyl chloride. *Biotechnol. Lett.*, 7(6), 383-388.

Hartmans, S., and de Bont, J.A.M. (1992). Aerobic vinyl chloride metabolism in *Mycobacterium aurum* L1. *Appl. Environ. Microbiol.*, **58**(4), 1220.

Scholtz-Muramatsu, H., Szewzyk, R., Szewzyk, U., and Gaiser, S. (1990). Benzoate can serve as a source of reducing equivalents for PCE reduction. *FEMS Microbiol Lett*(66), 81.

Sewell, G.W., and Gibson, S.A. (1991). Stimulation of the reductive dechlorination of tetrachloroethene in anaerobic aquifer microcosms by the addition of toluene. *Environ Sci Technol*, **25**(5), 982-984.

Zeikus, J.G. (1977). The biology of methanogenic bacteria. *Bacteriological Reviews*, **41**(2), 514-541.

ANAEROBIC OXIDATION OF (CHLORINATED) HYDROCARBONS

John A. Dijk and Jan M. de Bont (Wageningen University, Wageningen, The Netherlands)
Xiaoxia Lu (Peking University, Beijing, P.R. China)
Petra M. Becker, Tom N.P. Bosma, Huub H.M. Rijnaarts and Jan Gerritse (TNO Institute of Environmental Sciences, Energy Research and Process Innovation, Apeldoorn, The Netherlands)

ABSTRACT

Batch and soil column experiments were performed to investigate the potential of anaerobic oxidation processes for natural and stimulated degradation of chlorinated and non-chlorinated hydrocarbons. Model pollutants which included 1,2-dichloroethane (DCA), 1,1,1-trichloroethane (TCA), vinylchloride (VC), cis-1,2-dichloroethene (CIS), trans-1,2-dichloroethene (TRANS), trichloroethene (TCE), monochlororbenzene (MCB), 2-chloroethanol (CE) and octane were supplied as electron donor under iron (III), manganese (IV) or nitrate reducing conditions. Evidence was obtained that DCA, VC, MCB, CE and octane can be oxidised under anoxic conditions. Particularly nitrate-reducing conditions may have great potential for natural or stimulated remediation of sites contaminated with these particular hydrocarbons.

INTRODUCTION

Low-chlorinated and non-chlorinated hydrocarbons enter soil and groundwater systems through accidental spills, for example at industrial sites or petrol stations, and belong to the most important pollutants. In addition, they are formed by partial dechlorination of highly chlorinated hydrocarbons. In contaminated soil and groundwater systems that are anoxic, microbial degradation routes are based on processes for which oxygen is not needed. Reductive dechlorination is a well-established anaerobic biotransformation pathway for many highly chlorinated solvents such as tetrachloroethene (PCE) or hexachlorobenzene. Recent research indicates that contaminants such as VC, CIS, TCA, dichloromethane and n-alkanes (C_5-C_{12}) can also be removed through anaerobic oxidative pathways (Bradley et al., 1998; Bradley et al., 1997; Gerritse et al., 1999; Kohler-Staub et al., 1995; Rabus et al., 1999; Sherwood et al., 1997). Oxidised iron, manganese or nitrate may serve as alternative electron acceptors for the bacteria involved in these anaerobic oxidations. Knowledge of the role and potential of anaerobic oxidation processes for natural or stimulated degradation of aliphatic hydrocarbons is still limited.

Objectives. The aim of this paper is to identify the potential for anaerobic oxidation processes for natural and stimulated degradation of chlorinated and non-chlorinated hydrocarbons. Both batch and soil column experiments were performed with material obtained from five different polluted sites. Model-pollutants that were supplied as electron donors included: DCA, TCA, VC, CIS, TRANS, TCE, MCB, CE and/or octane. Biotransformation of these contaminants was investigated in the presence of nitrate, ferrihydrite ($Fe(OH)_3$) or manganese oxide (MnO_2) as electron acceptors.

MATERIALS AND METHODS

Batch cultures. Batches were prepared under N_2 atmosphere in 120 ml bottles which were crimp-sealed with butyl or viton rubber stoppers and contained 50 ml media. The growth media consisted of the following components: Na_2HPO_4/KH_2PO_4 buffer (20 mM, pH 7), 1.0 g/L NH_4Cl, 0.1 g/L $MgSO_4 \cdot 7H_2O$, 0.05 g/L $CaCl_2 \cdot 2H_2O$, 0.1 g/L yeast extract, 0.1 % resazurin, 1 ml/L trace elements (Gerritse and Gottschal, 1992) and vitamin solution (4-aminobenzoic acid, 0.1 mg/L; folic acid 0.05 mg/L; DT-α-lipoic acid, 0.01 mg/L; riboflavin, 0.1 mg/L; thiamine hydrochloride, 0.2 mg/L; nicotinic acid, 0.2 mg/L; pyridoxine hydrochloride, 0.5 mg/L; pantothenate, 0.1 mg/L; vitamin B_{12}, 0.1 mg/L; biotin 0.002 mg/L). For the batches grown with chloroethanol, NH_4Cl was replaced by $(NH_4)_2SO_4$ and 500 ml bottles were used containing 250 ml media. Vitamins were filter sterilised and added after autoclaving the media. Electron donors were added separately from autoclaved stock solutions to final concentrations of 1–10 mM. Electron acceptors were added to final concentrations of 5–10 mM ($NaNO_3$, MnO_2 or $Fe(OH)_3$). Amorphous $Fe(OH)_3$ was synthesised by neutralising a 0.4 M $FeCl_3$ solution with 1 M NaOH until a pH of 7 was achieved. Manganese oxide was prepared by mixing equal amounts of 0.4 M MnO_4 and 0.6 M $MnCl_2$ and adjusting the pH to 10 with 1 M NaOH. After preparing, the solutions were washed with milli-Q water by centrifugation. All solutions were prepared under N_2 atmosphere. Inoculum material was obtained from six different polluted sites in The Netherlands (Rotterdam, Arnhem, Tilburg, Uden, Groningen and Wageningen). Batches were incubated horizontally at 20°C under continuous shaking (100 rpm.).

Soil column Set-up. Four glass columns (35 cm length x 3.6 cm internal diameter) were packed in an anaerobic glove box with sediment from four different contaminated sites (Rotterdam, Arnhem, Tilburg and Uden). A layer of glass wool and glass beads were put at the bottom of the column to avoid clogging of the inlet. Sampling ports were sealed with both butyl rubber and viton rubber septa. The columns were percolated (up-flow) with media at a flow-rate of 5.8 ml/h (liquid retention time ± 18 hours) by a peristaltic pump (505S, Watson Marlow Ltd., UK). Influent medium was continuously stirred and flushed with N_2/CO_2 (90:10 v/v) and consisted of Na_2HPO_4/KH_2PO_4 buffer (2 mM, pH 7), 0.1 g/L NH_4Cl, 0.01 g/L $MgSO_4 \cdot 7H_2O$, 0.005 g/L $CaCl_2 \cdot 2H_2O$ and 0.1 ml/L trace elements. A substrate mixture with 10 mM DCA, 10 mM CIS, 10 mM TRANS, 5 mM TCE, 1mM MCB, 1 mM TCA and 1 mM VC was injected (58 µl/h) into the medium flow by a syringe pump. The substrate cocktail was mixed and diluted 100-fold with the medium in a stainless steel coil construction just before entering the column. $Fe(OH)_3$ (about 80 mmol) and MnO_2 (about 50 mmol) were mixed with the soil material before filling the columns. Nitrate was supplied by a syringe pump (final concentration 5 mM) with the same pumping-rate as the substrate pump. The columns were operated in a $20°C$ climate room.

Analytic methods. DCA, TCA, CIS, TRANS, TCE, MCB, VC, ethene, ethane and methane were determined by gas chromatography (GC). Liquid samples of 0.5 ml were taken from the column inlet or outlet and injected into 22 ml vials containing 7.5 ml milli-Q water and 0.1 ml $HgCl_2$ (50 g/L). $HgCl_2$ was added to prevent further microbial degradation. The vials were automatically sampled (Tekmar headspace autosampler) and 100 µl headspace was injected in a Varian Genesis GC equipped with a Porabond-Q column (Chrompack B.V., Middelburg, The Netherlands, i.d. 0.32 mm, length 25 m) and a flame ionisation detector ($300°C$).The temperature program of the column oven ranged from $35°C$ (3 min) to $250°C$ (5.5 min) and increased with a rate of $10°C$ /min. Helium was used as carrier gas. Concentrations of octane were determined by analysing 100 µl headspace samples on a Chrompack CP9000 gas chromatograph equipped with a Chrompack CP-Sil 8B column (length 25 m, i.d. 0.22 mm). The oven temperature was kept at $75°C$ and the detector (FID) temperature was $270°C$. Optical densities were determined at 660 nm in a Perkin-Elmer 55 A UV-VIS spectrophotometer. Chloride, nitrate, nitrite, manganese and iron were all determined colorimetrically. Chloride was measured according to Bergman and Sanik (1957). In batch cultures with chloroethanol, chloride production was used as a measure for the chloroethanol consumption. Nitrate was measured with a 6% brucine solution in methanol and sulfuric acid. Nitrite was measured with a 1% sulfanilamide solution in 2.5 N HCL and a 0.02% N-(1-naphtyl) ethylenediamine solution in water. A Dr. Lange Pipetting test (Dr. Lange Nederland B.V.) was used for measuring manganese. Iron was determined according to Lovley and Philips (1986)

Calculations. In batch experiments, concentrations of (chlorinated) hydrocarbons were determined by comparing integration areas with a calibration curve and ignoring headspace-liquid partitioning. In the column experiments, no headspace was present. Half-lives were calculated according to $t_{1/2} = \ln 2/k$, where $k = (\ln c_1 - \ln c_2)/(t_2-t_1)$ (day^{-1}). For the soil columns $\Delta t = (\pi d^2 \theta [l_2-l_1])/4K_p$ (Where l_1 and l_2 are the distances to the bottom of the column; d is the inner diameter; θ is the effective porosity and K_p the medium-pump rate). The effective porosity was assumed to be 0.3.

RESULTS

Batch cultures. To reveal the potential for anaerobic oxidation of chlorinated solvents, soil samples obtained from five polluted locations were incubated with different chlorinated hydrocarbons in the presence of either nitrate, Fe(OH)$_3$ or MnO$_2$. The studied compounds were VC, CIS, TRANS, TCE, DCA, MCB, CE and octane. Enrichment cultures were obtained with the following electron donor / electron acceptor combinations: CE plus nitrate, DCA plus nitrate, octane plus nitrate, and DCA plus MnO$_2$. In batch cultures with other electron donor / electron acceptor combinations no degradation was observed within 500 days.

The batch culture with CE and nitrate was obtained from a transfer of a batch culture growing on DCA and nitrate. The DCA enrichment culture was described previously (Gerritse et al., 1999), and the first order degradation rate constants (k-values) of DCA ranged from 0.15 to 1.68 day^{-1} (unpublished). Because the DCA-degrading capacity of this culture decreased after repeated transferring, CE (suspected intermediate) was used as electron donor instead of DCA to further enrich and isolate the DCA degrading bacteria. After a lag phase of 6 days, CE degradation started (Figure 1).

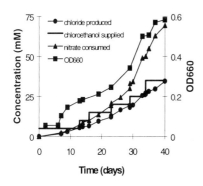 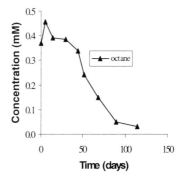

FIGURE 1. Degradation of 2-chloro-ethanol under nitrate reducing conditions.

Figure 2. Degradation of octane under nitrate reducing conditions.

The total amount of CE added was 35 mM, which resulted in a chloride production of 34 mM and a nitrate consumption of 69.6 mM. Nitrite accumulated temporarily to concentrations of about 1 mM but subsequently was further reduced, possibly to nitrogen. The mass balance calculation indicates that nitrate-coupled CE oxidation proceeded according to following reaction:

$$C_2H_4OHCl + 2HNO_3 \rightarrow 2CO_2 + N_2 + 3H_2O + HCl$$

The rate of chloride production showed that CE degradation proceeded at a first-order decay rate of 0.066 day^{-1}. This rate was relatively low compared to the k-values of the DCA degrading culture from which it originated. The increase of the OD_{660} indicated that CE degradation was coupled to growth, corresponding to a µ of approximately 0.04 day^{-1}. After 5 transfers, the CE degrading enrichment culture had lost the capacity to transform DCA in the presence of nitrate.

Under manganese(IV) reducing conditions, DCA was degraded with a maximum transformation rate of 0.012 day^{-1} (data not shown). As a result of manganese reduction, 0.46 mM Mn(II) was formed. However, according to a complete oxidation reaction 1 mol DCA would require as much as 5 mol manganese. Further research is needed to explain this discrepancy. Compared to denitrifying conditions, the DCA-transformation rate obtained under manganese reducing conditions was relatively low.

Octane was degraded under denitrifying conditions (Figure 2). The observed maximum transformation rate of 0.018 day^{-1}, was in the same order as the transformation rate obtained for nitrate coupled CE oxidation. Nitrate was reduced (2.1 mM) and no nitrite was detected. Theoretically, 10 mol nitrate is required to oxidise 1 mol octane. Since approximately 400 µM octane was consumed, 4 mM nitrate would be needed for complete oxidation to CO_2. The fact that only 2.1 mM nitrate was used suggests that octane was not completely oxidised to carbon dioxide and that microbial biomass and organic acids (e.g. acetate) may have been produced.

Soil Columns. Soil column experiments were performed with a mixture containing the following chlorinated compounds in the influent medium (concentrations ranging from 10 to 100 µM): VC, CIS, TRANS, TCE, DCA, TCA and MCB. To enhance the potential for oxidative degradation of the chlorinated compounds three different columns were supplied with either $Fe(OH)_3$ (± 0.22 mol/g of soil), MnO_2 (± 0.14 mol/g of soil) or NO_3^- (influent concentration 5mM) as external electron acceptors. One control column was run to which no external electron acceptor was added. The columns were operated at a hydraulic retention time of about 1 day, and the influent and effluent concentrations of the chlorinated compounds were monitored over a period of approximately 200 days. In the column supplied with $Fe(OH)_3$ only DCA was instantly removed (> 75 %) (Figure 3A). This DCA-removal could not be contributed to adsorption, because compared to some of the other compounds such as TCA and MCB, the adsorption potential of DCA is much lower. Both DCA and VC were degraded (75 % and 100 % respectively) in the column amended with MnO_2 (Figure 3B). The molar ratio of manganese (II) versus manganese (IV) increased from zero to 0.12 over the length of the column, indicating the occurrence of manganese reduction. The nitrate reducing soil column also showed removal of DCA and VC (>90%), and MCB concentrations were reduced to less than 50% of the influent (Figure 3C).

Nitrite was formed (about 6 µM) and the nitrate concentration decreased (from 5 to 3 mM). Interestingly, in the control column (i.e. no additional electron acceptors supplied) more than 70% of DCA, VC and MCB were removed (Figure 3D).

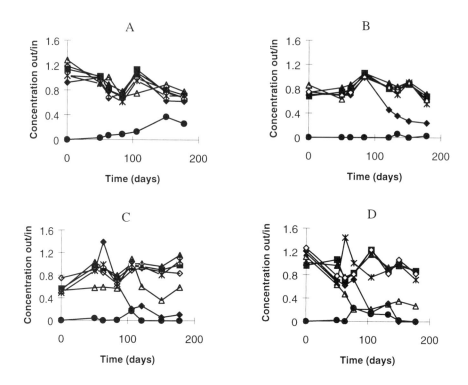

FIGURE 3. Degradation of a mixture of chlorinated compounds in soil columns supplied with A: $Fe(OH)_3$, B: MnO_2, C: NO_3^- and D: no external electron acceptor supplied. VC (♦), DCA (●), TCA (✕), cDCE (▲), tDCE (■), TCE (◇), MCB (△).

The lag-time for DCA degradation was about 2 weeks for all DCA-degrading columns while for VC it was about 3 months. Reductive dechlorination products such as ethane, ethene and benzene were not detected in any of the columns, indicating that VC, DCA and MCB degradation proceeded through anaerobic oxidation pathways. It is not clear which electron acceptors present in the soil may have been involved in the degradation of DCA, VC and MCB in the control column.

DISCUSSION

The results obtained both for the batch and soil column experiments reveal that it is possible to have anaerbic oxidation of DCA, CE, VC, MCB and octane (Table 1). DCA was degraded under all the three different electron accepting conditions studied (i.e. nitrate, iron and manganese-reduction) and the half-lives ranged from several hours to weeks. Chloroethanol and octane oxidation were only tested in

TABLE 1. Half-lives obtained for the oxidation of (chlorinated) hydrocarbons in batch cultures and soil columns

Pollutant	Abbreviation	Half-life (days) Electron acceptors		
		NO_3^-	MnO_2	$Fe(OH)_3$
monochlorobenzene	MCB	0.8^b	-	-
octane		17^a	n.d.	n.d.
vinyl chloride	VC	0.2^b	0.32^b	-
trans-1,2-dichloroethene	TRANS	-	-	-
cis-1,2-dichloroethene	CIS	-	-	-
trichloroethene	TCE	-	-	-
1,2-dichloroethane	DCA	0.14^a–4.6^a	0.09-25^a	0.25^b
1,1,1-trichloroethane	TCA	-		
2-chloroethanol	CE	11^a	n.d.	n.d.

^{a}obtained in batch experiments ^{b}obtained in soil column experiments

nitrate reducing batch cultures and proceeded with half-lives of 11 and 17 days, respectively. VC degradation was observed under both manganese and nitrate reducing conditions with half-lives of 0.2 and 0.3 days, respectively. Further research is required to definitely prove that the half-life of less than one day for MCB, observed in the control and the nitrate reducing columns, can be attributed to biological degradation processes.

Particularly nitrate-reducing conditions may have great potential for natural or stimulated remediation of sites contaminated with (low-chlorinated) hydrocarbons, since several of the chlorinated compounds were degraded under denitrifying conditions. Besides, denitrifiers are widely distributed in soil and sediments (Tiedje, 1988). Especially for stimulated remediation it is of big advantage that nitrate is highly soluble in water and can therefore be easily supplied to the soil or groundwater to enhance degradation.

ACKNOWLEDGMENT

This project is a co-operation between Wageningen University and TNO (Netherlands Organisation for Applied Scientific Research) and was financially supported by the Research Centre on Soil, Sediment and Groundwater Management and Remediation WUR/TNO, and the Royal Dutch Academy of Sciences.

REFERENCES

Bergman, J. G., and J. Sanik. 1957. "Determination of trace amounts of chlorine in naphta." *Anal. Chem.* **29**: 241-243

Bradley, P. M., J. E. Landmeyer, and R. S. Dinicola. 1998. "Anaerobic oxidation of [1,2-^{14}C]dichloroethene under Mn(IV)-reducing conditions." *Appl. Environ. Microbiol.* **64**: 1560-1562.

Bradley, P. M., and F. H. Chapelle. 1997. "Kinetics of DCE and VC mineralization under methanogenic and Fe(III)-reducing conditions." *Environ. Sci. Technol.* **31**: 2692-2696

Gerritse, J., A. Borger, E. van Heiningen, H. H. M. Rijnaarts, T. N. P. Bosma, J. Taat, B. van Winden, J. A. Dijk, and J. A. M. de Bont. 1999. "Assessment and monitoring of 1,2-dichloroethane dechlorination." In: A. Leeson and B. C. Alleman (Eds), *Engineered approaches for in situ bioremediation of chlorinated solvent contamination*, pp. 73-80. Batelle Press, Columbus, Richland

Gerritse, J., B. J. van der Woude, and J. C. Gottschal. 1992. "Specific removal of chlorine from the ortho-position of halogenated benzoic acids by reductive dechlorination in anaerobic enrichment cultures." *FEMS Microbiol. Lett.* **79**: 273-280

Kohler-Staub, D., S. Frank, and T. Leisinger. 1995. "Dichloromethane as the sole carbon source for *Hyphomicrobium* sp. strain DM2 under denitrification conditions." *Biodegradation* **6**: 229-235

Lovley, D. R., and E. J. P. Phillips. 1986. "Availability of Ferric iron for microbial reduction in bottom sediment of the fresh water tidal Potomac River." *Appl. Environ. Microbiol.* **52**: 751-757.

Rabus, R., H. Wilkes, A. Schramm, G. Harms, A. Behrends, R. Amann, and F. Widdel. 1999. "Anaerobic utilization of alkylbenzenes and n-alkanes from crude oil in an enrichment culture of denitrifying bacteria affiliating with the ß-subclass of *Proteobacteria*." *Environ. Microbiol.* **1**: 145-157

Sherwood, J. L., J. N. Petersen, and R. S. Skeen. 1997. "Biodegradation of 1,1,1-trichloroethane by a carbon tetrachloride-degrading denitrifying consortium." *Biotechnol. Bioeng.* **59**:393-399

Tiedje, J. M. 1988. Ecology of denitrification and dissimilatory nitrate reduction to ammonia. In: Zehnder (ed.). *Biology of Anaerobic Microorganisms*, pp. 245-303. Wiley-Liss, New York

IN SITU BIOWALL CONTAINING ORGANIC MULCH PROMOTES CHLORINATED SOLVENT BIOREMEDIATION

Patrick E. Haas, Technology Transfer Division, Air Force Center for Environmental Excellence, Brooks Air Force Base (AFB), Texas
Philip Cork, Civil Engineering, Environmental Flight, Offutt AFB, Nebraska
Carol E. Aziz, Ph.D., Mark Hampton, Groundwater Services Inc., Houston, Texas

ABSTRACT: In an effort to enhance in situ reductive dechlorination, an in situ permeable reactive treatment biowall (100 ft. (L) x 1 ft (W) x 23 ft (D)) was constructed which placed naturally-derived organic matter (i.e. a shredded tree compost and sand mixture) in direct flow-through contact with chlorinated solvents dissolved in ground water. The biowall was installed using a continuous trenching and backfill method. In addition, a surface amendment and control plot was constructed. A comparison of geochemical profiles before and 6 months after organic mulch emplacement indicated a progression toward more reducing conditions and significant shifts in cis-1,2dichloroethene (cis-DCE) to TCE ratios. For the in situ biowall, the mean cis-DCE/TCE ratio varied from 0.026 upgradient to 12 downgradient, or a mean increase in the ratio of a factor of 470. For the surface amendment plot, the cis-DCE/TCE ratio varied from 0.024 upgradient to 0.35 downgradient, with a mean increase in the ratio of a factor of 15. This work demonstrates that an in situ permeable biowall containing inexpensive organic material derived from shredded plant matter can enhance in situ reductive dechlorination (i.e. hydrogen releasing material).

INTRODUCTION:
A large body of evidence has been compiled that demonstrates that chlorinated ethenes like tetrachloroethene (PCE) and trichloroethene (TCE) are biodegraded under specific geochemical conditions. These conditions and biodegradation reactions (i.e. reductive dechlorination) are supported by the utilization of a carbon source (i.e. electron donor). Similarly, nitrate removal has been shown to occur in the presence of an acceptable carbon substrate (Boussaid, F. et. al., 1988; Robertson, W. D. et. al. 1995). The *Technical Protocol for Evaluating Natural Attenuation of Chlorinated Solvents In Ground Water, U. S. EPA, EPA/600/R-98/128* (Wiedemeier, T. H. et. al. 1998) provides descriptions of different plume behavior types based on the presence and source of a carbon substrate and the resultant biodegradation reactions. Specifically, a Type 2 behavior describes a zone within a plume where chlorinated solvents biodegradation is supported by the utilization of a naturally occurring organic carbon source. Type 2 behavior has been documented in coastal regions and wetland environments where chlorinated solvents migrate into organic carbon-rich zones. Thus, it follows that enhanced bioremediation could be facilitated by the establishment of similar conditions.

Objectives:
- Determine if an inexpensive, readily available source of natural organic carbon can sustain the in situ reductive dechlorination of chlorinated ethenes;
- Document performance, cost, and contaminant end products.

Site Description: Offutt AFB is located approximately five miles south of Omaha, Nebraska. The in situ biowall, surface amendment and control plots are situated near and west of the Base boundary where the depth to groundwater is only 3 to 10 bgs. (near MW-9S, 6 ft bgs). The hydraulic conductivity in the alluvial silt and clay near the test area is 1.8E-3 cm/sec and averaged 3.5 ft/day (mean of 5 slug tests in alluvial silt and clay). The hydraulic gradient is 0.01 ft/ft. Using an assumed effective porosity of 0.15, the computed groundwater seepage velocity is 0.34 ft/day or 124 ft/yr. At MW9S, the concentration of TCE is 372 ug/L. PCE is not a contaminant of concern, being detected only immediately downgradient of the source area at concentrations close to the quantitation limit (i.e., 1.5, 1.5, and 1.9 ug/L at MW7S, MW14, and MW18).

All three DCE isomers were detected in June-July 1996 groundwater samples, with cis-1,2-DCE being detected most frequently and at the highest concentrations (ranging up to 1,230 ug/L). Relatively low levels of trans-1,2-DCE and 1,1-DCE (ranging up to 9.4 and 28.6 ug/L, respectively) were detected, suggesting that cis-1,2-DCE is an intermediate of reductive dechlorination of TCE. The areal extent of the cis-1,2-DCE plume in June-July 1996 is significantly different than that of the TCE plume. The highest concentration of cis-1,2-DCE was detected near the source area. DCE concentrations appear to decrease west of the source area, but increase substantially near and west of the Base property boundary. Only two samples contained detectable concentrations of vinyl chloride, a reductive dechlorination product of DCE, and no ethene was detected at quantifiable concentrations. This data suggested that reductive dechlorination is generally not proceeding past the transformation of TCE to DCE.

MATERIALS AND METHODS:

An in situ permeable reactive treatment biowall (100 ft. (L) x 1 ft (W) x 23 ft (D)) which places naturally-derived organic matter (i.e. a shredded tree compost and sand mixture) in direct flow-through contact with chlorinated solvents dissolved in ground water has been constructed using a continuous trenching and backfill method. In addition, a surface amendment and control plot was constructed.

The trench was filled above the water table to approximately 2 feet below the ground surface (bgs) with a mixture of 50% mulch and 50% coarse sand. The mulch was generated by the Offutt AFB Civil Engineering Flight as part of a severe storm cleanup effort that was completed approximately eight months prior to biowall construction. Fallen tree limb and trunk material was passed through a tub grinder and stockpiled. The volume of material accumulated exceeded amounts needed for landscaping purposes. Thus, this material was temporarily a waste product. The material was actively composting upon mixing with sand and emplacement. The material was considered acceptable based on a visual inspection which verified that it contained more readily biodegradable material (i.e. partially composted leave and twig material) as well as more fibrous, less readily biodegradable material (i.e. cellulose and lignin-rich "chips").

The surface amendment plot was constructed as 30 ft long x 15-ft. wide plot of mulch with a 18-inch lift of mulch and sand mix located south of the trench location. The surface amendment plot was bermed to prevent run-off and to promote downward delivery of organic matter via infiltration to the contaminated ground water. Two monitoring wells (downgradient of the surface amendment area) were installed. The existing upgradient well is located 10 ft upgradient of the surface amendment area and the downgradient wells were spaced at 10 ft and 15 ft intervals.

Two additional monitoring wells, within the contaminated plume area were installed to act as control wells. These wells have been sampled for the same analytes as the treatment trench and surface amendment plot, but they will only be sampled at start-up and at the end of the test. Samples taken from these wells will be used to compare the rate and extent of chlorinated solvent degradation due to natural attenuation versus organic mulch addition.

Ground water sampling has been conducted for VOCs (PCE, TCE and degradation products such as DCE, chloroethane, vinyl chloride), alternate electron acceptors/by-products (NO_3^-, SO_4^{2-}, Fe^{3+}, CH_4, ethene, and ethane), total organic carbon, dissolved hydrogen, dissolved oxygen, pH, temperature, redox potential, and specific conductance. Sampling was conducted prior to organic mulch emplacement (i.e. January 1999), approximately six months later (i.e. June 1999), and is planned for approximately six months thereafter.

RESULTS AND DISCUSSION:

A comparison of geochemical profiles (Table 1) before and 6 months after organic mulch emplacement indicate a progression toward more reducing conditions as reflected in depressed dissolved oxygen levels and lower redox potential levels. Shifts toward more reducing conditions are not apparent in upgradient wells. There is some evidence that downgradient wells are displaying a change in geochemical profiles. Monitoring wells 31S and 32S exhibit an apparent lowering in dissolved oxygen and redox potential measurements and increase in dissolved methane. These results indicate that ground water geochemical conditions have not been affected at all locations 10 to 15 feet downgradient of the biowall. Given the 6-month timeframe between sampling events and the estimated ground water seepage velocity, one might expect to see dramatic changes in downgradient geochemical profiles. However, this may not be the correct conceptual model. Thus, it is possible that measurable geochemical changes may be localized near mulch material or may not be exhibited this far downgradient until more time has passed.

Significant shifts in cis-1,2dichloroethene (cis-DCE) to TCE ratios were exhibited as a result of both the in situ biowall and surface amendments (Table 1). For the in situ biowall, the mean cis-DCE/TCE ratio varied from 0.026

TABLE 1
Comparison of Baseline Sampling (January 1999) and 1st Sampling Round (June 1999)

	Units	UPGRADIENT MONITORING WELLS				DOWNGRADIENT MONITORING WELLS							
		Jan. 99 B301-MW23S (B301-2)	Jan. 99 B301-MW24S (B301-3)	Jun. 99 B301-MW23S (B301-2)	Jun. 99 B301-MW24S (B301-3)	Jan. 99 B301-MW31S (B301-10)	Jan. 99 B301-MW32S (B301-11)	Jan. 99 B301-MW33S (B301-12)	Jan. 99 B301-MW34S (B301-13)	Jun. 99 B301-MW31S (B301-10)	Jun. 99 B301-MW32S (B301-11)	Jun. 99 B301-MW33S (B301-12)	Jun. 99 B301-MW34S (B301-13)
Chlorinated Organics and Reduction By-Products													
TCE	mg/L	0.670	1.900	0.280	0.250	0.280	0.670	1.300	1.300	0.013	0.130	0.870	0.600
1,1-DCE	mg/L	0.0064	0.003	<0.001	<0.001	0.0032	0.0024	0.0023	0.0023	0.0012	0.0026	<0.001	<0.001
cis-1,2-DCE	mg/L	0.0082	0.020	0.0067	0.0068	0.27	0.07	0.045	0.02	0.550	0.73	0.067	0.040
trans-1,2-DCE	mg/L	0.0016	0.0041	0.0011	0.001	0.0083	0.0068	0.0041	0.0027	0.0045	0.0064	0.0033	0.0022
Vinyl chloride	mg/L	<0.001	<0.001	<0.001	<0.001	0.0023	0.0013	<0.001	<0.001	0.0061	0.0041	<0.001	<0.001
Ethene	ng/L	<3200*	<3200*	15	11	<3200*	<3200*	<3200*	<3200*	73	166	34	26
Ethane	ng/L	<2500*	<2500*	<5	<5	<2500*	<2500*	<2500*	<2500*	<5	<5	5	<5
cDCE/TCE ratio		0.012	0.011	0.024	0.027	0.964	0.104	0.035	0.015	42.308	5.615	0.077	0.067
Water Quality Parameters													
Temperature	°F	40.7	45.1	69.1	61.1	40.5	43.1	34.1	44.5	72.5	68.9	69.3	56.8
pH	pH units	6.9	6.88	7.14	7.05	6.89	7.64	7.91	6.97	7.07	7.06	7.05	7.03
Specific conductance	μmhos/cm	877	1,027	752	643	1,110	973	1,075	1,078	890	880	900	690
Total organic carbon	mg/L	<1.0	<1.0	3	3	24	13	15	5	3	3	3	3
Chloride	mg/L	13	19	17.6	17.2	9	10	13	10	--	17.4	18.6	19.1
Natural Attenuation Parameters													
Dissolved oxygen	mg/L	1.5	1.0	1.1	1.1	2.4	2.5	2.3	2.4	0.4	0.3	0.5	0.8
Redox potential	mV	166.5	174.8	127.2	128.8	133.2	159.6	177.1	182.6	30.6	79.9	159.1	84
Sulfate	mg/L	35	45.2	13	9	57.3	53.2	74.6	59	2 (11.8)	16 (23.3)	30	23 (27.4)
Nitrate	mg/L	4.9	5.7	2.06	2.22	<0.5	4.9	4.7	6.4	--	<0.1	2.05	1.69
Ferrous Iron	mg/L	0.04	<0.02	<0.2	<0.2	0.03	0.05	<0.02	0.06	<0.2 (0.25)	<0.2 (0.07)	<0.2	<0.20 (<0.02)
Methane	ug/L	2.8	<1.2	0.077	0.121	17	99	16	3	55.91	111.8	3.6	1.324
Hydrogen	nM	2.135	2.23	1.15	1.02	--	--	--	--	1.03	1.13	0.85	1.32
Alkalinity	mg/L	--	--	340	300	--	--	--	--	440 (360)	400 (362)	400	400 (298)

upgradient to 12 downgradient, or a mean increase in the ratio of a factor of 470. For the surface amendment plot, the cis-DCE/TCE ratio varied from 0.024 upgradient to 0.35 downgradient, with a mean increase in the ratio of a factor of 15. Vinyl chloride was non-detect in the same 5 of 9 wells before and after amendment addition. Vinyl chloride levels in the remaining 4 wells sampled did not display significant concentration increases. The analysis of a single sample of the organic mulch material and the above dissolved cis-DCE/TCE ratio data indicate that sorption is not a predominant mechanism. These results are considered a primary line of evidence in support of on-going reductive dechlorination.

CONCLUSIONS:

Recent work related to the natural and enhanced biodegradation of chlorinated ethenes has demonstrated that biodegradation requires the presence of a carbon substrate (electron donor). Anthropogenic activities as well as natural depositional processes have been shown to create the conditions conducive to chlorinated solvent biodegradation. However, the creation and maintenance of these conditions through the use of cost-effective engineered materials and distribution systems has met with more failure and overexpense than success. Success in terms of enhanced bioremediation appears to be a function of the quality, quantity, and distribution of the electron donor.

Quality can be measured in terms of the capacity to promote complete breakdown of contaminants to acceptable end products. The results from this project appear to support the concept that a wide variety of carbon-rich materials can satisfy the quality requirement.

Quantity is a function of having enough electron donor material to support enhanced biodegradation for an expended period of time so as to limit the need to repeatedly inject or emplace more electron donor. In addition, it would appear that a large excess of electron donor is needed per unit of chlorinated ethene biodegraded. Inexpensive organic mulch appears to meet this criteria.

Lastly, effective and cost-effective distribution systems must overcome the constraints imposed by porous media, heterogeneity, and the relatively "slow" rate of ground water migration versus the apparently "fast" rate at which microbes consume electron donors. Significant changes in geochemical conditions have not been observed at all monitoring well locations down gradient of the wall, but this may be observed in future monitoring events. The analysis of TCE to daughter product ratios (e.g. cis-DCE) is considered a useful technique to evaluate biodegradation as well as verify that downgradient wells are in the flowpath of ground water flowing through the biowall.

Future performance monitoring at this site will be aimed at evaluating the quality, quantity, and distribution methods of the electron donor material. Enhanced biodegradation has been stimulated. In summary, if reductive dechlorination continues without the accumulation of undesirable endproducts like vinyl chloride and these conditions are maintained for five years or more, then this in situ biowall approach is likely to be a cost-effective alternative for plume containment.

Cost-effective source reduction may be possible at "shallow" sites via in situ mixing (e.g. backhoe) of organic-rich material throughout source zones.

In situ biowall construction costs were approximately $70 US per linear foot (Organic mulch provided at no cost to the project; Sand and site work included in above unit cost).

This work demonstrates enhanced bioremediation of chlorinated ethenes can be created through the construction of an in situ permeable biowall containing inexpensive organic material derived from shredded plant material.

REFERENCES:

Wiedemeier, T. H. et. al. 1998. *Technical Protocol for Evaluating Natural Attenuation of Chlorinated Solvents in Ground Water*, U. S. EPA, EPA/600/R-98/128, http://www.epa.gov/ada/reports.html

Boussaid, F. et. al., 1988. "Denitrification In-situ of Groundwaters with Solid Carbon Matter". *Environmental Technology Letters*, Vol. 9, 803-816.

Robertson, W. D. et. al. 1995. "In Situ Denitrification of Septic-System Nitrate Using Reactive Porous Media Barriers: Field Trials", *Ground Water*, Vol. 33, No. 1. 99 – 111.

LABORATORY STUDIES USING EDIBLE OILS TO SUPPORT REDUCTIVE DECHLORINATION

Michael D. Lee (Terra Systems, Inc., Wilmington, DE, USA)
Ron J. Buchanan Jr. and David E. Ellis (DuPont, Wilmington, DE, USA)

Abstract: In laboratory microcosm tests, single additions of corn oil, beef tallow, melted corn oil margarine, coconut oil, corn oil, soybean oil, or partially hydrogenated soybean oil supported complete reductive dechlorination of perchloroethene (PCE) or trichloroethene (TCE) to ethene when a dechlorinating population was present. TOC production from the edible oils continued for up to 200 days. More than 250 mL of gas were produced in some microcosms. TCE concentrations as high as 236 mg/L were degraded. Coconut oil was added to a soil column that had been bioaugmented with a dechlorinating enrichment culture to evaluate edible oil-based TCE dechlorination in a flow-through system. Ethene and ethane were the only products found in the column effluent during the 88 days following the oil addition. The 10 mL of coconut oil (1,500 mg/kg) was projected to support TCE degradation for over two years. The edible oils are much cheaper than soluble substrates such as sodium lactate. Operation and maintenance costs for an edible oil based reductive dechlorination system should also be less than for a soluble substrate. The edible oil will need to be added only periodically. This technology should be applicable for source zone treatment.

INTRODUCTION

Some anaerobic microbial populations are capable of completely degrading chlorinated solvents such as perchloroethene (PCE), trichloroethene (TCE), or carbon tetrachloride to less chlorinated intermediates and innocuous products such as ethene, ethane, or methane (Lee et al. 1998). This process, known as reductive dechlorination, has been applied at several sites to degrade these chlorinated solvents. Recirculation based systems with a series of extraction and injection wells or trenches have generally been used to treat a source area or downgradient portion of a plume. Soluble substrates such as benzoate, methanol, and acetate have been added to the recirculated groundwater to generate anaerobic conditions in the groundwater and to provide the hydrogen necessary for reductive dechlorination. Bioaugmentation with a dechlorinating enrichment or isolate has been recently demonstrated for sites where the native microbial population is not capable of completely degrading the contaminants (Dybas et al. 1998).

Recirculation based systems suffer from several drawbacks including the relatively high costs for many of the soluble substrates, the frequent additions of substrate, and the need for on-going operation and maintenance program to maintain the recirculation system (Lee et al. 1998). Biofouling of the injection wells is a particular problem with recirculating reductive dechlorination systems that may result in decreased recirculation rates.

One approach that avoids the need for a recirculation system would be to use a long-lasting substrate that is injected only periodically into the subsurface. A polylactate ester, Hydrogen Release Compound from Regenesis, San Clemente, CA, has been developed as a slow-release source for hydrogen (Koenigsberg and Farone, 1999). Dybas et al. (1997) demonstrated that corn oil, solid food shortening, and hydrogenated cottonseed oil beads, would support carbon tetrachloride degradation by *Psedumonas stutzeri* strain KC. Other edible oils should also support reductive dechlorination. Mackie et al. 1991 reported that long and short chain fatty acids can be beta-oxidized to acetate by certain anaerobic organisms including the H_2-producing synotrophs.

MATERIALS AND METHODS

Microcosm Studies. A single addition of 6 mL of corn oil, beef tallow, melted corn oil margarine, and coconut oil were added to duplicate 280 mL serum bottles containing 50% by volume soil from the Victoria, TX site and 50% by volume water. The soil had been used in column studies fed benzoate or molasses and PCE for almost two years (Lee et al. 1997). The columns had been bioaugmented with a dechlorinating enrichment from another area of the site. A sterile control microcosm treatment (autoclaved for one hour on two successive days) amended with corn oil was also run. The microcosms were prepared in an anaerobic chamber with an atmosphere containing a mixture of 5% hydrogen, 5% carbon dioxide, and 90% nitrogen. Each microcosm was dosed with TCE on days 0, 2 and 71. Samples from the bottles were removed periodically over 200 days to be analyzed for volatiles by gas chromatography-mass spectrometry, end product gases (ethene, ethane, and methane) by head-space gas chromatography, and total organic carbon (TOC) by a Total Organic Carbon Analyzer using the methods described by Odom et al. (1995). The microcosms were incubated in the anaerobic chamber at 22 °C. Gas production was estimated by inserting a needle through the rubber septa and collecting the excess gas in the attached syringe.

Corn oil, soybean oil, and partially hydrogenated soybean oil were also tested as substrates to support reductive dechlorination in microcosm studies for three additional sites. The solids content of the microcosms ranged from 10 to 50 percent by volume site soil with the remaining volume being site groundwater. Some of the treatments were bioaugmented with the Pinellas Dechlorinating Enrichment (PDE) or the Dover Landfill enrichment (DLFE) two weeks after the studies were begun to evaluate whether bioaugmentation could stimulate complete dechlorination. The PDE was isolated from a site in Pinellas, FL (Harkness et al. 1999). The DLFE was isolated from a landfill at the Dover Air Force Base. The non-replicated microcosms were prepared, incubated, and sampled in an anaerobic chamber with 3% hydrogen, 5% carbon dioxide, and 92% nitrogen at 22 °C. These microcosm studies were analyzed for volatiles and metabolic gases by head-space gas chromatography in a modification of EPA

Methods 8021B and 8015. The microcosm studies lasted for between 84 and 189 days.

Column Study. An additional study involved the introduction of coconut oil to a soil column to evaluate edible oil-based reductive dechlorination in a flow-through system. The 7.6 cm diameter by 76 cm long steel Shelby tube held an intact core from Dover Air Force Base weighing 6 kg. The soil column had been bioaugmented with the PDE. Complete dechlorination of up to 90 μM TCE and 20 μM DCE to ethene had been observed when the column was fed lactate. Ten mL of coconut oil was added to the bottom of the up-flow column. Groundwater or tap water spiked with about 76 μM TCE (10 mg/L) was fed onto the column using a Ranin Rabbit® pump. The influent and effluent from the column were maintained at 4 °C to minimize changes in the substrate or chlorinated ethene concentrations between sampling events. The retention time on the column varied between 6 and 11 days.

RESULTS

Microcosm Tests. All of the oils and molasses supported complete reductive dechlorination of TCE to ethene over a period of greater than 49 days. Table 1 summarizes the results of the microcosm studies for four sites where edible oils have been tested as substrates.

Figure 1 shows the average concentrations of TCE and daughter products in μM concentrations, and methane and TOC concentrations in mg/L for a representative treatment from Site 1 (Victoria, TX) amended with corn oil. The TCE was degraded to a mixture of ethene and ethane over the 49 day incubation period. The autoclaved control treatment fed corn oil showed the production of TOC and ethene following the second TCE spike; autoclaving had not killed all of the microbes. Dissolved TOC production was highest initially with the maximum TOC of 670 mg/L recorded for the treatment with beef tallow at Day 29. TOC production continued in all oil based treatments over a period of 200 days (Figure 2) with coconut oil sustaining the longest TOC release. Dechlorination continued with the second spike of TCE on day 71. In excess of 50 mL of gas were produced in some of these microcosms; the septa was displaced in some microcosms. The high quantity of gas generated by the oils indicates that the oils may lead to plugging problems if gas bubbles accumulate in situ.

In the site 2 microcosms, the treatment with corn oil only showed dechlorination of TCE to DCE with no production of VC or ethene over the 84 day study. When the PDE was bioaugmented into a microcosm fed corn oil, TCE was dechlorinated completely to ethene. More than 290 mL of gas was produced with the treatment with the PDE and corn oil.

Treatments were prepared for the site 3 with both the till and the deeper, fractured bedrock soils and groundwaters. The bedrock samples were broken into small particles to place it into the 560 mL serum bottles and provide additional surface area. Groundwater from the same zone as the soil samples were added to the microcosms. The till groundwater contained an average of 215 μM TCE (28 mg/L) and 50 µM DCE (4.8 mg/L). The bedrock groundwater contained an average of 1,800 µM TCE (236 mg/L) and 78 µM DCE (7.5 mg/L). Both of the till microcosm treatments fed soybean oil and partially hydrogenated soybean oil were bioaugmented and showed complete conversion of the TCE and DCE to ethene. More than 200 mL of gas was generated in the till microcosms. The bedrock treatments with the soybean oil and partially hydrogenated soybean oil promoted the dechlorination of TCE to DCE. Some VC was produced in the treatment with soybean oil, but ethene or ethane was not generated. In the PDE-bioaugmented bedrock treatments with both the soybean oil and partially hydrogenated soybean oil, a mixture of VC and ethene were detected after 189 days. The Pinellas dechlorinating enrichment was able to biodegrade the very high concentrations of TCE. However, complete dechlorination to ethene was not observed possibly due to a substrate limitation or the need for a second inoculation with the PDE.

In the Site 4 microcosm studies, soybean oil alone promoted the essentially complete conversion of a mixture of PCE, TCE, cDCE, and VC to ethene. A dechlorinating enrichment had previously developed at this site and bioaugmentation with the Pinellas Dechlorinating Enrichment or the Dover Landfill Enrichment was not necessary. The maximum methane concentration observed in this study was 34 mg/L. Methane concentrations in excess of the aqueous solubility of about 25 mg/L at 22 °C were generated under the high pressures generated in the bottles.

Column Studies. Groundwater without substrate (TOC = 3-10 mg/L) plus TCE was fed onto the column. Ethene and low concentrations of ethane (<1.7 µM) were the only products found in the column effluent during the 88 days following the oil addition (Figure 3). The dissolved TOC in the column effluent ranged between 40 and 120 mg/L and the maximum methane concentration was 3.6 mg/L. One dose of oil supported complete dechlorination for almost 90 days in the flow-through system. A mass balance analysis of the column influent and effluent showed that less than 1% of the organic carbon present in the added coconut oil was used for TCE degradation, sulfate reduction, and methane production. Just under 9% of the added organic carbon was released from the column as dissolved organic carbon with over 90% unaccounted for. Assuming the unaccounted for organic carbon remained in the column and could be used for reductive dechlorination, the 10 mL of coconut oil (1.5 mg/kg) was projected to support TCE degradation for over two years.

DISCUSSION

The microcosm studies showed that edible oils including corn oil, beef tallow, corn oil margarine, coconut oil, soybean oil, and partially hydrogenated soybean oil can support reductive dechlorination at several sites. At one site, the native microbial population was able to dechlorinate TCE to ethene with just the addition of soybean oil. At the other four sites investigated, bioaugmentation with an dechlorinating enrichment allowed for complete dechlorination to occur. Very high concentrations of TCE were degraded at one site suggesting that this technology has potential applications for treatment of source areas with residual dense non-aqueous phase liquids. The column study showed that edible oil should be effective in aquifers.

The edible oils are much cheaper than soluble substrates such as sodium lactate. Operation and maintenance costs for an edible oil based reductive dechlorination system should also be less than for a soluble substrate since the edible oil will need to be added only periodically. The maintenance of a recirculation system of extraction and injection wells for the soluble substrate may also be avoided with an edible oil based system. However, the vegetable oils may cause excessive gas production in some soils potentially leading to gas bubbles partially plugging the subsurface.

REFERENCES

Dybas, M. J., G. M. Tatara, M. E. Witt, and C. S. Criddle. 1997. "Slow-Release Substrates for Transformation of Carbon Tetrachloride by *Pseudomonas* Strain KC." In *In Situ and On-Site Bioremediation: Vol. 3. Papers from the Fourth International In Situ and On-Site Bioremediation Symposium. New Orleans, LA. April 28-May 1, 1997.* p. 59. Battelle Press, Columbus, OH.

Dybas, M. J., M. Barcelona, S. Bezborodnikov, S. Davies, L. Forney, H. Heuer, O. Kawka, T. Mayotte, L. Sepulveda-Torres, K. Smalla, M. Sneathen, J. Tiedje, T. Voice, D. C. Wiggert, M. E. Witt, and C. S. Criddle. 1998. "Pilot-Scale Evaluation of Bioaugmentation for In-Situ Remediation of a Carbon Tetrachloride-Contaminated Aquifer." *Environmental Science and Technology.* 32(22): 3598-3611.

Harkness, M. R., A. A. Bracco, M. J. Brennan, Jr., K. A. DeWeerd, and J. L. Spivack. 1999. "Use of Bioaugmentation to Stimulate Complete Reductive Dechlorination of TCE in Dover Soil Columns." *Environmental Science and Technology.* 33(7): 1100-1109.

Koenigsberg, S. S. and W. A. Farone. 1999. "The Use of Hydrogen Release Compound (HRC) for CAH Bioremediation." In A. Leeson and B. C. Alleman, (Eds), *Engineered Approaches for In Situ Bioremediation of Chlorinated Solvent Contamination.* pp. 67-72. Battelle Press, Columbus, OH.

Lee, M. D., S. A. Bledsoe, S. M. Solek, D. E. Ellis, and R. J. Buchanan, Jr. 1997. "Bioaugmentation with Anaerobic Enrichment Culture Completely Dechlorinates Tetrachloroethene in Column Studies." In *In Situ and On-Site Bioremediation: Volume 3. Papers from the Fourth International In Situ and On-Site Bioremediation Symposium. New Orleans, LA. April 28-May 1, 1997.* p. 21. Battelle Press, Columbus, OH.

Lee, M. D., J. M. Odom, and R. J. Buchanan, Jr. 1998. "New Perspectives on Microbial Dehalogenation of Chlorinated Solvents. Insights from the Field." *Annual Reviews in Microbiology* 52: 423-452.

Mackie, R. I., B. A. White, and M. P. Bryant. 1991. "Lipid Metabolism in Anaerobic Ecosystems." *Critical Reviews in Microbiology* 17(6): 449-479.

Odom, J. M., J. Tabinowski, M. D. Lee, and B. Z. Fathepure. 1995. "Anaerobic Biodegradation of Chlorinated Solvents: Comparative Laboratory Study of Aquifer Microcosms." In R. E. Hinchee, A. Leeson, and L. Semprini (Eds), *Bioremediation of Chlorinated Solvents*. pp. 17-24. Battelle Press, Columbus, OH.

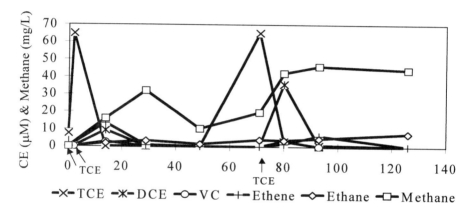

FIGURE 1. Chlorinated ethenes and methane concentrations for corn oil treatment of site 1 microcosms

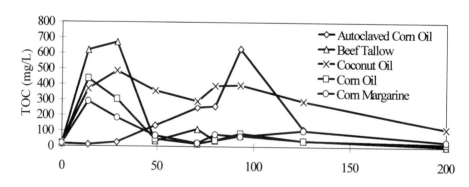

FIGURE 2. TOC concentrations for site 1 microcosms

Potential Electron Donors

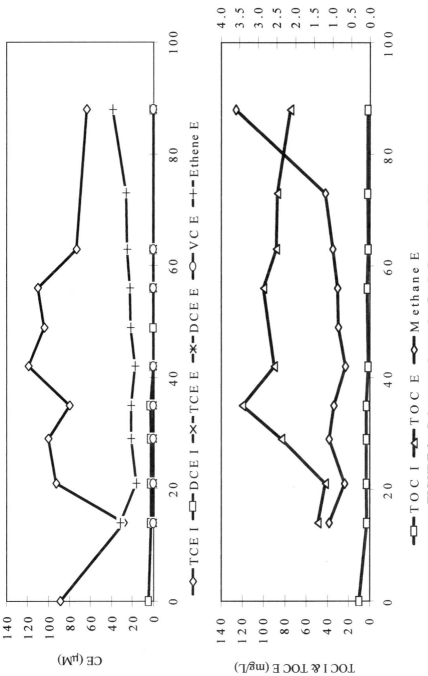

FIGURE 3. Column study results (I = Influent; E = Effluent)

TABLE 1. Summary of microcosm studies with edible oils

Treatment	Initial TCE μM	Initial cDCE μM	Final TCE μM	Final DCE μM	Final VC μM	Maximum Ethene μM	Maximum Ethane μM	Maximum Methane mg/L	Quantity Gas mL	Maximum TOC mg/L
Site 1 50% Soil 280 mL Microcosm Bottles Study 49 Days Long										
6 mL Corn Oil Autoclaved	94	<0.8	2.1	34	<1.2	10	5.5	29	>13	630
6 mL Corn Oil	65	<0.8	<0.6	<0.8	<1.2	13	1.4	32	>25	440
6 mL Beef Tallow	60	<0.8	<0.6	<0.8	<1.2	10	11	150	>51	670
6 mL Corn Oil Margarine	105	<0.8	<0.6	<0.8	<1.2	20	4.2	56	>30	290
6 ml Coconut Oil	31	<0.8	<0.6	<0.8	<1.2	4.3	4.3	46	>25	480
Site 2 50% Soil 560 mL Microcosm Bottles Study 84 Days Long										
2.8 mL Corn Oil	7.1	<0.1	<0.05	6.9	<0.1	<0.2	<0.2	>200	>106	
2.8 mL Corn Oil + PDE	14	<0.1	<0.05	<0.06	<0.1	32	<0.2	>290	>153	
Site 3 10% Soil 560 mL Microcosm Bottles Study 84-189 Days Long										
Bedrock 2.8 mL Soybean Oil	1990	80	8.6	1010	24	<18	<17	<0.5	0	
Bedrock 2.8 mL Partially Hydrogenated Soybean Oil	2060	76	<4	1390	<8	<18	<17	<0.5	0	
Bedrock 2.8 mL Soybean Oil + PDE	1450	82	4	<1.2	610	270	<8	8	2.2	
Bedrock 2.8 mL Partially Hydrogenated Soybean Oil + PDE	1700	72	2.4	<4	1000	190	<17	0.6	1.2	
Till 2.8 mL Soybean Oil + PDE	240	58	<0.2	<0.3	<0.4	420	<3	155	>255	
Till 2.8 mL Partially Hydrogenated Soybean Oil + PDE	190	42	<0.2	<0.3	<0.4	200	<3	67	>200	

Treatment	Initial PCE μM	Initial TCE μM	Initial cDCE μM	Initial VC μM	Initial Ethene μM	Final PCE μM	Final TCE μM	Final DCE μM	Final VC μM	Maximum Ethene μM	Maximum Methane mg/L
Site 4 50% Soil 560 mL Microcosm Bottles Study 84 Days Long											
2.8 mL Soybean Oil	29	35	29	14	2.5	<0.06	<0.08	<0.1	0.4	47	0.7
2.8 mL Soybean Oil + PDE	28	48	38	20	2.2	<0.04	<0.05	0.1	<0.1	85	34
2.8 mL Soybean Oil + DLFE	45	47	38	<0.4	0.9	<0.04	<0.05	<0.06	<0.1	34	21

BEHAVIOR OF NUTRIENTS IN THE BIODEGRADATION OF AZO DYE

Hiroaki Ozaki (Kyoto University, Japan)
Kusumakar Sharma (Asian Institute of Technology, Thailand)

ABSTRACT Reactive Red 22, a reactive azo dye was studied, in laboratory scale, to evaluate feasibility of simultaneous removal of color due to azo dye and nutrients. The effect of typical process variables such as influent dye concentration levels, influent nutrient concentration, nitrification and concentration of nitrate were also studied. Alternating anaerobic, aerobic and anoxic conditions were prevailed in single sludge sequencing batch reactor (SBR) to enhance the effective nutrient removal and biodegradation of azo dye. Influent concentrations of nutrients were varied to study the behavior of nutrients in the biodegradation of azo dye. The results indicate that complete cleavage of azo bond is easily accomplished in the anaerobic phase initiating the cycle of operation in the presence of co-substrate. Level of concentration of nitrate formed during nitrification plays inhibitory role in the biodegradation of azo dye. Influent azo dye concentration itself inhibits the biodegradation. In higher concentration of azo dye within the study range, azo dye cleaving process lowers the phosphorus release in the medium, as a result, phosphorus removal is decreased. In the simultaneous biodegradation process, chemical oxygen demand (COD) load is reduced to an acceptable level. The result thus indicate that simultaneous removal of organic matter, nutrient and color due to azo dye can be achieved in a modified operating pattern of sequencing batch reactor.

INTRODUCTION

Azo dye, widely used in dyeing and textile industries, is considered as a recalcitrant organic because of the stability of the dye molecule. Biodegradation of azo dye with corresponding reduction in color, regardless of the products and reductase, is achieved under anaerobic condition (Sheshadri *et al.*, 1994). Azo compounds are biodegraded via cleavage of bond under reducing conditions by facultative anaerobes (Wuhrmann *et al.*, 1980). However, unless the cleavage products of azo dyes are quickly consumed during the aerobic stage, the aromatic amines can self polymerize to form color and biologically recalcitrant compounds (Coughlin *et al.*, 1997). From different studies it is shown that alternating anaerobic and aerobic condition is necessary to enhance the biodegradation of azo dye which, is also a requirement for nutrient removal. However, the biodegradation of azo dye through cleavage of azo bond was found to be inhibited by various environmental factors (Wuhrmann *et al.*, 1980). Increased cell age and reduced food to microorganism ratio increase cell permeability and improve the rate of dye reduction (Roxon et al., 1967).

The studies conducted so far have shown relationship between nitrate and biodegradation of azo dye (Terashima et al. 1991, Whurmann et al. 1980). Simultaneous removal of color and nutrients and the behavior of nutrients in the biodegradation of azo dye have not been so far studied. The simultaneous removal in one way, would certainly reduce the cost and space of wastewater treatment processes and in other way could lay basis for establishing the dynamic relationship among the mixed culture bacterium to predict the effect of nutrients on the biodegradation of azo dye.

The purpose of this study is to understand the behavior of azo dye and nutrients in sequencing batch reactor process and to show the possibility of simultaneous removal of color (due to azo dye), nitrogen, phosphorus and organic matter using SBR. Based on the understandings of conventional continuous flow system for azo dye degradation and biological nutrient removal processes, anaerobic, anoxic treatments in addition to aerobic reactions were introduced into the operational schedule of laboratory scale SBR fed with artificially colored synthetic wastewater.

MATERIALS AND METHODS

A five-liter acrylic, cylindrical, laboratory scale tank (18.5 cm diameter and 18.6 cm liquid height) was used as the reactor for the operation of SBR in ambient laboratory of Asian Institute of Technology (AIT). In order to meet the requirement for anaerobic conditions inside, the reactor was fitted with screw lid. The reactor was equipped with inlet pipe, outlet pipe, and two peristaltic pumps. Overflow and sludge draining system were integrated to function by gravity.

Synthetic wastewater consisting of glucose and starch as carbon source, nutrients, and azo dye was used in this study approximating dye industry wastewater. The composition of the synthetic wastewater is given in Table 1 and the characteristics of the wastewater are presented in Table 2. The azo dye used to prepare synthetic wastewater is reactive red 22 (Figure 1).

TABLE 1. Composition of wastewater

Parameter	Quantity	Source of
Glucose	250-550 mg/l	Carbon
Starch	190-400 mg/l	Carbon
NH_4Cl	50-242 mg/l	Nitrogen
K_2HPO_4	160-660 mg/l	Phosphorus
Azo dye	5-35 mg/l	Color
0.5N-$NaHCO_3$	8-40 ml/l	Alkalinity
$CaCl_2$	2.7 mg/l	Trace element
$FeCl_3\ 6H_2O$	0.2 mg/l	Trace element
$MgSO_4\ 7H_2O$	2.25 mg/l	Trace element

TABLE 2. Characteristics of wastewater

Parameters	Quantity
SCOD	300-1100 mg/l
TKN	26-100 mg/l
Total PO_4-P	5-48 mg/l
Alkalinity	180-428 mg/l

A mixed culture acclimated to several days, capable of biodegrading monoazo, diazo and triazo dyes, was used in the research. The sludge used in the experiment was the excess sludge of sequencing batch reactor studying denitrification of colored effluent due to azo dye in the laboratory of Department of Chemical and Process Engineering, National University of Malaysia.

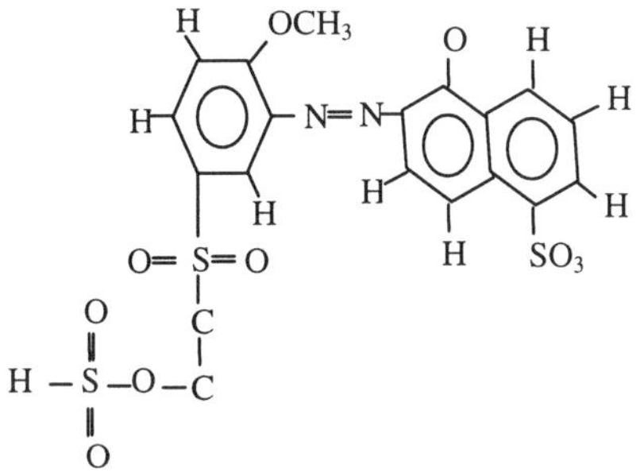

Bioremediation Microcosm and Laboratory Studies

FIGURE 1. Presumed molecular structure of Reactive Red 22 (Terashima et.al., 1991)

Acclimatization started by feeding the sludge obtained from another SBR removing carbon and nitrogen in the ambient laboratory of AIT with glucose and lower concentration of azo dye together with nutrients. To develop mixed population of bacteria, sludge containing heterotrophs and autotrophs was added from a biofilm reactor treating nitrogen in the same laboratory. After two weeks, sludge obtained from Malaysia was mixed with the sludge in the tank. Acclimatization of the sludge with azo dye concentration of 15 mg/l and COD 1000 mg/l took more than one month. Fresh synthetic wastewater was prepared daily for acclimatization as well as for actual experiment to avoid natural biodegradation of glucose. Stock substrate solution was stored in refrigerator at 4^0C and was used up within one week.

Operating Pattern. The SBR was operated in a cycle of 24 hour. Operation of schedule was adapted from the tested operating pattern by Park and Terashima (1988) for effective nutrient removal and improved sludge settleability. The operating pattern was divided in six phases of different conditions, each with four hour duration and the last two hour of the cycle were allotted for settling of sludge (Figure 2). Two hours of oxic phase before settling period is provided to achieve final nitrification and phosphorus uptake. The operation started with anaerobic

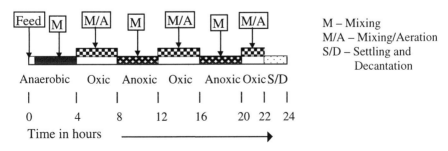

FIGURE 2. Operating pattern for one cycle and conditions in the SBR

condition followed by alternating aerobic and anoxic condition in one complete cycle. Mixing in all phases of operation was accomplished with the mechanical mixer and air was sparged through four porous stones. Hydraulic retention time was 24 hours.

Sampling and Analytical Method. Samples for influent and effluent were collected daily. All the samples were filtered through 1.2 μm GFC glass fiber filter. Supernatant samples for the determination of total phosphorus were collected separately and preserved in freezer below -10^0C. Samples for COD, NH_4-N, TKN, color due to Azo dye for the influent and effluent were analyzed daily unless steady state was reached. When steady state was achieved batch runs for concentration profile were conducted.

All the physical and chemical analyses were carried out in accordance with the Standard Methods (1995). NO_3 was analyzed by Hach spectrophotometer (DR 2000). Since the dye used in this study was color specific, a spectrophotometric technique was developed to quantify dye concentration in the influent and effluent streams using UV visible spectrophotometer (UV 106, Shimazdu) at the wavelength of 509 nm. Dissolved oxygen (DO), temperature of mixed liquor in the reactor were measured using DO electrode and a meter (YSI 68), and pH and ORP of mixed liquor in the reactor were measured using pH electrode (Knick SI 101) and ORP electrodes (Knick, Goldline) and meter (Knick).

RESULT AND DISCUSSION

Cyclic studies. The azo dye concentration was reduced from 33 mg/l to 4.5 mg/l at the end of anaerobic phase (Figure 3). In the absence of oxygen, an azo compound acts as a sole oxidant and its reduction rate is then governed exclusively by the rate of formation of electron donor (Wuhrmann et al., 1980). Complete reduction of azo dye and decolorization was achieved in the subsequent anoxic phases. Table 3 presents the biodegradation of azo dye during anaerobic phase. It is inferred that the degradation is dependent on the initial azo dye concentration. Within four-hour period of anaerobic phase the degradation rate is found to decrease on increasing the azo dye concentration. This implies that the higher concentration of azo dye might be inhibitory to microbial activity.

TABLE 3. Degradation of azo dye during anaerobic phase

Azo dye concentration		Percentage Removal
Influent	At the end of Anaerobic phase	
14.87	0	100
19.72	0	100
25.42	4.07	84
29.23	4.46	84
32.59	5.97	82

Ammonium, sole nitrogen source, was reduced from 44.8 mg/l to 33.3 mg/l (Figure 3) during anaerobic phase. Similar trend was observed throughout the experiment. This interesting result, observed in the experiment, of reduction in ammonium-nitrogen during anaerobic phase could be due to assimilation of ammonium for cell synthesis by biomass. However, the amount of ammonium reduced is very high than the budget required by cell biomass. This excess reduction in ammonium concentration may likely be due to the reaction between metabolites of azo dye and ammonium during anaerobic phase, which can be observed in the increased TKN value at the end of anaerobic phase.

There was fast reduction in ammonium-nitrogen concentration as soon as air was supplied in the reactor. The source of nitrogen in the experiment is

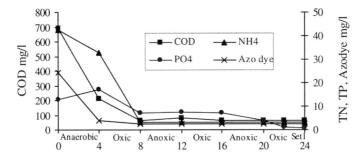

FIGURE 3. Cyclic profiles of Azo dye, COD, Total Nitrogen (TN) and Total Phosphorus (TP) during a complete cycle of operation.

ammonium chloride, hence denitrification during anaerobic phase is not observed. In the first oxic phase nitrification occurred which is supported by the reduction in alkalinity as shown in Figure 4 and increase in alkalinity during first anoxic phase indicating denitrification (Figure 5). The condition of pH 8.26-8.65 recorded

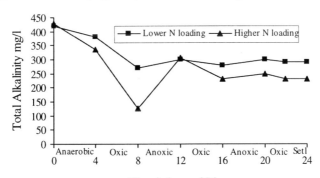

FIGURE 4. Total alkalinity during different phases of one cycle.

during nitrification falls within the nitrification range of 7 to 9 and during denitrification of value pH 7.59-7.69 also falls within the denitrification range of 7-9 (Henze et al., 1996). The DO in anoxic phases was below 0.1 mg/l.

The ORP readings observed, –223 mV in anaerobic phase, indicate the reduction condition whereas ORP of +39 mV in oxic phase represent the oxic condition. The ORP value during first oxic phase was +39mV and in following anoxic phase it was –180 mV that represent nitrification and denitrification in respective phases. Sores et al. (1991) found denitrification occurs at -80 mV to -330 mV. The ORP readings obtained are thus within range suitable for denitrification.

Effect of azo dye concentration on phosphate removal. Figure 3 shows the phosphorus release and uptake corresponding to azo dye concentration of 33 mg/l. In anaerobic phase, the phosphorus concentration in the solution increased from 12.9 mg/l to 17.7 mg/l. This indicates release of phosphorus in anaerobic condition and carbon storage by the cell-biomass. As soon as air was supplied, there was decrease in phosphate concentration from 17.7 mg/l to 9.6 mg/l. Phosphorus concentration in the following anoxic phase decreased from 9.6 mg/l to 8.1 mg/l. This decrease in phosphate concentration in anoxic phase is due to the consumption of phosphate by bacteria using nitrate as electron acceptor. Further decrease in phosphate occurred in subsequent phases. Table 4 illustrates the relationship between azo decolorizing rate and phosphorus release rate measured per unit of volatile suspended solids during four-hour period of anaerobic phase.

TABLE 4. Relationship between azo dye degradation rate and phosphate releasing rate in anaerobic phase of SBR operation

Azo dye concentration mg/l	Degradation rate mg Azo dye/hr-gVSS	Phosphate releasing rate mg TP/hr-gVSS
14.87	2.55	4.00
19.72	1.75	2.28
25.42	2.41	1.54
29.23	3.03	0.63

There is decrease in phosphate releasing rate and increase in azo dye degradation rate as the azo dye concentration is increased in an increased biomass medium. This stipulates the effect of azo dye dose on phosphorus removal. This result can be explained with the competition of azo cleaving bacteria and phosphate accumulating bacteria. From the result it can be stated that azo cleaving bacteria have higher affinity to glucose substrate than phosphate accumulating bacteria. Azo compound may induce inhibition on the activity of phosphate accumulating bacteria in substrate limited medium. Furthermore, the increase in degradation of azo dye is due to the increase in cell-biomass (data not shown), whilst this could not enhance phosphorus removal.

Effect of nitrogen concentration in the degradation of azo dye. Figure 3 and Figure 5 illustrate the nitrogen removal and azo dye degradation. In Figure 5 total nitrogen is almost same as TKN as nitrate concentration is negligible in the influent. Throughout all phases except at the end of anaerobic phase, in a cycle there is almost same difference in the concentration of ammonium and TKN indicating a fixed amount of non-biodegradable organic nitrogen. This means that azo compound containing N is not completely mineralized after cleavage and is reflected as TKN.

The effect of different concentrations of total nitrogen on the same concentration of azo dye can be seen in Figure 3 and Figure 5. When the nitrogen-loading rate in the influent was 68 mg NH_4/g COD (Figure 3), there is a very small production of nitrate and complete degradation was achieved. The nitrate produced could not show any effect in the degradation of azo dye. On increasing the nitrogen loading rate in the reactor with 82 mg NH_4/g COD (Figure 5), nitrate produced, as a result of nitrification in the first oxic phase showed effect on the degradation of azo dye. Azo dye concentration reduced from 32.6 mg/l to 8.3 mg/l during anaerobic phase. In subsequent alternating phases there is no change in concentration of azo dye. Hence, on increasing N-loading the degradation of azo dye is inhibited.

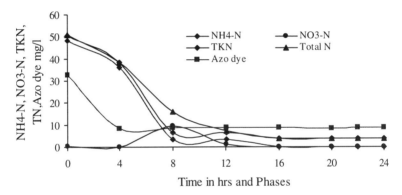

FIGURE 5. Relationship between total nitrogen, ammonium, nitrate concentration and azo dye

Due to presence of nitrate during anoxic phase, the degradation of azo dye is halted. The cleaving of azo bond might occur after all the nitrate has been denitrified (Wuhrmann et al., 1980). As there is least availability of co-substrate during anoxic phase the further biodegradation of azo dye is ceased even after complete denitrification. It must be admitted that the second oxic and anoxic periods are not always necessary. Some times phosphorus removal is enhanced in these phases.

CONCLUSION

The behavior of nutrient has been found influential in the biodegradation of azo dye. Nitrification has the influence in the azo dye degradation. As long as the presence of nitrate is observed, the degradation of azo dye ceases. Increasing concentration of nitrate shows the inhibition effect in the azo dye degradation. The phosphorus removal was dependent on the concentration of azo dye. Biodegradation of azo dye with simultaneous removal of nutrient was shown to be possible by modified operation of SBR.

REFERENCES

Coughlin, M.F., B.K. Tinkle, A. Tepper, and P. L. Bishop. 1997. "Characterization of Aerobic Azo Dye-Degrading Bacteria and their Activity in Biofilms." *Wat. Sc. Tech.* 36(1): 215-220.

Henze, M., P. Harremoes, J. La Cour Jansen, and E. Arvin. 1996. *Wastewater Treatment Biological and Chemical Processes.* Springer-Verlag, Heidelberg, Second Edition.

Park, D., and Y. Terashima. 1988. "New Indexes to Evaluate Substrate Feed and Anaerobic/Aerobic Sequencing Batch Activated Sludge." *Proc.of Second IAWPRC Asian Conference on Water.*

Roxon, J.J., A.J. Ryan, and S.E. Wright (1967). *Food Cosmet. Toxicol.* **5,** 645 as quoted by Dubrow et al., (1996). In Abraham Reife and Harold S. Freeman (Eds.), *Environmental Chemistry of Dyes and pigments.* John Wiley and Sons.

Seshadri, S., P.L. Bishop, and A.M. Agha. 1994. "Anaerobic/Aerobic Treatment of Selected Azo Dyes in Wastewater." *Waste Management 14:* 127-137.

Sharma, Kusumakar. 1998. *Simultaneous Removal of Carbon, Nitrogen, Phosphorus and Azo Dye Using a Sequencing Batch Reactor.* Master's Thesis. Asian Institute of Technology, Thailand.

Soares, M.I.M., S. Belkin, and A. Abeliovich. 1991. "Denitrification in laboratory sand columns: carbon regime, gas accumulation and hydraulic properties." *Wat. Res. 25* (3): 325.

Standard Method for Examination of Water and Wastewater (1995). 19th Edition. American Public Health Association/American Water Works Association/Water Pollution Control Federation, Washington, DC, USA.

Terashima, Y., H. Ozaki, and A.R. Rakmi. 1991. "Biodegradation of Azo Dye under Anoxic Condition and Characteristics of Formed Metabolics." *Proc. of Third International Symposium on Industry and Environment in the Developing World.*

Wuhrmann, K., Kl. Mechsner, and Th. Kappeler. 1980. "Investigation on Rate-Determining factors in the Microbial Reduction of Azo dyes." *European J. Appl. Microb. Biotechnol. 9:* 325-338.

ANAEROBIC BIOSTIMULATION FOR THE *IN SITU* TREATMENT OF CARBON TETRACHLORIDE

Mary F. DeFlaun and Robert J. Steffan
(Envirogen, Inc., Lawrenceville, NJ)

ABSTRACT: A microcosm study was used to test electron donor addition as an option for *in situ* biological treatment of carbon tetrachloride in a karst aquifer at the Upper East Fork Poplar Creek Site (the Site) at the U. S. Department of Energy's Y12 Plant in Oak Ridge, Tennessee. The study was implemented (February through July, 1999) after recommendation by participants in the Innovative Treatment Remediation Demonstration (ITRD) program to evaluate anaerobic biological *in situ* treatment technologies as an innovative remedy for this Site. Site data suggested that the aquifer was reducing and that intrinsic degradation of carbon tetrachloride was already occurring. Microcosms were constructed from aquifer sediment and groundwater removed from existing wells using a protocol designed to minimize exposure of the samples to oxygen. The microcosms, constructed in an anaerobic glove bag, were amended with different combinations of electron donor and nutrients (nitrogen and phosphorus) and were later spiked with carbon tetrachloride in a range that represented the approximate average concentration found at the Site. Results indicated rapid dechlorination of the carbon tetrachloride with the formation of chloroform, however, methylene chloride and chloromethane were never detected. Chloroform accumulated only in the microcosms without added electron donor indicating that it was completely degraded in the amended microcosms. Subsequent spikes of the microcosms to approximately 300 ppb carbon tetrachloride and sampling every 2 days over a period of 8 days resulted in rates of carbon tetrachloride degradation of approximately 48 ppb CCl_4 per day compared to 18 ppb/day in the unamended microcosms. These data will be used to assess anaerobic biostimulation on the basis of cost and efficacy as a potential remedial option at this Site.

INTRODUCTION

One alternative for treating carbon tetrachloride (CT) in groundwater and aquifer sediments is *in situ* biostimulation which involves enhancing bacterial degradation by adding electron donors, electron acceptors, nutrients or a combination of these (Semprini et al, 1992; Hooker et al, 1998). The assumption with this approach is that the indigenous microbial population is competent to degrade carbon tetrachloride, but lacks the appropriate electron donors or electron acceptors to maintain high levels of degradative activity. A treatability study was conducted with sediment and groundwater in order to determine whether the addition of electron donor with and without nutrients would be sufficient to stimulate degradation of the carbon tetrachloride at this Site. Electron donors were chosen based upon cost and ease of adding it to groundwater, past experience, and safety and regulatory considerations. Due to the limited timeframe for the study, some of the treatments were designed to provide accelerated results to allow an assessment of the capabilities of the indigenous microbial population under optimum conditions.

MATERIALS AND METHODS

There were three sets of microcosms that were tested (ORNL1, 2 and 3). All microcosms were constructed in an anaerobic hood in 125 mL serum vials fitted with Teflon-lined septa and aluminum crimp seals. Six microcosms were prepared for each treatment, three of these were analyzed at each time point and the others were kept as reserves. Liquid volume removed during sampling of the microcosms was replaced with sterile glass beads in the anaerobic chamber. ORNL 1 had 11 different treatments including six treatments that were designed to accelerate potential degradation. The treatments were: 1) no amendments; 2) killed control; 3) ethanol + lactate; 4) ethanol + lactate + N + P; 5) ethanol + corn syrup; 6) filter of 1 liter of groundwater; 7) filter + ethanol + lactate; 8) filter killed control; 9) 30^0C ethanol + lactate + RAM media; 10) 30^0C killed control; and 11) 30^0C no amendment. The filter treatments consisted of filters that had 1 L of groundwater filtered through them placed into the serum vial with groundwater filling the remaining volume in the vials. This accelerated treatment was designed to concentrate microbial biomass in the microcosms. For treatments with sediment (1 through 5 and 9 through 11), 50 mL of a sediment slurry was introduced into each bottle. The bottles were then completely filled with Site groundwater and incubated in the dark at either 22^0C (treatments 1 through 5) or 30^0C (treatments 6 through 8); the 30^0C incubations were an attempt to accelerate degradative activity. The amount of carbon in the amendments added was based on the concentration of VOCs in the Site groundwater. Per microcosm this calculation resulted in the addition of 6.2×10^{-5} moles of glucose (corn syrup), 1.25×10^{-4} moles of lactate, and 1.15×10^{-4} moles of ethanol. For killed controls, both sodium azide and mercuric chloride were added at 1 mg/mL.

The T=0 and T=25 day analysis of ORNL1 microcosm data indicated that low starting concentrations of carbon tetrachloride made it difficult to determine whether there was any degradation occurring. These samples were spiked with approximately 300 ppb of carbon tetrachloride 42 days after their initial set up. At this time it was also determined that contamination with methyl isobutyl ketone (MIBK) in ethanol had occurred during setup in the anaerobic glove bag. After decontamination of the anaerobic chamber, a second set of microcosms (ORNL2) were set up and spiked with approximately 300 ppb carbon tetrachloride. Data from ORNL1 suggested that accelerated treatment were not necessary for degradation therefore only treatments 1 through 5 were analyzed for ORNL2. ORNL2 microcosms were spiked at T=0 with 300 ppb carbon tetrachloride.

ORNL3 microcosms were a subset of ORNL1 [treatments 1 (no amendments), 2 (killed control) and 4 (ethanol + lactate + N + P)] and ORNL2 microcosms (treatments 1 and 4) that were respiked and analyzed at T= 0, 2, 4 and 8 days to compare rates in amended and unamended treatments.

RESULTS

The analysis of Site groundwater indicated anoxic to anaerobic conditions with a relatively high pH (8.15) and carbon tetrachloride concentrations of approximately 70 ppb. This value was judged to be high enough to construct the microcosms without addition of carbon tetrachloride for ORNL1. ORNL1 T=0 samples however had much lower concentrations of carbon tetrachloride than in the original groundwater sample (Table 1), likely as a result of volatilization during microcosm construction. Therefore, ORNL1 microcosms were spiked with 300 ppb carbon tetrachloride on day 42 and again at day 89. ORNL2 microcosms were spiked at T=0 and T=32 and ORNL3 microcosms were spiked at T=0.

For the eleven treatments in the ORNL1 microcosms, 2 weeks after addition of the 300 ppb carbon tetrachloride spike, the only microcosms with any appreciable carbon tetrachloride remaining were the three killed controls (treatments 2, 8 and 10; Table 1 accelerated treatments not shown), and the treatments in which the indigenous microbial population was added as filtrate of the groundwater rather than adding sediment (treatments 6 and 7). The no amendment control (treatment 1) had partial degradation of carbon tetrachloride and an accumulation of chloroform as product (chloroform data not shown). The 30^0C no amendment microcosm had complete degradation of carbon tetrachloride, but also showed the accumulation of chloroform. In the amended microcosms there was no chloroform indicating complete degradation. Chloroform was not produced in the killed controls, nor did they receive addition spikes of carbon tetrachloride because there was very little loss. The MIBK in the ORNL1 microcosms did not appear to inhibit degradation; the results for ORNL1 were very similar to those for ORNL2 (Table 1). Additional spikes of carbon tetrachloride into ORNL1 at T=89 days and ORNL2 at T=32 days show the same pattern, slow degradation in the no amendment microcosms with accumulation of chloroform, no degradation in the killed controls, and faster degradation in the amended microcosms with no accumulation of chloroform. The ORNL3 amended microcosms had from 94 to 100% removal of carbon tetrachloride over 8 days with a rate ranging from 44.5 to 48.1 ppb/day, while loss in the unamended microcosms ranged from 51 to 69%, with a degradation rate ranging from 17.9 to 18.5 ppb/day. There was only 4% loss in the killed control for a rate of 1.7 ppb/day (Table 1; Figure 1). As in ORNL1 and 2, subsequent degradation of the daughter product, chloroform, occurred in the amended ORNL3 microcosms, but accumulated in the no amendment controls. There was no chloroform detected in the killed controls.

Table 1. Average Carbon Tetrachloride Concentrations for ORNL1, ORNL2 and ORNL3 Microcosms

ORNL1-CONDITION / TIME (days)	0 ppb	sd	25 ppb	sd	56 ppb	sd	88 ppb	sd	96 ppb	sd	102 ppb	sd
no amendments (1)	11	1	0	0	24	6	0	0	147	6	63	8
killed (2)	7	6	9	1	207	15	233	15	177	6	213	31
EtOH + lactate (3)	8	1	2	3	0	0	0	0	5	1	2	2
EtOH + lactate + N + P (4)	8	1	0	0	1	2	0	0	8	5	0	0
EtOH + corn syrup (5)	7	2	0	0	0	0	0	0	4	4	0	0
ORNL2 -CONDITION / TIME (days)												
#2 no amendments (1)					0		31	0	39	0	45	3
#2 EtOH + lactate (3)					68		0	0	32	10	3	4
#2 EtOH + lactate + N + P (4)					67		0	0	1	2	0	0
#2 EtOH + corn syrup (5)					69	0	0	0	0	0	1	0

sd = standard deviation

ppb values are the average of triplicates

* ORNL spiked at T=42 and 89 days. ORNL2 spiked at T=0 and T=32 days

ORNL3-CONDITION / TIME (days)	0 ppb	sd	2 ppb	sd	4 ppb	sd	8 ppb	sd	ppb CCl4 removed /day	% removal
no amendments (1)	283	25	257	6	237	6	140	10	17.90	51
killed (2)	387	15	433	31	927	67	373	35	1.70	4
EtOH + lactate + N + P (4)	410	231	135	119	71	52	25	19	48.10	94
no amendments (1)[2]	213	50	114	28	220	80	65	48	18.50	69
EtOH + lactate + N + P (4)[2]	94	44	5	3	0	0	0	0	44.5*	100

[2] Microcosms from ORNL2, all others from ORNL1

sd = standard deviation

* Based upon data at day 2

Figure 1. ORNL3 results for carbon tetrachloride. Treatments with a superscript 2 are microcosms originally from ORNL2. The other treatments are microcosms originally for ORNL1.

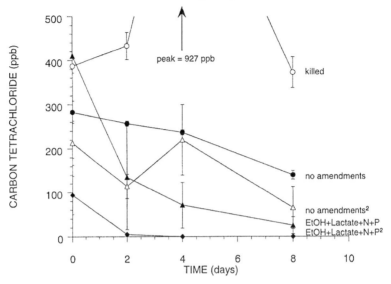

DISCUSSION

The results of these microcosm studies indicate that biostimulation by the addition of electron donor is a feasible option for remediation of carbon tetrachloride in a karst aquifer at the Upper East Fork Poplar Creek Site (the Site) at the U. S. Department of Energy's Y12 Plant in Oak Ridge, Tennessee. All treatments that included the addition of electron donor resulted in the complete degradation of carbon tetrachloride. Chloroform was formed as one of the degradation products, but was subsequently degraded in the microcosms with added electron donor. There was no detectable lag period prior to the onset of degradation, indicating that the indigenous microbial population was competent to degrade carbon tetrachloride. The lack of a lag period also indicates that the *in situ* biogeochemistry was not significantly altered during sampling and construction of the microcosms. Results at the first timepoint (25 days) for ORNL1 showed significant degradation of carbon tetrachloride in all the electron donor amended microcosms.

The agreement of the degradation rates for identical treatments from the two different sets of microcosms (ORNL1 and 2) lends considerable credibility to the rates obtained in the ORNL3 microcosm study. The slightly higher rate in the ORNL2 microcosms may be due to the higher inoculum in the form of a greater amount of aquifer solids in these microcosms. Similar rates for the two sets of microcosms also indicates that the presence of MIBK contamination in ORNL1 did not affect the degradation of carbon tetrachloride and that the storage of the sediment and groundwater at 4° C in the period of time between the construction of ORNL1 and 2 (57 days) did not obviously affect the biogeochemistry.

Although at least part of the carbon tetrachloride followed the stepwise dechlorination pathway to chloroform, dichloromethane or chloromethane were never detected. This same phenomenon was observed in the biostimulation field trial conducted by Semprini et al. (1992). It is likely that the rate of dichloromethane and chloroethane degradation were greater than the rate of chloroform degradation. Thus, the daughter products were degraded as soon as they were produced so they never accumulated in the microcosms. Degradation of carbon tetrachloride and chloroform most likely occurred by multiple pathways, in some cases going directly to CO_2. Direct degradation to CO_2 without the appearance of intermediate dechlorination products such as dichloromethane and chloromethane is the predominant pathway in some environments (Criddle and McCarty, 1991; Picardal et al., 1995; Van Eekert et al., 1998).

All of the electron donor and nutrient combinations tested were equally effective in stimulating the rapid degradation of carbon tetrachloride. Therefore, the selection of an additive for a full-scale remediation would be based upon the least expensive option. Based on the rapid degradation observed in this study, *in situ* biostimulation is an economical and relatively simple remedial approach for this Site.

The purpose of this study was to determine whether biostimulation would increase the rate of biodegradation observed at the Site. It was not intended to obtain degradative rates that could be used to predict the length of time for full-scale remediation. It is not known how the degradation rates obtained in these microcosm studies would relate to field rates, except that the rates after addition of electron donor would be higher than the natural degradation rates at the Site. The relative rates that we saw for the no amendment microcosms vs. the amended microcosms probably gives us the best indication of the magnitude of increase in the natural rates that we can expect if electron donor is added in the field. In this case rates were increased in the amended microcosms by a factor of three over the unamended microcosms, therefore, we would expect field rates after electron donor addition to be approximately three times higher than the natural biodegradation rate. A review of microcosm vs. field rates for biodegradation indicates that microcosm rates can range from being similar to the field rate to being 100 times faster (Rifai et al., 1995). The rates obtained in this study can be used in a qualitative manner to estimate rates that may be obtained in the field.

In order to obtain rates that can be used for scale-up to the field, there are experimental approaches that yield kinetic parameters that can be used to develop models for biodegradation at a particular Site. This approach has been used for carbon tetrachloride and the degradation of chlorinated ethenes (Skeen et al., 1995; Skeen et al., 1996).

Field Scale Application of Biostimulation. Results from this study present strong evidence that biostimulation can be a viable method for remediation of carbon tetrachloride and chloroform at the Upper East Fork Poplar Creek Site. Some of the advantages of this method as compared to conventional pump and treat include shorter remediation time periods and lower capital, maintenance, and operation costs. Using the transformation rates determined above, and assuming that amendments are injected into wells with 200-foot spacing, the time needed to provide sufficient amendment concentrations at a distance 100 feet from each well is approximately 10 years, assuming that these additives are carried by natural ground water flow. In contrast, a pump and treat system consisting of 30 wells at 200 foot spacing, each discharging at 4 gallons per minute, must be operated for 30 years to extract 5 pore volumes in the contaminated area.

Combining biostimulation with a pump and treat system or a recirculating well system will likely enhance the performance of biostimulation by enhancing the transport of amendments in the subsurface. By installing new wells or using selected existing wells, amendments can be injected into the aquifer and groundwater extracted from wells surrounding the injection point. Extracted water may be treated and re-injected at the injection well. This combination of injection-extraction should enhance the transport of amendments in the aquifer.

Costs for a biostimulation system lie mainly in the raw material needed to prepare the amendment slurry. The amendment slurry used in this study was based on a 10% utilization rate of the electron donor (i.e. corn syrup, lactate, and ethanol) towards contaminant breakdown. Using the same utilization rate for the field application, a slurry containing 300 lbs of corn syrup, 300 lbs of lactate, and 140 lbs of ethanol is needed to completely degrade 1 lb of carbon tetrachloride. These electron donor materials can be substituted with cheaper material (e.g. cotton waste) given that the ratio of moles of electrons donated towards one mole of contaminant is preserved. Nitrogen and phosphorous ratios in the amendment can be achieved in the slurry using a conventional agricultural fertilizer.

In order to move towards a full-scale application of biostimulation, additional data is needed regarding the extent of mixing and transport of injected amendments and the enhancement of the degradation rate in the subsurface. This can be achieved by a carefully designed pilot study. The primary goal of such a study would be to obtain detailed estimates for travel times and field biodegradation rates due to amendment addition.

REFERENCES

Criddle, C.S. and P.L. McCarty. 1991. "Electrolytic model system for reductive dehalogenation in aqueous environments". *Environ. Sci. Technol.* 25:973-978.

Hooker, B.S., R.S. Skeen, M.J. Truex, C.D. Johnson, B.M. Peyton, and D.B. Anderson. 1998. "*In situ* bioremediation of carbon tetrachloride: field test results". *Bioremediation Journal* 1: 181-193.

Picardal, F., R.G. Arnold, and B.B. Huey. 1995. "Effects of electron donor and acceptor conditions on reductive dehalogenation of tetrachloromethane by *Shewanella putrefaciens* 200". *Appl. Environ. Microbiol.* 61: 8-12.

Rifai, H.S., R.C. Borden, J.T. Wilson, and C.H. Ward. 1995. "Intrinsic bioattenuation for subsurface restoration". *In:* Intrinsic Bioremediation, R.E. Hinchee, J.T. Wilson, and D.C. Downey (eds.) Battelle Press, Columbus.

Semprini, L. G.D. Hopkins, P.L. McCarty, and P.V. Roberts. 1992. "In-situ transformation of carbon tetrachloride and other halogenated compounds resulting from biostimulation under anoxic conditions". *Environ. Sci. Technol.* 26: 2454-2461.

Skeen, R.S., J. Gao, and B.S. Hooker. 1995. "Kinetics of chlorinated ethylene dehalogenation under methanogenic conditions". *Biotechnol. Bioeng.* 48:659-666.

Skeen, R.S., J. Gao, B.S. Hooker, and R.D. Quesenberry. 1996. "Characterization of anaerobic chloroethene-dehalogenating activity in several subsurface sediments". *Pacific Northwest National Laboratory*, PNNL-11417, Richland, WA.

Van Eekert, M.H.A., T.J. Schroder, A.J.M. Stams, G. Schraa, and J.A. Field. 1998. "Degradation and fate of carbon tetrachloride in unadapted methanogenic granular sludge". *Appl. Environ. Microbiol.* 64:2350-2356.

SITE CLASSIFICATION FOR BIOREMEDIATION OF CHLORINATED COMPOUNDS USING MICROCOSM STUDIES

William A. Farone, Applied Power Concepts, Inc., Anaheim, CA
Stephen S. Koenigsberg, Regenesis, San Juan Capistrano, CA
Tracy Palmer, Applied Power Concepts, Inc., Anaheim, CA
Daniel Brooker, Applied Power Concepts, Inc., Anaheim, CA

ABSTRACT: Rate constants for the microbial reduction of chlorinated compounds determined in laboratory studies can rarely be applied to field sites due to the variation in microbial ecology and the differences in chemistry between the two regimes. The difference in rates is even greater when treating the site with materials intended to facilitate microbial dechlorination since these materials dramatically effect the microbial population in favor of dechlorinating bacteria. The natural selection of the desired organisms occurs much more easily in small, relatively homogeneous microcosms.

In studying a wide variety of soil samples from over 40 contaminated sites using a polylactate ester (HRC™) to dechlorinate PCE, TCE, TCA or PCP it was found that the microbial ecology during treatment in microcosm studies causes the soil samples from the field sites to segregate into three natural groups. In Group A sites the soil microbes reduce the chlorinated compounds and their secondary metabolites uniformly with time. With Group B sites the primary contaminant is reduced but the secondary metabolites build up before they are subsequently reduced. Class C sites show reduction of the primary contaminant, a build up of secondary metabolites with no clear indication that the secondary metabolites will be reduced with subsequent treatment in a reasonable time period. These microcosm studies will be compared with actual field experience to determine whether the classification system is predictive of field behavior. Such predictability provides a useful tool for estimating the difficulty of remediation of a site.

INTRODUCTION

Over the last several years dozens of microcosm tests have been conducted in our laboratory using soil from sites contaminated with various chlorinated compounds. The purpose of these tests were to determine whether the soil contained the appropriate microbiological population to accomplish dechlorination when augmented over a long time with slow release source of lactic acid and other nutrients released from a polylactate ester (HRC™).

Kinetic studies on the rates of degradation in these systems indicated that the most important factor is the number and type of microbes (Farone et al., 1999). The dependence of bioremediation on the microbial population implies that it is difficult to assess the rate at which bioremediation will occur in contaminated sites.

TCE was used as a model compound in the testing although several other compounds have been used also. Since TCE occurs widely in contaminated sites, the testing in microcosm tests may provide a means of classifying sites for their ease of remediation.

Objective. The objective of this study is to determine whether microcosm testing can be used to predict success at actual field sites. Due to the wide variations in rates of remediation reported in the literature it is not clear that the rates of degradation in microcosm tests are related to actual field rates in any useful manner.

EXPERIMENTAL METHODS

Test tube microcosms were established in sealed 200 ml test tubes. Ten grams of soil from the contaminated sites were used in each test tube along with 130-160 ml of a solution that contained the contaminant. For TCE the concentration levels of 10 mg/l and 25 mg/l were used. Approximately 1.5 grams of the polylactate ester was added. Previous studies had established that the rates are independent of the amount of the ester over a wide range of concentration due to the fact that the release of lactic acid occurs in response to the microbial population.

The systems are run in triplicate with a single control having soil and water only. The purpose of the control is to judge the background microbial population during the period of the test. The tests are run for 28 days with sampling at approximately 7 days intervals. Analysis of TCE, DCE, and VC is performed by gas chromatography using both PID and FID detectors.

The field data has been collected from eight sites where HRC™ has been used over an extended period of time. In order to make a reasonable comparison the initial rates in the test tube experiments were compared to the incremental rates in the field study. The initial rates were determined simply from the amount of TCE reduced during the first week in the test tubes. The field rates were determined after the introduction of HRC when the first two data points spaced in time became available. The rates are expressed as linearly (decrease in concentration divided by the time interval) even though the rate expressions are much more complicated. There is insufficient data to evaluate the rates in a more accurate manner.

RESULTS AND DISCUSSION

Figure 1 shows the microcosm data from 35 tests performed on soil using TCE as the contaminant selected from 45 sites tested. The average remediation rate for the initial period of the tests was 0.0575 mg/(l-day). The wide horizontal bands in Figure 1 are the 99% confidence levels around that average.

The wide distribution of values is expected. For the purposes of classifying sites it is suggested that values that are more rapidly remediated will have rates that are above the 99% confidence intervals (Class A sites). These sites also have the characteristic that there is less accumulation of either DCE or VC when TCE is degraded. That is, the degradation rates for the intermediate metabolites are also much higher.

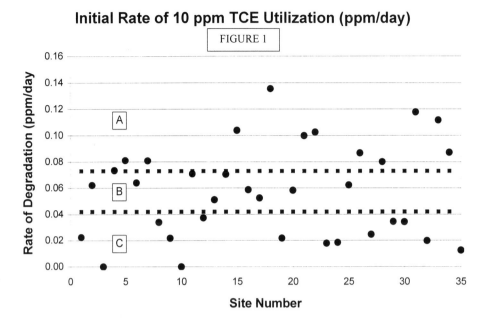

FIGURE 1. Classification Based on 35 TCE Microcosm Studies at 10ppm.

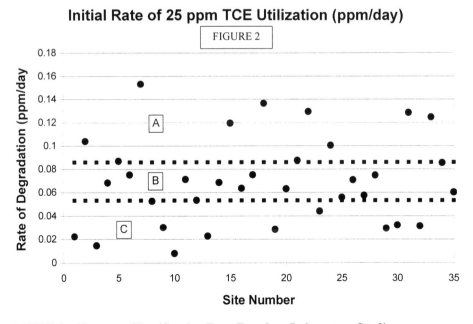

FIGURE 2. Alternate Classification Data Based on Laboratory Studies.

Sites within the 99% confidence limits of the average (Class B sites) have slightly lower rates of TCE degradation. These sites also show higher accumulation of DCE and VC before these metabolites are also remediated. For example, soil from these sites inoculated with 10 mg/l of TCE will show accumulations of as much as 2-3 mg/l of DCE and 1-2 mg/l of VC before these are also reduced.

Finally, sites below the 99% confidence limits are the "slowest" sites (Class C sites). In these sites DCE can accumulate up to 5-7 mg/l before it is degraded at a slower rate than TCE. Vinyl chloride can reach 3 mg/l.

Figure 2 shows that the ranking of sites does not change much if we use a larger amount of TCE in the experiments. In this case the test tubes were inoculated with 25 mg/l of TCE. The average initial rate of degradation of the TCE was 0.0696 mg/(l day) for the 35 soil samples. This is slightly higher than the rates with a lower inoculation concentration as would be expected.

Figure 3 shows the results of actual rates measured in seven field sites correlated with the laboratory results.

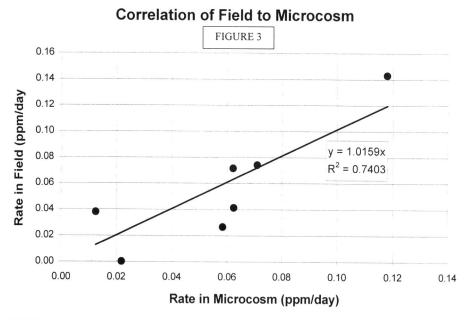

FIGURE 3. Correlation of Field Degradation Rates to Microcosm Studies

The filed sites were chosen from those available at the time of writing for which the laboratory data was available. Other field sites that are related to the laboratory measurements are in progress or in planning. Clearly the correlation is not very good in a statistical sense. The correlation line in Figure 3 was forced though the coordinate axis. The results do not change appreciably ($R^2 = 0.75$) if this constraint is removed.

If we take a slightly less quantitative approach, however, we can use the microcosm data to categorize our expectations for actual field behavior. In Table 1 eight sites are compared for which we have complete data. In the "Site Rating A-C" column the sites are listed by the rating that was assigned based on Figure 1. The sites are listed in increasing order of the laboratory test rates (Lab Rate). The field rates are given in the Field Rate column. All of the sites except site 21 had initial contamination levels that were in the range of 2-15 mg/l. Site 21 had contamination levels much less than 1.0 mg/l and thus the rates are misleading even though the contamination was essentially completely removed. For this reason it is listed as Not Relevant (NR). The fact that contaminants are being remediated to the detection limits indicates that its position as a Class A is justified even if the rate seems low because the starting concentration is low.

Table 1. Classification of Field Sites Based on Microcosm Data Compared to Both Microcosm and Field Results

Site Number	Lab Rate ppm/day	Field Rate ppm/day	Site Rating A-C
35	0.0126	0.0381	C
9	0.0218	0.0003	C
20	0.0584	0.0267	B
2	0.0622	0.0717	B
25	0.0624	0.0414	B
11	0.0710	0.0743	B
21	0.1000	NR	A
31	0.1178	0.1431	A

The only site that does not seem to fit the trend in Table 1 is site 25 where the laboratory testing would have indicated a better ranking than is apparent from the field results. Site 25 was also contaminated with a significant amount of PCE and TCA. Since PCE passes though TCE in the degradation pathway the amount of TCE as a function of time is elevated and it is difficult to get an accurate picture of the TCE rate. If we base the rate on both PCE and TCE reductions the field rate increases to 0.052 mg/(l day). The TCA degradation rate reaches a value of 0.5 mg/(l day) as the site has as much as 120 mg/l of TCA. The VC concentration never exceeds 5 µg/l. The bacteria on this site are very busy and very effective. The site justifies its position in the area of expectation based on laboratory results.

CONCLUSIONS

Laboratory test tube microcosms can be used to semiquantitatively develop a rating system for contaminated field sites. When the laboratory tests indicate better performing sites the actual sites will mirror the performance. While the actual rates in microcosm tests may not correlate with the field rates for a wide variety of reasons, the ranking of sites will be consistent.

Entering a contaminated site with an expectation of the result is helpful for all concerned. Although it is often difficult to define exact resources required for site remediation, microcosm tools such as presented here offer a means of understanding the site in greater.

The single largest factor in bioremediation is the concentration and type of the microorganisms relied upon to perform the task. We have yet to find a soil that with some help in the way of nutrients and time did not remediate chlorinated compounds. The range of rates of remediation varies widely due to the variability in soil microbiology.

The laboratory results reported here were based on the use of TCE as the contaminant. Tests based on other contaminants could be used to serve a similar purpose. Table 2 provides some data on several other materials that have been used in similar studies with HRC™. As more data is collected these contaminants could be used in a similar manner.

Table 2. Rates of Degradation of Several Compounds in Microcosm Studies

Chemical	Initial conc. ppm	Initial Rate ppm/day	Overall Rate ppm/day
Chromate	10	0.021	0.020
Chromate	25	0.011	0.023
TCA	2	0.032	0.017
TCA	2	0.033	0.017
PCP	10	0.027	0.011
PCP	10	0.028	0.010
PCE	10	0.111	0.042
PCE	25	0.111	0.042
TCA	25	0.051	0.030
TCA	250	0.010	0.042

It will be interesting to determine if site rankings are "generic" or whether they differ by compound. One would expect that similar patterns would be seen even though the specific rates for each compound are different. The ranking system is based on a statistical distribution of the rates and should be independent of the specific values. This would provide valuable evidence to test the hypothesis that the microbes to degrade most chlorinated compounds are either the same or are available in virtually all soils.

REFERENCE

Farone, W.A., S.S. Koenigsberg, and J. Hughes. 1999. "A Chemical Dynamics Model for CAH Remediation with Polylactate Esters." Engineered Approaches of In Situ Bioremediation of Chlorinated Solvent Contamination, Andrea Leeson and Bruce Alleman, Editors, Batelle Press, 1999, 287-292

MICROBIAL DEGRADATION OF 1,2-DICHLOROPROPANE AND 2,2'-DICHLORODIISOPROPYL ETHER IN CONTINUOUS REACTORS

Hauck, Regine (Technical University Berlin, Germany)
Hegemann, Werner (Technical University Berlin, Germany)

ABSTRACT: For investigations on the microbial degradation of 1,2-dichloropropane (DCP) and 2,2'-dichlorodiisopropyl ether (DDE) different mixed cultures were enriched. An anaerobic culture able to dechlorinate DCP to propene could be established in a laboratory scale fluidized bed reactor with polyurethane (PU) foam as carrier material. An aerobic mixed culture was immobilized on a porous sinter glass and established in a fixed bed reactor prior to adaption to DDE degradation. After a lag phase of about 50 days during continuous reactor operation, DDE degradation could be detected. Furthermore, determination of the released chloride amount indicated complete dechlorination of DDE.

INTRODUCTION

DCP and DDE are formed as by-products in a magnitude of about 10% and 5%, respectively during high-temperature chlorination of propene to allyl chloride, which is further used for the production of epichlorohydrine. Furthermore, they occur as a by-product (about 9% and 2.7%, respectively) in the industrial 1,2-propylene oxide (PO) production. In 1992 these processes yielded over 78,000 metric tons DCP and 23,000 tons DDE in Germany (Plinke et al., 1994).

DCP as well as DDE are still used by industry and for research, where they have applications as solvents, extractants, paint and varnish removers.

Until 1994 DCP was a production-related contaminant in fumigants to control root parasitic nematodes and due to its high mobility in soil, its moderate solubility in aqueous systems and the slow biodegradation by microorganisms in natural habitats, DCP can be found in groundwater reaching concentrations of up to 9 mg/m^3 in samples from areas in the Netherlands (Boesten et al, 1992) and up to 1.2 g/m^3 in California (Holden, 1986).

DDE has been detected in effluents from industrial plants, in several rivers (e. g., the Ohio (0.5–5.5 mg/L), Kanawha, and Mississippi Rivers in the US and the Rhine and Scheldt Rivers in the Netherlands) and in tap water (0.8 mg/L) from the Ohio River (IARC, 1986). Regarding the microbial transformation of DDE, the National Toxicology Program (1982) indicated that DDE is stable in aqueous media and nonbiodegradable in river water.

There is a high level of carcinogenicity concern for DCP and DDE since tumors were observed in animal tests and they have been classified as group 3 carcinogens (IARC, 1987, Parker et al., 1982).

Some investigations concerning the biological degradation of DCP (Löffler et al., 1997, Vandenbergh and Kunka, 1988) and DDE (Kawamoto and Urano, 1990, Patterson and Kodukala., 1981) in have been made, but no studies with continuous reactors were carried out.

MATERIALS AND METHODS

Anaerobic Cultivation. For enrichment of an anaerobic dechlorinating mixed culture sediment samples from the river Saale were taken near an introduction of a waste water stream of a chemical plant (Buna-Werke, Schkopau, near Halle, Germany). At Buna besides other chemicals also PO was produced, so that the autochthon culture could already be adapted to DCP.

The sediment was filled in serum flasks right after sampling, overlaid with autoclaved mineral medium (Shelton and Tiedje, 1984) and closed airtight with rubber septa. The flasks were incubated in the dark at 28°C without shaking.

Immobilization of the mixed culture on PU foam cubes (volume: 1 cm^3) was carried out prior to the start of the continuous reactor. The foam cubes were incubated with the mixed culture for two month in a serum flask. During the immobilization period DCP and co-substrates were added to the sample regularly after depletion. The PU foam particles were than transferred to the reactor and continuously supplied with model water, containing DCP and other co-substrates.

Aerobic Cultivation. For investigations of the aerobic degradation of DDE a mixed culture existing in a groundwater contaminated with different chloroorganic and organic compounds, like benzene, DCP, diethyl ether, was enriched. 25 ml of the groundwater were mixed with 125 ml mineral medium (Oldenhuis et al., 1989) and transferred to a flask closed with a teflon coated rubber septum, that was equipped with a canula and a membrane filter for oxygen supply. The flask was shaken at room temperature and after beginning of bacterial growth porous sinter glass (diameter: 0.12 mm) was added. No additional substrates were supplied. For a period of one month the carrier material was transferred to a mixture of fresh groundwater and mineral medium for several times. The porous siran glass overgrown with a biofilm was transferred into the reactor and for the period of one year it was continuously supplied with a medium, that contained mineral medium with various DCP concentrations (3 –10 mg/L) and different co-substrates like methanol and benzene. To this medium DDE was supplied additionally, to investigate DDE transformation.

Reactor configuration. The aerobic and anaerobic laboratory scale reactor systems consisted of a stainless-steel reactor vessel (volume: 5 L) containing the carrier material with the immobilized culture (Figure 1). Both systems were independent from each other.

Complete mixing of the containing liquid was realized by pumping the liquid upward through the reactor. For the anaerobic DCP degradation PU foam particles were used as carrier material and through the internal recycle the fluidized bed was realized. For aerobic DDE-transformation a fixed bed reactor with siran glass as carrier material was employed. The synthetic substrate was fed continuously from a feeding tank. The tank contained mineral medium with organic substrates and was equipped with an aluminum coated nitrogen filled balloon to minimize loss of volatile chloroorganics through evaporation. Temperature was steady at 25-27 °C. The reactor performance was monitored by measuring pH, temperature and redoxpotential or O_2 with electrodes. Collection of the data and, for DDE degradation performance, regulation of the oxygen concentration (5 mg/L) were performed online.

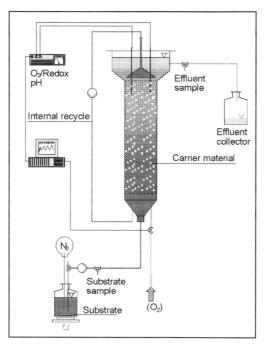

FIGURE 1. Scheme of bioreactor configuration

Analytical methods. The chlorinated aliphatics were quantified by gaschromatography using flame ionization detection (HP 5890 series II). Probes were injected directly from the gas phase using an headspace autosampler (HS 40, Perkin Elmer). The amount of chloride released during dechlorination of DDE was detected photometrically using a turbimetric method.

RESULTS AND DISCUSSION
Anaerobic DCP degradation. In batch-tests with fresh sediment an initial DCP concentration of 20 mg/L was depleted after 14 days, with 50 mg/L methanol supplied additionally as co-substrate.
Also in batch cultivation 1- as well as 2-monochloropropane were determined as metabolites of DCP transformation and a stochiometric formation of propene was observed.
      Investigations of Löffler et al. (1997) with an enriched mixed culture from river sediment also showed a stepwise dechlorination of DCP to propene under anaerobic conditions.

Due to a dehydrohalogenation step that is carried out either with sodium hydroxide or calcium hydroxide the real waste water of a PO production site contains besides different short chain chloroorganics high salt concentration of up to 50 g/L NaCl or $CaCl_2$ at a pH value of 13.

In order to investigate the influence of these factors on the DCP degradation, batch tests with the sediment containing mixed culture at different pH values and various salt concentrations were carried out.

No dechlorination could be detected at pH values above 9 and below 4 through the sediment containing mixed culture. Concerning the influence of chloride concentration the batch tests showed, that a $CaCl_2$ concentration of 5 g/L and a NaCl concentration over 10 g/L inhibited DCP dechlorination.

These results are only partially transferable to the continuous reactor, since immobilized bacteria are more stable against the extreme conditions in real waste water (Rehm, 1990).

Continuous anaerobic reactor operation. For development of a continuous process, reactors with immobilized biomass were chosen, not only because of the higher stability against unfavorable conditions, but also because the retention time of the specialized bacteria is disconnected from the hydraulic retention time (HRT) of the reactor. Furthermore, a higher cell density compared to suspended biomass and a continuous teamwork in the biocenoses through short transportation ways can be established (Diekmann et al., 1990). PU foam cubes as carrier material were successfully employed in reactor experiments concerning microbial degradation of chlorobenzenes and an on site treatment of a contaminated groundwater (Nowak, 1994, Meierling, 1997). In order to make the initial start up of the reactor system easier the mixed culture was immobilized onto the inert support in advance.

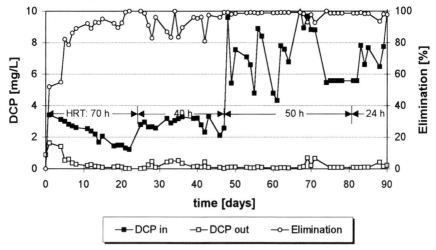

FIGURE 2: Anaerobic dechlorination of DCP in a continuous fluidized bed reactor

In Figure 2 the reactor performance in the first 90 days of operation is shown. DCP degradation through the mixed culture could be detected about 10 days after the beginning of continuous operation. With an influent DCP concentration of 3 mg/L a 90% elimination was achieved; as co-substrates 100 mg/L methanol and 75 mg/L sodium acetate were added.

Parallel to the reduction of HRT to 24 h DCP feed concentration was increased to a concentration of about 7 mg/L after 50 days of reactor operation, which equals a space loading of 50 µmol/(L*d). Under these conditions an elimination of over 95% was reached.

Through further increase of DCP feed concentration the capacity of the system was determined. An increase of space loading above 700 µmol/(L*d) resulted in a decrease in elimination from 90% to 60%.

Under the same conditions a variation of different co-substrates were examined for their ability to support DCP degradation in the reactor at variable space loads, to optimize the reactor operation parameters. In **Table 1** the results of these investigations are summarized.

TABLE 1: Reactor performance with various co-substrates at different space loads

Co-substrate [mg/L]	DCP$_{in}$ [mg/L]	Space loading [µmol/L*d]	Elimination [%]
Methanol, 75 Acetate, 50	29	250	92,4
Methanol, 75 Acetate 50	82	700	90,1
Acetate 100	29	250	49,3
Methanol 100	42	370	88,3
Pyruvate 100	33	290	92,8
Pyruvate 100	81	700	67,1

A space loading of 700 µmol/(L*d) could be eliminated to 90% with a co-substrate combination of methanol and sodium acetate. However, for a real application of a treatment system, besides a high efficiency and stability the costs are also an important factor. Since methanol is relatively cheap, it could serve as the single co-substrate also achieving high elimination.

Aerobic DDE degradation. The original inoculum was enriched from a groundwater contaminated with organic and chloroorganic compounds and was able to transform monochlorobenzene, dichloromethane and benzene (Meierling, 1997). For establishment of a DDE transforming mixed culture a continuous reactor was used that was previously used for investigations in the degradation of DCP.

In order to determine if this mixed culture was also able to degrade DDE, it was supplied additionally to the mineral medium already containing DCP, benzene, methanol, 2-propanol and diethyl ether. The initial DDE influent concentration was not significantly transformed in the first 47 days of reactor operation and could be quantitatively found in the collected effluent (Figure 3). Complete recovery of the supplied DDE concentration proved, that the reactor system was closed and that stripping effects could be neglected.

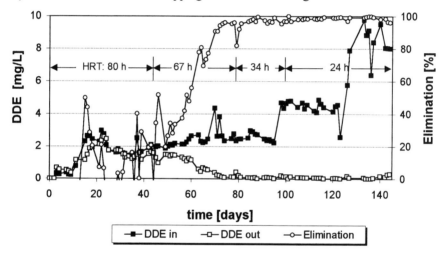

FIGURE 3: Aerobic DDE transformation in a fixed bed reactor

Beginning at day 47 the effluent concentration decreased and within the next 30 days the elimination rose to over 90%, showing clearly, that an adaption of the culture during reactor operation took place. The shortage of HRT from 80 to 24 h and increase of DDE influent concentration from 2 mg/L to 8 mg/L was accompanied with a reduction of co-substrate variety. After 140 days of reactor operation with 30 mg/L methanol and 30 mg/L benzene as co-substrates an elimination was achieved, that remained stable at over 97%.

In further reactor experiments the influent concentration was further increased up to 25 mg/L and with methanol (75 mg/L) as co-substrate still an elimination of 98% was detected in the fixed bed reactor.

With a *Rhodococcus* strain, that was isolated from the mixed reactor culture a stochiometric chloride release could be detected, that indicated, that DDE was completely dechlorinated. Metabolites of the degradation pathway have not yet been found, but since there are several ways to cleave ether bonds (White, et al., 1996), it is likely, that DDE is mineralized.

ACKNOWLEGDGEMENTS. These investigations were carried out within the framework of a co-operative research program Sonderforschungsbereich 193 „Biological treatment of industrial wastewater", sponsored by the Deutsche Forschungsgemeinschaft.

REFERENCES

Boesten, J. J. T. I., L. J. T. van der Pas, M. Leistra, J. H. Smelt, and N. W. H. Houx. 1992. "Transformation of 14 C-labeled 1,2-dichloropropane in water-saturated subsoil materials." *Chemosphere*. 24: 993-1011.

Diekmann, R.; M. Naujoks, M. Gerdes-Kühn, and D. C. Hempel. 1990. "Effects of suboptimal environmental conditions on immobilized bacteria growing in continuous culture." *Bioprocess. Engineering.* 5: 13-17

Holden, P.W. 1986. *Pesticides and groundwater quality-issues and problems in four states*. National Acadamy Press. Washington, D.C.

International Agency for Research on Cancer (IARC). 1986. *IARC monographs on the evaluation of carcinogenic risk of chemicals to humans*. Volume 41. pp. 149-160, IRAC, Lyon

International Agency for Research on Cancer (IARC). 1987. *IARC monographs on the evaluation of carcinogenic risk of chemicals to humans; Overall evaluation of carcinogenicity: An updating of IARC monographs Volumes 1 to 42*, Supplement 7, 59, IRAC, Lyon

Kawamoto, K., and K. Urano. 1990. "Parameters for predicting fate of organochlorine pesticides in the environment (III) Biodegradation rate constants." *Chemosphere*, 21(10-11): 1141-1152.

Löffler, F. E., J. E. Champine, K. M Ritalahti, S. J. Spargue, and J. M. Tiedje. 1997. "Complete reductive dechlorination of 1,2-dichlorpropane by anaerobic bacteria." *Appl. Environ. Microbiol.* 63: 2870-2875

Meierling, L. 1997. "Untersuchungen zur mikrobiellen Reinigung organisch belasteter Grundwässer in Kombination mit chemischen und physikalischen Verfahren." PhD Thesis, Technical University Berlin, Germany.

Nowak, J. 1994. "Umsatz von Chlorbenzolen durch methanogene Mischkulturen aus Saalesediment in batch- und Reaktorversuchen." PhD Thesis, Technical University Berlin, Germany.

National Toxicology Program (NTP). 1982. "Carcinogenesis bioassay of bis(2-chloro-1-methyethyl)ether (~70%) (CAS No. 108-60-1) containing 2-chloro-1-methylethyl(2-chloropropyl)ether (~30%) (CAS No. 83270-31-9) in B6C3F$_1$ mice (Gavage study)." Technical Report No. 239, US Department of Health and Human Services, Research Triangle Park, NC.

Oldenhuis, R., R. L. J. M. Vink, D. B. Janssen, and B. Witholt. 1989. "Degradation of chlorinated and non-chlorinated aromatic solvents in soil suspensoins by pure bacterial cultures." *Appl. Environ. Microbiol.* 30: 211-217.

Parker, C. M., W. B. Coate, and R. W. Volker. 1982. "Subchrinic inhaloation toxicity of 1,3-dichloropropene/1,2-dichloropropane (D-D) in mice and rats." *J. Toxicol. Environ. Health.* 9: 899-910

Patterson, J. W., and P. S. Kodukala. 1981. "Emission and Effluent control: Biodegradation of hazardous organic pollutants." *Chem. Eng. Prog.* April, 48-55

Plinke, E., R. Schüssler, and K. Kämpf. 1994. "Konversion Chlorchemie." Hessisches Ministerium für Umwelt, Energie und Bundesangelegenheiten, Eigendruck, Wiesbaden.

Rehm, H.-J. 1990. "Besondere Eigenschaften immobilisierter Mikroorganismen beim Abbau von Problemsubstanzen." *gwf-Wasser/Abwasser.* 131: 9

Shelton, D. R., and J. M. Tiedje. 1984. "General method für determing anaerobic biodegradation potential." *Appl. Environ. Microbiol.* 48: 850-857

Vandenbergh, P. A., and B. S. Kunka. 1988. "Metabolism of volatile chlorinated aliphatic hydrocarbons by *pseudomonas fluorescens*." *Appl. Environ. Microbiol.* 54: 2578-2579

White, G. F., N. J. Russell, and E. C. Tidswell. 1996. "Bacterial scission of ether bonds." *Appl. Environ. Microbiol.* 60. 216-232

LEACHATE PERMEATION IN SOILS: ITS IMPLICATIONS FOR CLOGGING

Naresh Singhal and Jahangir Islam
University of Auckland, Auckland, New Zealand

Abstract: Contamination of groundwater by landfill leachate represents a major environmental concern associated with landfilling of waste. In New Zealand most landfills constructed prior to 1970 are unlined and potentially leak leachate to underlying soils and groundwater. Leachate ponding of up to 1 meter at the landfill base have been observed in a number of test pits at the Gisborne landfill. Laboratory analysis has shown that the soil structure has undergone changes. Laboratory and field studies in other countries also suggest that hydraulic conductivity of soils subject to leachate infiltration decrease due to metal precipitation and microbial growth.

This paper reviews the literature on field and laboratory evidence of plugging of soils. The paper shows that there is growing evidence that compacted soils can under appropriate conditions be plugged to reduce the hydraulic conductivity of compacted soils to the levels specified for engineered landfill liners. The reduced hydraulic conductivity appears to be resistant to increase after exposure to acidic and basic solutions or under wet-dry cycles. The literature review shows that biological growth coupled with bioprecipitation may potentially be a feasible technology for creating waste containment barriers in soils. However, a number of challenges remain to be overcome before the engineering of such systems can be done satisfactorily.

INTRODUCTION

Comparative studies in several countries on the various possible means of eliminating solid urban waste (landfilling, incineration, composting, etc.) have shown landfilling to be the cheapest. In New Zealand, until recently, few municipal landfills were lined. The older landfills typically have no liner and a great reliance is placed upon the natural attenuation of contaminants in the underlying unsaturated soils and the groundwater aquifer by biological and physico-chemical processes.

Field observations have shown that leachate permeation may lead to clogging of soils resulting in significant decrease in hydraulic conductivity of the matrix. Nelson (1995) reports that ponding of leachate up to depths of 1 meter has been observed in numerous pits at the Gisborne landfill. Although the responsible mechanisms were not identified, laboratory analysis revealed that the soil structure had undergone changes. Clogging of leachate collection systems has been extensively reported (for example, see Rowe, 1998; Rowe et al., 1997a, b, 1995; Rittmann et al., 1996; Giroud, 1996; Brune et al., 1991). Some field studies involving leachate contaminated soils at landfills have shown the formation of new mineral phases via precipitation of metals under aerobic to anaerobic conditions (see Heron et al., 1998; Gade et al., 1997).

In addition to the field studies, various laboratory investigations have provided information on the composition of minerals formed by metal precipitation under different redox conditions and the effect of microbial growth on soil permeability. This paper discusses the field and laboratory studies to investigate the possibility of soil plugging in leachate contaminated soils and the potential for creating waste containment barriers in soils by promoting bioprecipitation processes.

FIELD STUDIES

Field investigations of landfills in Germany by Brune et al (1991) showed significant clogging of soils. Clogging was extensive in areas exposed to leachate with high concentrations of organics (measured as chemical oxygen demand, COD) and calcium. The Altwarbüchen landfill with COD and calcium concentrations on the order of 50,000 mg/l and 3,500 mg/l, respectively, in the leachate showed greatest clogging. Conversely, the Venneberg landfill with COD and calcium concentrations of approximately 1,000 mg/l and 300 mg/l, respectively, in the leachate showed limited clogging. The encrusted material, on the average, contained 20.5% calcium (by mass of dry solid matter), 33.6% carbonate, 8% iron, and 15% silica. Rowe et al. (1995) reported similar findings for 2 to 4 year old sections of the leachate collection system at the Keele Valley landfill in Canada. The clog has a soft (organic) and hard (predominantly $CaCO_3$) component (Rowe, 1998).

Heron et al. (1998) investigated the geochemistry at a landfill leachate contaminated aquifer in Grindsted, Denmark, and showed that minerals of iron-calcium carbonates precipitated in the reduced part of the contaminated zone.

Gade et al. (1997) investigated two Bavarian hazardous waste landfills, Raindorf and Gallenbach, with different ages of landfilling. The waste areas researched were 4 years old at Raindorf and 15-20 years old at Gallenbach. The Gallenbach leachate contained about 15% organic substances and similar numbers of aerobic, anaerobic, and sulfate reducing bacteria, while the Raindorf leachate contained 4% organic substances and no anaerobic or sulfate reducing bacteria. Calcite formation was observed in the leachate drainage systems at both landfills. In addition several newly formed minerals incorporating heavy metals, typically phosphates and sulfates of heavy metals, were observed.

Rowe et al. (1997b) state that exhumation of the leachate collection system after a 1 to 5 year exposure to leachate at the Keele Valley Landfill in Canada showed that the stone drainage blanket contained significant quantities of soft black slime coating the stones and occupying most of the void space. The pH of the slime ranged from 7.5 to 8.8. The hydraulic conductivity of a sample taken from the collection system using leachate permeant under anaerobic conditions was estimated to be 10^{-4} m/s, implying a 3 order decrease over 4 years relative to hydraulic conductivity in similar unclogged stone. The clog material included a significant fraction of fine gravel, sand, and other material that was not originally present in the clear stone. Based on the findings of various microbiological and chemical studies (Brune et al., 1991; Rowe et al., 1997a, b, 1995; Rittmann et al., 1996) Rowe (1998) propose that clogging is a result of mobilization processes involving fermentative and iron and manganese reducing bacteria followed by precipitation processes primarily involving methanogenic and sulfate reducing bacteria.

LABORATORY INVESTIGATIONS

The reduction of permeability in soils due to microorganisms is well documented in the literature. Allison (1947) reported significant reductions in the hydraulic conductivity of saturated soils from microbial growth during prolonged flows. Frankenberger et al. (1979) showed that the hydraulic conductivity initially decreased due to bacterial clogging and eventually stabilized at a constant value. Reductions in aquifer permeability have been observed at wastewater injection sites and it is common practice to chlorinate waters to avoid clogging at injection wells (Taylor and Jaffé, 1990a). Jenneman et al. (1984) and Raiders et al. (1986) showed that in situ microbial growth can selectively plug high permeability zones and drive an initially heterogeneous permeability field towards homogeneity.

Rowe et al. (1997a, b) examined the effect of temperature, flow rate and particle size on the clogging process using field obtained (from the Keele Valley Landfill) and synthetic leachates. The lag time of 80 to 90 days at 21°C was practically reduced to zero at 27°C. However, the chemical oxygen demand (COD) removed was the same at both temperatures for columns with the same bead size (Rowe et al., 1997b). The drainable porosity was smallest near the inlet and greatest at the outlet end of the column due to the greater amount of substrate that was available to support biological growth near the inlet. Reduction in drainable porosity was greater for small bead size than for larger bead sizes suggesting that clogging occurs faster in a porous media that has small pores and larger surface area. Although this result appears to contradict those reported by Jenneman et al. (1984) and Raiders et al (1986), it is an artifact of the manner in which the flows through the columns were maintained in these studies. Rowe et al. (1997b) maintained equal flows through columns with different particle size while Jenneman et al. (1984) and Raiders et al. (1986) had higher flows through the column with more porous soil. As the leachate flow in soils under field conditions will be governed by the leachate head, greater flows will result in soils that are more porous. Therefore, the observations of Jenneman et al. (1984) and Raiders et al. (1986) are likely to prevail in the field.

Methanogenesis plays an important role in the clogging of soils by converting organic acids into methane and inorganic carbon (total dissolved carbonate), which raises the pH and results in carbonate supersaturation (Rowe et al., 1997b; Rittmann et al., 1996). Isotopic studies at three different landfills by Rowe et al. (1997b) showed that of the two pathways for methanogenesis – fermentation of volatile fatty acids and reduction of carbon dioxide – acetate fermentation dominates methane production in all cases. Although carbon dioxide reduction is secondary, it does seem to increase as the landfill matures.

Clogging of soils can also be caused by precipitation of insoluble salts resulting from anaerobic bacterial activity (Bagchi, 1994; Brune et al., 1991).

Bisdom et al. (1983) studied the precipitation of heavy metals in column experiments simulating conditions in soils underlying landfills. Sectioning of the sandy column showed that brown and black heavy metal containing incrustations could be detected on soil surfaces. The brown incrustations contained smaller amounts of heavy metals while the black ones contained the largest amounts. The authors suggest that the black incrustations probably represent brown incrustations in which the crystallization of heavy metal sulfides has occurred to a degree at which FeS and PbS are formed. The black color can then be explained by the presence of considerable quantities of very small crystallites of FeS and PbS in the incrustation.

Experiments using compacted soils in flexible wall permeameter showed that three order reductions in hydraulic conductivity to 10^{-8} cm/s following bacterial treatment from initial values of 10^{-5} to 10^{-6} cm/s (Dennis and Turner, 1998). The plugged zone was resistant to various inorganic chemical permeants. Soil Specimens with well established biofilm barriers showed essentially no change in hydraulic conductivity when permeated with a 0.5 normal saline solution, a pH 3 acidic solution, and a pH 11 basic solution. The biofilm barrier was also fairly resistant to wet-dry cycles, but such conditions can cause a moderate to significant increase in the hydraulic conductivity. Table 1 shows that reductions in hydraulic conductivity of 3 orders can be achieved in soils. Final hydraulic conductivities corresponding to engineered liners are likely not achievable in coarse soils but may be achieved in compacted lesser permeable soils.

CURRENT UNDERSTANDING OF BIOCLOGGING PROCESSES

Reduction of the hydraulic conductivity of porous media has been associated with bacterial growth, production of large amounts of extracellular polymers by bacteria and the precipitation of metals by bacteria. The mechanisms most often advanced to account for the observed reduction in hydraulic conductivities involve the production of slime (Allison, 1947; Shaw et al., 1985) and the accumulation of these gummy substances (Nevo and Mitchell, 1967). Other causal mechanisms that have invoked include the precipitation in soil pores of FeS and MgS produced by SO_4 reducing bacteria (Van Beek, 1984), the deposition of Fe hydroxides or Mg oxides produced by bacteria (Kuntze, 1982; Van Beek, 1984), precipitation of calcium carbonate (Rowe et al., 1997b), or simply the physical plugging of pores by bacterial cells (Gupta and Schwartzendruber, 1962).

The processes by which bacteria affect the hydraulic conductivity of soils are poorly understood (Okubo and Matsumoto, 1983). Although in recent times many contributions have been made in the area of bacterial transport (see Bhowmick (2000) for a comprehensive literature review of bacterial transport in soils), only a few of these have advanced our understanding of bacterial clogging of soils. The most significant papers in this area are by Taylor et al. (1990), Taylor and Jaffé (1990b), Clement et al. (1996), and Sarkar et al. (1994a, 1994b). Each of these studies has studied bacterial clogging of soils and put forward mathematical models to describe the process. However, Vandevivere et al. (1995)

show that such models do not predict satisfactorily the saturated hydraulic conductivity reductions that are observed in fine-textured material.

TABLE 1. Summary of Observed Hydraulic Conductivity Reductions in Porous Media.

Porous Media	Hydraulic Conductivity (cm/s)		Bacteria	Reference
	Initial	Lowest		
Hanford loam	6×10^{-4}	4×10^{-5}	Indigenous	Allison (1947)
Exeter sandy loam	3×10^{-3}	7×10^{-5}	Indigenous	Allison (1947)
Sandy loam	4×10^{-3}	2×10^{-4}	Indigenous	Allison (1947)
Berea sandstone	5×10^{-4}	3×10^{-5}	Aerobic (Indigenous)	Raiders et al. (1986)
Berea sandstone	6×10^{-4}	2×10^{-5}	Anaerobic (Indigenous)	Raiders et al. (1986)
Berea sandstone	3×10^{-4}	4×10^{-5}	Bacillus #47 (Injected)	Raiders et al. (1986)
ASTM C-190 sand	2.5×10^{-1}	2.5×10^{-4}	Aerobic, sewage and sludge	Taylor and Jaffé (1990a)
Silty sand	2×10^{-5}	1.4×10^{-8}	*Beijerinckia indica*	Dennis and Turner (1998)
Silty sand	1×10^{-5}	1×10^{-8}	*Beijerinckia indica*	Dennis and Turner (1998)
Silty sand	6×10^{-6}	9×10^{-9}	*Beijerinckia indica*	Dennis and Turner (1998)
Silty sand	1×10^{-6}	5×10^{-9}	*Beijerinckia indica*	Dennis and Turner (1998)
Silty sand	2×10^{-7}	1.5×10^{-8}	*Beijerinckia indica*	Dennis and Turner (1998)
Silty sand	5×10^{-6}	1.2×10^{-8}	*Beijerinckia indica*	Dennis and Turner (1998)

The typical inorganic constituents in municipal leachate are presented in Table 2. The inorganic constituents that can precipitate in significant amounts and lead to soil plugging are manganese, iron and calcium. Bioprecipitation occurs under anaerobic conditions due to the formation of insoluble sulphides resulting from reaction of hydrogen sulphide (H_2S) with the trace metals. Additionally under anaerobic conditions, the iron- (Fe(III)) and manganese- (Mn(IV)) reducing bacteria use these cations as electron acceptors and precipitate several metals as carbonates or oxides (Gorby and Lovely, 1992; Smith et al., 1994). Other precipitation reactions involve the complexation of calcium, magnesium and other metals with hydrogen carbonate (HCO_3^-) to form a precipitate (Christensen et al., 1994; Smith et al., 1994). Soil solutions provide great opportunity for variety in speciation, as they contain organic ligands (fulvic acid), HCO_3^-, CO_3^{2-}, OH^-, and numerous other anions that are capable of forming insoluble complexes with metal cations (McBride, 1994). Precipitation of crystals induces co-precipitation of other metals in the system (Ehrlich et al., 1973). Additional research is required to exploit the potential benefits from these microorganisms (Baldi et al., 1997).

TABLE 2. Inorganic Constituents in Municipal Waste Leachate.

Parameter	Concentration (mg/L)
Calcium	<300-4,000
Copper	0.1-9
Sodium	0.2-4,000
Potassium	35-4,000
Magnesium	3-15,000
Iron	200-5,500
Manganese	0.6-41
Zinc	0.6-200
Aluminium	<10-200
Chloride	<100-50,000
Sulphate	25-2,000
Total phosphorus	0.1-300

CONCLUSIONS

Bioclogging is prevalent in practice. Field and laboratory studies show that bioclogging could be stimulated to reduce the hydraulic conductivity of some soils to levels corresponding to engineered liners. Some evidence also exists that soils once plugged resist changes in hydraulic conductivity upon contact with acids, bases, and saline solutions. As bioplugged soils have a tendency to reseal the leaking areas their use may offer some benefits over the use of engineered liners as waste containment barriers. However as the literature review shows, our present understanding of the governing processes is limited and there is need for additional research before the bioclogging processes can be engineered to create waste containment barriers.

REFERENCES

Allison, L. E. 1947. "Effect of microorganisms on permeability of soil under prolonged submergence." *Soil Sci.* 63:439-450.

Bagchi, A. 1994. *Design, construction and monitoring of sanitary landfills.* Pp. 178-179. Wiley, New York.

Baldi, F., V. P. Kukhar, and Z. R. Ulberg. 1997. Bioconversion and Removal of Metals and Radionuclides. In J. R. Wild et al. (Eds.), *Perspectives in Bioremediation*, pp. 75-91.

Bhowmick, S. 2000. "Physical and chemical factors affecting bacterial transport in porous media." M.E. Thesis, University of Auckland, Auckland, New Zealand.

Bisdom, E. B. A., A. Boekstein, P. Curmi, P. Lagas, A. C. Letsch, J. P. G. Loch, R. Nauta, and C. B. Wells. 1983. "Submicroscopy and chemistry of heavy-metals-contaminated precipitates from column experiments simulating conditions in soil beneath a landfill." *Geoderma*. 30:1-20.

Brune, M., R. B. Ramke, H. Collins, and H. H. Hanert. 1991. "Incrustation processes in drainage systems of sanitary landfills." *In Proceedings of 3rd International Landfill Symposium, Sardinia,* pp. 999-1006.

Christensen, T. H., P. Kjeldsen, H. Alberchtsen, G. Heron, P. H. Nielsen, P. L. Bjerg, and P. E. Holm. 1994. "Attenuation of Landfill Leachate Pollutants in Aquifers." *Crit. Rev. Env. Sci. Technol.* 24(2):119-202.

Clement, T. P., B. S. Hooker, and R. S. Skeen. 1996. "Macroscopic models for predicting changes in saturated porous media properties caused by microbial growth." *Groundwater.* 34(5):934-942.

Dennis, M. L. and J. P. Turner. 1998. "Hydraulic conductivity of compacted soil treated with biofilm." *J. Geotech. Geoenv. Eng.* 124(2):120-127.

Ehrlich, H. L., S. H. Yang, and J. D. Mainwaring. 1973. "Bacteriology of Manganese Nodules. Fate of Copper, Nickel, Cobalt, and Iron During Bacterial and Chemical Reduction of the Manganese (IV)." *Z. Allg. Mikrobiol.* 13:39-48.

Frankenberger, W. T., Jr., F. R. Troeh, and L. C. Dumenil. 1979. "Bacterial effects on hydraulic conductivity of soils." *Soil Sci. Soc. Am. J.* 43:333-338.

Gade, B., H. Westermann, A. Heindl, R. Hengstmann, H. Pöllmann, J. Riedmiller, and G. Wiedemann. "Geochemistry and equilibrium models in Hazardous waste landfills." *In Proceedings of the 6th International Landfill Symposium, S. Margherita di Pula, Cagliari, Italy, October,* 2:163-180.

Giroud, J. P. 1996. "Granular filters and geotextile filters." *In Proceedings of Geofilters '96,* pp. 565-680.

Gorby, Y. A., and D. R. Lovely. 1992. "Enzymatic Uranium Precipitation." *Env. Sci. Technol.* 26:205-207.

Gupta, R. P. and D. Schwartzendruber. 1962. "Flow-associated reduction in the hydraulic conductivity of quartz sand." *Soil Sci. Soc. of Am. Proc.* 26:6-10.

Heron, G., P. L. Bjerg, P. Gravesen, L. Ludvigsen, and T. H. Christensen. 1998. "Geology and sediment geochemistry of a landfill leachate contaminated aquifer (Grindsted, Denmark)." *J. Contam. Hydrol.* 29:301-317.

Jenneman, G. E., R. M. Knapp, M. J. McInerney, D. E. Menzie, and D. E. Revus. 1984. "Experimental studies of in-situ microbial enhanced oil recovery." *Soc. Pet. Eng. J.* 24:33-37.

Kuntze, H. 1982. "Iron clogging in soils during saturated and unsaturated flow." *J. Wat. Pollut. Control Fed.* 42:1495-1500.

McBride, M. B. 1994. *Environmental Chemistry of Soils.* Oxford University Press Inc.

Nelson, A. 1995. "Landfill Leakage and Biofilms – Can we rely on self clogging mechanisms." *In Proceedings of the 7th Annual Conference of the Waste Management Institute NZ Inc., Auckland, New Zealand,* pp. 431-442.

Nevo, Z. and R. Mitchell. 1967. "Factors affecting biological clogging of sand associated with groundwater recharge." *Water Res.* 1:231-236.

Okubo, T. and J. Matsumoto. 1983. "Biological clogging of sand and changes of organic constituents during artificial recharge." *Water Res.* 17:813-821.

Raiders, R. A., M. J. McInerney, D. E. Revus, H. M. Torbati, R. M. Knapp, and G. E. Jenneman. "Selectivity and depth of microbial plugging in Berea sandstone cores." *J. Ind. Microbiol.,* 1:195-203.

Rittmann, B. E., I. Fleming, and R. K. Rowe. 1996. "Leachate chemistry: Its implications for clogging." *North American Water and Environment Congress '96, Anaheim, CA, June,* Paper 4 (CD Rom) 6p, Session GW-1, Biological Processes in Groundwater Quality.

Rowe, R. K. 1998. "Geosynthetics and the minimization of contaminant migration through barrier systems beneath solid waste." *In Proceedings of the Sixth International Conference on Geosynthetics, Atlanta*, pp. 27-102.

Rowe, R. K., I. R. Fleming, M. D. Armstrong, S. C. Millward, J. VanGulck, and R. D. Cullimore. 1997a. "Clogging of leachate collection systems: Some preliminary experimental findings." *In Proceedings of the 50th Canadian Geotechnical Conference, Ottawa, October*, 1:153-160.

Rowe, R. K., I. R. Fleming, M. D. Armstrong, A. J. Cooke, R. D. Cullimore, B. E. Rittmann, P. Bennett, and F. J. Longstaffe. 1997b. "Recent advances in understanding the clogging of leachate collection systems." *In Proceedings of the Sixth International Landfill Symposium, S. Margherita di Pula, Cagliari, Italy, October*, 3:383-392.

Rowe, R. K., I. Fleming, R. Cullimore, N. Kosaric, and R. M. Quigley. 1995. "A research study of clogging and encrustation in leachate collection systems in municipal solid waste landfills." Geotechnical Research Center, University of Western Ontario, Report submitted to Interim Waste Authority Ltd., June, 128 p.

Sarkar, A. K., G. Georgiou, and M. M. Sharma. 1994. "Transport of bacteria in porous media: I. An Experimental Investigation." *Biotechnology and Bioengineering*. 44:489-497.

Sarkar, A. K., G. Georgiou, and M. M. Sharma. 1994. "Transport of Bacteria in Porous Media: II. A model for convective transport and growth." *Biotechnology and Bioengineering*. 44:499-508.

Shaw, J. C., B. Bramhill, N. C. Wardlaw, and J. W. Costerton. 1985. "Bacteria fouling in a model core system." *Appl. And Envir. Microbiol*. 49(3):693-701.

Smith, L. A., B. C. Alleman, and L. Copley-Graves. 1994. "Biological Treatment Options." In Means and Hinchee (Eds.), *Emerging Technology for Bioremediation of Metals*, pp. 1-12. CRC Press. Boca Raton.

Taylor, S. W. and P. R. Jaffé. 1990a. "Biofilm growth and the related changes in the physical properties of a porous medium. 1. Experimental Investigation." *Water Resour. Res*. 26(9):2153-2159.

Taylor, S. W. and P. R. Jaffé. 1990b. "Substrate and biomass transport in a porous medium." *Water Resour. Res*. 26(9):2181-2194.

Taylor, S. W., P. C. D. Milly, and P. R. Jaffé. 1990. "Biofilm growth and the related changes in the physical properties of a porous medium. 2. Permeability." *Water Resour. Res*. 26(9):2161-2169.

Van Beek, C. G. E. M. 1984. "Restoring well yields in the Netherlands." *J. Am. Water Works Assoc*. 76(10):66-72.

Vandevivere, P., P. Baveye, D. S. de Lozada, and P. DeLeo. 1995. "Microbial clogging of saturated soils and aquifer materials: Evaluation of mathematical models." *Water Resour. Res*. 31(9):2173-2180.

DETERMINATION OF FLUORANTHENE BIODEGRADATION METABOLITES AND THEIR TOXICITY

Ester Šepič ("Jožef Stefan" Institute, Ljubljana, Slovenia)
Mihael Bricelj (National Institute of Biology, Ljubljana, Slovenia)
Hermina Leskovšek ("Jožef Stefan" Institute, Ljubljana, Slovenia)

ABSTRACT: Analytical procedures for isolating and identifying the biodegradation products of fluoranthene were developed using the pure bacterial strains *Pasteurella* sp. IFA and *Mycobacterium* sp. PYR-1. The stable degradation products were determined over an 8, 10 and 14 day incubation period. Samples for metabolite isolation were extracted with a two step liquid liquid extraction. The chromatographic properties of the metabolites were improved by derivatisation (sylilation and oxime formation) and were identified using GC-MS (SCAN and SIM mode). In total seven stable fluoranthene metabolites were identified and quantified.
Toxicological (growth inhibition) tests for fluoranthene and its 9 metabolites were made using algae *Scenedesmus subspicatus*. Algae growth was determined by counting cells and the amount of chlorophyll a formed.
A comparison of calculated EC_{50} values showed that there is no direct correlation between algae growth and chlorophyll production and therefor a simpler method of measuring chlorophyll can not replace counting algae. Toxicological tests using the algae *Scenedesmus subspicatus* revealed that with the exception of 9-hydroxyfluorene - which is in the same toxicity range as fluoranthene, all the metabolites were in the order of 1000 times less toxic than fluoranthene.

INTRODUCTION

Polycyclic aromatic hydrocarbons (PAHs) are considered as serious environmental pollutants since they can be either acutely toxic or genotoxic depending upon the number and configuration of the benzene rings and the presence and position of their substituents. Microbial degradation is a major factor affecting the persistence of PAHs in the environment. The ability of bacteria in water, soil or sediment to degrade PAHs depends on the complexity of its chemical structure and the extent of enzymatic adaptation in response to chronic exposure to aromatic hydrocarbons.
Microbial degradation of lower molecular weight PAHs with two to three aromatic rings (e.g. naphthalene, fluorene, phenanthrene) with a microbial consortia or pure strains has been studied extensively and proven to be relatively easy. PAHs with four or more aromatic rings (fluoranthene, pyrene and benzo(a)anthracene) are more recalcitrant to microbial attack and are degraded with greater difficulty.
Fluoranthene, a four ring PAH, is one of the most frequent and abundant PAH pollutants of pyrogenic origin in the environment and was chosen as a

model compound for studying the bioremediation potential. The present study concentrates on using a new bacterial isolate from an activated sludge (*Pasteurella sp.* IFA) and compares the results with the degradation of fluoranthene by *Mycobacterium sp.* PYR-1 (known to be able to degrade fluoranthene) under the same experimental conditions. The aim of this study is to elucidate the microbial activity of different bacterial strains, to isolate and identify metabolites of the model compound (fluoranthene), to determine metabolic pathways and toxicity of stable intermediates.

MATERIALS AND METHODS

Biodegradation. The bacterial strains *Pasteurella* sp. IFA and *Mycobacterium* sp. PYR-1 were chosen for fluoranthene biodegradation studies in aqueous media. Both strains were previously isolated from an oil-contaminated soil and an estuarine sediment. The preparation of the substrate, mineral media with trace elements and the inoculumn are described in full elsewhere (Šepič et al, 1996, 1997, 1998).

All experiments were set up in aqueous media at room temperature. In the first experiment using *Pasteurella* sp. IFA, four cotton wool-stoppered flasks (1 litre) were filled with mineral media containing 20 mg l^{-1} fluoranthene and incubated for 10 days. A similar experiment was set up for *Mycobacterium* sp. PYR-1 with 15 flasks using the same concentration of fluoranthene (20 mg l^{-1}). Ten flasks were incubated for 8 days and five flasks for 14 days, at room temperature. The optimal times of incubation (10 days for *Pasteurella* sp. IFA and 8 days for *Mycobacterium* sp. PYR-1) were found from previous studies (Šepič et al, 1997). A 14 day incubation period for *Mycobacterium* sp. PYR-1 was chosen to determine whether different metabolic products were accumulated after prolonged incubation.

At set intervals, the cultures were extracted three times with equal volumes of ethylacetate (EtAc, 50 ml) and combined to give a neutral extract. The cultures were then acidified to pH 2,5 with H_2SO_4 and extracted once more with another three volumes of ethylacetate. These were combined to give acidic extract. Both extracts were dried (anhydrous Na_2SO_4) and reduced in volume to 1 ml (N_2). The extracts were analysed using gas chromatography with mass selective detection (GC-MSD, Hewlett Packard 6890-5972, Waldbronn, Germany). The chromatograph was equipped with a HP-MS5 crosslinked 5 % phenylmethyl silicone capillary column (30 m length, 0,25 mm diameter, 0,25 mm film thickness)

Commercially available standard compounds (Aldrich Chemical Co., Milwaukee, WI, USA) were dissolved in organic solvents and screened using GC-MS. The structure of these compounds is presented in Figure 1. The elucidation of the structure of the metabolites E, F, H and I was only possible after derivatisation when characteristic spectra were obtained.

The various metabolites differ in their chemical structure and functional groups and no one derivatising agent is suitable for all the metabolites. In our work sylilation with N-methyl-N-(trimethylsilyl)trifluoracetamide (MSTFA,

Aldrich) was applied for those metabolites containing R-OH and R-COOH groups (Šepič *et al*, 1999) and derivatisation (oxime formation) with O-(2,3,4,5,6-pentafluorobenzyl)-hydroxylamine hydrochloride (PFBHA.HCl; Fluka Chemie AG, Buchs, Switzerland) (Šepič *et al*, 1999). All sample extracts were derivatised with both derivatising agents.

To estimate the amount of totally degraded fluoranthene, the evolved CO_2 was measured (Šepič *et al*, 1998). The apparatus consisted of three dreschel bottles connected in series (first bottle: 5 mg fluoranthene/250 ml, second: standardised BaOH solution, third: pure water - airlock). The amount of evolved CO_2 was calculated from the residual amount of BaOH measured by titration (0,05 M HCl).

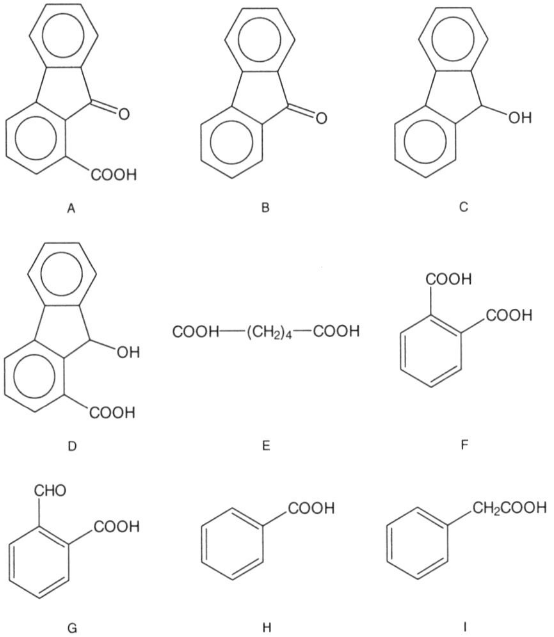

Figure 1: **Structural formula of expected fluoranthene metabolites** *(A:* **9-fluorenone-1-carboxylic acid,** *B:* **9-fluorenone,** *C:* **9-hydroxyfluorene,** *D:* **9-hydroxy-1-fluorene-carboxylic acid,** *E*: **adipic acid,** *F*: **phtalic acid,** *G:* **2-carboxybenzaldehyde,** *H:* **benzoic acid and** *I:* **phenylacetic acid)**

Ecotoxicity. Growth inhibition test for fluoranthene and its intermediates was made with algae *Scenedesmus subspicatus* CHODAT according to modified ISO standard (ISO 8692-1989). Algae growth was determined with counting algae cells after 7 day incubation with 12 hour day/night rhythm of lighting at 100 µE/m2/sec. Each sample contained app. 1×10^4/ml of algae culture.

According to standard requirements, sample concentrations must follow geometric ratio. Concentrations were carefully chosen in the way that 4-5 of them covered 10 – 90 % inhibition. For each compound at the selected concentration 4 samples and 4 blanks were prepared. Algae cells were counted at the beginning of the experiment (day 0) and after 7 day incubation.

At the same time formed chlorophyll was measured. The algae samples were first centrifuged, then the precipitated algae was dissolved in hot methanol and the extinction for chlorophyll a was measured (Vollenveider, 1974). Our task was also to find out if measuring chlorophyll a could replace the time consuming counting of algae cells.

RESULTS AND DISCUSSION

Biodegradation. Neutral and acidic extracts of each biodegraded sample from both bacteria were first analysed using GC-MS operating in the SCAN mode (m/z=50-500). Fluoranthene metabolic products were identified on the basis of the molecular and fragment ions in the mass spectra and the chromatographic retention time (Rt) of authentic compounds. Because of the low metabolite concentrations, only GC-MS/SIM was sufficient for metabolite determination. Derivatisation was applied to make the compounds more accessible to gas chromatography. Detailed derivatisation procedure and mass spectra of derivatised compounds, showing the advantages of applied derivatisation, are described in full elsewhere (Šepič *et al*, 1999).

In the fluoranthene biodegradation experiment with *Pasteurella* sp. IFA only two metabolites (9-fluorenone-1-carboxylic acid and 9-fluorenone) were identified by GC-MS directly, while 5 metabolites (9-fluorenone-1-carboxylic acid, 9-hydroxy-1-fluorene-carboxylic acid, 2-carboxybenzaldehyde, benzoic acid and phenylacetic acid) were identified in their derivatised forms. In the case of *Mycobacterium* sp. PYR-1 9-fluorenone-1-carboxylic acid and 9-fluorenone were determined directly and 6 metabolites after derivatisation. With *Mycobacterium* sp. PYR-1 metabolite 9-hydroxyfluorene was also identified in its derivatised form.

On the basis of the identified metabolites a biodegradation pathway for *Pasteurella* sp. IFA is proposed (Figure 2). The presence of 9-fluorenone-1-carboxylic acid represents an initial attack on the fused aromatic ring portion of the fluoranthene molecule, most likely on the 1,2-positions *via dioxygenases*, which almost certainly leads to the formation of a dihydroxylated fluoranthene intermediate. With subsequent *meta* cleavage 9-fluorenone-1-carboxylic acid is formed, which can further undergo decarboxylation to form 9-fluorenone. The presence of 9-hydroxy-1-fluorene carboxylic acid indicates an alternative pathway, which occurs simultaneously with the 9-fluorenone-1-carboxylic route. Transition from one to another pathway is also possible *via* biological hydrogenation of 9-fluorenone-1-carboxylic acid to 9-hydroxy-1-fluorene carboxylic acid. Further decarboxylation of this stable metabolite leads to the formation of 9-hydroxyfluorene and benzoic derivatives (2-carboxybenzaldehyde, benzoic acid, phenylacetic acid). On the basis of the identification of 7 metabolic products the same biodegradation pathway can be assumed for *Mycobacterium* sp. PYR-1 confirming part of a biodegradation scheme previously proposed by Kelly *et al* (1993).

Biodegradation results show that *Pasteurella* sp. IFA degraded 25 % of total fluoranthene, of which approximately two thirds were mineralised to CO_2 and one third was identified as stable metabolic products. For *Mycobacterium* sp. PYR-1 46 % of total fluoranthene was degraded after 14 days of incubation, of which approximately four fifth were mineralised to CO_2 and one fifth was identified as stable metabolites. A comparison of the mass balance for each bacterium shows that *Pasteurella* is not as efficient as *Mycobacterium* sp. PYR-1 in degrading and mineralising fluoranthene. This was confirmed by measuring the biochemical oxygen demand and CO_2 evaluation. Overall, *Pasteurella* sp. IFA required two months to degrade the same amount of 3 and 4 ring polycyclic aromatic hydrocarbons (80 %), that *Mycobacterium* sp. PYR-1 was able to degrade within the first month of incubation. *Pasteurella* sp. IFA is more efficient in degrading initial metabolic products (e.g. 9-fluorenone-1-carboxylic acid), which resulted in the accumulation of higher amounts of one-ring metabolites (2-carboxybenzaldehyde, benzoic acid). Biodegradation experiments also showed that *Pasteurella* sp. IFA, although slower than *Mycobacterium* sp. PYR-1 is active over a longer period.

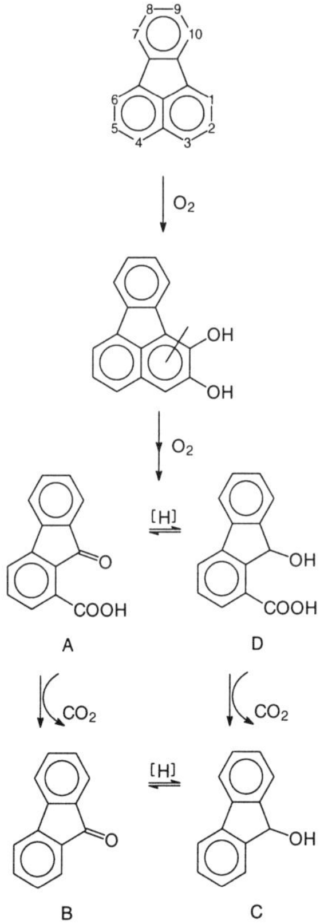

Figure 2: Proposed fluoranthene biodegradation pathway

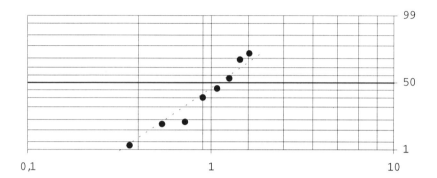

Figure 3: Graphical determination of EC_{50} for 9-hydroxyfluorene (growth rate, interpolation: $EC_{50}=1,18$)

Ecotoxicity. EC_{50} values were determined graphically separately for each compound (fluoranthene and 9 metabolites). Figure 3 shows the EC_{50} value for intermediate 9-hydroxyfluorene ($EC_{50}=1,18$) obtained with interpolation using biomass integral (area under the growth curve). Abscissa (X axis) represents compound concentration (mg l^{-1}) and ordinate (Y axis) growth inhibition (%).

The Tables 1 and 2 show calculated EC_{50} values for algae cells (Table 1) and formed chlorophyll (Table 2) applying both growth rate and biomass integral with interpolation and the regression method.

Table 1 and 2: EC_{50} values for fluoranthene and its intermediates determined with Growth inhibition biotest (algae cells counting: Table 1, formed chlorophyll: Table 2) applying growth rate and biomass integral with interpolation and regression method (SS = saturated solution)

Compounds	Growth rate				Biomass		integral	
	EC_{50} interpol. gl^{-1}	mgl^{-1}	EC_{50} regress. gl^{-1}	mgl^{-1}	EC_{50} interpol. gl^{-1}	mgl^{-1}	EC_{50} regress. gl^{-1}	mgl^{-1}
A	0,18		0,29		0,55		0,56	
B	30,0%SS		26,3%SS		47,2%SS		44,1%SS	
C		0,82		0,17		1,18		1,18
D	>0,75		>0,75		>0,75		>0,75	
E	0,51		0,61		0,52		0,89	
F	0,45		0,20		0,54		0,49	
G	0,63		0,59		0,80		0,78	
H	0,67		0,52		0,59		0,84	
I	0,64		0,40		0,68		0,58	
Fluoranthene		0,19		0,19		0,23		0,23

Compounds	Growth rate		Biomass integral	
	EC_{50} interpolation gl^{-1} mgl^{-1}	EC_{50} regression gl^{-1} mgl^{-1}	EC_{50} interpolation gl^{-1} mgl^{-1}	EC_{50} regression gl^{-1} mgl^{-1}
A	0,23	0,21	0,74	1,56
B	26,2%SS	24,8%SS	48,6%SS	57,8%SS
C	0,56	0,75	1,14	0,79
D	>0,75	>0,75	>0,75	>0,75
E	0,55	0,68	0,63	0,64
F	0,63	0,58	0,76	0,96
G	0,67	0,71	1,10	1,10
H	0,81	0,81	0,94	0,99
I	0,71	0,69	0,79	0,79
fluoranthene	0,19	0,19	0,23	0,23

Table 1 shows calculated EC_{50} values for counting algae cells for fluoranthene and its metabolites applying growth rate and biomass integral with interpolation and regression method. From calculated EC_{50} values for biotest growth inhibition of algae cells *Scenedesmus subspicatus* CHODAT it is seen that the most toxic compound for tested cells is fluoranthene and its calculated EC_{50} value is 0.19mg/l (growth rate) and 0.23 (biomass integral). Only one intermediate (9-hydroxy fluorene) has a EC_{50} value in the same concentration range (mg/l), e.g. growth rate: EC_{50}=0.82 (interpolation) and 0.71 mg/l (regression) and biomass integral: EC_{50}=1.18 mg/l with both interpolation and regression method). 9-hydroxy fluorene was determined in the process of fluoranthene metabolite detection in only one experiment and in very low concentrations. Therefor we estimated that this metabolite is unstable (at selected incubation times) and harmless for the algae population. For the rest of the metabolites all the EC_{50} values were in the order of 1000 times higher concentration range (g/l).

The aim of our work was also to find out whether the simpler and less time consuming method of measuring chlorophyll could replace counting algae cells. Considering ± 10 % error for both calculated EC_{50} values, quotient f between EC_{50} for counting cells and EC_{50} for measuring chlorophyll should be in the range between 0,90 and 1,10.

Table 3 represents factors f obtained with method applying growth rate and biomass integral and calculated with interpolation and regression method. From Table 3 it can be seen that agreement within 10 % error is achieved only for metabolite E when growth rate method (interpolation) was applied and for metabolites B and C when growth rate (regression) and biomass integral (interpolation) were applied. From a direct comparison of calculated EC_{50} values it is clearly seen that there is no direct correlation between algae growth and chlorophyll production and therefor a simpler method of measuring chlorophyll can not replace counting algae.

To confirm this unsuitable correlation between measuring chlorophyll and counting algae cells, analysis of variance was also carried out. 144 pairs of values of counted algae cells and formed chlorophyll were compared with linear and exponent model. To achieve suitable agreement, R^2 should be > 0.80. In both cases values are much lower (linear model: 0.380, exponent model: 0.445), which proves that there is no suitable correlation between chlorophyll production and counting cells when testing growth inhibition of algae *Scenedesmus subspicatus*.

Table 3: Factors EC_{50} values for fluoranthene metabolites at growth inhibition test with algae *Scenedesmus subspicatus*

Compounds	Growth rate		Biomass	integral
	EC_{50} interpol.	EC_{50} regression	EC_{50} interpol.	EC_{50} regression
A	1,27	0,50	1,34	2,28
B	0,87	0,94	1,02	1,31
C	0,68	1,05	0,96	0,71
E	1,07	1,11	1,21	0,71
F	1,40	2,90	1,40	1,90
G	0,90	1,20	1,30	1,41
H	1,21	1,56	1,59	1,18
I	1,11	1,73	1,16	1,36

CONCLUSIONS

A combination of a two step liquid liquid extraction procedure and derivatisation by silylation with MSTFA and oxime formation with PFBHA.HCl proves to be a powerful isolation technique for quantitative GC-MS analysis of stable fluoranthene metabolites. The method resulted in identification of six metabolites with *Pasteurella* sp. IFA (9-fluorenone-1-carboxylic acid, 9-fluorenone, 9-hydroxy-1-fluorene-carboxylic acid, 2-carboxybenzaldehyde, benzoic acid and phenylacetic acid). In the case of *Mycobacterium* sp. PYR-1 also 9-hydroxy fluorene was identified. Derivatisation greatly improved the chromatographic and spectrometric properties of both authentic compounds and metabolites and this, combined with the low detection limits operated in SIM mode, allowed the identification of previously unobserved metabolic products.

On the basis of identified metabolites, a fluoranthene biodegradation pathway was for the first time proposed for *Pasteurella* sp. IFA and was partly in agreement with degradation pathway published by Kelly *et al* (1993). According to the same identified metabolites (with an addition of 9-hydroxy fluorene in the case of *Mycobacterium* sp. PYR-1), it can be assumed that both bacterial strains undergo the same degradation pathway.

Results from toxicological (growth inhibition) tests using algae *Scenedesmus subspicatus* revealed (Tables 1 and 2) that all the metabolites, with the exception of 9-hydroxyfluorene, which is in the same toxicity range as fluoranthene, were in order of 1000 times less toxic than fluoranthene. Thus, the high biodegrading potential for fluoranthene - a recalcitrant polycyclic aromatic hydrocarbon and the fact that *Pasteurella* sp. IFA and *Mycobacterium* sp. PYR-1 were isolated from natural environment makes them ideal candidates to be incorporated in an effective bioremediation technique.

REFERENCES

Kelley, I., J.P. Freeman, F.E. Evans, and C.E. Cerniglia. 1993. "Identification of metabolites from the degradation of fluoranthene by *Mycobacterium* sp. strain PYR-1". *Appl. Environ. Microbiol. 59:* 800-806.

Šepič, E., C. Trier, and H. Leskovšek. 1996. "Biodegradation studies of selected compounds from diesel oil." *Analyst. 121*: 1451-1456.

Šepič, E., M. Bricelj, and H. Leskovšek. 1997. "Biodegradation studies of polyaromatic hydrocarbons in aqueous media." *Jour. Appl. Microbiol. 83:* 561-568.

Šepič, E., M. Bricelj, and H. Leskovšek. 1998. "Degradation of fluoranthene by Pasteurella sp. IFA and Mycobacterium sp. PYR-1: Isolation and identification of metabolites." *Jour. Appl. Microbiol. 85:* 746-754.

Šepič, E,. and H. Leskovšek. 1999. "Isolation and identification of fluoranthene biodegradation products." *Analyst. 124*: 1765-1769.

Volenveider, R.A.. 1974. "Primary production in aquatic environments." In *Int. Biol. Prog.,* Handbook 12, pp. 225. Blackwell Sc. Pub., Oxford, UK.

TRANSFORMATION OF CHLOROBENZENE IN A LABORATORY AQUIFER COLUMN

Carsten Vogt, Helmut Lorbeer, Lothar Wuensche and Wolfgang Babel
(Centre for Environmental Research Leipzig-Halle, Permoserstrasse 15, 04318 Leipzig, Germany)

ABSTRACT: The possibility of an anaerobic biotransformation of chlorobenzene (CB) under nitrate- or sulfate-reducing conditions was examined in a laboratory glass column filled with sediment from a CB-polluted aquifer. Under all conditions, CB and nitrate disappeared partially and nitrite was formed, whereas sulfate concentrations remained stable. Small amounts of 3-chloro-catechol (3-CC) were detected in the influent of the column, and bacteria growing aerobically on CB as sole source of carbon and energy were isolated from the column, as well as bacteria growing anaerobically with succinate and nitrate. These results suggest that the disappearance of CB in the column system was initiated partially by aerobic CB-degrading bacteria, and that degradation products of CB were probably consumed by nitrate-reducing bacteria.

INTRODUCTION

During the last 100 years, lignite mining and chemical industry were widespread and dominated the landscape of the region Bitterfeld/Wolfen, Saxonia-Anhalt, Germany. Consequently, the groundwater was planarly and seriously contaminated with pollutants, mainly halogenated aliphatic and aromatic hydrocarbons (Peters et al., 1995). For this reason, an area in the southeast of the city Bitterfeld was choosen as a model area for testing novel *in-situ* groundwater remediation technologies within the German groundwater research project SAFIRA. CB is the main pollutant in the model area aquifer system in depths of 15 to 20 m, reaching concentrations of 8 to 51 mg/L (Weiß et al., 1997). Oxygen concentrations in these depths are nearly zero. Therefore, we focussed our research on the possibility of an anaerobic degradation of CB by authochtonous bacteria. A part of this work was to study the biotransformation of CB under different electron-acceptor conditions, mainly nitrate-reducing conditions, in a laboratory glass column which was filled with original aquifer sediment material and which was continuously percolated with anoxic, bicarbonate-buffered mineral salt medium and CB added as sole source of carbon and energy.

MATERIAL AND METHODS

The glass column (lenght: 120 cm; inner diameter: 4.2 cm; volume: about 1200 ml) was filled under a stream of N_2 with aquifer sediment to a volume of approximately 1000 ml. The aquifer sediment was taken from a reactor which was previously percolated for 300 days with anoxic groundwater supplemented with nitrate (62 mg/L). A bicarbonate buffered anoxic mineral salt medium with added

CB (15 – 30 mg/L) was pressured with N_2 and pumped through the column in upflow direction by means of a peristaltic pump, in flow-rates between 6 and 12 ml/h (see Results); fresh medium was connected once per week. The mineral salt medium contained (in g/L): NaCl, 0.5; KH_2PO_4, 0.5; NH_4Cl, 0.4; KCl, 0.4; Na_2SO_4, 0.2; $CaCl_2$, 0.1; $MgCl_2$, 0.5; $NaHCO_3$, 2.52. The medium was supplemented with trace element solution SL-10 (1 ml/L; Trueper and Pfennig, 1992) and vitamin solution (5 ml/L; Pfennig, 1965), the pH was adjusted to 6,7. CB was added as pure substance (99.5%). To dissolve CB, the medium was agitated over night before using. For adjusting anoxic conditions, the medium was purged with N_2 for 30 minutes and reduced with Na_2-dithionite (5-10 mg/L) before adding CB. The inlet and outlet of the glass column were connected to 100 ml flasks via tubes, high-grade steel tubing connectors and teflon-coated, valve-containing plugs (Serto Jacob Ltd., Germany); from these flasks, samples for influent and effluent analyses were taken. To prevent oxygen diffusion from behind the column, a 1-L flask was inserted between the effluent sample flask and the flask in which the outflow was collected. All used tubes were made of oxygen-impermeable material (Iso-Versinic; Ochs Ltd., Germany).

Sampling and Analysis. Samples from the influent and effluent were taken from the sample flasks by a pipette inside an anaerobic chamber. CB was immediatly analyzed after sampling. Samples for measuring inorganic ions and 3-CC were usually frozen at $-20°C$ and analyzed later.

CB was measured by automatical headspace gas chromatography on a CombiPal autosampler and a Varian 3800 gas chromatograph equipped with a 0.25 mm (inner diameter) x 25 m (length) CP SIL 5 CB capillary column (DF 0.12 μm) and a flame ionization detector. Chromatographic conditions were as follows: injector temperature: $250°C$ (split 1:2); detector temperature: $300°C$; oven temperature program: $35°C$ (3 min), $10°C$ / min $\rightarrow$ $65°C$, $30°C$ / min $\rightarrow$ $260°C$. Liquid test samples (diluted 1:10 or 1:20 in 1.6 mM H_2SO_4, end volume: 10 ml) were prepared in 20 ml glass vials. Samples were incubated for 20 minutes at $35°C$ in an agitator (rotation regime: 250 rpm for 5 sec, no rotation for 2 sec) prior to analysis. 1 ml headspace of the samples was injected. For calibration, diluted standards of CB (0.005-2 mg/L) prepared from stock solutions were treated in the same way as described for the samples. Stock solutions of CB (11,12 mg/L) were prepared in pure methanol.

3-chlorocatechol (3-CC) was quantified by high-performance liquid chromatography (HPLC), using a Shimadzu LC-6A Liquid Chromatograph equipped with a Nucleosil-100 (4 mm ID) column and a Shimadzu SPD-6AV UV-VIS detector (wave lenght 283 nm). Samples were eluted isocratically in 40 % (v/v) acetonitrile / 60% phosphate (pH 2.8), at a flow rate of 1 ml/min.
Nitrate and sulfate were measured using a Dionex DX-100 ionchromatograph equipped with an IonPac AS4A Guard (4 x 50 mm) precolumn, an IonPac AS4A (2 x 250 mm) column and a conductivity detector. Samples were eluted isocratically in Na_2CO_3 (190.8 mg/L)/$NaHCO_3$ (122.4 mg/L) buffer, at a flow rate of 2 ml/h.

Nitrite was measured photometrically using Merck test Spectroquant (detection range: 0.02 – 3 mg/L).

Preparation of microcosms. Anoxic microcosms under nitrate-reducing conditions with CB as sole source of carbon and energy were prepared from the aquifer sediment before starting the column experiment. The microcosms consisted of 20 g sediment suspended in 30 ml anoxic, CB- (11 mg/L) and nitrate- (1000 mg/L) amended mineral salt medium (see above) in 116 ml flasks sealed with teflon-coated butyl septa and aluminium crimps, and were incubated statically at room temperature in an anaerobic chamber.

Isolation and characterization of pure bacterial cultures. Liquid influent or effluent samples were spread onto plates containing modified phosphate-buffered ‚Brunner' mineral salt medium (DSMZ medium No. 457) with added agar. The Brunner medium contained (in g/L destilled water): Na_2HPO_4, 2.44; KH_2PO_4, 1.52; $(NH_4)_2SO_4$, 0.5; NaCl, 0.5; $MgSO_4$ x 7 H_2O, 0.2; $CaCl_2$ x 2 H_2O, 0.05. The medium was supplemented with trace element solution SL-10 (1 ml/L); Trueper and Pfennig, 1992) and vitamin solution (5 ml/L; Pfennig, 1965) and adjusted to pH 6.9.

For isolation of CB-degraders, plates were incubated in an air-filled, closed jar. CB was given as sole source of energy and carbon by means of a glass plate filled with 200 μl CB (99.5%) which was placed at the top of the stacked plates so that the bacteria were supplied with the substrate via the gas-phase. Plates were incubated at room temperature until growth of colonies was visible; single colonies were subcultured on dilution plates and fed with CB for at least 3 times until purity was ensured microscopically.

For growth and isolation of denitrifying succinate degraders, the modified Brunner medium was supplemented with succinate (295 mg/L) and nitrate (620 mg/L), and plates were incubated in an anaerobic atmosphere (Anaerocult® A; Merck, Germany) at room temperature.

Bacterial strains were identified by primary tests such as Gram staining, morphology, motility and oxidase test, and by means of identification systems such as analyses of methylated cellular fatty acids (Haertig et al., 1999), and the BIOLOG (BIOLOG Inc., Hayward, USA) Automated Microbial Identification System (Wuensche and Babel, 1996).

For CB degradation experiments with *Rhodococcus* strain UFZ B528 under microaerobic conditions, mineral salt medium was purged with N_2 as long as oxygen concentrations were lower than 0.5 mg/L. The medium was spiked with CB (11-22 mg/L) in an anerobic chamber and inoculated with CB-pregrown *Rhodococcus* cells. Experiments were done in 116 ml flasks filled with 30 ml cell suspension and sealed with teflon-coated butyl septa and aluminium crimps. The flasks were incubated at 14°C on a rotary shaker (111 rpm).

RESULTS AND DISCUSSION

Column experiment. Successively, three peroids with different electron acceptor conditions were established: (1) nitrate (5 mM) and sulfate (2 mM); (2) sulfate (2 mM); (3) nitrate (1 mM) and sulfate (2 mM). In Figure 1 and 2, the results of all periods are summarized.

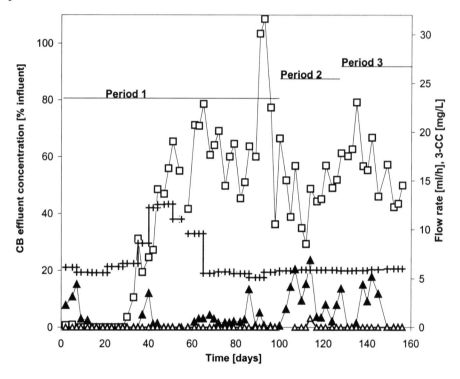

Figure 1. Time course of CB disappearance, 3-CC production, and flow rate during the column experiment. + flow rate; □ CB effluent concentration (% of influent) ; ▲ 3-CC (influent); △ 3-CC (effluent);

In period 1, during the first 28 days of operation, no CB was detected in the effluent. Beginning with day 30, CB concentrations in the effluent started to raise. At flow rates between 5 to 6 ml/h, the effluent concentration reached 65.3 % of the influent concentration in average, accompanied by a small decrease of nitrate and production of nitrite. When nitrate supply was stopped (period 2), sulfate concentrations remained stable, but the average concentration of CB in the effluent decreased slightly (average effluent concentration: 46.1 %), indicating that the decrease in the CB concentration was due to abiotic losses or to biotic transformations independent of the presence of nitrate and sulfate. Under low nitrate conditions (period 3), CB further disappeared partially (average effluent concentration: 56.8 %), and both nitrite production and nitrate removal were observed again. During all column flow phases, active bacteria were observed in the influent and effluent sample flasks, and small amounts of 3-CC, a typical intermediate of aerobic microbial breakdown of CB (van der Meer, 1997), were detected irregularly in the influent (Fig. 1). This may indicate that traces of oxygen have penetrated the system (see below). Since 3-CC could not observed in the effluent (with the exception of day 114), it was transformed or degraded inside the column.

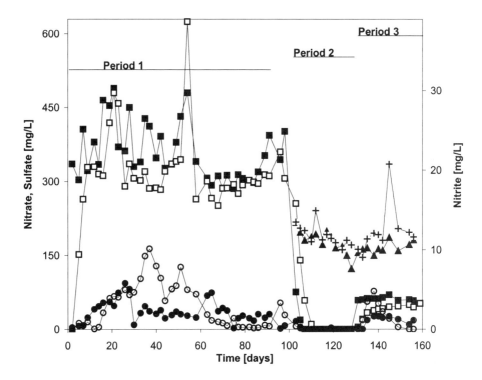

Figure 2. Time course of nitrate, nitrite and sulfate concentrations during the column experiment (sulfate concentrations are shown only for periods 2 and 3). ■ nitrate (influent); □ nitrate (effluent); ● nitrite (influent); ○ nitrite (effluent); ▲ sulfate (influent); + sulfate (effluent)

Establishment of CB-amended microcosms under anaerobic, nitrate-reducing conditions. Establishment of batch microcosms made of column material capable of degrading CB under anaerobic, denitrifying conditions failed; CB concentrations in the microcosms remained stable for months. Therefore, under the experimental conditions used, bacteria of the aquifer were seemingly not able to transform CB anaerobically. Only a few indications exist that CB is degradable without oxygen (Fathepure and Vogel, 1991; Nowak et al., 1996).

Isolation of aerobic CB-degrading and denitrifying succinate-degrading bacteria. From the influent and effluent of the column, five bacterial strains were isolated growing on CB as sole source of carbon and energy, four of them belonging to the genus *Rhodococcus*, known for bearing members with the capacity to degrade CB (Zaitsev et al., 1995). One strain was identified as *Acidovorax facilis*, recently detected as highly abundant bacterial group in activated sludge of wastewater treatment plants (Schulze et al., 1999), but not described for CB degradation to date. *Rhodococcus* UFZ B528, isolated from the influent of the column, degraded CB and accumulated 3-CC when incubated with CB under microaerobic (oxygen content lower than 0.5 mg/L) conditions, indicating that the organism is at least partly responsible for the observed production of 3-CC in the influent. 3-CC accumulation during microbial degradation of CB under low-oxygen (< 0.5 mg/l) conditions was recently described (Ferreira Jorge and Livingston, 1999). Three bacterial strains were isolated growing anaerobically on succinate and nitrate, but yet not further characterized. Succinate is an intermediate of the usual aerobic degradation pathway of CB or 3-CC.

CONCLUSIONS

The results from the column and microcosms experiments demonstrate that, under the experimental conditions used, the autochthonous aquifer bacteria were not able to degrade CB anaerobically under nitrate- or sulfate-reducing conditions. However, the results suggest that (a) traces of oxygen might have penetrated the column system, and the disappearance of CB in the column was initiated at least partially by aerobic CB-degrading bacteria, (b) degradation products of CB were consumed by nitrate-reducing bacteria. In the future, attempts will be made to mineralize CB in the column by adding small amounts of oxygen concomitantly with nitrate in excess.

ACKNOWLEDGEMENTS

This work was carried out as part of the project SAFIRA (acronym of the German „Sanierungsforschung in regional kontaminierten Aquiferen" = remediation research in regionally contaminated aquifers) which is under the management of Dr.Holger Weiß (Centre for Environmental Research Leipzig-Halle, Department of Industrial and Mining Landscapes, Leipzig). The authors thank Mrs. Renate Boetz, Mrs. Elke Haeusler and Mrs. Rita Remer for excellent technical assistance, and Dr. Claus Haertig for analysis of cellular fatty acids.

REFERENCES

Fathepure, B.Z., and T.M. Vogel. 1991. "Complete degradation of polychlorinated hydrocarbons by a two- stage biofilm reactor." *Appl. Environ. Microbiol. 57*, 3418-3422.

Ferreira Jorge, R.M., and A.G. Livingston. 1999. "A novel method for characterisation of microbial growth kinetics on volatile organic compounds." *Appl. Microbiol. Biotechnol. 52*: 174-178.

Haertig, C., N. Loffhagen, and W. Babel. 1999. "Glucose stimulates a decrease of the fatty acid saturation degree in Acinetobacter calcoaceticus." *Arch. Microbiol. 171*: 166-172.

Nowak, J., N.H. Kirsch, W. Hegemann, and H.-J. Stan, 1996. "Total reductive dechlorination of chlorobenzenes to benzene by a methanogenic mixed culture enriched from Saale river sediment." *Appl. Microbiol. Biotechnol. 45*: 700-709.

Peter, H., Großmann, J., and G. Schulz-Terfloth. 1995. "Remediation outline concept for the Bitterfeld/Wolfen region" (Rahmensanierungskonzept des Großprojektes Bitterfeld/Wolfen). In H.-P. Luehr (Ed.), *Groundwater remediation (Grundwassersanierung) 1995*. IWS Schriftenreihe 23, pp. 123-128. Erich Schmidt, Berlin, Germany.

Pfennig, N. 1965. "Enrichment cultures for red and green sulfur bacteria" (Anreicherungskulturen fuer rote und gruene Schwefelbakterien). In Schlegel H.G., E. Kroeger (Eds.), *Enrichment culture and mutant selection (Anreicherungskultur und Mutantenauslese)*, Zentralbl. Bakteriol., I. Abt., Suppl. 1, pp. 179-189.

Schulze, R., S. Spring, R. Amann, I. Huber, W. Ludwig, K.H. Schleifer, and P. Kaempfer. 1999. "Genotypic diversity of Acidovorax strains isolated from activated sludge and description of Acidovorax defluvii sp. nov." *Systematic and Applied Microbiology. 22*: 205-214.

Trueper, H.G., and N. Pfennig. 1992. "The family Chlorobiaceae." In Balows A., H.G. Trueper, M.. Dworkin, W. Harder, and K.-H. Schleifer (Eds.), *The Prokaryotes*, 2nd Edition, pp. 3583-3592. Springer, New York.

Van der Meer, J.R. 1997. "Evolution of novel metabolic pathways for the degradation of chloroaromatic compounds." *A. van Leeuwenhoek. 71*: 159-178.

Weiß, H., G. Teutsch, and B. Daus (Eds.). 1997. *"Remediation research in regionally contaminated aquifers (SAFIRA)" (Sanierungsforschung in regional kontaminierten Aquiferen)*. Internal UFZ report to the feasability study for the model location Bitterfeld.

Wuensche, L., and W. Babel. 1996. "The suitability of the BIOLOG automated microbial identification system for assessing the taxonomic composition of terrestrial bacterial communities." *Microbiol. Res. 151*: 133-143.

Zaitsev, G.M., J.S. Uotila, I.V. Tsitko, A.G. Lobanok, and M.S. Salkinoja-Salonen. 1995. "Utilization of halogenated benzenes, phenols, and benzoates by *Rhodococcus opacus* GM-14." *Appl. Environ. Microbiol. 61*: 4191-4201.

BIODEGRADATION OF MTBE BY COMETABOLISM IN LABORATORY-SCALE FERMENTATIONS

Pascal Piveteau, Françoise Fayolle, Yann Le Penru and *Frédéric Monot*
Institut Français du Pétrole, Département de Microbiologie,
1-4, avenue Bois-Préau, 92852 Rueil-Malmaison Cedex, France.

ABSTRACT: A mixed culture of two defined strains has been used for studying the cometabolic biodegradation of methyl *tert*-butyl ether (MTBE). The first strain, *Gordonia terrae* IFP 2007, is able to degrade MTBE to *tert*-butyl alcohol (TBA) and formate in presence of another growth substrate such as ethanol. The second one, *Burkholderia cepacia* IFP 2003, was selected for its ability to mineralize TBA to CO_2. Growth parameters (growth rate, biomass yields), as well as the effect of some important factors such as MTBE/ethanol ratio and TBA concentration were determined using individual strains. Using the mixed culture, MTBE degradation was studied in laboratory scale fed-batch fermentations, in the presence of varying concentrations of ethanol. At a MTBE/ethanol ratio equal to 0.5 (w/w) MTBE was totally degraded, without any accumulation of TBA, and the mixed culture was efficient during more than 200 hours.

INTRODUCTION

Methyl *tert*-butyl ether (MTBE) is an octane enhancer commonly used in reformulated gasoline. In Europe, ETBE is also used for the same purpose. As a consequence of its extensive use and of its high water solubility, numerous aquifers have been contaminated with MTBE (Squillace et al., 1996).

First studies on MTBE biodegradability pointed out the recalcitrance of this compound. Salanitro et al. (1994) obtained the first microbial consortium growing on MTBE as a sole carbon and energy source. Thereafter, cometabolic biodegradation of MTBE to *tert*-butyl alcohol (TBA) by pure strains was reported in the presence of short-chain alkanes (Garnier et al., 1999; Hardison et al., 1997; Steffan et al., 1997). More recently, Hanson et al. (1999) isolated a pure bacterial strain, PM1, able to grow on MTBE.

At IFP, we isolated a strain of *Gordonia terrae* IFP 2001 able to grow on ETBE as the sole carbon and energy source (Fayolle et al., 1998). It grew at the expend of a C2-compound liberated after cleavage of the ether-bond, leading to the concomitant accumulation of TBA. A constitutive variant (strain IFP 2007) of the former strain was also isolated. Resting cells of *G. terrae* IFP 2007 were able to degrade MTBE to TBA, showing that the enzymatic system involved in the initial breakdown of the ether bond could attack ETBE and MTBE (Hernandez et al., 1998). However, MTBE did not support growth and addition of a carbon and energy source such as ethanol was necessary for MTBE degradation (Hernandez et al., 2000). A strain of *Burkholderia cepacia* IFP 2003 was also isolated for its capacity to grow on TBA as the sole carbon and energy source and a mixed culture of both strains was used to completely degrade ETBE even at high concentrations (450 mg.L^{-1}) (Fayolle et al., 1999).

Several aspects aiming at determining the capacity of the mixed culture consisting of *G. terrae* IFP 2007 and *B. cepacia* IFP 2003 to mineralize MTBE were studied: 1) optimization under batch conditions of the ethanol/MTBE ratio required for a complete and fast mineralization of MTBE degradation by *G. terrae* IFP 2007, 2) determination of the growth parameters of each strain, 3) utilisation of the mixed culture in a 6 L-fermentor for continuous MTBE degradation in a culture with ethanol.

MATERIALS AND METHODS

Microorganisms. *G. terrae* IFP 2007 was a variant strain of *G. terrae* IFP 2001 initially isolated from an activated sludge for its ability to grow on ETBE as the sole carbon and energy source (Hernandez et al., 2000). Compared with *G. terrae* IFP 2001, *G. terrae* IFP 2007 was constitutive regarding catabolism of ETBE. *B. cepacia* IFP 2003 was isolated from an activated sludge for its ability to grow on TBA. They were identified and deposited at the CNCM, Pasteur Institute, Paris, France, and maintained on glycerol (20%, w/w) at –80°C for preservation.

Medium. Both strains, *G. terrae* IFP 2007 and *B. cepacia* IFP 2003, as well as the reconstituted mixed culture were cultivated on a mineral medium (MM) containing in g per liter of deionized water : KH_2PO_4, 1.4; K_2HPO_4, 1.7; MgSO4. $7H_2O$, 0.5; $NaNO_3$, 1.5; $CaCl_2$. $2H_2O$, 0.04; $FeCl_3$. $6H_2O$, 0.0012 and 1 mL of a vitamin solution (Fayolle et al., 1999). The pH was 7.0. 10 mL of a solution of trace elements containing in mg per liter of deionized water : nitrilotriacetic acid, 1.5; $Fe(NH_4)_2(SO_4)_2$, $6H_2O$, 0.2; Na_2SeO_3, 0.2; $CoCl_2$, $6H_2O$, 0.1; $MnSO_4$, $2H_2O$, 0.1; Na_2MoO_4, $2H_2O$, 0.1; $ZnSO_4$, $7H_2O$, 0.1; $AlCl_3$, $6H_2O$, 0.04; $NiCl_2$, $6H_2O$, 0.025; H_3BO_3, 0.01; $CuSO_4$, $5H_2O$, 0.01, was added to the MM medium. The culture purity was checked on solid Luria-Bertani (LB) medium (15 g.L^{-1} pure agar). Specific counts of *B. cepacia* IFP 2003 were also performed on Petri dishes containing solid MM medium (15 g.L^{-1} highly pure agar, Sigma) supplemented with TBA (1 g.L^{-1}).

Batch cultures. The ability of both strains to degrade MTBE, TBA, formate and ethanol was tested in gas-tied sealed flasks (250 mL) containing 50 mL of MM medium. MTBE, TBA and ethanol were tested at a concentration of about 100 mg.L^{-1} whereas the initial formate concentration was 2 g.L^{-1}. Each flask was inoculated with a centrifuged and washed cell suspension to reach an initial biomass concentration of about 44 mg.L^{-1} cell dry weight. Flasks were incubated aerobically at 30°C under constant agitation. Samples were withdrawn over time, filtered (0.22 μm) and analyzed for substrate consumption.

The effect of ethanol on the degradation of MTBE by *G. terrae* IFP 2007 was also studied in gas-tied sealed flasks (250 mL) containing 50 mL of MM medium. About 100 mg.L^{-1} MTBE were introduced and varying concentrations of ethanol, from 0 to 500 mg.L^{-1}, were added to the media. All flasks were seeded with the same centrifuged and washed cell suspension of *G. terrae* IFP 2007 obtained from a 24 h preculture on LB broth. The initial cell concentration was 44 mg.L^{-1} in each flask. Samples were withdrawn over time, filtered (0.22 µm) and analyzed for residual ethanol, MTBE and for TBA production.

The effect of the initial TBA concentration on its degradation rate by *B. cepacia* IFP 2003 was measured in non-sealed flasks containing 100 mL of MM medium. Culture samples were withdrawn, filtered (0.22 µm) and analyzed.

In order to assess possible losses of substrates, an abiotic control was prepared for each experiment and incubated under the same conditions. The decrease of MTBE was less than 5% and there was no loss of the other substrates.

Fermentation set-up and mode. 24-hour precultures of *G. terrae* IFP 2007 and *B. cepacia* IFP 2003 were prepared on 200 mL of LB and MM medium supplemented with glucose (1 g.L^{-1}), respectively. Cells were collected by centrifugation (10,000 g for 10 minutes at 4°C) and resuspended in 50 mL of MM medium. The cell suspensions were used for inoculating the fermentor in order to reach an initial population of approximately 1x10^7 colony forming units (cfu)/mL for each strain. Fed-batch fermentations were performed on MM medium (4 L) in a 6-liter fermentor (Biolafitte, France), at 30°C, and with a constant aeration rate of 1 L.L^{-1}.h^{-1}. Dissolved oxygen was monitored with an Ingold probe. MTBE and ethanol were injected at the bottom of the fermentor via a volumetric pump (Pharmacia) at the desired relative concentrations and rates. CO_2 was measured in the effluent gas stream at the outlet of a condensor used to avoid MTBE loss in the gas phase. Samples were withdrawn from the culture for ethanol, TBA and MTBE analysis and for bacterial counts.

Analytical procedures. Culture samples were filtered (0.22 µm) and analyzed for MTBE, TBA and ethanol using a VARIAN 3300 gas chromatograph fitted with a flame ionization detector on a 0.32 mm x 60 m DB 1 (1-µm-thick film) capillary column. The temperature of the column was initially 100°C during 0.5 min, then increased to 150°C at 10°C.min^{-1} in a first step and up to 250°C at 50°C.min^{-1} in a second step. The carrier gas was helium (1.6 mL.min^{-1}).

Total organic carbon (TOC) was determined on culture samples using a DC-80 carbon analyzer (Dorhmann).

Formate in the culture samples was assayed using an enzymatic kit (Boerhinger, Mannheim, Germany).

The cell dry weight was determined from 50 to 100 mL of culture using a Mettler infra-red dryer (LP16) and weighing scale (PM100).

Chemicals. MTBE, TBA, ethanol were purchased from Sigma.

RESULTS AND DISCUSSION

Degradation of MTBE, ethanol, TBA and formate by both isolates. *G. terrae* IFP 2007 was able to partially degrade MTBE, but no growth was observed and the degradation stopped after 50 h (Table 1). TBA accumulated stoichiometrically during MTBE degradation and was not utilized by *G. terrae* IFP 2007. Formate, another product of MTBE degradation was not used as a growth substrate. These results show that the enzymatic system cleaving the ether bond remained active in the cells but, as no intermediates could be metabolized, energy could not be produced thereby limiting the MTBE degradation. In order to sustain MTBE degradation activity, different substrates were considered in order to provide energy to *G. terrae* IFP 2007 cells without inhibiting the breakdown of the MTBE molecule. Ethanol met these prerequisites and MTBE degradation in presence of ethanol was further studied.

To complete MTBE degradation, another strain, able to degrade both TBA and formate, the MTBE degradation products, was required. *B. cepacia* IFP 2003 utilized TBA and formate as growth substrates without accumulation of any end-products, but did not degrade MTBE (Table 1). The biomass yield during TBA degradation was similar to those observed for readily degradable substrates such as glucose. This was quite unexpected considering the structure of the TBA molecule which is probably the cause of its recalcitrance. The methylotrophic nature (growth on C1-compound such as formate) of *B. cepacia* IFP 2003 is another interesting aspect of this microorganism as all enzymes needed for MTBE mineralization are present except those catalyzing the first reaction in the degradation pathway of MTBE, *i.e* attack of the ether bond and release of a TBA molecule. Moreover, *B. cepacia* IFP 2003 did not metabolize ethanol, thus substrate competition between *G. terrae* IFP 2007 and *B. cepacia* IFP 2003 was avoided during mixed cultivation on a MTBE and ethanol mixture.

TABLE 1. Growth and kinetic parameters of *G. terrae* IFP 2007 and *B. cepacia* IFP 2003 on different substrates.

Microorganism	Substrate	Substrate degradation	μ_{max} (h^{-1})	$Y_{x/s}$ (g biomass (dry weight).g^{-1} substrate)	Rate of degradation ($g.g^{-1}$ biomass (dry weight) .h^{-1})
G. terrae	MTBE**	+	nd	nd	8.10^{-3}
IFP 2007	Ethanol*	+	0.21	0.4	0.14
	TBA	-	nd	nd	nd
	Formate	-	nd	nd	nd
B. cepacia	MTBE	-	nd	nd	nd
IFP 2003	Ethanol	-	nd	nd	nd
	TBA*	+	0.03	0.5	0.02
	Formate*	+	nd	0.02	nd

nd : not determined.
* : substrates supporting growth.
** : degradation without growth.

Degradation of MTBE by *G. terrae* IFP 2007 in presence of ethanol. Degradation of MTBE by *G. terrae* IFP 2007 was studied in presence of different concentrations of ethanol. Results are presented in Table 2.

TABLE 2. Effect of the MTBE/ethanol ratio on MTBE degradation by *G. terrae* IFP 2007.

MTBE/ Ethanol ratio	Initial MTBE $(mg.L^{-1})$	Final ethanol $(mg.L^{-1})$	Final MTBE $(mg.L^{-1})$	MTBE degraded (%)	MTBE degradation rate $(mg.g^{-1}$ (dry weight)$.h^{-1})$	Time for complete MTBE degradation (h)
MTBE alone	113	-	57	49.6	7.6	> 170 *
1/0.65	110	0	32	70.9	12.4	> 170 *
1/1.5	111	0	0	100	29.4	170
1/3.2	95	0	0	100	41.5	60
1/4.1	114	0	0	100	32.3	25

*Experiments were carried out for 170 h.

G. terrae IFP 2007 was able to degrade more than 50 $mg.L^{-1}$ MTBE in the absence of any other substrate probably because the enzymatic system responsible for the breakdown of MTBE into TBA was constitutively expressed. The products of MTBE degradation could not support growth of *G. terrae* IFP 2007 and they accumulated in the culture media, the degradation stopped after 50 h. When no ethanol was added the degradation stopped after 50 hours of incubation (data not shown). When ethanol was added at a 1/0.65 MTBE/ethanol ratio, the degradation rate was higher and 71% of the initial MTBE was degraded. At higher ethanol concentrations, MTBE degradation was complete and the incubation time needed for complete MTBE degradation decreased. This shorter degradation time was due to a higher production of biomass.

Degradation of TBA by *B. cepacia* IFP 2003. The ability of *B. cepacia* IFP 2003 to degrade TBA was assessed at various initial TBA concentrations (Table 3). Up to 1 $g.L^{-1}$, TBA was degraded at a rate of about 0.02 $g.g^{-1}$ (dry weight)$.h^{-1}$.

TABLE 3. Effect of initial TBA concentration on the growth of *B. cepacia* IFP 2003.

Initial TBA $(mg.L^{-1})$	Final TBA $(mg.L^{-1})$	Initial $O.D._{600nm}$	Final $O.D._{600nm}$	Degradation time (h)	TBA degradation rate $(mg.g^{-1}$ (dry weight)$.h^{-1})$
375 ± 15	0	0.11 ± 0.01	0.54 ± 0.01	69	20 ± 1.5
517 ± 49	0	0.07 ± 0.02	0.72 ± 0.06	69	22 ± 2.0
717 ± 44	0	0.12 ± 0.01	0.93 ± 0.01	93	23 ± 1.3
1019 ± 48	0	0.10 ± 0.01	1.24 ± 0.09	165	26 ± 7.4
2865 ± 270	0	0.14 ± 0.01	2.85 ± 0.37	165	7 ± 2.2

TBA mineralization remained total even at very high TBA concentrations, although it occurred at a lower rate.

MTBE degradation by the mixed culture at different MTBE/ethanol feeding ratios. The suitability of the mixed culture composed of *G. terrae* IFP 2007 and *B. cepacia* IFP 2003 for the MTBE degradation was then assessed in fed-batch fermentations at two different MTBE/ethanol ratios. The feeding rate was fixed at 1 mL.h^{-1} and the MTBE/ethanol ratio varied between experiments.

The time courses of the concentrations of substrates (ethanol and MTBE) and products (TBA, biomass) obtained with a feeding rate of 50 mg.h^{-1} for both MTBE and ethanol are presented in Figure 1A. Ethanol injected was totally consumed by *G. terrae* IFP 2007 and the strain was under ethanol limitation after about 40 hours. Ethanol was not fed at a sufficient concentration to allow complete MTBE degradation as residual MTBE was measured all over the experiment (more than 200 hours). TBA was transiently produced and then *B. cepacia* IFP 2003 was under TBA limitation after 30 hours of incubation. *B. cepacia* IFP 2003 grew to a final population of 3×10^9 cfu/mL. *G. terrae* IFP 2007 did grow but accurate bacterial counts were not possible due to its tendency to aggregate. Nevertheless, the culture density increased in the fermentor and reached a final O.D. greater than 4. Total organic carbon was measured and its value corresponded to the concentration of residual MTBE obtained by GC.

Different results were obtained when increasing the ethanol concentration in the feeding solution to 100 g.L^{-1} with a feeding rate for MTBE and ethanol of 50 and 100 mg.h^{-1}, respectively (Figure 1B). During the first 24 h, residual concentrations of ethanol, MTBE and TBA were detected. Then, MTBE, as well as ethanol, were totally degraded and TBA was transiently produced. After 50 hours, MTBE, ethanol and TBA concentrations remained equal to zero. Total organic carbon was nil confirming the total degradation of the 3 compounds and the absence of accumulation of any other intermediates. For instance, formate did not accumulate and less than 1 mg.L^{-1} was detected throughout the fermentation, showing the ability of *B. cepacia* IFP 2003 to utilize formate produced during MTBE degradation. *B. cepacia* IFP 2003 population reached 2×10^9 cfu/ml after 200 h and the final O. D. was greater than 8. Again, aggregation of *G. terrae* IFP 2007 prevented bacterial counts during growth.

In both cases, the cultures lasted more than 200 h without any loss of activity, and a total amount of 10 g MTBE was degraded.

CONCLUSION

In presence of ethanol as a carbon and energy source, complete MTBE degradation by the bacterial consortium constituted of *G. terrae* IFP 2007 and *B. cepacia* IFP 2003 was obtained. In addition, this performance was stable during more than 200 hours. As *B. cepacia* IFP 2003 was able to grow on TBA and formate, the consortium completely mineralized MTBE without any accumulation of degradation products. The MTBE/ethanol ratio was crucial to obtain a total MTBE degradation. The results show that this defined culture could be utilized

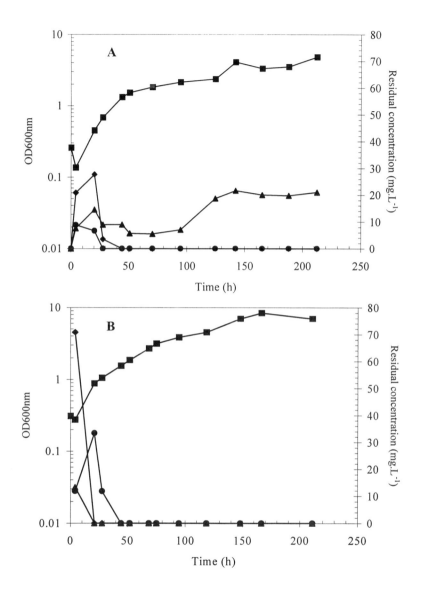

FIGURE 1. MTBE degradation by a mixed culture composed of *G. terrae* IFP 2007/*B. cepacia* IFP 2003 during fed-batch fermentations with a feeding rate of 1 mL.h^{-1} of: (A) a 1/1 (w/w) mixture of MTBE (50 g.L^{-1}) and ethanol (50 g.L^{-1}); (B) a 1/2 (w/w) mixture of MTBE (50 g.L^{-1}) and ethanol (100 g.L^{-1}). (■) total biomass, (▲) MTBE, (♦) ethanol, (●) TBA.

for remediation of polluted aquifers. This consortium is now studied in biotrickling filter experiments.

REFERENCES

Fayolle, F., G. Hernandez, F. Le Roux and J.-P. Vandecasteele. 1998. "Isolation of Two Aerobic Bacterial Strains that Degrade Efficiently Ethyl *t*-Butyl Ether (ETBE)". *Biotechnol. Lett. 20*: 283-286.

Fayolle, F, F. Le Roux, G. Hernandez and J.-P. Vandecasteele. 1999. "Mineralization of ethyl *t*-butyl ether by defined mixed bacterial cultures", p.25. *In* . B.C. Aleman and A. Leeson (ed.), In Situ Bioremediation of Petroleum Hydrocarbon and Other Organic Compounds. Battelle Press, Columbus, OH.

Garnier, P., R. Auria, M. Magana, and S. Revah. 1999. "Cometabolic biodegradation of methyl *t*-butyl ether by a soil consortium", p.31. *In* . B.C. Aleman and A. Leeson (ed.), In Situ Bioremediation of Petroleum Hydrocarbon and Other Organic Compounds. Battelle Press, Columbus, OH.

Hanson, J.R., C.R. Ackerman, and K.M. Scow. 1999. "Biodegradation of Methyl *tert*-butyl ether by a bacterial pure culture". *Appl. Environ. Microbiol. 65*: 4788-4792.

Hardison, L.K., S.S. Curry, L.M. Ciuffetti and M.L. Hyman. 1997. "Metabolism of Diethyl Ether and Cometabolism of MTBE by a Filamentous Fungus, a *Graphium* sp.". *Appl. Environ. Microbiol. 63*: 3059-3067.

Hernandez, G., F. Fayolle, F. Le Roux and J.-P. Vandecasteele.1998. "Procédé de traitement bactérien d'effluents contenant au moins un éther". French patent N° 98/16520.

Hernandez, G., F. Fayolle and J.-P. Vandecasteele. 2000. "Biodegradation of ethyl *t*-butyl ether (ETBE), methyl *t*-butyl ether (MTBE) ant *t*-amyl methyl ether (TAME) by *Gordonia terrae*". Submitted for publication.

Salanitro, J.P., L.A. Diaz, M.P. Williams and H.L. Wisniewski. 1994. "Isolation of a Bacterial Culture that Degrades Methyl *t*-Butyl Ether". *Appl. Environ. Microbiol. 60*: 2593-2596.

Squillace, P.J., Zogorski, J.S., Wilber, W.G. and Price C.V. 1996. "Preliminary assessement of the occurrence and possible sources of MTBE in groundwater of the United States, 1993-1994". *Environ. Sci. and Technol. 30*: 1721-1730.

Steffan, R.J., K. Mac Clay, S. Vainberg, C.W. Condee and D. Zhang. 1997. "Biodegradation of the Gasoline Oxygenates MTBE, ETBE and TAME by Propane-Oxidizing Bacteria". *Appl. Environ. Microbiol. 63*: 4216-4222.

COMETABOLIC DEGRADATION OF MTBE BY *ISO*-ALKANE-UTILIZING BACTERIA FROM GASOLINE-IMPACTED SOILS

Michael Hyman (North Carolina State University, Raleigh, North Carolina)
Christine Taylor (North Carolina State University, Raleigh, North Carolina)
Kirk O'Reilly (Chevron Research and Technology Co. Richmond, California)

INTRODUCTION

Methyl *tertiary* butyl ether (MTBE) is frequently added to gasoline as an oxygenate to reduce automobile emissions of CO and NO_x. There is currently considerable interest in the potential use of biological processes for the remediation of MTBE contaminated ground water resulting from surface fuel spills and leaking underground fuel storage tanks. Several microorganisms have been described that can grow on MTBE as a sole source of carbon and energy (Salanitro *et al.*, 1994, Hanson *et al.*, 1999). In contrast, our research focuses on cometabolic processes in which MTBE is fortuitously degraded by organisms that grow on substrates other than MTBE.

We have previously demonstrated that several microorganisms can rapidly degrade MTBE after growth on a variety of low molecular weight alkanes, including both straight chain (*e.g.* propane, *n*-butane and *n*-pentane) and simple branched alkanes (*e.g. iso*-butane and *iso*-pentane) (Hyman *et al.*, 1998). We are particularly interested in the possible connection between MTBE oxidation and the microbial degradation of branched alkanes for two reasons. First, simple branched alkanes are reasonable structural analogs of MTBE. Second, simple branched alkanes are some of the most major components of gasoline itself. Our current hypothesis is that microbial enzymes predisposed to oxidizing branched hydrocarbon structures are likely to be consistent in their ability to oxidize the branched MTBE molecule. If this hypothesis is correct the abundance of branched alkanes in gasoline raises the possibility that cometabolic processes may contribute to degradation of MTBE in certain aerobic environments where MTBE and *iso*-alkanes are present simultaneously.

In the present study we have further investigated the connection between MTBE-degrading activity and the ability of microorganisms to metabolize *iso*-alkanes. Whereas our previous studies have examined well-characterized alkane-oxidizing organisms obtained from culture collections, in this study we have investigated the MTBE-degrading activity of newly isolated microorganisms that have been isolated from gasoline-impacted soil. These organisms were obtained from enrichment cultures that were provided *iso*-butane as a sole source of carbon and energy. Our results confirm that there is a strong correlation between the ability of bacteria to grow on branched alkanes and their ability to cometabolically degrade MTBE.

MATERIALS AND METHODS

Separate enrichment cultures were initially established using gasoline-impacted surface soil from two gasoline stations in Raleigh, NC. Soil samples (~2 g) were mixed with sterile minerals salts medium (2 ml) in a sterile pyrex test tube. The soil/buffer mixture was brief sonicated in an ultrasonic water bath (3 x 10s) and the soil particles were then allowed to sediment for 10 min. Samples of the supernatant (100 µl) were then removed and added to mineral salts medium (30 ml) in glass serum vials (120 ml). The vials were sealed with butyl rubber stoppers and crimp seals and *iso*-butane (3 ml) was added as an overpressure. The vials were incubated for 3 weeks in the dark at 30° C in a temperature-controlled shaking incubator (150 rpm). Samples (100 µl) of the enrichment culture were then transferred to a fresh incubation vial containing mineral salts medium (30 ml) and *iso*-butane (3 ml). After further incubation for 2 weeks, samples (100 µl) of these cultures were spread on mineral salts agar plates. The inoculated plates were then incubated in a glass dessicator jar with *iso*-butane (~10% gas phase) as the sole source of carbon and energy. Individual colonies were subsequently picked from these plates and were streaked out on mineral salts agar plates and reincubated in the presence of *iso*-butane. This procedure was repeated until a single colony type was obtained for each isolate. The morphological and structural characteristics of each isolate were determined by standard techniques. 16S rRNA genes were amplified using universal primers 515F and 1492R.

The isolates described in this study were cultivated in a variety of methods. To determine which alkane substrates supported the growth of these organisms, small scale cultures were grown in 12-well microtiter plates. Each well contained mineral salts medium (2 ml) and was inoculated with a single colony transferred from a mineral salts plates using a sterile toothpick. The microtiter plates were incubated in the dark at 30° C in glass dessicator jars. Gaseous hydrocarbons were added to the jars to a final concentration of ~10% (v/v gas phase). Liquid hydrocarbons were supplied through the vapor phase. In these cases an open test tube containing each liquid hydrocarbon was placed in the center of the incubator jar. Growth was determined by A_{600} measurements. Larger scale liquid cultures of each isolate were grown at 30° C in minerals salts medium (100 ml) in glass bottles (600 ml) sealed with screw cap tops fitted with butyl rubber septa. All gaseous hydrocarbon growth substrates (60 ml) were added to the bottles as an overpressure. All liquid hydrocarbon growth substrates were added to an initial concentration of 0.1% vol/vol. After incubation for 5 days the bottles were opened under sterile conditions to replenish oxygen. The vials were then sealed again and the corresponding hydrocarbon was added to the culture at the concentrations described above. After an additional 24 h incubation the cells were harvested by centrifugation (10,000 x g for 10 min). The cell pellet was resuspended in sodium phosphate buffer (50 mM, pH 7.0) and was centrifuged again. The supernatant was discarded and the cell pellet was finally resuspended with buffer at a final protein concentration of ~ 10 mg protein ml^{-1}. These cell suspensions were stored on ice and were used in experiments within 4 h after harvesting.

The oxidation on MTBE by each isolate was investigated in a standard assay using small scale reactions maintained at 30° C in a shaking water bath (150 rpm). The reactions were conducted in glass serum vials (10ml) stoppered with butyl rubber septa and aluminum crimp seals. The reaction vials contained buffer (900μl) and MTBE (1100 nmoles). The reactions were initiated by the addition of cells (0.2 to 1.2 mg total protein) to the vials. Samples (2μl) were removed from the reaction and were directly injected into a gas chromatograph equipped with a flame ionization detector. The reactants and products were separated using a 2m stainless column packed with Porapak Q (60-80 mesh) that was operated at 160° C. Nitrogen was used as carrier gas at a flow rate of 20 ml/min.

RESULTS

A total of nine distinct bacterial strains isolated from *iso*-butane enrichment cultures seeded with soil samples obtained from two sites in Raleigh, NC. Five strains (CT-1, -5, -7, -8A and -8B) were isolated from a sampling site on Avent Ferry Road (AFR) while the remaining four other strains were obtained from a second sampling site on Hillsborough Street (HS). The isolates include 4 Gram-positive and 5 Gram-negative strains. Our preliminary identifications of these strains based on 16SrRNA sequence analyses suggest the Gram-positive strains include examples of well-known hydrocarbon utilizing genera (*e.g. Rhodococcus* and *Nocardia*) while the Gram-negative isolates also include several common and metabolically diverse soil bacteria (*e.g. Pseudomonas*, *Alcaligenes* and *Rhizobium*).

TABLE 1. Characteristics of*iso*-butane-utilizing strains

Strain	Source	Gram Stain	Color	16SrRNA
CT-1	AFR	+	Bright Orange	ND
CT-2	HS	+	Orange	*Rhodococcus*
CT-3	HS	-	Cream	*Pseudomonas*
CT-4	HS	-	Tan	*Alcaligenes*
CT-5	AFR	-	Bright Orange	*Rhizobium*
CT-6	HS	-	Peach	ND
CT-7	AFR	+	Milk White	*Nocardia*
CT-8A	AFR	-	Yellow	ND
CT-8B	AFR	+	White	*Rhodococcus*

The MTBE degrading activity of each isolate was determined after growth on *iso*-butane. Our results (Table 2) demonstrate that each isolate was capable of oxidizing MTBE under these conditions. In each case MTBE degradation involved the production of both *tertiary* butyl alcohol (TBA) and *tertiary* butyl formate (TBF) as reaction products. In each case the oxidation of MTBE and the accumulation of TBA and TBF was also inhibited by the presence of acetylene (10%: v/v gas phase). We also investigated the kinetic constants (K_s and V_{max}) associated with the oxidation of MTBE by several of these isolates. In these experiments *iso*-butane-grown cells were exposed to varying concentrations of MTBE (0 to 8.2.µmoles) for a fixed period of time (30 min). The rate of MTBE oxidation was determining from the combined concentrations of TBA and TBF found in the reaction medium. Our results (Table 2) derived from Lineweaver-Burke plots indicate the K_s values for the five strains tested to date range from ~120 µM to ~500 µM while the V_{max} values ranged from ~2 to ~14 nmoles/min/mg protein.

TABLE 2. Kinetics of MTBE oxidation by*iso*-butane-grown isolates.

Strain	MTBE Oxidation Assay (nmoles/min/mg protein)	K_s (µM)	V_{max} (nmoles/min/mg protein)
CT-1	4.9	ND	ND
CT-2	4.6	ND	ND
CT-3	2.8	ND	ND
CT-4	1.9	ND	ND
CT-5	3.8	482	14.1
CT-6	2.2	265	6.4
CT-7	2.6	180	4.6
CT-8A	1.6	210	5.2
CT-8B	1.8	119	2.0

We have previously demonstrated that organisms such as *Mycobacterium vaccae* JOB5 can grow on a wide range of alkanes and that these organisms retain the ability to oxidize MTBE after growth on many of these substrates (Hyman *et al.*, 1998). We have conducted similar studies with many of the isolates described in this study. Our results (Figure 1) demonstrate that all of the isolates are capable of growth on propane, *n*-butane, *n*-pentane, *n*-hexane, *n*-octane, as well as *iso*-butane and *iso*-pentane. We also investigated whether these isolates retained the ability to oxidize MTBE after growth on a selection of these hydrocarbon substrates. In these experiments cells were grown on each hydrocarbon substrate and were then assayed under standard conditions in small scale, short term reactions, as described earlier. Of the eight strains examined in these experiments, all were able to oxidize MTBE after growth on propane. Strains CT-1, -2, -3, -4, -5, -7 and -8A degraded MTBE after growth on *n*-pentane and substantially similar results were observed for the the majority of these organisms after growth on *n*-hexane. In all but one case (CT-2), substantial rates of MTBE oxidation were also observed for cells grown on *n*-octane. In most cases the rate of MTBE oxidation in our standardized assay were similar, irrespective of the hydrocarbon substrate used to initially grow the organisms. However, in some cases (*e.g.* CT-2, CT-4 and CT-8A) the rates of MTBE oxidation were generally faster with the shorter *n*-alkanes (*e.g.* propane and *n*-pentane) than with the longer *n*-alkanes (e.g. *n*-hexane and *n*-octane).

DISCUSSION

The overall aim of this study was to test the hypothesis that organisms that have the ability to grow on simple branched alkanes are consistent in their ability to cometabolically degrade MTBE. The underlying rationale for this hypothesis rests on two issues. First, to date the microbial oxidation of MTBE under aerobic conditions has been consistently associated with oxygenase enzymes. Second, there is the expectation that organisms capable of growth on simple branched alkanes such as *iso*-butane are likely to express oxygenase enzymes that are predisposed to the oxidation of branched hydrocarbon structures. Our observation that all of the *iso*-butane-metabolizing strains isolated during this study are capable of cometabolically degrading MTBE clearly supports our hypothesis. The fact that we have such obtained consistent results with newly isolated strains suggests that the association between *iso*-alkane metabolism and MTBE degradation is not an artifact limited to well-established laboratory strains and occurs widely in hydrocarbon-oxidizing bacteria.

Further support for our suggestion that *iso*-alkane metabolising activity is a key feature of microorganisms that can cometabolically degrade MTBE could possibly have been provided if this study had identified MTBE-degrading strains that that have the ability to grow on *iso*-alkanes but not *n*-alkanes. However, our results (Table 2), indicate that all of the strains examined in this study have a broad hydrocarbon growth substrate ranges and that all of the organisms tested have the ability to cometabolically degrade MTBE after growth on hydrocarbon substrates other than *iso*-alkanes. In many cases the rates of MTBE degradation seem with cells grown on *n*-alkanes (Table 3) also exceeds the rate of MTBE oxidation observed for *iso*-butane-grown cells (Table 2). It is not clear at this stage whether these differences in rates are attributable to different oxygenase enzymes with different kinetic features or whether these organisms express different levels of the same oxygenase in response to different hydrocarbon growth substrates. These issues are being addressed in our ongoing research with these isolates.

An important issue raised by our observations is the potential role of gasoline hydrocarbons as stimulants for cometabolic MTBE degradation in the environment. Our results again demonstrate that a wide variety of *n*- and *iso*-alkanes can support the growth of MTBE degrading bacteria. This raises the possibility that a natural attenuation of MTBE contamination can occur in environments where dominant gasoline hydrocarbons such as *iso*-pentane are simultaneously present with MTBE and oxygen. The high vapor pressure of *iso*-pentane and MTBE suggests that the aerobic unsaturated zone above point sources of gasoline contamination would be a good example of this type of environment. Our results also support the idea that decreases in MTBE concentration can be expected as a result of treatments such as bioventing or air

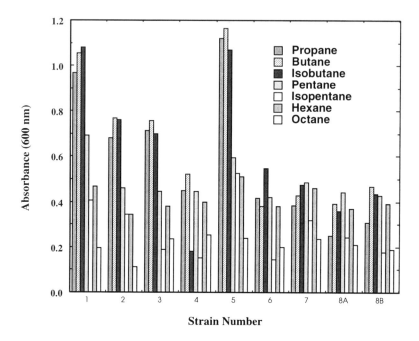

FIGURE 1. Growth of isolates on selected n-alkanes and iso-alkanes in microtiter plate cultures

TABLE 3. Oxidation of MTBE by isolates grown on selected n-alkanes

Strain	MTBE oxidation rate after growth on n-alkane (nmoles products (TBA + TBF)/min/mg protein)			
	propane	n-pentane	n-hexane	n-octane
CT-1	6.9	7.3	4.9	4.5
CT-2	11.7	18.3	3.3	0.3
CT-3	6.3	5.7	6.3	4.8
CT-4	7.5	10.6	6.2	1.6
CT-5	8.5	5.0	5.6	6.0
CT-6	2.0	ND	0.7	2.9
CT-7	5.9	11.1	8.1	5.4
CT-8A	10.7	11.9	ND	4.2

sparging that introduce oxygen into contaminated environments to promote gasoline hydrocarbon degradation.

REFERENCES

Hanson, J. R., C. E. Ackerman, and K. M. Scow. 1999. "Biodegradation of Methyl *tert*-Butyl Ether a Bacterial Pure Culture". *Applied and Environmental Microbiology*, 65 (11): 4788-4792.

Hyman, M. R., P. Kwon, K. Williamson, and K. O'Reilly. 1998 "Cometabolism of MTBE by Alkane-Utilizing Microorganisms". In, Wickramanayake, G. B. and R. E. Hinchee (Eds.), *Natural Attenuation: Chlorinated and Recalcitrant Compounds*, pp. 321-326, Battelle Press, Columbus, OH

Salanitro, J. P., L. A. Diaz, M. P. Williams, and H. L. Wisiewski. 1994 "Isolation of a Bacterial Culture that Degrades Methyl *t*-Butyl Ether". *Applied and Environmental Microbiology* 60 (7): 2593-2596.

IN SITU APPLICATION OF PROPANE SPARGING FOR MTBE BIOREMEDIATION

Robert J. Steffan, Charles Condee, Joseph Quinnan, Matthew Walsh, Stewart H. Abrams, and Jonathan Flanders, Envirogen, Inc., 4100 Quakerbridge Road, Lawrenceville, NJ 08648

ABSTRACT: In earlier work we demonstrated that propane oxidizing bacteria can mineralize MTBE to CO_2 after growth on propane (Steffan et al., Appl. Environ. Microbiol., 63:4216-4222, 1997; US Patent #5,814,514, Sept. 29, 1998). Consequently, propane injection can be used to promote MTBE biodegradation by indigenous microorganisms in MTBE-contaminated aquifers, or to maintain the activity of exogenous propane oxidizers used as seed cultures. We are preparing for a field demonstration of the technology at a service station site in New Jersey, and have performed treatability testing for a second site located in northern California. Operation of an air sparging system at the New Jersey site had little effect on groundwater MTBE concentrations (~100 mg/L MTBE). Laboratory treatability testing demonstrated that more than 60 days of propane stimulation may be required to achieve measurable levels of propane and MTBE degradation, presumably because of the low pH at the site (~pH5.6). An adjustment of pH and addition of a seed culture of the propane oxidizing bacterium ENV425 allowed rapid degradation of MTBE. Field-scale biological treatment will involve adding a seed culture of ENV425 and injecting a mixture of 0.2% propane in air (10% of the propane LEL) into a set of three existing air sparging wells. Treatability testing with samples from the California site demonstrated that indigenous organisms responded rapidly to the addition of propane and degraded MTBE (~20 mg/L) to below detection limits (<50 µg/L).

INTRODUCTION

The widespread use of MTBE as a gasoline oxygenate to meet the demands of the Clean Air Act has resulted in it becoming a common groundwater contaminant at gasoline stations and terminals. Because MTBE is highly soluble in water (~43,000 mg/L) and has a low tendency to adsorb to soils (K_{owc} ~1.05), it is extremely mobile in groundwater and results in the formation of large contaminant plumes. Few remedial technologies have been developed remediate such plumes, and most current methods require pumping the water to above ground treatment systems.

In our earlier work we identified a group of microorganisms, the propane oxidizing bacteria (POB), that can degrade MTBE completely to CO_2 (Steffan et al., 1997). Because these organisms are widely distributed in nature, their population numbers in a selected environment may be amplified by adding propane and oxygen to the environment. The amplified population can then facilitate biodegradation of the target contaminant (e.g., MTBE) in situ. This in situ biostimulation approach to remediation has been applied previously to degrade chlorinated solvents in aquifers (Hazen et al., 1994; Semprini et al., 1999).

There are several potential advantages to use biostimulation for degrading MTBRE In situ. For example, the technology can be applied in a number of configurations depending on site characteristics and treatment needs including: 1) re-engineered or modified multi-point air sparging (AS) / soil vapor extraction (SVE) systems that deliver propane and air throughout a contaminated site (suitable for use with existing AS/SVE systems or specially designed systems); 2) a series of air/propane delivery points arranged to form a permeable treatment wall to prevent off site migration of MTBE; 3) permeable treatment trenches fitted with air and propane injection systems; 4) in situ recirculating treatment cells that rely on pumping and reinjection to capture and treat a migrating contaminant plume; and, 4) through bubble-free gas injection devices to minimize off-gas release and contaminant stripping. Furthermore, propane is widely available, transportable even to remote sites, and relatively inexpensive. In situ approaches also may be used to treat an entire plume simultaneously, potentially allowing more rapid site clean-up.

Objective. The objective of this work was to evaluate the use of in situ biostimulation for remediating MTBE contaminated aquifers. The work involved performance of laboratory treatability studies to test our ability to stimulate the growth and activity of POB in two aquifers. We also sought to evaluate the addition of exogenous POB seed cultures for situations where indigenous organisms are not rapidly stimulated. The work ultimately led to the design of and permitting for a field demonstration system at a gasoline station site in New Jersey.

METHODS

Treatability testing. Treatability samples consists of aquifer samples incubated in glass serum vials. Aquifer sediment (saturated zone) and groundwater samples were collected from a gasoline service station site in central New Jersey and from a fuel terminal site in northern California. In each case, the samples were emediately placed on ice and shipped overnight or transported directly to Envirogen. Subsamples of the sediment (50 g) were placed into sterile 150 ml serum vials, and 60 ml of groundwater were then added. Three treatment scenarios were evaluated: 1) the addition of propane and oxygen only; 2) the addition of propane, oxygen, and nutrients (N+P); and 3) the addition of propane, oxygen, nutrients (N+P), and strain ENV425 (~1 x 10^6 cells/ml of slurry). For the New Jersey site a fourth treatment consisting of propane, oxygen, nutrient, and pH adjustment (from pH 5.6 to pH 7.0) was included. Control samples were poisoned with $HgCl_2$ to inhibit microbial activity. Each treatment was evaluated in triplicate. Propane (4 mL) was added to the sample headspace just prior to sealing the vials. The microcosms were sealed with Teflon-lined crimp seals and incubated on a shaker at 15°C. Periodically, subsamples of the headspace gas were removed and analyzed for MTBE on a gas chromatograph equipped with a flame ionization detector. Alternatively, subsamples of the slurry were removed and analyze by purge and trap gas chromatography/mass spectroscopy by using EPA Method 8260. Method 8260 allowed for an MTBE detection limit of only 5 µg/L (ppb), whereas direct injection had a detection limit of approximately 300 µg/L (ppb). Additional MTBE, propane, and/or oxygen were added periodically as needed.

To evaluate the combined use of bioaugmentation (adding microorganisms) and biostimulation, microcosms were seeded with the propane oxidizing bacterium strain ENV425 (Steffan et al., 1997). The strain was grown in basal salts medium (BSM; Hareland et al., 1975) with propane or isopropanol as a carbon suorce, washed, and added to the microcosms to a final concentration of ~1 x 10^6 cells/ml. Propane and oxygen were added to the microcosms as described above.

RESULTS

California aquifer. Results of the treatability testing are presented in Figures 1 and 2. Figure 1 shows the degradation of MTBE in microcosms amended with propane or propane + nutrients (nitrogen and phosphorous). MTBE degradation (initial concentration ~ 8 mg/L) was initiated with a limited lag period of 10 to 20 days. In samples amended with nutrients, MTBE concentrations were <300 ppb by day 30. Additional MTBE (~10 mg/L) was added to all of the biostimulation microcosms on day 33. The added MTBE was all degraded by day 42 in the nutrient amended microcosms, indicating an apparent removal rate of ~ 1 mg/L/day. Less of the added MTBE was degraded in the microcosms that did not receive nutrients. The degradation rate in these microcosms was approximately 0.6 mg/L/day. Nutrient amended microcosms were spiked again with ~7 ppm MTBE on day 43, and most of the added MTBE was degraded by day 57. On day 58, the propane was removed from the microcosm headspace by purging the samples with air. Removal of the propane resulted in a significant decline in MTBE degradation activity between days 58 and 76.

Propane utilization was constant throughout the first 60 days of incubation, with some fluctuation in propane concentrations during the first 20 days (data not shown). Propane was removed from the sample headspace on day 58, and it resulted in a decline in MTBE degradation.

In a separate set of experiments, strain ENV425 was added to the bioaugmentation microcosms to a final concentration of 1 x 10^6 cells/ml (Figure 2) on day 3 of incubation, and again to a final concentration of 1.8 x 10^6 on day 35 of the incubation. MTBE was rapidly degraded in the microcosms that received strain ENV425, and on day 35 additional MTBE (~7 ppm) was added to the microcosms. All of the added MTBE was degraded in the microcosms by day 59, with an apparent degradation rate of approximately 1 mg/L/day during the first 5 days after adding the spike. A second spike of MTBE (~7 ppm) on day 50 was degraded at a rate of approximately 0.3 mg/L/day. No degradation of MTBE or propane was observed in poisoned (killed) control samples.

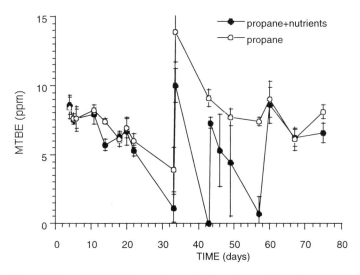

Figure 1. Degradation of MTBE in northern California aquifer microcosms. Open circles, microcosms that received only propane; closed circles, microcosms that received propane and nutrients (N+P). Additional MTBE was added to all of the microcosms on day 33, and to the nutrient amended microcosms on day 43 and day 60. Values represent means (n=3) and error bars represent one standard deviation.

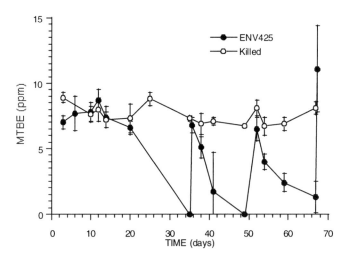

Figure 2. Effect of bioaugmentation on MTBE degradation in aquifer microcosms. Open circles, microcosms poisoned with $HgCl_2$ and sodium azide; closed circles, microcosms augmented with strain ENV425, propane, and nutrients (N+P). Augmented microcosms received 1×10^6 cells/ml of ENV425 on day 3, and 1.8×10^6 cells/ml of ENV425 on day 35. Additional MTBE was added to the bioaugmented microcosms on days 35, 52, and 67. Values represent means (n=3) and error bars represent one standard deviation.

New Jersey aquifer. The New Jersey test aquifer had an ambient pH of ~5.6. Prolonged propane biostimulation of aquifer microcosms from this site, with or without pH adjustment, did not result in active MTBE degradation (data not shown). Furthermore, we were not able to enrich for MTBE degrading propanotrophs from this site even using traditional enrichment culturing techniques. Because of this apparent lack of an indigenous propanotroph population we tested the use of strain ENV425 as a seed culture to promote activity. Even when added at low cell density (~10^6 cells/ml) the strain promoted MTBE degradation in the aquifer microcosms (Figure 3), and degradation was not inhibited by BTEX in the groundwater. The addition of propane appeared necessary to maintain degradation. When the microcosms were starved for propane on days 34 through 40, the rate of MTBE degradation decreased. The repeated doses of MTBE were readily degraded in the microcosms without the further addition of strain ENV425.

The biostimulated and bioaugmented microcosms were analyzed for MTBE by purge and trap gas chromatography on days 57 and 49, respectively. MTBE concentrations in all of the bioaugmented microcosms were below the method detection limit (~10 ppb). Two of the nutrient amended biostimulation microcosms also had MTBE concentrations below 10 ppb, and the third replicate contained approximately 1900 ppb MTBE on day 57.

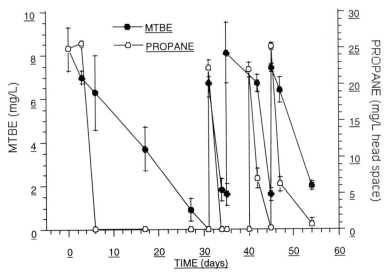

FIGURE 3. MTBE and propane degradation in New Jersey aquifer microcosms inoculated with strain ENV425. Values represent means (n=3) and error bars represent one standard deviation. The pH of the aquifer material was adjusted to ~7, and nitrogen and phosphorous were added as nutrients. Additional MTBE was added to the microcosms on days 31, 36, and 45, and propane was added as needed. No losses occurred in killed control samples.

Field application system. Field-scale application of in situ biostimulation with bioaugmentation is scheduled to begin in the summer of 200 at the New Jersey site the project is currently in the permitting process. The site is contaminated with MTBE and BTEX at average concentrations of 100 mg/L and 10 mg/L, respectively. The system will utilize three existing air sparging wells at the site. Approximately 20 L of ENV425 (~10^9 cells/ml) will be added to each well. Air will be added to the sparge wells by using a compressor, and propane will be metered into the air stream to achieve a final propane concentration of 0.2% (1/10 of the lower explosive limit [LEL] of propane). A soil vapor extraction (SVE) system with an activated carbon trap will be used to remove any fugitive propane and capture hazardous chemicals (e.g., BTEX).

DISCUSSION

Few remedial technologies are available for in situ remediation of MTBE contaminated sites. In one notable case, an MTBE-degrading mixed culture was injected into an aquifer and supplied oxygen to support MTBE degradation (Salanitro et al., 1999). Field results suggested that the culture began degrading MTBE within about 30 days after injection, and low levels of MTBE were achieved. The addition of such cultures may be limited, however, because of the need to grow them on substrates like MTBE which generate poor yields, thereby increasing fermentation costs and times. Likewise, obtaining permits for injecting poorly characterized microbial mixtures into aquifers may be difficult and costly. As an alternative, we evaluated the use of biostimulation of indigenous MTBE-degrading organisms and the use of a fast-growing pure bacterial seed culture that can be grown on innocuous and inexpensive growth substrates and supported in situ through the addition of propane.

Results of our initial treatability studies demonstrated that the addition of propane and oxygen to some sites supports rapid degradation of MTBE by indigenous organisms (Figure 1). In one site, however, a natural population of propane/MTBE degrading microbes was not present or it did not respond rapidly to the addition of the growth substrates. In the latter case, the addition of a seed culture of a known MTBE-degrading propanotroph, strain ENV425, was sufficient to support prolonged in situ MTBE degradation (Figure 3). In each case MTBE was degraded to very low levels (<10 µg/L), and the organisms could degrade repeated additions of MTBE. Because ENV425 is a pure bacterial culture that grows rapidly and to high cell density on many typical fermentation substrates, it will be relatively inexpensive to produce for large-scale aquifer seeding for MTBE remediation. Efforts are underway to isolate adhesion deficient variants of the strain that will be efficiently distributed upon injection into aquifers (Steffan et al., 1999).

Application of propane injection in the field also may be limited in some cases because of concerns about creating explosive mixtures of propane and air in situ. To address these concerns during early demonstrations, we have proposed to add propane to sites at concentrations of only 10% of the LEL, and to use SVE to prevent in situ accumulation of propane. Likewise, in a related ongoing propane injection project for TCE remediation at Lakehurst Naval Air Engineering

Station, New Jersey, we are using plugged silicone tubing as a bubble-free propane delivery system, thereby further minimizing fugitive emissions of propane (unpublished). As more demonstrations and full-scale applications of this technology are performed, concerns about the hazards of propane injection will likely diminish.

In summary, the results of these studies suggest that propane biostimulation will be a viable in situ remediation technology for MTBE-contaminated sites. The technology can be applied using existing air sparging systems or by using any of a variety of systems designed around specific site conditions. In cases where the natural propanotroph population is insufficient or does not respond to propane injection, exogenous strains like ENV425 can be added as seed cultures to facilitate MTBE degradation.

ACKNOWLEDGEMENTS

The authors acknowledge the excellent analytical support of Allen Thomas and Jamie Latham. This work was supported in part by Small Business Innovative Research grant DMI-9661329 from the National Science Foundation and funding from Kinder Morgan Energy Partners, L.P. Propane biostimulation for MTBE remediation is protected by US Patent #5,814,514, Sept. 29, 1998.

REFERENCES

Hareland, W., R. L. Crawford, P. J. Chapman, and S. Dagley. 1975. "Metabolic function and properties of 4-hydroxyphenylacetic acid 1-hydroxylase from *Pseudomonas acidovorans". J. Bacteriol.* 121:272-285.

Hazen, T.C., K. H. Lombard, B. B. Looney, M. V. Enzien, J. M. Dougherty, C. B. Fliermans, J. Wear, and C. A. Eddy-Dilek. 1994. "Summary of in situ bioremediation demonstration (methane biostimulation) via horizontal wells at the Savannah River Site Integrated Demonstration Project". *In Situ Remediation: Scientific Basis for Current and Future Technologies*, Battelle Press, Richland, WA. pp. 137-150.

Salanitro, J.P., G. Spindler, P. Maner, H. Wisniewski, and P. Johnson. 1999. "1Potential for MTBE bioremediation-in situ inoculation of specialized cultures". *Proceedings of the API/NGWA Conference on Petroleum Hydrocarbons and Organic Chemicals in Ground Water: Prevention Detection and Remediation Conference.* Houston, TX.

Semprini, L., M. Dolan, A. Tovannabootr, V. Magar, A. Leason, A. Lightner, and E. Becvar. 1999. "Microcosm protocol". *Partners in Environmental Technology-Technical Symposium and Workshop.* Arlington, VA.

Steffan, R. J., K. McClay, S. Vainberg, C. W. Condee, and D. Zhang. 1997. "Biodegradation of the gasoline oxygenates methyl *tert*-butyl ether, ethyl *tert*-butyl ether, and *tert*-amyl ether by propane-oxidizing bacteria". *Appl. Environ. Microbiol.* 63:4216-4222.

Steffan, R. J., K. L. Sperry, M. T. Walsh, S. Vainberg, and C. W. Condee. 1999. "Field-scale evaluation of in situ bioaugmentation for remediation of chlorinated solvents in groundwater". *Environ. Sci. Technol.* 33:2771-2781.

BIOTREATMENT OF MTBE WITH A NEW BACTERIAL ISOLATE

Robert J. Steffan, Simon Vainberg, Charles Condee, Kevin McClay and Paul Hatzinger, Envirogen, Inc., 4100 Quakerbridge Road, Lawrenceville, NJ 08648

ABSTRACT: We isolated an MTBE-degrading bacterium, *Hydrogenophaga flava* ENV735, that grows slowly on MTBE, but grows rapidly on MTBE or tertiary butyl alcohol (TBA) in the presence of a small amount of yeast extract (0.01%). The strain mineralizes uniformly labeled [^{14}C]MTBE to $^{14}CO_2$, and exhibits an initial MTBE oxidation rate of ~46 nmole/min/mg cell protein. A culture of ENV735 (OD_{550}=1) degraded 25 mg/L MTBE to below analytical detection limits in ~15 min. MTBE degradation by the strain is not affected by the presence of BTEX, and the strain does not degrade BTEX. The strain can be grown rapidly to high cell densities on rich media for bioaugmentation and bioreactor applications. We used ENV735 as a catalyst for in situ bioaugmentation and as a seed for laboratory-scale membrane and fluid bed bioreactor systems. In aquifer microcosms the strain rapidly degraded 22 mg/L MTBE when added at cell densities of 1 x 10^7 to 1 x 10^9 cells/mL. In an 85-L membrane bioreactor (MBR) seeded with the strain, ~1200 mg/L MTBE was degraded to <100 µg/L, and degradation was not inhibited by 30 mg/L BTEX. In a fluid bed bioreactror (FBR) seeded with the culture, 10 mg/L MTBE was degraded to <100 µg/L.

INTRODUCTION

Methyl *tert*-butyl ether (MTBE) has been used extensively in the United States as a gasoline oxygenate to meet the vehicle emissions requirements of the 1990 Clean Air Act Amendments. Because MTBE is highly soluble in water (~43,000 ppm) and has a low tendency to adsorb to soils (K_{owc} ~1.05), it is now often found in groundwater near service stations, storage facilities, and filling terminals throughout the United States. As little as 4 liters of reformulated gasoline (~11% MTBE) can contaminate 16 million liters of groundwater to above 20 µg/L.

Objective. The objective of this work was to isolate a microorganism that can use MTBE as a growth substrate, and to apply that organism as a catalyst for in situ and ex situ MTBE remediation. In this report, we describe a new MTBE-degrading bacterium, *Hydrogenophaga flava* strain ENV735, and the application of the strain as a catalyst for in situ bioaugmentation and a seed for ex situ reactor systems.

METHODS

Strain isolation. Strain ENV735 was isolated by classic enrichment culture techniques. MTBE-contaminated groundwater and colonized activated carbon from a laboratory fluid bed bioreactor treating MTBE were added to basal salts medium (BSM; Hareland et al., 1975) containing 25 mg/L MTBE. The culture was incubated at 25 °C with shaking until an increase in turbidity was observed, a portion of the culture liquid was added to a new flask of BSM with MTBE, and the process was repeated. After 3 to 4 rounds of enrichment, a subsample of the culture was grown on R2A agar plates. Individual colonies were picked from the plates and added to fresh BSM with MTBE. Cultures that grew were streaked again on R2A plates and evaluated microscopically to assess culture purity. One isolate was selected for further studies and designated strain ENV735. To characterize strain ENV735, a subculture of the strain was plated on R2A agar plates and sent to Acculabs, Inc. (Newark, DE) for fatty acid and rRNA sequence analysis.

MTBE degradation assays. MTBE biodegradation assays were performed essentially as described in Steffan et al., 1997. Cells grown in rich medium (LB or 0.4% YE in BSM), BSM + TBA (100 mg/L), or BSM + sucrose (0.5% w/v) were collected by centrifugation, washed, and suspended in BSM to an optical density at 550 nm (OD_{550}) of 1, unless otherwise indicated. Subsamples of the cultures were placed in 60 mL serum vials, and MTBE was added to the culture as either neat compound or an aqueous solution depending on the desired final concentration. For high concentration MTBE assays, cultures were placed in 160-mL serum vials to insure oxygen availability. Vials were sealed with Teflon-lined septa and shaken at 25 °C.

To measure the amount of MTBE in the vials, a subsample of the culture liquid was centrifuged and analyzed by gas chromatography (GC) as previously described (Steffan et al., 1997). This method had a detection limit of ~300 µg/L. When a lower detection limit was desired, the samples were analyzed by using EPA Method 8260. The detection limit of Method 8260 was approximately 5 µg/L.

Mineralization of MTBE was confirmed by measuring the production of $^{14}CO_2$ from uniformly labeled [^{14}C]MTBE. To perform the assay, 10 mL of cell suspension was placed in a 30-mL serum vial, amended with 1.36 µCi uniformly labeled [^{14}C]MTBE (>99%; 10.1 mCi/mmol; Dupont NEN) in 2.5 µL ethanol and 2 mg/L MTBE, and incubated at 25 °C. At timed intervals, air was passed through the samples and $^{14}CO_2$ in the effluent air were trapped and quantified (Marinucci and Bartha, 1979). Oxosol C^{14} (National Diagnostics) was used as a CO_2 trapping solution and liquid scintillation cocktail. The amount of $^{14}CO_2$ produced was quantified by liquid scintillation counting.

Aquifer microcosms. Aquifer microcosms were prepared using saturated aquifer soils and groundwater from an MTBE-contaminated fuel terminal site located in northern California. Each microcosm contained 15 g of soil and 30 mL of groundwater (22 mg/L MTBE; pH 6.7) in 160 mL serum vials sealed with Teflon-lined septa. Triplicate microcosms received either 1×10^7, 1×10^8, or 1×10^9, cells/mL of TBA-grown ENV735. Control microcosms received no ENV735. Microcosms were incubated at 20 °C with shaking.

Bioreactor systems. MTBE degradation by ENV735 was tested in two MBR systems made from 85-L PCV tanks (16 inch diameter) and internal microporous hollow fiber membranes (Zenon, Inc.). The first reactor (MBR1) was seeded with ENV735 on Oct. 5, and it has a feed stream of MTBE in 0.4X BSM. The second reactor (MBR2) was inoculated on November 21 with a culture from MBR1, and was fed MTBE in tap water with NH_4Cl and H_3PO_4 as nutrients. Flow into the reactors was controlled to achieve a hydraulic retention time (HRT) of one to 3 days, and the influent MTBE concentration was maintained at between 1000 and 2000 mg/L. The pH of the reactors was controlled by adding 5 N NaOH. A calculated solids retention time (SRT) of 160 days was maintained by removing a portion of the reactor contents each day. The dissolved oxygen concentration in the reactor was maintained at ~2 mg/L by adding air. MTBE stripping was evaluated by analyzing samples of the effluent air, and it was always <5% of the MTBE loss.

A laboratory-scale FBR was constructed from a glass column with Teflon and stainless steel tubing and fittings to minimize abiotic losses. The system had a total liquid volume of 4.5 L. Granulated activated carbon (settled bed volume ~ 800 mL) was used as a support medium. The carbon was fluidized to an expanded bed height of 125% of the original bed height by using a gear pump in the recycle line. Contaminated water was fed at a rate of 5.6 to 17 mL/min using a peristaltic pump, corresponding to hydraulic retention times (HRTs) of approximately 3 to 1 h, respectively. The HRT was calculated based on expanded bed volume and feed flowrate (assumes no significant biodegradation occurred in other wetted areas of the reactor).

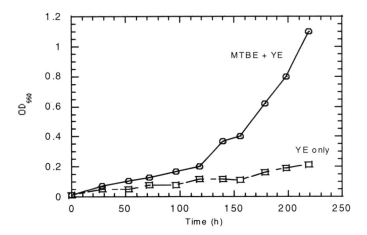

FIGURE 1. Growth of ENV735 on MTBE + yeast extract (YE). ENV735 was grown in BSM with YE (0.01% v/v) or YE (0.01% v/v) + MTBE (185 mg/L). Additional YE (0.01%) was added to each flask, and additional MTBE (185 mg/L) was added to the MTBE + YE flask, at T= 118, 156, 178, and 198 h.

RESULTS

Isolation of ENV735. After several weeks of enrichment culturing on MTBE as a sole carbon source, small clumps of cells and cell growth on the enrichment flask walls were observed. These cells were streak plated on R2A media, and individual colonies were added to fresh BSM media containing 100 mg/L MTBE. Once culture purity was achieved, an individual colony was selected from the plate, streaked onto a fresh R2A agar plate, and designated ENV735. Strain ENV735 was a Gram negative rod-shaped bacterium. Fatty acid analysis indicated that the strain was most closely related to bacteria of the genus *Hydrogenophaga* (similarity index = 0.720), and 500 bp 16s rRNA analysis indicated that the strain is most closely related to *Hydrogenophaga flava* (0.58% difference from library strain). The strain grew readily on H_2 as a sole energy source, but did not grow on BTEX, propane or isopentane.

Growth of strain ENV735 on MTBE as a sole source of carbon was slow and resulted in the production of dense bacterial clumps and cells attached to the container surface at the air/water interface, thereby making it difficult to collect representative samples for measuring cell yield on MTBE. If a small amount of YE (0.01% w/v) was added to the media, however, the cells grew more rapidly and were dispersed throughout the media (Figure 1). Cell yields on 0.01% YE in the absence of MTBE or TBA were low, but the strain did grow well on high concentration (0.4% w/v) YE. Cell yields on BSM with MTBE + 0.01% YE were at least 0.20 to 0.26 g ENV735 (dwt)/g MTBE after subtracting the dry weight of cells produced during growth on 0.01% YE alone.

MTBE degradation assays. To measure the ability of strain ENV735 to degrade MTBE, the cells were grown in either BSM + MTBE + 0.01% YE, BSM + 0.4% (w/v) YE, or BSM + Sucrose (0.5%) and incubated with MTBE or uniformly-labeled [^{14}C]MTBE. Results of one such assay are presented in Figure 2. Regardless of the growth substrate used, MTBE degradation occurred without a lag period. Initial maximum MTBE oxidation rates achieved were 46 nmole/min./mg total cell protein at 25 mg/L MTBE. MTBE degradation rates were higher with cells grown on rich medium (LB or 0.4% YE) than with cells grown on either MTBE or TBA. Strain ENV 735 was able to degrade MTBE at concentrations as great as 3000 mg/L, and the pH optimum for MTBE degradation by strain ENV735 was 7.0. TBA degradation occurred after a lag period of about 5 hr. If cells were grown on MTBE or TBA, however, there was no accumulation of TBA during MTBE degradation, and no lag period before TBA degradation. MTBE degradation was not affected by the presence of H_2 nor the gasoline components benzene, ethylbenzene, toluene and zylenes (BTEX) (Figure 3).

Microcosm studies. Results of initial microcosm studies with ENV735 are shown in Figure 4. MTBE concentrations in the microcosms seeded with 10^9 or 10^8 ENV735/mL were reduced to below the limit of detection (50 µg/L) within the first 100 hours of incubation, and they were below detection limits in microcosms inoculated with 10^7 cells/mL after 6 days of incubation (data not shown).

Bioremediation of MTBE   169

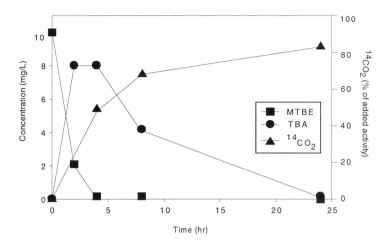

FIGURE 2. Degradation of [u-^{14}C]MTBE by ENV735. Symbols are as indicated the graph and represent means of triplicate samples. Standard errors were within the size of the symbols.

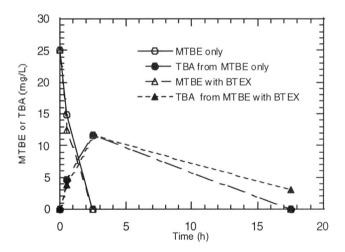

FIGURE 3. Degradation of MTBE by ENV735 in the presence of BTEX (25 mg/L). Values represent means (n=2). In each case the range of data was within the size of the symbols and is not shown.

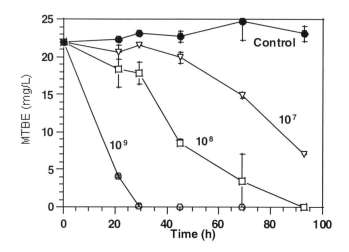

FIGURE 4. Degradation of MTBE in aquifer microcosms by TBA-grown ENV735 added at 10^9, 10^8, or 10^7 cells/mL as indicated. Values represent means (n=3). Error bars represent one standard deviation.

Bioreactor studies. Results of membrane bioreactor studies are shown in Figure 5. MTBE influent concentrations were either ~1000 mg/L (MBR1) or ~2000 mg/L (MBR2) in the two MBR reactors, and hydraulic retention times were maintained at ~3 days (1.2 L/hr into the 85L reactors). MTBE removal rates in MBR1 reached 42 mg/L/hr during operation with a 1 d HRT. MTBE removal in MBR2 reached 28 mg/L/hr while BTEX removal was 0.4 mg/L/hr BTEX (>99%) during the same time. After an initial evaluation period during which solids retention times (SRT) were infinite (i.e., no solids removal) the SRTs were maintained at a 160 days (calculated value) with MTBE as the only carbon source. Biomass in MBR1 increased from ~5000 mg/L on day 1 to ~12,000 mg/L on day 125. The feed to biomass ratio in the reactor was approximately 0.03 mg MTBE/h/mg biomass. Plate count analysis of MBR1 performed after 2 months of operation revealed that ENV735 comprised >70% of the reactor microbial population. Both MBRs recovered well from spikes in the influent MTBE concentration, and from periods of no feed caused by feed pump malfunctions.

A fluid bed bioreactor inoculated with the ENV735-containing culture from MBR1 degraded 10 mg/L MTBE to <100 µg/L at a hydraulic retention time of 3 h (Figure 7). This represented a removal efficiency of ~99%. When the flow rate into the reactor was increased 3-fold (1 hr HRT), the reactor continued to remove >90% of the added MTBE.

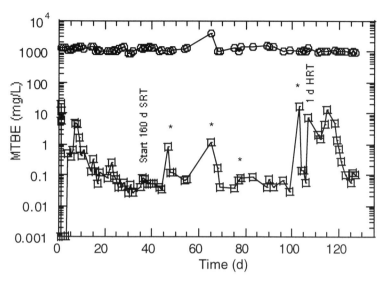

FIGURE 5. MTBE biodegradation in MBR1. The reactor was inoculated with a 16 L culture of ENV735 on day 1, and it received a constant feed of MTBE in 0.4 X BSM. The reactor was operated with an HRT of 3 days until day 106 when the HRT was decreased to 1 day. Circles indicate influent MTBE concentration, and squares indicate effluent concentration. The symbol "*" indicates three feed pump malfunctions and a spike in feed MTBE concentration (>4000 mg/L on day 65).

DISCUSSION

The isolation of a pure culture that degrades MTBE provides many advantages over the use of mixed cultures for bioaugmentation or reactor seeding. Most notably, pure cultures can be grown rapidly and to high cell density on high-yield substrates, thereby reducing the cost of producing biomass for such applications. Consortia must be grown on the target compound to maintain integrity of the consortium, and sometimes, as in the case of MTBE, these target compounds produce low cell yields and slow growth. Strain ENV735 can be grown on minimal media and sucrose, or on rich media like YE. Additionally, the use of a pure culture for in situ bioaugmentation allows increased certainty about the safety or pathogenicity of the microbes that will be injected into the aquifer. It is difficult to evaluate the pathogenicity of each organisms in a consortium.

The results of this work demonstrate that the newly-isolated strain ENV735 mineralizes MTBE to CO_2, and that it is an efficient catalyst for degrading MTBE. Preliminary bioaugmentation studies suggest that the strain can be added to contaminated aquifers at a relatively low cell density (10^7 cells/mL) and that it can degrade at least 20 mg/L MTBE to below detection

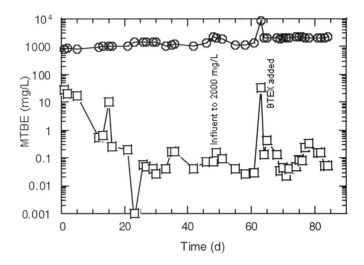

FIGURE 6. MTBE Biodegradation in MBR2. MBR2 was inoculated with a culture from MBR1 and fed MTBE in tap water. BTEX (30 mg/L) addition was initiated on January 20. The HRT was 3 d, and the calculated SRT was maintained at 160 d. Circles are influent MTBE concentration and squares are effluent concentration.

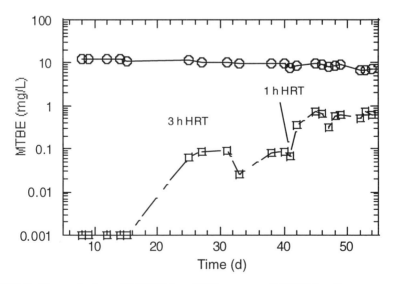

FIGURE 7. Degradation of MTBE in a FBR system. MTBE loading rates at 3 h and 1 h HRTs were 3.3 and 10 mg MTBE/h/L, respectively. Circles represent influent MTBE concentration, and squares represent effluent concentration.

limits (Figure 3). When added to an MBR or FBR system, the strain survived and degraded MTBE for extended periods without the addition of co-substrates (Figures 4 to 7). Cell densities in the MBRs continued to increase with time, even though ENV735 does not grow well on MTBE alone (Figure 1). Of particular importance in this work is the observation that MTBE degradation by ENV735 is not inhibited by BTEX (Figure 3 and 6) which are common co-contaminants in groundwater.

Preliminary operating cost analyses performed with data from this work and data provided by suppliers of alternative treatment technologies indicate that, at MTBE concentrations >1mg/L, biological treatment in an MBR or FBR with ENV735 should be considerably less expensive than other currently-available technologies including carbon absorption. The cost advantage of biotreatment is further accentuated if the contaminated water also contains BTEX.

ACKNOWLEDGEMENT

This work was supported in part by a Small Business Innovative Research Grant (# DMI-9960886) from the National Science Foundation. We are grateful to Allen Thomas and Jamie Latham for their excellent analytical support.

REFERENCES

Hareland, W., R. L. Crawford, P. J. Chapman, and S. Dagley. 1975. "Metabolic function and properties of 4-hydroxyphenylacetic acid 1-hydroxylase from *Pseudomonas acidovorans*". *J. Bacteriol.* 121:272-285.

Marinucci, A.C. and R. Bartha. 1979. "Apparatus for monitoring the mineralization of volatile ^{14}C-labeled compounds". *Appl. Environ. Microbiol.* 38:1020-1022.

Steffan, R. J., K. McClay, S. Vainberg, C. W. Condee, and D. Zhang. 1997. "Biodegradation of the gasoline oxygenates methyl *tert*-butyl ether, ethyl *tert*-butyl ether, and *tert*-amyl ether by propane-oxidizing bacteria". *Appl. Environ. Microbiol.* 63:4216-4222.

FACTORS INFLUENCING BIOLOGICAL TREATMENT OF MTBE IN FIXED FILM REACTORS

William T. Stringfellow (Lawrence Berkeley National Laboratory, Berkeley, CA)
Robert D. Hines (Envirex/U. S. Filter, Waukesha, WI)
Dirk K. Cockrum (Camp, Dresser & McKee Inc., Denver, CO)
Scott T. Kilkenny (Kinder Morgan Energy Partners, Orange CA)

ABSTRACT: Data from fluidized bed bioreactors treating contaminated ground water at two field sites have been collected and compared to laboratory studies with the objective of improving the reliability of methyl *tert*-butyl ether (MTBE) biotreatment in the field. Laboratory studies demonstrated that MTBE biodegradation was inhibited by a broad range of compounds, including o-xylene, methanol, toluene, and trichloroethylene (TCE). The general inhibition of MTBE degradation is similar to effects previously observed with nitrifying bacteria. Field data was examined to determine if two inhibitors, toluene and TCE, could be shown to effect MTBE treatment in fluidized bed reactors. It was found that there is a higher probability of poor MTBE removal efficiency during periods of higher toluene loading, but inhibition by TCE was not conclusively demonstrated. Results also show that periods of poor treatment also occur independently of effects attributable to toluene loading alone. These results illustrate the complexity of MTBE treatment and the limitations of using laboratory results to predict results in the field.

INTRODUCTION

MTBE has been used as a gasoline additive since 1979. As a consequence, MTBE is now a widespread environmental contaminant. Many gasoline and fuel transfer stations have MTBE contaminated ground water that must be recovered and treated before either re-injection or discharge. Activated carbon adsorption is currently the most widely used technology for the treatment of MTBE contaminated water.

Activated carbon is an effective treatment regime for MTBE, but has draw-backs. Activated carbon does not have a large sorption capacity for MTBE. It is also a phase transfer technology that does not result in the ultimate destruction of the MTBE. The spent carbon must be shipped off site for disposal or other treatment. Thus, there is a strong interest in developing alternative treatments for MTBE contaminated groundwater.

Our research has focused on the development of biological treatment as a viable, field-ready alternative for MTBE treatment at larger MTBE contaminated sites. We are developing biotreatment as both a stand alone technology and as a technology to be used in conjunction with carbon filters. Biological treatment also has potential for the treatment of *tert*-butyl alcohol and other contaminants that are not efficiently treated by activated carbon filtration.

The focus of this work has been the evaluation and application of up-flow, fluidized-bed bioreactor technology for MTBE treatment under real world conditions. Fluidized-bed bioreactors have been widely applied for the treatment of ground water contaminated with gasoline hydrocarbons. The objective of our research is to understand the mechanism of MTBE degradation in these types of bioreactors and to delineate the parameters controlling MTBE treatment efficiency. In this paper we examine MTBE degradation efficiency in the presence of other groundwater contaminants, specifically toluene and TCE.

MATERIALS AND METHODS
Field sites. Data were collected from two field sites for this study. The Sparks Solvent Fuel Site (SSFS) is a fuel transfer terminal located in Sparks, NV. The Mission Valley Terminal (MV) is a smaller fuel transfer terminal located in San Diego, CA. Both sites have fluidized-bed bioreactors designed by Envirex/U. S. Filter that contain granular activated carbon (GAC) as a support media. Envirex/U. S. Filter bioreactors were installed at SSFS in 1995 for gasoline hydrocarbon treatment and began degrading MTBE in 1996. MTBE removal was demonstrated to be due to biological degradation (Stringfellow 1998). SSFS has a pair of 183 cm diameter reactors that are operated in parallel. MV has one 51 cm diameter reactor. The effective reactor volume at SSFS is approximately 17,600 liters and the MV reactor is approximately 680 liters. At SSFS, the reactors are an integral component of the MTBE control strategy. Fluidized bed bioreactors are being piloted at MV. Data from SSFS was collected as part of the requirements for regulatory compliance. Data from MV was collected as part of two separate pilot studies conducted at the site. For this paper, data from both MV studies have been pooled as one data set. All analyses were conducted using EPA approved methods at a contract analytical laboratory. Table 1 summarizes the operational conditions for the two sites.

TABLE 1: Summary of reactor flow and loading conditions for Envirex/U.S. Filter fluidized-bed bioreactors at Sparks Solvent Fuel Site, NV and Mission Valley, CA.

	SSFS Mean	MV Mean
Influent Flow Rate, liters per minute	700	16
Hydraulic Retention Time, hour	0.2	0.8
MTBE Concentration, $\mu g/L$	245	6,970
MTBE Load, mg/L-reactor volume/day	18	259

Laboratory studies. Samples of bed material from SSFS were collected and shipped on ice overnight to Lawrence Berkeley National Laboratory for testing. The bed material consisted of GAC coated with a microbial biofilm. Samples of GAC were placed in 40 mL glass vials, supplemented with a mineral salts buffer, and spiked with a solution of MTBE in water to give the appropriate final MTBE concentration required for each experiment. The vials contained 10 mL of liquid, approximately 30 mL of headspace, and were capped with Teflon vial caps. Compounds tested as inhibitors were added through the vial caps using a 10 µL syringe. MTBE was monitored by analysis of 100 µL head-space samples using a flame ionization detector after gas chromatographic separation. Samples were tested in triplicate for inhibition studies. Kinetic analysis was conducted using single vials for each concentration point. Initial degradation rates were measured after allowing for rapid equilibrium of the MTBE with the GAC sample.

RESULTS AND DISCUSSION

In order to examine the effects of other ground water constituents on MTBE degradation potential, batch experiments were conducted using GAC from SSFS. Initial MTBE degradation rates were reduced in vials receiving an additional compound in comparison to those vials receiving MTBE alone (Table 2). Most of the inhibitors tested (toluene, p-xylene, and methanol) were degraded over time, typically within three days, and MTBE degradation continued until all MTBE was degraded. Analysis of the headspace showed the vial still contained significant amounts of oxygen at the end of the experiment. Inhibition by TCE followed a different pattern in that the TCE was not degraded and MTBE degradation did not go to completion.

TABLE 2: Inhibition of initial MTBE biodegradation rates by bacteria grown as a biofilm on granular activated carbon. Data presented are means of three replicates.

Inhibitor	Initial MTBE Degradation Rate, % of Control	Final Inhibitor Concentration, % of Initial Concentration	Final MTBE Concentration, % of Initial Concentration
Toluene (430 mg/L)	25	0	0
TCE (1000 mg/L)	29	13	57
p-Xylene (170 mg/L)	44	0	0
Methanol (500 mg/L)	33	0	0

These results suggest that MTBE degradation rate can be influenced by the presence of many other compounds. The broad sensitivity of MTBE degradation activity is reminiscent of observations made with nitrifying bacteria. Nitrifying bacteria and ammonia removal in water treatment plants are well known to be sensitive to a broad range of inhibitors, including solvents and metals (Eckenfelder 1980). The exact reason for this sensitivity is not well understood.

The inhibitory effect of toluene and TCE on MTBE degradation was investigated further by conducting kinetic experiments using a constant toluene (520 mg/L) and TCE (400 mg/L) concentration and varying the initial MTBE concentration between 2 and 400 mg/L. Results from these experiments are presented as Lineweaver-Burk plots in Figure 1a and 1b. Lineweaver-Burke analysis allows the mechanism of inhibition to be investigated by comparing the slopes of linear fits of reaction rate data collected with MTBE plus an inhibitor and with MTBE only. In the case of both TCE and toluene, the lines intercept at the x-axis, suggesting the inhibition is due to a non-specific mechanism (non-competitive inhibition).

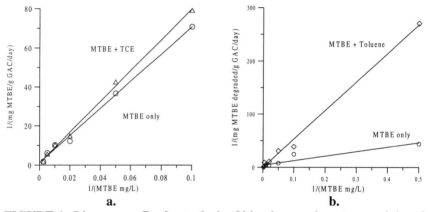

FIGURE 1. Lineweaver-Burke analysis of kinetic experiments examining the concentration dependent rates of MTBE degradation in the presence of toluene and trichloroethylene (TCE).

Given laboratory results, the question arises as to what relevance these results may have on understanding the operation of the field reactors. Both the full-scale reactors (at SSFS) and the pilot-scale reactor (at MV) have MTBE biodegrading populations present. In laboratory studies, GAC samples from both sites have consistently high MTBE degrading activity, even for samples collected during periods when the field systems have poor MTBE removal efficiency. Both sites have MTBE removal efficiencies that can fluctuate greatly, but also have demonstrated stable MTBE removal (greater than 90%) for extended periods. SSFS exhibited greater than 90% removal for periods longer than 100 days and 80% removal consistently for over 200 days. Both sites have maintained benzene removal efficiencies greater than 99% during their entire history of operation, even during periods when MTBE removal was completely lost. Based on these observations and our laboratory results, we postulated that toluene or TCE inhibition could be a contributing factor in the loss of MTBE removal efficiency in field reactors.

One approach to answering this question is to plot removal efficiency data as a function of plant loading conditions. The SSFS treatment system receives very little toluene (mean < 2 µg/l), but some TCE (mean = 6 µg/L). In contrast the MV system receives significant amounts of toluene (mean = 1,150 µg/L) and does not receive TCE (non-detectable on all analysis). These differences can allow us to examine the influence of these parameters independently at the two sites.

Toluene and MTBE loading and removal rate data for MV are presented in Figure 2a and 2b. Toluene removal rates remained high during the complete course of this study. The maximum toluene loading capacity for this system has not been reached (Figure 2a). Unlike toluene, MTBE removal rate is not simply a direct function of MTBE loading at MV (Figure 2b). Figure 2b indicates that MTBE removal rate may reach a maximum of approximately 300 mg MTBE per liter reactor volume per day in this system.

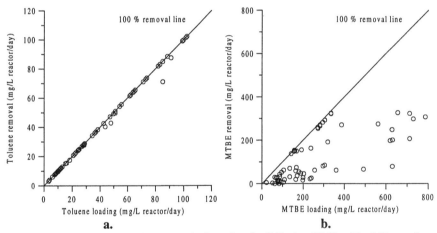

FIGURE 2. Loading and removal plots for the Mission Valley Fuel Transfer Station, CA.

MTBE removal efficiency as a function of toluene loading at MV was examined. Toluene loading and MTBE loading are not correlated in this system ($r^2 = 0.075$, n = 67), so the influence of toluene loading can be examined independently. Figure 3 is a plot of MTBE removal efficiency as a function of toluene loading. At toluene loading above 40 mg per liter reactor volume per day, MTBE removal efficiency was 60% or less (Figure 3). However, there were days where MTBE removal efficiency was low when toluene loading was low, indicating that toluene loading is only one of many factors that influence MTBE removal in this type of fluidized bed bioreactor.

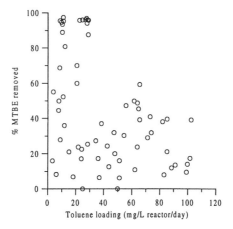

FIGURE 3. Methyl *tert*-butyl ether removal efficiency as a function of toluene loading at Mission Valley Terminal, CA.

A similar analysis has been conducted using data from SSFS. TCE and MTBE loading and removal efficiency data for MV are presented in Figure 4a and 4b. TCE removal averaged approximately 56% for the period included in this study. MTBE loading rates at SSFS are less than at MV (Table 1 and Figures 2b and 4b). The SSFS reactors exhibit a more consistent relationship between MTBE loading rates and MTBE removal rates. MTBE removal appears to be more stable at higher loading rates (Figure 4b).

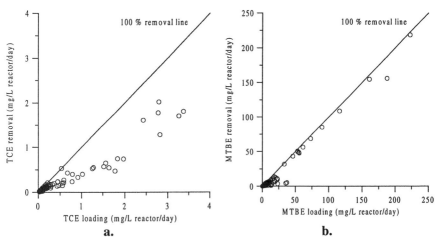

a.   b.

FIGURE 4. Loading and removal plots for the Sparks Solvent Fuel Site, NV. TCE removal rates are stable, but consistently less than 100%.

MTBE removal efficiency is plotted as a function of TCE loading in Figure 5. MTBE removal efficiency is consistently below 20% when TCE loading exceeds 2 mg per liter reactor volume per day, however there are very few days when TCE loading is at this level. The data may indicate that TCE has a negative influence on MTBE removal efficiency, but the information is hardly conclusive. In any case, the analysis indicates that there are few days when TCE loading could be high enough to warrant concern as an operating variable for MTBE removal control.

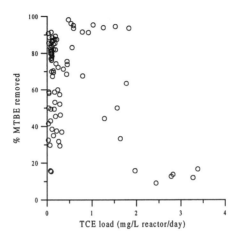

FIGURE 5. Methyl *tert*-butyl ether removal efficiency as a function of trichloroethylene (TCE) loading at Sparks Solvent Fuel Site, NV.

CONCLUSIONS

It can be concluded from this study that experiments examining factors influencing MTBE biodegradation conducted in the laboratory are useful for defining broad issues. However, laboratory studies have limitations in their ability to guide operation of field reactors. It appears that co-occurring compounds can influencing the treatment of MTBE in fluidized bed bioreactors, but the relationship is complicated by multi-variable interactions. Toluene and TCE removal rate varies as a direct function of loading in these reactors. In contrast, MTBE removal is not obviously a function of any one loading variable alone. This analysis indicates that MTBE removal in these systems is more difficult to predict and maintain than either toluene or TCE removal.

ACKNOWLEDGMENTS

This research was funded by a grant from Kinder Morgan Energy Partners to the Department of Energy, Lawrence Berkeley National Laboratory.

REFERENCES

Eckenfelder, W.W. 1980. Principals of Water Management. P. 315-325. CBI Publishing Company, Inc., Boston, MA.

Stringfellow, W.T. 1998. Biodegradation of methyl *tert*-butyl either by microorganisms found in a groundwater treatment system. Abstracts of the 98th Annual Meeting of the American Society for Microbiology, Atlanta, GA, May 17-21.

COMETABOLIC DEGRADATION OF MTBE BY A CYCLOHEXANE-OXIDISING BACTERIA

Diego Corcho, Robert J. Watkinson and David N. Lerner
(University of Sheffield, Sheffield, UK)

ABSTRACT: A cyclohexane-oxidising bacterial culture has been tested for its ability to degrade gasoline oxygenates, including methyl tert-butyl ether (MTBE), ethyl *tert*-butyl ether (ETBE) and *tert*-amyl methyl ether (TAME) and di-isopropyl ether (DIPE). The initial oxidation of MTBE resulted in the production of nearly stoichiometric amounts of *tert*-butyl alcohol (TBA). The maximum MTBE transformation rate that the mixed culture exhibited at 20°C was 6.4 mg MTBE/g dry cells/h. Inhibition studies with 1-aminobenzotriazole (1-ABT) suggested that the enzyme responsible for the oxidation of MTBE is a cytochrome P-450 monooxygenase, most likely the cyclohexane monooxygenase (CMO). The cyclohexane-oxidising culture was capable also of degrading benzene and toluene efficiently, and had the ability to degrade MTBE in the presence of benzene or toluene.

INTRODUCTION

Fuel oxygenates are organic additives designed to increase the octane content in gasoline and to reduce vehicle emissions (Peaff, 1994). Methyl *tertiary* butyl ether (MTBE) is the most commonly used oxygenate and is currently added at concentrations of up to 15% (vol/vol) to reformulated gasolines (Kirschner, 1995). Other fuel oxygenates include ethyl *tert*-butyl ether (ETBE), *tert*-amyl methyl ether (TAME), di-isopropyl ether (DIPE) and *tert*-butanol (TBA). The widespread use of MTBE coupled with incidents of leaking underground storage tanks has led to MTBE contamination of surface waters, groundwater, soil and sediments. The relatively high water solubility and poor adsorption characteristic of MTBE, combined with its persistence, makes it the second most common groundwater pollutant in the USA (Squillace *et al.*, 1996).

A few reports have been published addressing the biodegradability of MTBE under controlled laboratory conditions by microorganisms in either mixed or pure culture. In the first report showing degradation of MTBE, an aerobic bacterial consortium composed of at least six different bacteria, transformed MTBE into tert-butyl alcohol (TBA) and subsequently to CO_2 and H_2O (Salanitro *et al.*, 1994). Other reports have addressed the biodegradability of MTBE under both aerobic (Cowan and Park, 1996; Mo *et al.*, 1997) and anaerobic conditions (Mormile *et al.*, 1994; Suflita and Mormile, 1993). To date, two studies described pure bacterial cultures capable of using MTBE as a sole carbon and energy source, however low MTBE removal rates and low biomass yield were reported (Mo *et al.*, 1997; Hanson *et al.*, 1999).

Since more than 80% of compounds present in reformulated gasoline are alkanes and aromatic hydrocarbons, cometabolic biodegradation of MTBE has received recent attention. Specifically interest has focussed on gaining an understanding of the mechanisms of natural attenuation of MTBE in soil and sediments. Reports have described the potential of microorganisms growing on propane (Steffan *et al.*, 1997; Hardison *et al.*, 1997), pentane (Garnier *et al.*, 1999) or butane (Hardison *et al.*, 1997) to efficiently degrade MTBE.

In this study, we evaluated the ability of a cylohexane-utilising culture to metabolise gasoline oxygenates, including MTBE, ETBE, DIPE, TAME and gasoline compounds including benzene and toluene. This study suggests that MTBE oxidation is initiated by a cytochrome P-450 monooxygenase, probably an inducible cyclohexane monooxygenase (CMO).

MATERIALS AND METHODS

Chemicals. All chemicals (TAME 94%, ETBE 99%, TBA, 99.5%, DIPE 99%, MTBE 99.8%, toluene 99.8% and benzene 99%) were purchased from Sigma-Aldrich Co. Ltd. (Dorset, UK). Cyclohexane was obtained from BDH (Merck Ltd., Dorset, UK).

Growth of the cyclohexane-oxidising culture. The mixed bacterial culture used in this study was taken from a bioscrubber treating air contaminated with cyclohexane. The mixed culture was grown in a 2L bench aerated bioreactor on mineral media. Cyclohexane, as the sole source of carbon and energy, was supplied to the reactor in the gas phase. The temperature was maintained at 25°C and the pH at 6.5

Degradation assays. Cells were grown as described above on cyclohexane to an OD_{600} of ≥ 1.2 and then collected by centrifugation, washed twice with mineral media, and suspended to an optical density of 1.5 to 2.0. Sub-samples of the cell suspensions (10ml) were placed in 60-ml serum vials, and amended with different organic compounds. MTBE, ETBE, TAME, or DIPE dissolved in distilled water were added to give final concentrations of 60 to 350 mg/L. Benzene was added to microcosms amended with MTBE (to give benzene final concentrations of 5, 15 and 120 mg/L). Toluene was added to microcosms amended with MTBE (to give toluene final concentrations of 2.5, 7.5 and 110 mg/L). The vials were sealed with Teflon-lined septa and crimp seals and incubated at 23°C on a rotary shaker at 200 rpm for at least 72 hours. Control samples were prepared without the addition of the mixed culture. Depletion of the target compounds was measured by gas chromatography (GC).

Inhibition assays. Inhibition assays were performed by incubating cyclohexane-grown cells with either TBA or 1-aminobenzotriazole (ABT) (Sigma Chemical Co., Dorset, UK). Resting cells were prepared as in the degradation assays. Subsamples of the cell suspension (10ml) were then placed in 70-ml serum vials and dosed with TBA (20, 50 and 300 mg/L) or ABT (0.1 mM final concentration). The reaction mixtures were incubated at 23°C for 10 min before MTBE was added to a final concentration of 100 mg/L. After 6 and 24 h of incubation samples were taken from the microcosms and analysed by GC for MTBE degradation.

Analysis. The measurement of ether oxygenates (MTBE, ETBE, TAME and DIPE), their catabolic metabolites and aromatic hydrocarbons (benzene and toluene) in the aqueous phase were carried out by direct injection using a gas chromatograph (GC) with a photo-ionisation detector (FID). The analysis was performed by first removing cells from samples by centrifugation, and injecting 1 µl of the supernatant directly into a Varian 3400 gas chromatograph equipped with a 30-m DB-MTBE capillary column 0.45-mm ID and 2.55 µm film thickness (J& W Scientific, Folsom, CA, USA).

RESULTS AND DISCUSSION

The MTBE degradation experiments using resting cells initially showed the transformation of MTBE into a single metabolite, t-butanol (TBA), which accumulated in the microcosms supernatant. Experimental results (Figure 1.A) showed a stoichiometric transformation of MTBE into TBA. The MTBE transformation rate of the mixed culture was 4.25 mg MTBE/g dry cells/h. Repetition of the biodegradation experiments were performed showing a rate of MTBE transformation ranging between 6.4 and 4.5 mg MTBE/g dry cells/h.

The cyclohexane-utilising culture was also able to efficiently degrade other ether oxygenates including TAME, ETBE and DIPE (Table 1). The data showed a very fast maximum degradation rate for TAME (9.18 mg TAME/g of cells/h).

TABLE 1. Summary of biodegradation experiments of MTBE and other ether oxygenates.

Substrate	Mean % substrate remaining (SD)		Maximum degradation rate (mg/g cells/h) (µmol/ g cells/h)
	High concentration (320 mg/L)	Low concentration (60 mg/L)	
MTBE	33.13 (3.5)	0 (0)	6.4 (73.2)
TBA	75.20 (6.8)	40.43 (10.4)	1.02 (13.8)
TAME	6.42 (0.1)	0 (0)	9.18 (89.8)
ETBE	66.25 (9.8)	5.37 (7.5)	1.91 (18.7)
DIPE	49.19 (4.7)	20.58 (2.8)	1.36 (13.3)

The bacterial culture transformed TAME into a single product, probably *tert*-amyl alcohol, equivalent to the transformation of MTBE into TBA. Figure 1.B shows the disappearance of TAME with time and the concomitant appearance of the primary degradative metabolite.

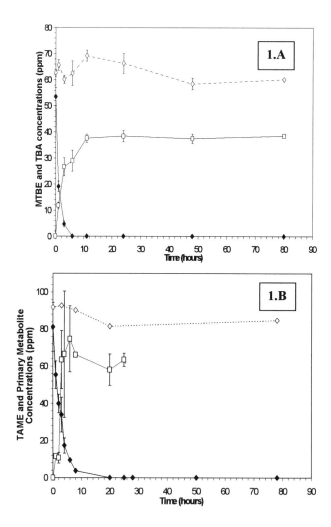

FIGURE 1. Degradation experiments in liquid microcosms at 20°C using the cyclohexane-oxidising culture. (A) ◊, **MTBE concentration in negative control microcosm;** ♦, **MTBE concentration in active microcosms;** ☐, **TBA concentration on active microcosms. (B)** ◊, **TAME concentration in negative control microcosm;** ♦, **TAME concentration in active microcosms;** ☐, **Primary degradative metabolite concentration in active microcosms. Values are means (n=2) and error bars represent the spread of the duplicate values.**

To confirm the role of a P-450 enzyme in MTBE degradation, the cyclohexane-utilising culture was incubated with 1-aminobenzotriazole (1-ABT), a highly specific inhibitor of the cytochrome P-450 activity (Ortiz de Montellano et al., 1981). In the presence of ABT the MTBE degradation was reduced by 89.9% compared with the microcosms without the inhibitor. This result suggests that the enzyme involved in the degradation of MTBE is likely to be a P-450 enzyme, most likely the cyclohexane monooxygenase (CMO). Other researchers have shown a P-450 propane monooxygenase involved in the degradation of MTBE by propane-oxidising bacteria (Steffan, 1997).

Inhibition experiments were also carried out to determine the effect of TBA on the degradation of MTBE. The presence of TBA (20, 50 and 300 mg/L) proved to have an inhibitory effect on MTBE removal with inhibitory values ranging from 10.4 to 32.1 %.

Resting cells from the cyclohexane-oxidising culture were capable of oxidising benzene and toluene. No lag phase was observed in the degradation of benzene or toluene although the cells were not exposed to those contaminants in the growing phase. The maximum degradation rate for benzene ranged from 0.025 mg benzene/g cells/h at the lowest concentration studied (5 mg/L) to 0.375 mg benzene/g cells/h (15 mg/L and 120 mg/L). After 24 hours of incubation benzene was only observed in the 120 mg/L benzene microcosm, with 11% of initial benzene present. Toluene was more difficult to transform than benzene. The maximum degradation rate for benzene ranged from 0.0125 mg toluene/g cells/h at 2.5 mg/L toluene to 0.035 mg toluene/g cells/h at 110 mg toluene/l. After 24 hours of incubation, 69% of the initial toluene was present in the microcosms containing the maximum concentration (110 mg/L)

Table 2 summarises the effect of benzene and toluene on MTBE degradation. The data suggest a strong inhibitory effect on MTBE degradation by toluene, even at very low concentrations (2.5 mg/L). On the other hand, only benzene at high concentrations exhibited an inhibitory effect on the removal of MTBE (56.5% inhibition at a concentration of 120 mg benzene/L).

TABLE 2. Summary of biodegradation experiments of MTBE in the presence of benzene or toluene.

Substrates Initial concentration (mg/L)	Maximum degradation rate (mg/g cells/h)	Inhibition of MTBE degradation (%)
MTBE (80)	3.5	0
MTBE (80) + Benzene (5)	3.5	11.8
MTBE (80) +Benzene (15)	2.2	6.8
MTBE (80) + Benzene (120)	1.1	56.5
MTBE (80) + Toluene (2.5)	1.9	48.1
MTBE (80) + Toluene (7.5)	1.25	77.6
MTBE (80) + Toluene (110)	0.62	70.3

CONCLUSIONS

A cyclohexane-oxidising culture showed the capability of degrading efficiently a range on ether oxygenates (MTBE, TAME, ETBE and DIPE) and other organic compounds present in gasoline including benzene and toluene.

This study suggested that the enzyme responsible for initiating the oxidation of MTBE is a P-450 cyclohexane monooxygenase. This result is in agreement with other reports on MTBE degradation, where the enzyme involved was proved to be a cytochrome P-450 (Hardison, 1997; Steffan, 1997).

Microbial cometabolic processes have the potential to play a key role in bioremediation schemes for MTBE contaminated sites. An understanding of the mechanisms involved in the cooxidation of MTBE may be crucial to predicting the fate and persistence of MTBE in the environment.

REFERENCES

Cowan, R. M., and K. Park. 1996. "Biodegradation of the Gasoline Oxygenates MTBE, ETBE, TAME, TBA and TAA by Aerobic Cultures". In *Proceedings of the Mid-Atlantic Conference*, Buffalo, NY. pp 523-539.

Garnier, P. M., R. Auria, C. Auger, and S. Revah. 1999. "Cometabolic biodegradation of Methyl t-Butyl Ether by *Pseudomonas aeruginosa* grown on pentane". *Applied Microbiology and Biotechnology 51*, 498-503.

Hanson, J. R., C. E. Ackerman, K. M. Scow. 1999. "Biodegradation of Methyl t-Butyl Ether by a Bacterial Pure Culture". *Applied and Environmental Microbiology 65*, 4788-4792.

Hardison, L. K., Curry, S. S., Ciuffetti, L. M., Hyman, M. R. 1997. "Metabolism of Diethyl Ether and Cometabolism of Methyl tert-Butyl Ether by a Filamentous Fungus, a *Graphium sp*". *Applied and Environmental Microbiology 63*, 3059-3067.

Kirschner, E. M. 1995. "Production of top 50 chemicals increased substantially in 1994". *Chemical Engineering News 73*, 16-20.

Mo, K., C. O. Lora, A. E. Wanken, M. Javanmardian, X. Yang, and C. F. Kulpa. 1997. " Biodegradation of Methyl t-Butyl Ether by Pure Bacterial Cultures". *Applied Microbiology and Biotechnology 47*, 69-72.

Mormile, M. R., Liu, S. & Suflita, J. M. 1994. "Anaerobic Degradation of Gasoline Oxygenates: Extrapolation of Information to Multiple Sites and Redox Conditions". *Environmental Science and Technology 28*, 1727.

Ortiz De Montellano, P. R., and M. Mathews. 1981. "Autocatalytic alkalation of the cytochrome P-450 prosthetic haem group 1-aminobenzotriazole". *Biochemical Journal 195*, 761-764.

Peaff, G. 1994. "Court ruling spurs continued debate over gasoline oxygenates". *Chemical Engineering News* 72, 8-13.

Salanitro, J. P., Diaz, L. A., Williams, M. P. & Wisniewski, H. L. 1994. "Isolation of a Bacterial Culture That Degrades Methyl t-Butyl Ether". *Applied and Environmental Microbiology 60*, 2593-2596.

Squillace, P. J., Pankow, J. F., Korte, N. E. & Zogorski, J. S. 1996. "Environmental Behaviour and Fate of Methyl tert-Butyl Ether (MTBE)". USGS.

Steffan, R. J., McClay, K., Vainberg, S., Concee, C. W., Zhang, D. L. 1997. "Biodegradation of the Gasoline Oxygenates Methyl tert-Butyl Ether, Ethyl tert-Butyl Ether, and tert-Amyl Methyl Ether by Propane-Oxidizing Bacteria". *Applied and Environmental Microbiology* 63, 4216-4222.

Suflita, J. M. & Mormile, M. R. 1993. "Anaerobic Biodegradation of Known and Potential Gasoline Oxygenates in the Terrestrial Subsurface". *Environmental Science & Technology 27*, 976-978.

MINERALIZATION OF MTBE WITH VARIOUS PRIMARY SUBSTRATES

Gregory J. Wilson, Amy P. Richter, and Makram T. Suidan
(University of Cincinnati, Cincinnati, OH)
Albert D. Venosa (U.S. Environmental Protection Agency, Cincinnati, OH)

ABSTRACT: Five specialized bioreactors have been operated for over a year to evaluate the biodegradability of the fuel oxygenate methyl-t-butyl ether (MTBE) under different substrate/co-substrate conditions. One bioreactor has been fed MTBE at an influent concentration of 150 mg/L (1.7 mM) as the only organic carbon source. In the other reactors, MTBE has been fed at an influent concentration of 75 mg/L (0.85 mM) with the balance of the COD fed as either ethanol, diethyl ether, diisopropyl ether (DIPE), or benzene/toluene/ethylbenzene/xylene (BTEX) compounds. The reactors were seeded with a mixed culture from several sources. A 10-L porous pot has been placed inside each reactor to facilitate exit of the treated effluent while retaining the biomass. Effluent from each bioreactor had been monitored by measuring the concentration of MTBE and its intermediate breakdown products in both the liquid effluent and in the exhaust gas. The concentration of MTBE in the effluent from all five reactors has been consistently below 1.0 to 5.0 µg/L. No intermediates or end-products from the biodegradation of MTBE have been found in the effluent. These findings demonstrate that MTBE can be effectively mineralized biologically in an *ex-situ* reactor to extremely low levels if a high enough sludge age and biomass concentration are maintained.

INTRODUCTION

An oxygenate that has grown substantially as the additive of choice in gasoline to increase its octane rating is methyl-t-butyl ether (MTBE). MTBE and the other oxygenates are mixtures of process chemicals containing polar compounds and thus have a higher water solubility than most gasoline hydrocarbons. For that reason, they can occur in higher concentrations than the benzene/toluene/ethylbenzene/xylene (BTEX) compounds contaminating groundwater from underground storage tank (UST) spills. Polar compounds, especially ethers, adsorb poorly to soil organic matter, adding to the mobility of oxygenates in a spreading plume. In addition, MTBE has the potential to decrease the sorptive retardation of BTEXs via co-solvent effects, which would further enhance the mobility of BTEXs in groundwater. Eventually, MTBE and BTEX plumes separate within the aquifer due to their differential solubilities. The presence of high concentrations of MTBE could reduce the biodegradability of BTEXs if MTBE is preferentially attacked by the degrading microbial populations present in groundwater or if it is toxic to the degraders.

Previously, researchers have studied MTBE biodegradation under both anaerobic and aerobic conditions (Hardison et al., 1997; Mormile et al., 1994; Salinitro et al., 1994); Steffan et al., 1997; Suflita and Mormile, 1993; Yeh and Novak, 1994 and 1995). Generally, the findings have shown that MTBE degrades either slowly or not at all, although some differences have been noted. The conflicting information may be due in part to the lack of sufficient time for microbial enrichment of degrading activity during startup of reactor operations.

This study was initiated to determine the extent to which MTBE is biodegradable under aerobic and anaerobic continuous flow conditions in both the presence and absence of other carbon sources. Since it was known that biomass yield is low when microorganisms are grown on MTBE (Salinitro et al., 1994), a specialized bioreactor was used to retain as high a concentration of biomass as possible to provide the highest probability of success. This paper will discuss the aerobic conditions only. Insufficient progress has been made with the anaerobic research for an adequate data presentation at this time.

MATERIALS AND METHODS

Description of Reactor. The type of bioreactor used was a porous pot reactor, which consisted of a stainless steel outer casing housing an inner porous barrier made of polyethylene. The water flowing into the reactor system flowed out under gravity through the porous sides of the inner barrier, thereby allowing total control of the biomass solids remaining within the polyethylene pot. The total inner volume of the stainless steel container was 12 L, while the aeration volume was 6 L. The influent feed rate was maintained at 2.4 L/d, giving a hydraulic retention time of 2.5 d. Initially, during the first 115 days of operation, biomass solids were wasted from all reactors at a rate of 5%/day (20-day sludge age). Following that period, the wasting rate was reduced to almost zero; i.e., the only time solids were removed from the reactor was for sampling. An abiotic reactor was also set up in which 0.2% sodium azide was included in the feed water to inhibit aerobic growth of respiring microorganisms.

Operating Conditions. Influent concentrations of MTBE were either 75 mg/L or 150 mg/L depending on whether the MTBE was either in combination with another carbon source or the sole carbon source, respectively. Total influent chemical oxygen demand (COD) was 417 mg/L. Temperature was held constant at 20°C, and pH was maintained between 7.5 to 8.1. The dissolved oxygen was always greater than 3 mg/L.

Analytical Measurements. MTBE and *t*-butyl alcohol (TBA) were measured by gas chromatography using a flame ionization detector (GC/FID) either by the direct injection method or using the purge and trap system. Column conditions were: 30 m J&W DB1 (0.53 mm bore x 3 μm film); temperature program of 35°C for 6 min followed by 12°C/min ramp rate to 190°C, held for 6 min). The detection limit for the direct injection method was 0.1 mg/L, but for the purge and trap method, it was 0.1 μg/L.

Seed Culture. All porous pot reactors were initially inoculated with a mixture of four different sources of microorganisms: 2 L mixed liquor from the Cincinnati Mill Creek Wastewater Treatment Plant (half industrial and half municipal wastewater), 600 mL mixed liquor from Shell Development Corporation's refinery in Houston, TX (kindly provided by J. Salinitro), 140 mL aquifer material wash water from the U.S. Navy's Port Hueneme site in California, and 2 L biomass solids actively growing on diethyl ether in a biofilter at the University of Cincinnati.

RESULTS

MTBE-Fed Reactor. Figure 1 presents the performance data for the MTBE-fed bioreactor. As stated previously, during the first 115 days of operation, biomass solids were wasted at a rate of 5% per day, giving an SRT of 20 days. During this time period, good biodegradation of MTBE was observed (upwards of 95%), but the results were unstable and erratic (see data to the left of the vertical line at the 115-day mark). After that time period, wasting of biomass solids was ceased.

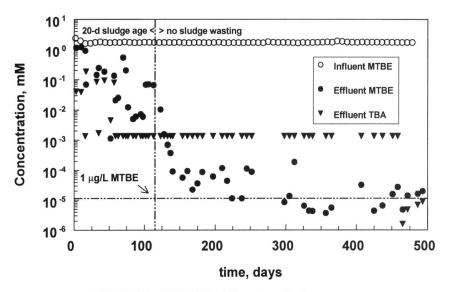

FIGURE 1. MTBE-Fed Reactor Performance

Within a very short time, the effluent MTBE concentration declined to undetectable levels and remained stable for the next 350 days close to the detection limit (about 1 µg/L). TBA was undetectable during this entire time period. Note that at 470 days, the purge and trap method was used instead of the direct injection method, and the detection limit for TBA declined to the same level as MTBE. Thus, both MTBE and TBA were near 1-5 µg/L for an extended period of time. No other intermediates were ever detected in the effluent from the porous pot reactor.

Figure 2 is a plot of the concentration of MTBE as a function of biomass solids during this time period. Clearly, as biomass increased, the MTBE in the

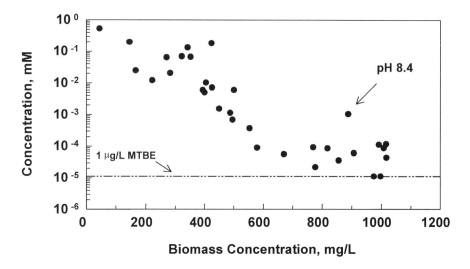

FIGURE 2. MTBE Biodegradation as a Function of Biomass Concentration

effluent decreased. Above about 600 mg/L biomass, MTBE concentrations were near the detection limit.

MTBE/DIPE-Fed Reactor. Figure 3 shows the performance of the MTBE/DIPE-fed reactor over a period of 500 days of operation. Results were nearly identical to the MTBE-fed reactor: prior to cessation of sludge wasting, good but erratic biodegradation of MTBE was observed. After biomass solids wasting was stopped,

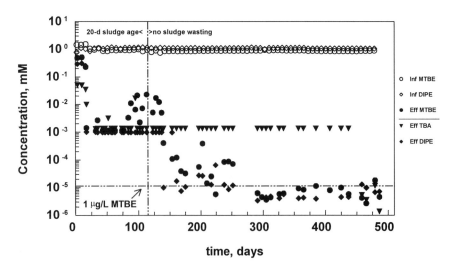

FIGURE 3. MTBE/DIPE-Fed Reactor Performance

MTBE, DIPE, and TBA were all at or below the detection limit for well over 300 days of operation.

MTBE/BTEX-Fed Reactor. Figures 4 and 5 summarize the performance of the reactor fed MTBE and BTEX. This reactor has not operated as long as the other reactors. However, results were again similar to the other reactors. Effluent MTBE has averaged about 14 µg/L, TBA about 2.1 µg/L, benzene about 0.10 µg/L, toluene about 0.39 µg/L, ethylbenzene about 4.2 µg/L, and xylenes about 8.0 µg/L.

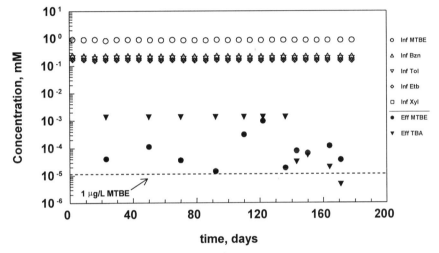

FIGURE 4. MTBE/BTEX-Fed Reactor Performance: MTBE/TBA in Effluent

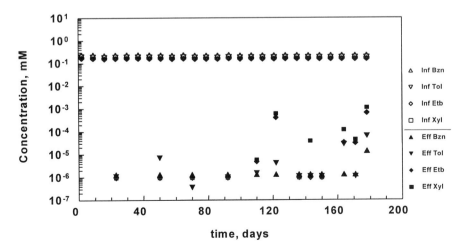

FIGURE 5. MTBE/BTEX-Fed Reactor: BTEX Effluent Concentrations

Abiotic Reactor. The reactor fed MTBE along with sodium azide showed virtually identical effluent concentrations of MTBE as influent for a period of 170 days (data not shown). This demonstrates good system integrity. Batch serum bottle experiments have recently been started to define the kinetics of mineralization of MTBE and TBA. Results were not available to include in this presentation.

DISCUSSION AND CONCLUSIONS

Results from the ethanol-fed reactor and the diethyl ether-fed reactor (data not shown) were similar to those from the MTBE, DIPE, and BTEX reactors. Clearly, MTBE is rapidly and effectively mineralized either as sole carbon and energy source or in the presence of other carbon sources provided the biomass is maintained at high levels and the pH is not less than 6.7 nor more than 8.2.

The porous pot reactor is a good model for laboratory research but is not an effective reactor system for field use. The reason is that the flow rate to the porous pot system is restricted to gravity feed, and only a limited amount of flow can be put through the system. One way to attain the high biomass solids needed for effective biodegradation of MTBE in groundwater in an *ex-situ* pump and treat operation is through the use of a membrane bioreactor (MBR). An MBR is a system that uses a normal aeration tank for the biodegradation reactions, but instead of a clarifier to separate the biomass solids from the effluent, it uses a concentration device composed of a ceramic membrane whose pore size is <0.02 mm. In this membrane device, all biomass solids are retained while the water flows through the membrane to exit to the outside. A major feature of the MBR is that it can be pressurized to handle high flow rates, a major advantage for treatment of MTBE-contaminated drinking water at the wellhead. Such a reactor system will be tested at pilot scale for cleaning up MTBE from groundwater.

REFERENCES

Hardison, L.K., S. S. Curry, L.M. Ciuffetti, and M.R. Hyman. 1997. "Metabolism of Diethyl Ether and Cometabolism of Methyl *tert*-Butyl Ether by a Filamentous Fungus, a *Graphium* spp." *App. Env. Microbiol.* 63:3059-3067.

Mormile, M.R., S. Liu, and J.M. Suflita. 1994. "Anaerobic Biodegradation of Gasoline Oxygenates: Extrapolation of Information to Multiple Sites and Redox Conditions." *Env. Sci. Technol.* 28:1727-1732.

Salanitro, J.P., L.A. Diaz, M.P. Williams, and H.L. Wisniewski. 1994. "Isolation of a Bacterial Culture that Degrades Methyl *tert*-Butyl Ether." *Appl. Env. Microbiol.* 60:2593-2596.

Steffan, R.J., K. McClay, S. Vainberg, C.W. Condee, and D. Zhang. 1997. "Biodegradation of Gasoline Oxygenates Methyl *tert*-Butyl Ether, Ethyl *tert*-Butyl Ether, and *tert*-Amyl Methyl Ether by Propane-Oxidizing Bacteria." *Appl. Env. Microbiol.* 63(11):4216-4222.

Suflita, J.M. and M.R. Mormile. 1993. "Anaerobic Biodegradation of Known and Potential Gasoline Oxygenates in the Terrestrial Subsurface." *Env. Sci. Technol.* 27:976-978.

Yeh, C.K. and J.T. Novak. 1994. "Anaerobic Biodegradation of Gasoline Oxygenates in Soil." *Water Env. Res.* 66:744-752.

Yeh, C.K. and J.T. Novak. 1995. "The Effect of Hydrogen Peroxide on the Degradation of Methyl and Ethyl *tert*-Butyl Ether in Soil." *Water Env. Res.* 67:828-834.

DECHLORINATION OF TETRACHLOROETHYLENE BY A MEMBRANE-ASSOCIATED DEHALOGENASE FROM *CLOSTRIDIUM BIFERMENTANS* DPH-1

Benedict C. Okeke, Young C. Chang, Masahiro Hatsu and *Kazuhiro Takamizawa* (Gifu University, Yanagido 1-1, Gifu 501-1193, Japan)

ABSTRACT: We studied a tetrachloroethylene (PCE) dehalogenase from cell-free extracts of *Clostridium bifermentans* DPH-1. The dehalogenase catalyzed PCE degradation at a Vmax of 73 nmol/mg.protein and Km of 12 nmol/ml. Maximal activity was recorded at 30°C and pH 7.5. Enzymatic activity was not stimulated by addition of metal ions. The molecular weight of the native enzyme was estimated to be approximately 70 kDa. SDS-PAGE and MALDI-TOF/MS analyses revealed a dimeric structure. A mixture of the alkylating agent, propyl iodide, and the reducing agent, titanium citrate, caused a light-reversible inhibition of enzymatic activity. A broad spectrum of chlorinated aliphatic compounds (PCE, trichloroethylene, *cis*-1,2-dichloroethylene, *trans*-1,2-dichloroethylene, 1,1-dichloroethylene, 1,2-dichloropropane and 1,1,2-trichloroethane) were degraded. The highest rate of degradation was achieved with PCE, and PCE was principally degraded via trichloroethylene to *cis*-dichloroethylene. Twenty-seven N-terminal amino acids of the enzyme were determined and the gene was cloned. Partial sequence of the gene showed striking homology with some outer membrane microbial proteins and no homology with any dehalogenase, suggesting a distinct PCE dehalogenase.

INTRODUCTION

Environmental pollution by toxic xenobiotic substances is a serious problem of worldwide importance. Halogenated aliphatic substances such as tetrachloroethylene (perchloroethylene, PCE), trichloroethylene (TCE), isomers of dichloroethylene (*cis*-1,2-DCE, *trans*-1,2-DCE, and 1,1-DCE), vinyl chloride, dichloromethane, 1,1-dichloroethane, 1,2-dichloroethane, 1,2-dichloropropane, and 1,1,2-trichloroethane occur in significant concentrations in the environment: waste and natural water courses, soils and the atmosphere. PCE and TCE, in particular, are excellent solvents, widely used in dry cleaning and in the textile industry, in the scouring of machines, in the extraction of fats (Neuman et al., 1994) and paint stripping (Distafano, 1999). These chemicals pose serious public health problem and are therefore considered priority pollutants. This has spurred both physicochemical and biological research for their remediation. Biological strategies (bioremediation) offer an attractive option which is relatively cheap with concern for the environment (Okeke et al., 1997).

PCE is an important model for the study of biodegradation of chlorinated aliphatic compounds because of its high halogen content and toxicity. PCE is recalcitrant under aerobic condition but can be dechlorinated by a few microbes under anaerobic condition. Thus there is growing interest in anaerobic biological systems for PCE decontamination. Mixed anaerobic enrichment cultures have been frequently reported to effect reductive dechlorination of PCE. However, only a few PCE degrading monocultures, e.g. *Dehalospirillum multivorans* (Neumann et al., 1994; Miller et al., 1997), *Desufomonile tiedjei* (Fathepure et al., 1987), *Dehalobacter restrictus* (Schumacher and Holliger, 1996), *Dehalococcoides ethenogenes* strain 195 (Maymó-Gatell et al., 1997), and *Desulfitobacterium* sp. PCE-S (Miller et al., 1998) have been studied in detail and their dehalogenases characterized. Most monocultures degrade PCE to principally *cis*-DCE, however,

D. ethenogenes 195 (Maymó-Gatell et al., 1997) can reductively dechlorinate PCE to ethylene (Figure 1). *cis*-DCE is the principal intermediate product of anaerobic dechlorination of PCE. Aerobic degradation of *cis*-DCE by *Rhodococcus rhodochrous* (Malachowsky et al., 1994) and *Nitrosomonas europaea* (Vanneli et al., 1990), have been reported.

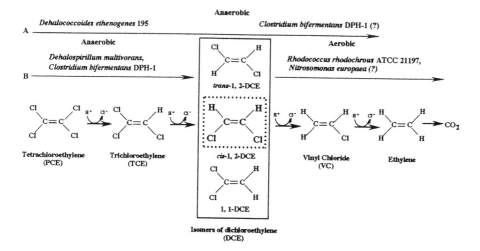

FIGURE 1. **Possible pathway for transformation and mineralization of PCE A: Single-stage process (anaerobic), B: Two-stage process (anaerobic and aerobic). Structures of chlorinated hydrocarbons were adapted from Vogel and McCarty, 1985; Maymó-Gatell et al., 1997; and Garant and Lynd, 1998.**

Corrinoids (protein-bound cobalamin) from some anaerobic microorganisms e.g.: *Dehalospirillum multivorans* (Neumann et al., 1994), *Sporomusa ovata* (Terzenbach and Blaunt, 1994) and *Methanosarcina thermophila* (Jablonski and Ferry, 1992) have been reported to be involved in reductive dechlorination of PCE. Moreover, the abiotic dehalogenation of PCE by cyanocobalamin (vitamin B_{12}) has been reported (Burris et al., 1998).

This study presents the degradation of PCE and other chlorinated aliphatic substances by a PCE dehalogenase from *C. bifermentans* DPH-1. The enzyme was identified by purification and the gene was cloned.

MATERIALS AND METHODS

Microorganism and Chemicals. *Clostridium bifermentans* DPH-1 was isolated from ditch sludge, contaminated with wastes from an electric company, in Japan (Chang et al., 2000). PCE was obtained from the Kanto (Tokyo, Japan). TCE, *cis*-DCE, *trans*-1,2-DCE, 1,1-DCE, dichloromethane (DM), 1,3-dichloropropene (1,3-DP), 1,2-dichloropropane (1,2-DP), 1,1,1-trichloroethane (1,1,1-TE), 1,1,2-trichloroethane (TE) and vinyl chloride (VC) were obtained from GL Sciences Inc. (Tokyo, Japan).

Cultivation of *C. bifermentans* and enzyme extraction. The culture was maintained in 26 ml serum bottles containing MMY medium under anaerobic condition, essentially as described by Chang et al. (1998). PCE was added to a final concentration of 6 µM to sustain PCE degrading activity. For enzyme preparation, cells were grown in 120 ml serum bottles containing 100 ml MMY medium at 30°C until the peak of PCE degradation (46 to 56 h). *C. bifermentans* cells were then harvested by centrifugation after adding DTT to a final concentration of 2 mM. Cells (0.8 to 1.6 g) resuspended in 50 mM Tris-HCl pH 7.5, 2 mM dithiothreitol (DTT) and 5% glycerol, were lysed by ultrasonic disruption. Unbroken cells and debris were separated by centrifugation and the extraction procedure repeated with the unbroken cells and debris. Supernatant filtrate served as the cell-free enzyme.

Enzyme assay. PCE dehalogenase assay was essentially according to the methods of Neuman et al. (1996) and Magnuson et al. (1998). Assay was performed in 26-ml serum bottles using either 0.61 mM ethanol or reduced methyl viologen as electron donors, in 4 ml Tris-HCl pH 7.5. The bottles were purged with oxygen-free nitrogen and sealed with Teflon-coated stoppers before 12 µM PCE was added. One unit (U) was defined as nmol of products (TCE+*cis*-DCE) produced under the assay condition.

Analysis of chlorinated aliphatic compounds. Chlorinated aliphatic compounds were quantified by the headspace method (Gossett, 1987), using a gas chromatograph equipped with an electron capture detector (ECD) or a flame ionization detector (FID).

Degradation of halogenated chemicals by cell free enzyme extract. The influences of various potential electron donors (ethanol, methanol, glucose, pyruvate, fumarate, acetate, lactate, formate and reduced methyl viologen) were determined at a final concentration of 1.22 mM as described under enzyme assay. The effects of temperature and pH were examined. Time course of PCE degradation and product formation was performed using an initial PCE concentration of 12 µM. PCE biotransformation and product formation (TCE and cDCE) were continuously monitored for for 70 min at 10 min interval, using the gas chromatograph equipped with ECD.

Purification and characterization of PCE dehalogenase. PCE dehalogenase was firstly partially purified from the cell-free enzyme extract, by ultrafiltration (50 kDa nominal molecular weight limit ultrafiltration membrane), ion-exchange and molecular exclusion column chromatography. Further purification was by ion-exchange high performance liquid chromatography (IEX-HPLC) and Size-exclusion high-performance liquid chromatography (SE-HPLC). Molecular weight of the native enzyme was determined by SE-HPLC. The purity and also molecular weight of the enzyme were examined by SDS-PAGE and Matrix-Assisted Laser Desorption Ionization-Time of Flight/Mass spectrometry (MALDI-TOF/MS). The influences of temperature, pH and metal ions on enzyme activity were determined. Degradation of chlorinated aliphatic compounds was subsequently examined. The effect of propyl iodide on PCE dehalogenase activity was examined essentially according to the method described by Neuman et al. (1995).

Based on 27 N-terminal amino acids sequence of the enzyme, degenerate primers were designed and used for PCR amplification of the N-terminal sequence region of the gene. The PCR product was confirmed by sequencing and was used as a probe to clone the gene.

RESULTS AND DISCUSSION

PCE Dechlorination. The rate of PCE degradation varied with different potential electron donors tested. Reduced methyl viologen, ethanol and glucose significantly enhanced PCE degradation. Reduced methyl viologen was the most effective, followed by ethanol. PCE degradation was also observed with other potential electron donors but at lower levels. Reductive dechlorination of PCE by *C. bifermentans* dehalogenase using a variety of potential electron donors would be advantageous for biotreatment in complex environments polluted with PCE and mixtures of anthropogenic and natural compounds.

Time course studies of PCE dechlorination and products formation, showed that PCE was rapidly degraded within 30 min. PCE was degraded to principally cis-DCE via TCE as the intermediate product. TCE accumulated within 10 min and was thereafter dechlorinated to *cis*-DCE. After about 60 min, a trace of vinyl chloride was detected. Neither ethylene nor CO_2 was detected during the 70 min reaction period. The rapid rate of PCE dechlorination to *cis*-DCE via TCE, observed with *C. bifermentans* dehalogenase is important (Figure 1). Two strategies have been proposed for the biotreatment of PCE and TCE contaminated sites: complete degradation by reductive dechlorination (de Bruin et al. 1992; Mayomó-Gatell et al. 1997); and by a combination of anaerobic and aerobic systems, in which PCE or TCE is converted to *cis*-DCE by anaerobic reductive dechlorination followed by complete aerobic metabolism of *cis*-DCE (Koziollek et al., 1999).

Properties of PCE Dehalogenase. The Km and Vmax for the PCE dehalogenase were 12 nmol and 73 nmol/mg.protein respectively. Maximal activity was recorded at $30^{\circ}C$ and pH 7.5. Enzymatic activity was not stimulated by addition of metal ions. The molecular weight of the native enzyme was estimated to be approximately 70 kDa. SDS-PAGE analysis and MALDI-TOF/MS revealed a molecualar weight of approximately 35 kDa, indicating a homo-dimer. The enzyme was slightly oxygen sensitive. The N-terminal amino acid sequence (AEVYNKDGNKLDLYGKVDGLHYFSNDT) of *C. bifermentans* dehalogenase showed striking homology with the sequences of maily outer membrane microbial proteins. No homology with the N-terminal sequence of any known PCE degrading bacterium suggesting a distinct dehalogenase. These properties are significantly different from those reported for other dehalogenases (Neuman et. al., 1996; Magnuson et al., 1998).

Enzymatic activity was inhibited upon incubation with a mixture of an alkylating agent (propyl iodide) anda reducing agent (titanium citrate), in the dark. Subsequent exposure to light restored the activity. Titanium citrate in the absence of propyl iodide did not have any inhibitory effect and no inhibition was recorded when propyl iodide was supplied singly. This observation is an indication that the enzyme may be a corrinoid protein. Such proteins from a number of anaerobic organisms similarly degrade PCE. Little is known about the exact mechanism of PCE dehalogenation by corrinoid proteins, however, it has been postulated that the supperreduced state of the corrinoid (Co^+) can bind to the alkyl residue of a halogenated hydrocarbon or an alkylating agent such as propyl iodide, achieving either dechlorination or activity inhibition respectively (Neuman et al., 1995).

Spectrum of chloroaliphatics degraded. *C. bifermentans* dehalogenase significantly degraded various chlorinated aliphatic compounds. The levels of degradation achieved with the cell-free dehalogenase extract are presented in Figure 2. Similarly the purified enzyme degraded multiple chlorinated aliphatic substances but to a lower extent. In general, the highest rate of degradation was achieved with PCE. Vinyl chloride was recalcitrant to *C. bifermentans* dehalogenase. Surprisingly, *cis*-DCE was degraded when added as the initial compound for degradation. The product(s) of *cis*-DCE degradation and the reasons why *cis*-DCE as an intermediate product is recalcitrant, are not clear. This aspect requires indepth studies to understand the mechanism and product(s).

As mixtures of chlorinated aliphatic substances are often found in polluted environments, the ability of *C. bifermentans* dehalogenase to degrade various chlorinated aliphatic substances is interesting. Only a few studies had described the anaerobic transformation of a variety of chlorinated hydrocarbons (Gerritse et al., 1996; Bouwer and McCarty, 1983; Miller et al., 1997). In methanogenic cultures, the transformation of halogenated methanes, 1,2-dichloroethane, 1,1,2,2-tetrachloroethane and PCE at low concentrations were observed (Bouwer and McCarty, 1983).

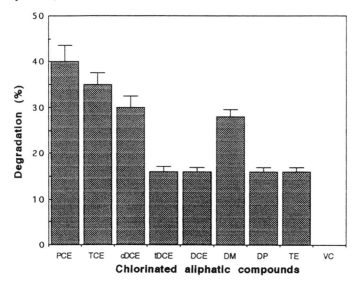

FIGURE 2. Degradation of various chlorinated aliphatic compounds by cell-free extract. Tetrachloroethylene (PCE), trichloroethylene (TCE), *cis*-dichloroethylene (cDCE), *trans*-1,2-dichloroethylene (tDCE), 1,1-dichloroethylene (DCE), dichloromethane (DM), 1,2-dichloropropane (DP), 1,1,2-trichloroethane (TE) and vinyl chloride (VC). Initial concentration of each compound was 12 µM.

Results indicate that *C. bifermentans* dehalogenase could play some important role in the initial breakdown of PCE (Figure 1) and other chlorinated aliphatic compounds, in sites contaminated with mixtures of halogenated substances. Furthermore, the striking homology with mainly outer membrane proteins, exhibited by the dehalogenase N-terminal amino acid sequence and partial sequence of the encoding gene is an indication that a novel group of outer membrane proteins may be involved in PCE degradation. Studies on expression of *C. bifermentans* dehalogenase using strong prokaryotic expression systems could yield active recombinant dehalogenase for biotreatment in cases where use of the live organism would be a drawback.

REFERENCES

Bouwer, E. J., and P. L. McCarty. 1983. "Transformation of 1- and 2- Carbon Halogenated Aliphatic Organic Compounds under Methanogenic Conditions". *Appl. Environ. Microbiol.* 45: 1286-1294.

de Bruin, W. P., M. J. J. Kotterman, M. A. Posthumus, G. Schraa, and A. J. B. Zehnder. 1992. "Complete Biological Reductive Transformation of Tetrachloroethene to ethane". *Appl. Environ. Microbiol.* 58: 1996-2000.

Burris, D. R., C. A. Delcomyn, B. Deng, L. E. Buck, and K. Hatfield. 1998. "Kinetics of Tetrachloroethylene-Reductive Dechlorination Catalyzed by Vitamin B_{12}." *Environ. Toxicol. Chem.* 17: 1681-1688.

Distafano, T. D. 1999. "The Effect of Tetrachloroethylene on Biological Dechlorination of Vinyl chloride: Potential Implication for Natural Bioattenuation." *Wat. Res.* 33: 1688-1694.

Chang, Y. C., M. Hatsu, K. Jung, Y. S. Yoo, and K. Takamizawa. 1998. "Degradation of a Variety of Halogenated Aliphatic Compounds by an Anaerobic Mixed Culture". *J. Ferment. Bioeng.* 86: 410-412.

Chang, Y. C., M. Hatsu, K. Jung, Y. S. Yoo, and K. Takamizawa. 2000. "Isolation and Characterization of a Tetrachloroethylene Dechlorinating Bacterium, *Clostridium bifermentans* DPH-1". J. Biosc. Bioeng. In Press.

Fathepure, R., J. P. Nengu, and S. A. Boyd. 1987. "Anaerobic bacteria that Dechlorinate Perchloroethylene". *Appl. Environ. Microbiol.* 53: 2671-2674.

Garant, H., and L. Lynd. 1998. "Applicability of Competitive and Noncompetitive Kinetics to the Reductive Dechlorination of Chlorinated Ethenes". *Biotechnol. Bioeng.* 57: 751-755.

Gerritse, J., V. Renard, T. M. P Gomes, P. A. Lawson, M. D. Collins, and J. Gottschal. 1996. "*Desulfitobacterium* sp. strain PCE1, an Anaerobic Bacterium that can Grow by Reductive Dechlorination of Tetrachloroethene or *ortho*-Chlorinated Phenols". *Arch. Microbiol.* 165: 132-140.

Gossett J. M. 1987. "Measurement of Henry's Law Constants for C1 and C2 Chlorinated Hydrocarbons". *Environ. Sci. Technol.* 21: 202-208.

Jablonski, P. E., and J. G. Ferry. 1992. "Reductive Dechlorination of Trichloroethylene by the CO-reduced CO Dehydrogenase Enzyme Complex from *Methanosarcina thermophila*.." *FEMS Microbiol. Lett.* 96: 55-60.

Koziollek, P., D. Bryniok, and H-J. Knackmuss. 1999. "Ethene as an Auxiliary Substrate for the Cooxidation of *cis-1,2-dichloroethene* and Vinyl Chloride." *Arch. Microbiol.* 172: 240-246.

Magnuson, J. K., R. V. Stern, J. M. Gossett, S. H. Zinder, and D. R. Burris. 1998. "Reductive Dechlorination of Tetrachloroethene to Ethene by a Two-Component Enzyme Pathway." *Appl. Environ. Microbiol.* 64: 1270-1275.

Malachowsky, K. J., T. J. Phelps, A. B. Teboli, D. E. Minnikin, and D. C White. 1994. "Aerobic mineralization of Trichloroethylene, Vinyl Chloride and Aromatic Compounds by *Rhodococcus* species". *Appl. Environ. Microbiol.* 60: 542-548.

Maymó-Gatell, X., Y. Chien, J. M. Gossett, and S. H. Zinder. 1997. "Isolation of a Bacterium that Reductively Dechlorinates Tetrachloroethene to Ethene." *Science* 276: 1568-1571.

Miller, E., G. Wohlfarth, and G. Diekert. 1997. "Comparative Studies on Tetrachloroethene Reductive Dechlorination Mediated by *Desulfitobacterium* sp. Strain PCE-S". *Arch. Microbiol.* 168: 513-519.

Miller, E., G. Wohlfarth, and G. Diekert. 1998. "Purification and Characterization of the Tetrachloroethene Reductive Dehalogenase of Strain PCE-S." *Arch. Microbiol.* 169: 497-502.

Neumann, A., H. Scholz-Muramatsu, and G. Diekert. 1994. "Tetrachloroethene Metabolism of *Dehalospirillum multivorans*." *Arch. Microbiol.* 162: 295-301.

Neumann, A., G. Wohlfarth, and G. Diekert. 1995. "Properties of Tetrachloroethene Dehalogenase of *Dehalospirillum multivorans*." *Arch. Microbiol.* 163: 276-281.

Neumann, A., G. Wohlfarth, and G. Diekert. 1996. "Purification and Characterization of Tetrachloroethene Reductive Dehalogenase from *Dehalospirillum multivorans*." *J. Biol. Chem.* 271: 16515-16519.

Okeke, B. C., A. Paterson, J. E. Smith, and I. A. Watson-Craik. 1997. "Comparative Biotransformation of Pentachlorophenol in Soils by Solid Substrate Cultures of *Lentinula edodes* ."*Appl. Microbiol. Biotechnol.* 48: 563-569.

Schumacher W., and C. Holliger. 1996. "The Proton Electron Ratio of the Menaquinone-Dependent Electron Transport from Dihydrogen to Tetrachloroethene in "*Dehalobacter restrictus*". *J. Bacteriol.* 178: 2328-2333.

Terzenbach, D. P., and M. Blaut. 1994. "Transformation of Tetrachloroethylene to Trichloroethylene by Homoacetogenic Bacteria". *FEMS Microbiol. Lett.* 123: 213-218.

Gossett J. M. 1987. "Measurement of Henry's Law Constants for C1 and C2 Chlorinated Hydrocarbons." *Environ. Sci. Technol.* 21: 202-208.

Vanneli, T., M. Logan, D. M. Arciero, and A. B. Hooper. 1990. "Degradation of Halogenated Aliphatic Compounds by the Ammonia-Oxidizing Bacterium *Nitrosomonas europaea* ." *Appl. Environ. Microbiol.* 60: 542-548.

Vogel, T. M., and P. L. McCarty. 1985. "Biotransformation of Tetrachloroethylene to Trichloroethylene, Dichloroethylene, Vinyl Chloride and Carbon Dioxide under Methanogenic Conditions." *Appl. Environ. Microbiol.* 49: 1080-1083.

ENHANCED BIOLOGICAL REDUCTIVE DECHLORINATION AT A DRY CLEANING FACILITY

Michael Lodato, IT Corporation, Tampa, FL.
Duane Graves, Ph.D. IT Corporation, Knoxville, TN.
Judie Kean, Florida Department of Environmental Protection, Tallahassee, FL.

ABSTRACT: A significant level of enhanced biological reductive dechlorination was demonstrated at a commercial dry cleaning facility in Orlando, Florida. Under the auspices of the Florida Department of Environmental Protection, and in accordance with the Dry Cleaning Solvent Cleanup Program, the upper portion of the surficial aquifer at the site was treated with an experimental source of time-release hydrogen (HRC®, Regenesis Bioremediation Products, Inc.). The hydrogen, which is produced by fermentation from HRC derived organic acids, serves as an electron donor that mediates the reduction of chlorinated hydrocarbons. The upper portion of the surficial aquifer at the site generally consists of sand and silty sand to a depth of approximately 29 feet below land surface (bls). It is underlain by approximately 3 feet of sandy clay and clay which separates it from the lower portion of the surficial aquifer. The average linear groundwater velocity in the upper portion of the surficial aquifer was calculated as 16 feet/year.

The site was extensively characterized with state-of-the-art direct push diagnostic protocols. It was determined that an area of approximately 14,600 square feet was situated within the 1 mg/L isopleth for perchloroethylene (PCE); in some wells contamination levels approached 9 mg/L. Approximately 6,810 pounds of HRC were injected into the area as described via 144 direct-push points spaced 10 feet on center. The total PCE contaminant mass was reduced by 96 percent after 152 days. This was calculated for a larger area that was bounded by wells which included both 1) a series of proximal up gradient wells and 2) a down gradient well series that could have been impacted by the advection and diffusion of the applied hydrogen releasing compound in the 152 days. This designated area, approximately 240' X 180', began with a mass of 19,183 g PCE and rose to 24,378 g by Day 43, presumably through physical desorption related to the injection activity. Subsequent to that point in time, mass was reduced to 17,925 g by Day 77, 12,869 g by Day 110 and then to 822 g by Day 152. Additionally, the daughter products trichloroethene (TCE), dichloroethene (DCE) and vinyl chloride all declined.

INTRODUCTION

The dry cleaning industry uses tetrachloroethene (PCE) as a degreaser and waterless cleanser for clothes. The use of PCE has resulted in the accidental release of this chlorinated solvent at many dry cleaning shops. The facilities are often located in strip malls and shopping centers and near residential areas. Even small towns are likely to have one or more dry cleaners. Many dry cleaning businesses were independently owned with little regulatory oversight regarding the disposal of spent solvent or the storage of fresh solvent. As a result, PCE contamination at dry cleaner sites is very common.

The field investigation evaluated parameters that give qualitative and quantitative indications of the occurrence of reductive dechlorination of chlorinated solvents. The combined evidence generated from several different aspects of the evaluation suggests that natural attenuation by the process of reductive dechlorination was occurring and was significantly affecting the fate of chlorinated compounds in the aquifer. Measurable levels of cis-1,2-DCE and vinyl chloride supported the conclusion that reductive dechlorination of PCE and TCE affected the chemical composition of dissolved contaminant groundwater plume. Using the evaluation approaches advanced by Region 4 EPA (undated) and the US Air Force (Weidemeier et al., 1996), evidence from the site indicates that naturally-occurring intrinsic biodegradation was making a measurable contribution to the restoration of the aquifer.

Reductive dechlorination rates for the site were found to range from 13 to 27 years for PCE, 5 to 11 years for TCE, and 17 to 35 years for vinyl chloride. An attenuation rate for DCE could not be calculated with the available data. Unfortunately, the rate of groundwater restoration was inadequate to provide a suitable level of groundwater protection. In order to expedite the reductive dechlorination of chlorinated solvents, a hydrogen releasing compound (HRC®, Regenesis Bioremediation Products, Inc.) was applied to the site as a large scale pilot demonstration under the Florida Dry Cleaning Solvent Cleanup Program.

Site Hydrogeology. The shallow subsurface generally consists of layers of tan to brown fine quartz sand to a depth of approximately 24 to 30 feet below land surface. The surficial sand is underlain by a gray clay which ranges in thickness from approximately 1 to 12 feet. Below this clay is a tan to white and gray silty sand and sandy clay unit which ranges in thickness from 20 to 25 feet. This is underlain by a pale to dark olive green sandy clay to clay with trace phosphate nodules at approximately 50 to 55 feet below land surface which represents the Hawthorne Group sediment. The groundwater seepage velocity the aquifer systems ranged from 16 ft per year in upper surficial aquifer, 2.6 ft per year in the lower surficial aquifer, and 5.8 ft per year in the intermediate aquifer.

HRC APPLICATION AND PERFORMANCE MONITORING

The application of HRC was designed to treat the upper surficial aquifer because the greatest concentrations of chlorinated solvents were observed in this part of the aquifer. Thus this hydrogeologic unit was targeted for the most aggressive treatment. HRC was directly injected into the upper surficial aquifer using geoprobe direct push technology at 144 locations within the area bounded by the 1,000 µg/L chlorinated hydrocarbon concentration contour (Figure 1). The injection points were located on 10-foot centers within an 80-ft by 180-ft grid. Each point received 2.45 gallons of HRC (22.5 lbs. or 0.9 lbs. per foot) between a depth of 5 to 30 ft BLS. A total of 6,800 lbs. of HRC was injected in the upper surficial aquifer.

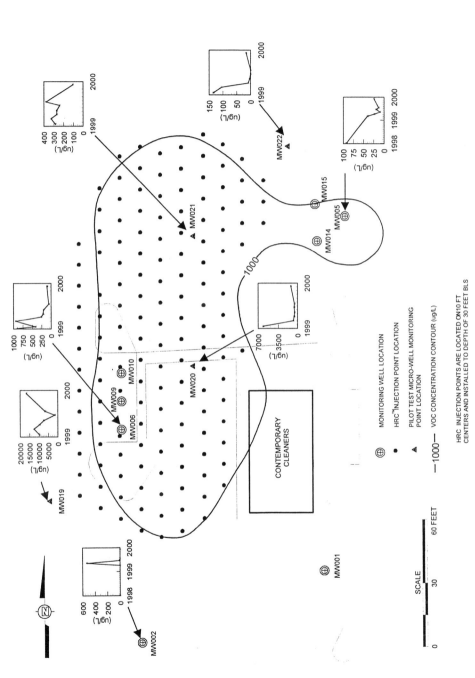

FIGURE 1. Site Map, HRC Injection Grid, and Concentration Reduction in Monitoring Wells

The effect of HRC on groundwater geochemistry and chlorinated solvent concentrations were determined by periodically sampling and analyzing groundwater from seven monitoring wells. Analysis included chlorinated solvents, dissolved oxygen, oxidation-reduction potential, pH, conductivity, temperature, ferrous iron, nitrate and nitrite, sulfate, methane, ethene, and ethane, manganese, and phosphorus. Groundwater samples were collected for six months following the HRC application to monitor progress of the treatment.

RESULTS AND DISCUSSION

The application of HRC resulted in an observable change in the concentration of chlorinated solvents. An area approximately 240 feet by 180 feet was affected by the HRC application. The mass of PCE and its dechlorination products before HRC application and at various time points after the application is shown in Table 1.

TABLE 1. Mass of chlorinated hydrocarbons at various times after HRC application.

Compound	Initial (g)	43 Days (g)	77 Days (g)	110 Days (g)	152 Days (g)
PCE	19,183	24,378	17,925	12,869	822
TCE	2,548	1,261	1,108	1,222	1,254
Cis-1,2-DCE	6,309	3,144	3,946	3,705	4,012
Vinyl Chloride	2,350	1,287	670	572	1,016

The PCE mass increased from the initial mass to the mass estimated after 43 days. This change was presumably due to physical desorption related to the injection activity. Overall the PCE mass was reduced by 96 percent after 152 days of treatment. The TCE mass was reduced by approximately 51 percent. The cis-1,2-DCE mass was reduced by 36 percent and the vinyl chloride mass was reduced by 58 percent. The dramatic reduction in PCE mass and the less dramatic reduction in the mass of the lesser chlorinated ethenes suggests that the PCE was being dechlorinated to TCE, DCE, and vinyl chloride. Because these compounds were being formed and dechlorinated concurrently, the overall reduction in mass was not as great than observed for PCE.

In addition to the overall mass reduction, the concentration in individual wells was observed to change over time. Figure 1 shows the change in the total chlorinated hydrocarbon concentration before HRC application and thereafter. With the exception of MW002, all other monitoring wells indicated on Figure 1 were located either within the treated area or close enough to the treated area to be influenced by the application. Concentration reductions were observed in MW006, MW021, MW022, MW020, and MW005. The concentration of chlorinated hydrocarbons appeared to be unaffected by treatment in MW019. This is due to recently discovered additional PCE source resulting from a broken sewer line leading from the dry cleaner to the city sewer system.

The behavior of specific chlorinated hydrocarbons in the HRC amended groundwater is shown for MW006 and MW020 in Figure 2. The presence and persistence of each chlorinated compound matches the expected behavior of that compound based on the level of chlorination and the reported biological dechlorination of PCE. Since the application of HRC, the PCE has declined to a very low concentration. TCE concentrations were also greatly reduced during treatment. Cis-1,2-DCE and vinyl chloride are the most prevalent compounds in both wells. This supports their position in the dechlorination sequence of PCE. As indicated in Figure 2, cis-1,2-DCE is the dominant DCE isomer in the groundwater although low concentrations of trans-1,2-DCE were also observed. This suggests the biological production of DCE. A continued decline in the concentrations of both DCE and vinyl chloride is expected. The apparent rate of concentration reduction is expected to increase as the production rate declines due to the complete dechlorination of residual of PCE and TCE.

HRC stimulated, biologically mediated reductive dechlorination of PCE was confirmed by changes in groundwater geochemistry that are typically catalyzed by biological activity. The spider chart shown in Figure 2 highlights a few groundwater characteristics that indicate the influence of HRC on groundwater chemistry. The thin line crosses each axis of the spider chart at the parameter value observed before HRC addition. The heavy line represents the values of the same parameters six months after HRC application. As indicated, pH tended to decline from neutral to acidic, conductivity was essentially unchanged, dissolved oxygen was depleted, the oxidation reduction (redox) potential changes from a positive potential to a very negative potential, sulfate was depleted, sulfide was produced, low concentrations ethene and ethane accumulated, much higher concentrations of methane were produced, and the chloride concentration was unchanged. Many of these changes in groundwater geochemistry were indicative of an increased level of anaerobic microbial activity in the aquifer.

The overall results from the HRC application and continued monitoring indicated that HRC appears to be an effective remedial alternative for the restoration of groundwater impacted with chlorinated ethenes. The cost for this large-scale demonstration was favorable and should encourage the application of the technology at appropriate sites. The overall cost of the project was $127,000. HRC product cost was $27,197. Additional project costs included the preparation of a detailed work plan, a sampling and analysis plan, a health and safety plan, preparation of an underground injection permit, Geoprobe subcontracting, labor, monthly reports and meetings, contractor oversight, and field and laboratory analyses. This project represents the successful collaboration of the Florida Department of Environmental Protection, Regenesis Bioremediation Products, and IT Corporation, the state cleanup contractor.

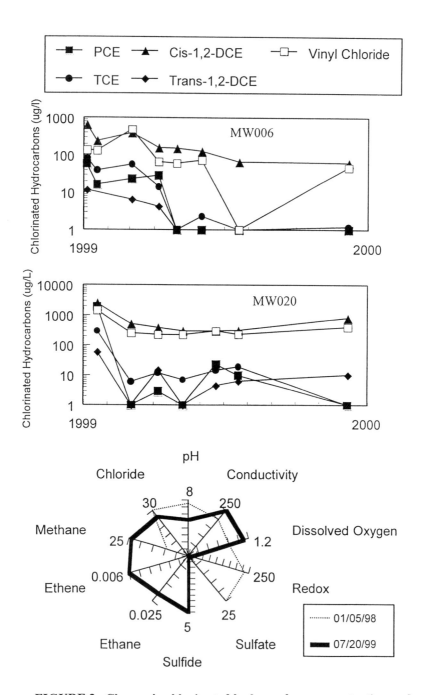

FIGURE 2. Change in chlorinated hydrocarbon concentration and groundwater geochemistry during treatment.

REFERENCES

Draft Region 4 Approach to Natural Attenuation of Chlorinated Solvents (Undated)

Wiedemeier, T. H., M. A. Swanson, D. E. Moutoux, E. K. Gordon, J. T. Wilson, B. H. Wilson, D. H. Kampbell, P. E. Haas, R. N. Miller, J. E. Hansen, F. H. Chapelle, 1998, Technical Protocol for Evaluating Natural Attenuation of Chlorinated solvents in Ground Water. US EPA Office of Research and Development, Washington, DC. EPA/600/R-98/128

REMEDIAL ACTION USING HRC UNDER A STATE DRY CLEANING PROGRAM

David Anderson (Oregon DEQ, Portland, Oregon, USA)
Mark Ochsner (Ecology and Environment, Inc., Portland, Oregon, USA)
Craig Sandefur (Regenesis, Inc., San Clemente, California, USA)
Steve Koenigsberg, (Regenesis, Inc., San Clemente, California, USA)

ABSTRACT: Hydrogen Release Compound (HRC™) was installed at a dry cleaning site in May 1999 to help accelerate and promote biodegradation of chlorinated compounds under the Oregon Dry Cleaning Fund. HRC was chosen because it is cost effective, is easy to install, and requires little or no operation and maintenance. Cost effectiveness is important due to limited funding available from the Oregon Dry Cleaning Fund. Groundwater monitoring has taken place at the site to ensure that chlorinated compounds are degrading. The monitoring has revealed that chlorinated compounds are degrading, but slower than originally anticipated. The slow degradation rate may be attributed to the relatively flat groundwater gradient across the site for approximately 6 months following HRC injection and for approximately six months until fall, when wet weather returned. Closure of the site within a year had been anticipated by the Department of Environmental Quality, but additional monitoring is still necessary. It is too early to determine if additional remedial measures may be necessary, or if additional HRC product is required. However, initial evaluation and positive mass removal and degradation of chlorinated solvents may have been enough for DEQ to consider HRC use as a state dry cleaner presumptive remedy.

INTRODUCTION

In 1995, Oregon's Environmental Cleanup Laws were revised to provide specific requirements and exemptions from cleanup liability for dry cleaning owners and operators. The State Dry Cleaner Fund is used to investigate and clean up contamination at dry cleaners. After DEQ determines a facility is eligible for funding, the order of investigations or cleanup efforts for all participating sites is determined. Priority is based on the potential risks to human health or the environment, and the availability of funds. DEQ funds and manages a group of contractors that provide the dry cleaner environmental services to DEQ. All of the funding for investigation and cleanup of the facilities comes from the Dry Cleaner Fund. Hayden Island Cleaners is one of the facilities accepted into the Dry Cleaner Program.

Hayden Island Cleaners is an active dry cleaning facility located on Hayden Island, an island located in the Columbia River, north of the city of Portland, Oregon. The groundwater flow direction and gradient at the site change depending on time of year and because the shallow groundwater at the site is in direct communication with the river. Responses related to river stage and tidal changes are observed. The result is that groundwater beneath the site fluctuates in direction and magnitude.

Investigations at the site indicate that soil and groundwater have been impacted by chlorinated solvent contamination near the dry cleaning building. Tetrachloroethene (PCE) contamination occurs in soil at concentrations of 1,500 micrograms per kilogram (μg/kg) beneath the facility. Shallow groundwater has been impacted by PCE contamination above the Federal Drinking Water Maximum Contaminant Level (MCL) of 5 micrograms per liter (μg/L) and the United States Environmental Protection Agency (EPA), Region IX, Preliminary Remediation Goal (PRG) of 1.1 μg/L. Trichloroethene (TCE) and cis-1,2-dichlorothene (cis-DCE) have also been detected in the groundwater. The highest concentrations of PCE have historically been detected in MW-1, located near the northwest corner of the site (Figure 1).

Hydrogen Release Compound (HRC), manufactured by Regenesis, of San Clemente, California was used and evaluated as an Interim Remedial Action Measure (IRAM) to help accelerate and promote biodegradation of chlorinated compounds. HRC was chosen because it is cost-effective, is easy to install, and requires little or no operation and maintenance. HRC is also being considered by the DEQ as a presumptive remedy in the State of Oregon Dry Cleaning Fund.

MATERIALS AND METHODS

Description of Technology. HRC is a passive, treatment option for in-situ bioremediation of chlorinated aliphatic hydrocarbons (CAHs). HRC is a proprietary, environmentally safe, food quality, polylactate ester specially formulated for slow release of lactic acid upon hydration. Bioremediation with HRC is a multi-step process. Microbes metabolize the lactic acid released by HRC, and produce hydrogen, which can be used by reductive dehalogenators, which are capable of dechlorinating CAHs.

Bench Scale Testing. To determine if indigenous bacteria at the site were capable of reducing PCE, soil samples were collected for bench scale testing. A representative soil sample was collected from the saturated zone beneath the dry cleaners site at approximately 25 feet below ground surface (BGS), at an interval where HRC could be applied.

The soil sample was shipped to Applied Power Concepts, Inc., (APC) laboratory where the sample was split into six discrete samples for the bench scale testing. The samples were tested using a protocol designed by Regenesis and outlined in their *Bench Scale Experiments* manual. The bench scale test provides an accelerated response to anaerobic remediation. The focus of the test was to determine whether the soil contains a population of bacteria that are not only suitable to perform the remediation, but could also respond to an increase in the carbon compound biochemical energy and the hydrogen generated from HRC.

Field Application. Based on the results of the bench scale testing, the site configuration, the groundwater flow direction and gradient, and the groundwater sampling data, two barrier arrays were determined to be able to provide the best coverage of HRC at the site. HRC was installed at the Hayden Island Cleaners site in May 1999. A standard van mounted Geoprobe® rig equipped with a 1.25-inch outside diameter drill rod was used to inject the HRC. The HRC was pumped from buckets, through an application hose, and injected through the drill rods using a specially designed pump provided by Regenesis. The HRC was placed along the entire vertical saturated interval of the aquifer from approximately 25 to 40 feet BGS.

An 80-foot barrier array was installed downgradient of MW-1 along the northern property boundary. The array consisted of 34 HRC delivery points, offset in two rows, and spaced at 5-foot intervals (Figure 1). Forty-eight pounds of HRC was injected per hole (3.2 pounds per vertical foot [lbs./vertical foot]) for a total of 1,680 pounds. The 3.2 lbs./ vertical foot was at the low end of an application range of 2 to 10 lbs./vertical foot as defined by Regenesis. Monitoring well MW-1 was used as a monitoring point along this northern barrier array.

The second barrier array was oriented along the south portion of the property. This array consisted of eight injection points, spaced at 5-foot intervals (Figure 1). Forty-eight pounds of HRC was injected per hole (3.2 lbs./vertical foot) for a total of 624 pounds. Monitoring well MW-3 was used as a monitoring point along this southern barrier array.

Groundwater Monitoring. Baseline groundwater samples were collected from six upgradient and downgradient monitoring wells prior to HRC installation. Groundwater samples were analyzed using EPA Method 8260 for Volatile Organic Compounds (VOCs). Natural attenuation parameters were also analyzed and included total and dissolved iron, nitrate, nitrite, sulfate, sulfide, methane, ethane/ethene, chloride, phosphorous, and manganese. In addition, field chemical parameters were collected for dissolved oxygen (DO), oxidation/reduction (redox) potential, pH, and temperature.

Figure 1. Site Layout, HRC Probe Points

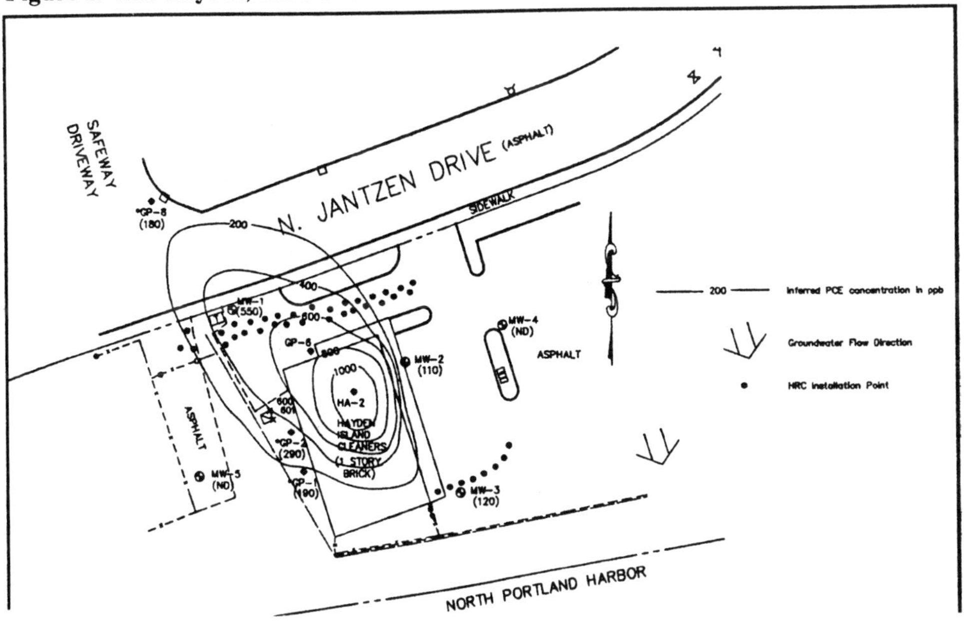

RESULTS AND DISCUSSION

Bench Scale Testing Results. The bench scale samples were run from 11/9/98 to 12/4/98. TCE was added to the soil samples at concentrations of 10 mg/L and 25 mg/L along with 1.5 grams of HRC (sorbitol polylactate). The results of the test indicated that significant reductions of TCE were observed in all of the samples nine days into the test, including the production of cis-DCE and vinyl chloride (VC). By the end of the test, the sample set that began with a low TCE value (10 mg/L) had been reduced to an average of 1.00 mg/L. The sample set with the highest TCE value (25 mg/L) was reduced to an average value of 5.29 mg/L. However, no degradation of TCE was indicated after the first 15 days of the test. Cis-DCE and VC still were produced and apparently degraded for the length of the test.

Results of organic acid analysis (lactic and pyruvic) indicated an increase in acid concentration, especially lactic acid, indicating that HRC was continuing to be available for degradation. Bacterial counts of the samples indicate a healthy population of aerobic, anaerobic, and sulfate reducing bacteria in the subsurface soils at the site.

Based on comparisons of the bench scale testing data to other test data, the Hayden Cleaners site was determined to have suitable microbes in the subsurface to be a good candidate for a pilot field test.

Groundwater Monitoring Results. Groundwater samples were collected from site monitoring wells prior to and following placement of the HRC product.

PCE concentrations have decreased approximately 75% from the initial sampling event in March 1998 in monitoring wells MW-1 and MW-3 (Figure 2). The initial decrease in MW-1 and MW-3 may be attributed to dilution based on the fluctuations of the shallow groundwater. However, it is believed that biodegradation/natural attenuation of the parent PCE compound was occurring at a slower rate before installation of HRC. Concentrations in MW-2 have decreased approximately 30% and represent the slower natural background aquifer conditions.

Following injection of HRC, significant degradation of PCE were observed in the wells, particularly at MW-3, where the daughter degradation products, TCE and cis-DCE concentrations have increased and appear to be accumulating (Figure 3). Monitoring wells MW-1 and MW-2 showed a similar decrease in PCE, but a slower increase in TCE concentrations and no production of cis-DCE. Vinyl chloride has not been detected in any of the site wells.

A review of selected geochemical parameters measured at the site confirms that degradation is occurring (Table 1). Baseline conditions at the site indicate an aquifer with low DO (<1.5 mg/l) and redox conditions (<100 mV).

Nitrate concentrations at the site have been decreasing with time, indicating that the site aquifer is depleted with respect to nitrate. Total iron and total manganese concentrations have been increasing with time, indicating iron and manganese reducing environments. Sulfate concentrations have increased with time, have stabilized, and are beginning to decrease, indicating possible sulfate reduction. The presence of methane beginning approximately three months after HRC injection may indicate that the terminal electron donor processes for nitrate, iron, and sulfate have begun to be depleted and that a move toward methanogenic conditions is occurring.

Figure 2. PCE Concentrations in Monitoring Wells

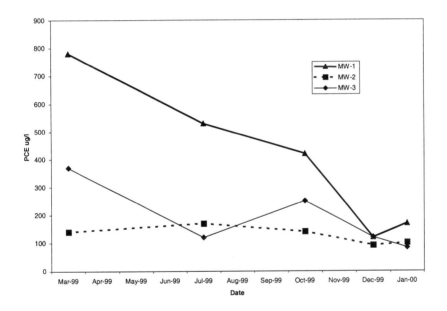

Figure 3. Volatile Organic Compounds Detected in MW-3

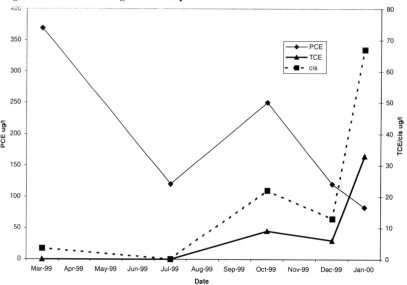

Table 1. Selected Geochemical Parameters in Groundwater (mg/l)

Well	Date	Nitrate	Iron (Total)	Mn (Total)	Sulfate	Methane
MW-1	12/28/98	2.1	0.315	NA	18	NA
	07/12/99	9.2	0.288	0.0635	24	ND
	10/07/99	6	0.012	0.0458	22	ND
	12/17/99	0.7	0.124	1.39	23	0.0042
	01/18/00	ND	0.667	3.04	24	0.0005
MW-2	12/28/98	2.3	NA	NA	6	NA
	07/12/99	4.5	0.081	0.0233	17	NA
	10/07/99	5.2	0.077	0.0159	14	NA
	12/17/99	0.4	0.148	0.0526	8.5	0.0224
	01/18/00	1.2	0.02	0.0481	7.7	ND
MW-3	12/28/98	2.1	NA	NA	4	NA
	07/12/99	0.5	0.997	0.0688	36	ND
	10/07/99	ND	3.95	1.27	22	0.11
	12/17/99	0.2	7.91	1.29	30	0.504
	01/18/00	0.25	8.74	1.28	25	0.612

NA = not analyzed. ND = Not detected

Project Costs. The costs associated with the HRC installation are outlined below. HRC was selected as an IRAM for the Hayden Island site with the primary objective to increase contaminant mass removal and reduce remaining concentrations to below risk-based concentrations. Also, DEQ was interested in reviewing HRC as a cost-effective means compared to other, more common remedies such as soil vapor extraction (SVE) and air sparging. Following are the actual costs incurred at the site, excluding site investigation activities, contractor costs, and DEQ oversight:

HRC product	$6.00 per pound x 2,310 pounds = $14,000
Regenesis Pump Rental	$900
Bench Scale Testing	$2,000 per sample
Drilling Subcontractor	$4,000
4 quarters of monitoring	$10,000
Total Costs	**$31,000**

Because of the limited amount of funding the state receives to investigate and cleanup dry cleaner sites, DEQ is evaluating the use of HRC as a presumptive remedy to help manage site prioritization, characterization, and eventual cleanup or no further action determination. At a similar dry cleaning site considered for HRC use, DEQ plans to spend an estimated $150,000 for installation of a multiphase extraction system. If successful in achieving site closure at the Hayden Island site, HRC would save approximately $120,000.

CONCLUSIONS. Specific conclusions that can be drawn from the use of HRC at the Hayden Island site are as follows:

- Shallow groundwater contaminated with PCE has shown degradation to TCE and cis-DCE;
- A general decline in PCE was observed across the site before installation of HRC, probably because of groundwater flushing and dilution from the adjacent river;
- Approximately seven months following HRC installation, a reduction of PCE by 75 % at the site was observed in wells MW-1 and MW-3 adjacent to the HRC injection arrays;
- Degradation rates observed in MW-2 are indicative of natural degradation rates for the site;
- Approximately seven months following HRC installation, an increase in associated production of daughter products (TCE and cis-DCE) is observed at the site;
- Based on the apparent success of HRC to reduce contaminant mass and stimulate degradation of chlorinated solvents, DEQ likely will continue to use HRC as a presumptive remedy.

ENHANCED REDUCTIVE DECHLORINATION OF ETHENES LARGE-SCALE PILOT TESTING

James A. Peeples (Metcalf & Eddy, Inc., Columbus, Ohio)
Joseph M. Warburton (Metcalf & Eddy, Inc., Columbus, Ohio)
Ihsan Al-Fayyomi (Metcalf & Eddy, Inc., Columbus, Ohio)
James Haff (Meritor Automotive, Heath, Ohio)

ABSTRACT: A shallow sand and gravel aquifer at an industrial site in central Ohio is contaminated with cis-1,2-dichloroethene (cDCE) and vinyl chloride (VC). Multi-phase field-scale pilot testing of enhanced reductive dechlorination began in November 1996. Pilot testing results indicated that reduction of cDCE and VC to concentrations at or near MCLs occurred within six months. Methods to expand and control the distribution of amendments in the shallow aquifer were developed. During the most recent pilot testing, a single amendment injection location was used to provide 90% or greater dechlorination within an aquifer volume of approximately 1000 yd^3 (763 m^3) with 50% or greater dechlorination achieved in 17,400 yd^3 (13,282 m^3) of aquifer.

INTRODUCTION

Chlorinated solvents are a common constituent of groundwater contaminant plumes. In situ reductive dechlorination of PCE/TCE to ethene by iron-reducing, sulfate-reducing, and methanogenic bacteria has been shown to be an effective remediation method in laboratory and field-scale studies (Beeman et al., 1994, de Bruin *et al.*, 1996, Bradley and Chappelle 1997). The presence of relatively high concentrations of cDCE and VC in groundwater at an industrial site in Ohio indicated that partial dechlorination of PCE and/or TCE had occurred. The primary source areas for chlorinated solvents and their daughter products are four former wastewater lagoons (Figure 1). The lagoons contained wastewater generated by metal machining operations from the early 1950s through 1985. The wastewater contained PCE and TCE, cutting oils, and other compounds. The liquid level in the lagoons was higher than the potentiometric surface of the underlying sand and gravel aquifer, and downward movement of lagoon waters and dissolved constituents occurred through the base of the lagoons. The lagoons were closed and filled in 1987.

A remedial investigation completed in 1993 described the subsurface hydrogeology, defined probable source areas, and delineated an area of approximately 38 acres (15.4 hectares) that was impacted by chlorinated solvents (Figure 1). Groundwater flow within the shallow sand and gravel aquifer is generally to the east, with an average gradient of 2 X 10^{-3} in the lagoon area (Figure 1). Depth to groundwater is 11 feet (3.4 m) onsite, the saturated thickness of the unit is 25 feet (7.6 m), and the hydraulic conductivity is approximately 4 X 10^{-2} cm/sec.

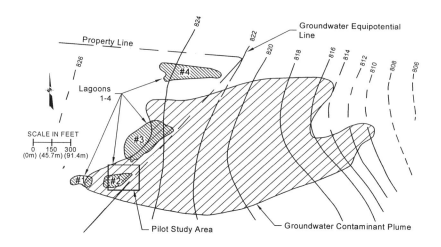

FIGURE 1. Site Area Layout

Groundwater samples were obtained to determine if reductive dechlorination of VC to ethene was occurring naturally within the aquifer. The analytical results provided evidence that reductive dechlorination to ethene was occurring at a low rate. Microcosm studies were conducted, using soil and groundwater from the site, to confirm the presence of the complete reductive dechlorination pathway and to evaluate the potential to enhance the process with amendments. Molasses and sulfate were chosen as amendments based on performance and cost. Field-scale pilot tests were then conducted to evaluate the potential to enhance the reductive dechlorination process in situ, to develop methods to increase the volume of treated aquifer, and to manage well biofouling and other issues associated with the injection process. The pilot testing conducted at this site is the subject of this paper.

MATERIALS AND METHODS

A multi-phase approach was used to implement field-scale testing of enhanced in-situ reductive dechlorination. The Phase I test confirmed the capability of enhancing the reductive dechlorination process in situ. The Phase II test evaluated amendment distribution strategies, and the Phase III test evaluated the maximum volume of aquifer that could be treated from a single injection location.

Phase I. The first phase of field-scale testing was conducted from November 1996 through August 1998. The purpose of this test was to demonstrate the ability to enhance reductive dechlorination in-situ, and to determine if treatment to low concentrations could be achieved. The Phase I pilot system consisted of an extraction well (EW-1), an injection well (IW-1) and an observation well OW-1

(Figure 2). Groundwater was extracted from EW-1, amended with sodium sulfate, molasses, ammonium chloride, and monopotassium phosphate, and reinjected into IW-1. The amended groundwater traveled through the aquifer, distributing nutrients and a reduced carbon source to the microbial community. This groundwater was partially captured by EW-1 and recirculated. The effectiveness of the system was assessed in terms of the degradation of cDCE and VC, the formation of ethene and ethane, and the appropriate distribution of nutrients and reducing conditions.

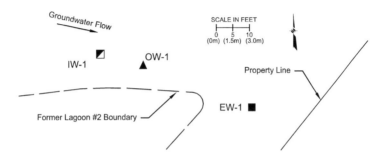

FIGURE 2. Phase I Test, System Layout

Phase II. A second phase field-scale pilot test was implemented from September 1998 through April 1999. The objectives of the test were to apply the enhanced reductive dechlorination process to a larger aquifer volume, to develop methods to alleviate well clogging, and to evaluate treatment efficiency within a broader area. Injection wells IW-3 and IW-4, and observation wells OW-3 through OW-11 were installed (Figure 3). The injection wells were located 6 ft (1.8 m) apart, along a line in the approximate direction of groundwater flow. Sodium sulfate

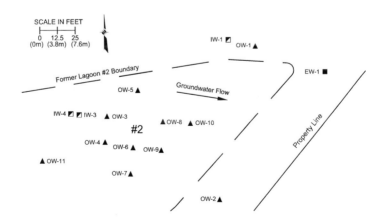

FIGURE 3. Phase II Test, System Layout

and molasses were added to the downgradient injection well (IW-3) and groundwater was injected into the upgradient well (IW-4). The injection of groundwater into IW-4 was used to move the injected nutrients away from IW-3, controlling amendment distribution and reducing biofouling in this well. Groundwater from RRW-2 (average concentration of cDCE 78 µg/L, VC 140 µg/L) and from EW-1 (average concentration of cDCE 109 µg/L, VC 160 µg/L) was injected into IW-3 at 6 gpm (23 L/min) and into IW-4 at 3 gpm (11 L/min). Nutrient injection and groundwater monitoring continued for 8 months.

Phase III. To evaluate the application of the two-well injection system to a full-scale operation, the injection method was modified for the Phase III field-scale test. During this phase, amendments were injected in varying proportions to IW-3 and IW-4, and the relative rate of injection was varied. The relative rate of groundwater injection into IW-3 and IW-4 was used to control the areal distribution and concentration of amendments within the treatment zone, and to reduce biofouling in the injection wells. A combined average injection rate of 10 gpm (38 L/min) was utilized. Groundwater used for injection into IW-3 and IW-4 was pumped from RW-6, a well located approximately 560 ft (171 m) northeast of EW-1. Pumping from RW-6 began in May 1999. The injected groundwater contained average cDCE and VC concentrations of 412 µg/L and 224 µg/L.

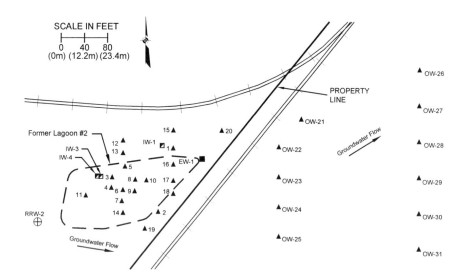

FIGURE 4. Phase III Test - System Layout

Wells OW-12 through OW-20 were installed in June 1999 (Figure 4), expanding the monitoring network used in the Phase II study to accommodate the expected increase in the size of the treatment area. Wells OW-21 through OW-31 (Figure 4) were installed in the off-site area east of the former Lagoon 2 in

September 1999 to evaluate the effects that the on-site reductive dechlorination process will have on downgradient locations in the aquifer.

IN SITU PILOT TESTING - RESULTS AND DISCUSSION

Phase I. Figure 5 provides a summary of the relative concentrations of cDCE, VC, and ethene and ethane at monitoring well OW-1 during the Phase I test. The concentrations are plotted as a percentage of the total micromolar concentrations of these constituents. The concentrations of cDCE and VC in groundwater collected from OW-1 were initially 265 µg/L (2.73 µmolar) and 15 µg/L (0.24 µmolar), respectively. Prior to the start of the pilot test, 91.9 percent of the total micromolar concentration was cDCE and 8.1 percent was VC. No detectable concentrations of ethene or ethane were present at the start of the test. The relative contributions of cDCE, VC, and ethene/ethane were 6.1, 1.3, and 92.6 percent, respectively, 251 days after the start of the test, indicating a nearly complete conversion of the chlorinated ethenes to ethene/ethane. Concentrations of cDCE were reduced to less than 25 µg/L, and VC concentrations to less than 4 µg/L after 380 days.

The declining concentrations of cDCE and VC at OW-1 occurred despite the continued introduction of groundwater to IW-1 with an average cDCE concentration of 413 µg/L and an average VC concentration of 150 µg/L. VC and cDCE were largely degraded in the aquifer between the injection well (IW-1) and the monitoring well (OW-1).

The Phase I test demonstrated that reductive dechlorination could be enhanced in the shallow aquifer beneath the site and that the process could be

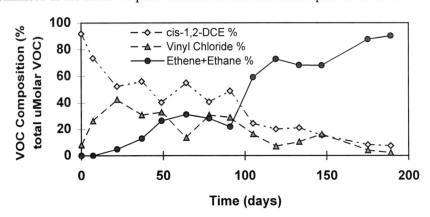

FIGURE 5. Relative Concentrations of cDCE, VC, and Ethane/Ethene, Phase I Test

applied in situ to reduce the concentrations of cDCE and VC to below their respective MCLs. The test was limited to a relatively small treatment area, and the effectiveness of treatment was measured in only one location (OW-1). The test indicated that batching nutrients slowed the process of injection well clogging, and that recirculation aggravated well clogging. Management of injection well clogging during the test was accomplished by regular cleaning using pump and surge techniques.

Phase II. Total Organic Carbon (TOC) analysis was used to track the distribution of an electron donor supplied by the injected molasses. The area where TOC increased in groundwater during the Phase II injections is shown in Figure 6. The area was approximately 180 ft (55 m) long, and 110 ft (34 m) wide at a distance of 110 ft (34 m) downgradient from IW-3. Increases in concentrations of ferrous iron and methane, and decreases in concentrations of sulfate occurred within this area, indicating that methanogenic, iron-reducing, and sulfate-reducing bacteria were being stimulated by the injected amendments.

Reductive dechlorination of cDCE and VC occurred in most of portions of the aquifer reached by amendments (Figure 6). Treatment proceeded to near detection limits in portions of the treatment zone, while some degree of dechlorination occurred over most of the amendment distribution area. At least 90% dechlorination occurred over an area of approximately 1200 ft^2 (111 m^2), equivalent to 1100 yd^3 (841 m^3) of the aquifer. At least 50% dechlorination was achieved in approximately 5,300 yd^3 (4052 m^3) of the aquifer.

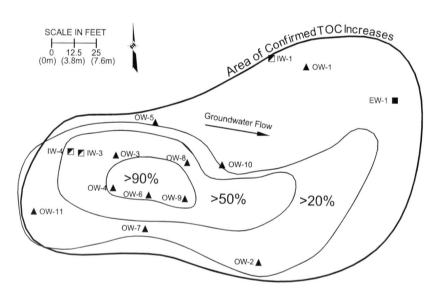

FIGURE 6. Distribution of TOC and Percent Dechlorination, Phase II Test

Clogging of IW-3 was minimized by the injection of groundwater into IW-4. The Phase II test ran for 26 weeks without shutdown, and IW-3 remained open and operable throughout the test. Amendment injection was discontinued from April 26, 1999, to June 29, 1999, while groundwater from RW-6 was injected into IW-3 and IW-4. It was assumed that the injection of this ground water, which contained cDCE (average 420 µg/L) and VC (average 220 µg/L), would recreate a VOC plume within the area downgradient of IW-3 and IW-4. This did occur in areas on the periphery of the Phase II injection zone, but within the area of the Phase II treatment zone, efficient VOC dechlorination continued without amendments.

Phase III. The area treated by the Phase III injections (Figure 7) was at least 180 ft (55 m) downgradient from IW-3, and was at least 120 ft (37 m) wide at a

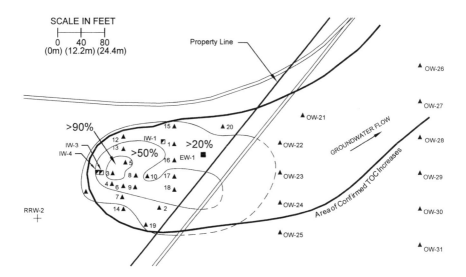

FIGURE 7. TOC Distribution and Percent Dechlorination, Phase III Test

distance of 40 ft (12 m) downgradient of the injection wells. This compares with a width of approximately 60 ft (18 m) achieved at this distance during the Phase II study. Reductive dechlorination of cDCE and VC occurred in most of the aquifer where amendments had been delivered (Figure 7). Treatment occurred to near detection limits in portions of the treatment zone, with 50% or greater dechlorination occurring over an area of approximately 18,700 ft^2 (1738 m^2), a volume of 17,400 yd^3 (13,282 m^3). This is more than 3 times the volume treated to this level in the Phase II test.

TOC concentrations in wells OW-21, OW-22, OW-23, OW-24, OW-26, and OW-27 indicate that amendments are moving northeast in the immediate off-site area. Concentrations of cDCE are generally lower in the north portion of the off-site well grid, with levels less than 10 µg/L in wells OW-21, OW-22, OW-26, OW-27, and RM-16. Vinyl chloride concentrations vary, but the general trend is for lower levels in the northern part of the grid, with the exception of OW-23.

Preliminary analysis of ethene distribution in the off-site area suggests that active reductive dechlorination enhanced by on-site amendment injection is occurring in the off-site area.

CONCLUSIONS

The following conclusions have been made based on the pilot testing conducted to date:

- Biologically mediated reductive dechlorination can be enhanced in the impacted aquifer by delivering the appropriate amendments to the aquifer at the appropriate concentrations;
- Following the creation of favorable conditions in the aquifer, the reductive dechlorination process is rapid and effective;
- Removal of cDCE and VC to levels at or near the MCLs is possible at this site using biologically mediated reductive dechlorination;
- A two-well injection system can be used to increase and control the areal distribution of the treatment zone;
- The efficiency of injection can be maximized by operating a two-well injection system, batching nutrients, and obtaining injection water from outside the treatment area;
- Full-scale implementation of enhanced reductive dechlorination at this site is feasible.

REFERENCES

Beeman, R. E., J. E. Howell, S. H. Shoemaker, E. A. Salazar, and J. R. Buttram, 1994. A field evaluation of in situ microbial reductive dehalogenation by the biotransformation of chlorinated ethenes. pp 14-175 In: Bioremediation of Chlorinated and Polycyclic Aromatic Hydrocarbon Compounds, R. Hinchee, *et al.* 1994 Lewis Publishers, Boca Raton, Florida.

Bradley, P.M. and F.H. Chapelle. 1997. Kinetics of cis-1,2-DCE and vinyl chloride mineralization under methanogenic and Fe(III) reducing conditions. Environ. Sci. Technol. 31:2693-2696.

deBruin, W.P., J. Michiel, J. Kotterman, M.A. Posthumus, G. Schraa, and A.J.B. Zehnder. 1992. Complete biological reductive transformation of tetrachloroethene to ethene. Appl. Environ. Microbiol. 58:1996-2000.

ENHANCED ANAEROBIC *IN SITU* BIOREMEDIATION OF CHLOROETHENES AT NAS POINT MUGU

Daniel P. Leigh (the IT Group, Concord, California, USA)
Christian D. Johnson, Rodney S. Skeen, and Mark G. Butcher (Battelle Pacific Northwest Division, Richland, Washington, USA)
Lisa A. Bienkowski (the IT Group, Irvine, California, USA)
Steve Granade (U.S. Navy, Point Mugu, California, USA)

ABSTRACT: A pilot-scale test of accelerated *in situ* bioremediation (ISB) of chlorinated ethenes is underway at a former underground storage tank site at the Naval Air Station Point Mugu in southern California. Groundwater at the site had been affected by a discharge of chloroethenes, predominantly TCE. Lactic acid was circulated throughout the test portion of the contaminated aquifer in regular, high concentration pulses to stimulate sulfate reduction and chloroethene biodegradation. A final large pulse of lactic acid was injected after 57 days of groundwater recirculation. Recirculation was halted at day 64 and long-term monitoring has been occurring since then; to date, the pilot test has been operating for 16 months. Complete reduction of the sulfate in the treatment area occurred within the first 60 days. Upon removal of the sulfate TCE and DCE were rapidly dechlorinated to VC. VC data are somewhat different than expected based on laboratory microcosms. Dissolved ethene indicates that the VC transformation rate is two orders of magnitude lower than expected based on laboratory microcosms with site sediment and groundwater. Other treatment alternatives, such as aerobic *in situ* bioremediation, are being considered to more quickly remediate the remaining VC.

INTRODUCTION

A former underground storage tank (UST) site at the Installation Restoration Program (IRP) Site 24 at Naval Air Station (NAS) Point Mugu in southern California has been the subject of a pilot-scale demonstration of accelerated *in situ* bioremediation for treatment of chlorinated ethenes in the groundwater. The initial work at UST Site 23 was described previously [Johnson et al., 1999]. For convenience, a brief overview of the site and the pilot-scale design will be repeated here.

Site 23 Characterization. UST Site 23 is located approximately 1 mile (1.6 km) inland from the Pacific Ocean. The water table is located about 5 feet (1.5 m) below ground surface (bgs) and the regional groundwater gradient is approximately 0.001 m/m to the south. Figure 1 is a cross section of the pilot-scale test site showing the approximate lithology. The wells used in this pilot test are screened in the sandy Zone B from 20 to 30 ft (6 to 9 m) bgs.

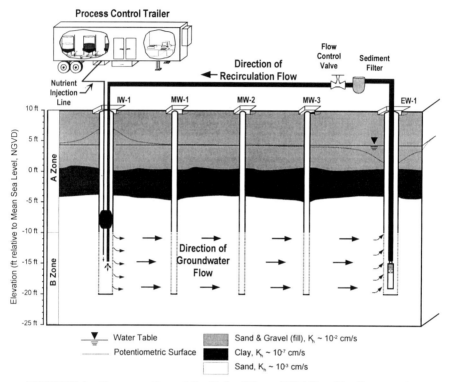

FIGURE 1. Cross-section of the Point Mugu UST Site 23 pilot test site depicting the approximate lithology and the pilot-scale system. K_h denotes hydraulic conductivity.

Historical records show that UST Site 23 was used for oil/water separation. The contaminants of concern at UST Site 23 are trichloroethene (TCE), dichloroethene (DCE; mostly cis-1,2-DCE), and vinyl chloride (VC). Sampling in August of 1997 found maximum concentrations in Zone B of approximately 1700 µg/L, 750 µg/L, and 1 µg/L for TCE, cis-1,2-DCE, and VC, respectively. Because the site is so near the ocean, seawater intrusion has resulted in chloride and sulfate concentrations in Zone B of approximately 5,000 mg/L and 700 mg/L, respectively.

A remediation alternative analysis in 1997 [Johnson et al., 1998] determined that UST Site 23 had too little organic carbon relative to the large mass of chlorinated ethenes and that natural attenuation would take too long to remediate the plume. A pilot test of accelerated *in situ* bioremediation (ISB) was implemented for UST Site 23 in the Fall of 1998 to demonstrate the effectiveness of the technology and to provide operating and cost information for evaluating a full-scale system.

Pilot Test Design. The pilot test design has been previously described [Johnson et al., 1999]. The RT3D code [Clement, 1997; Clement et al., 1998] was used to simulate reactive flow and transport to determine well placement and amendment addition strategy. The pilot-scale system consists of an extraction/injection well pair (EW-1 and IW-1, respectively) and five monitoring wells (MW-1 to 5). Three monitoring wells are in-line between the extraction and injection wells and two monitoring wells are offset. Figure 2 shows the well locations in a plan view of the portion of IRP Site 24 where the pilot-scale test is occurring. As shown in Figure 1, groundwater was circulated in a closed loop from EW-1 to IW-1 and through the contaminated aquifer. For this pilot test, groundwater was recirculated at a flowrate of 10 gpm (0.63 L/s).

Operation of the accelerated ISB pilot test included two nutrient injection phases. Initial laboratory microcosm experiments performed at Oregon State University using site sediments and groundwater had indicated that reductive dechlorination at this site will not occur under sulfate reducing conditions [Semprini, 1998, personal communication]. Hence, the objective of the first phase was to reduce the sulfate to a concentration where reductive dehalogenation would effectively proceed. From December 23, 1998 to February 18, 1999 (day 57), approximately 17 gallons per day of 88% lactic acid was injected with the recirculated groundwater.

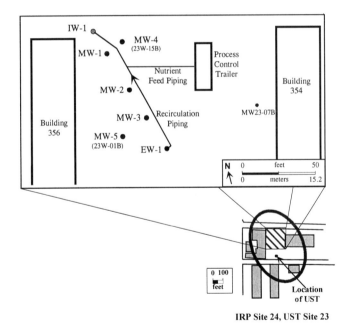

FIGURE 2. Plan view of IRP Site 24 with enlargement of the portion of UST Site 23 where the ISB pilot test is being conducted. Contour lines on the view of IRP Site 24 indicate the approximate extent of chlorinated ethene plumes in the groundwater.

The objective of the second nutrient injection phase of the pilot test was to provide the necessary conditions for rapid reductive dechlorination. The second phase began on February 18, 1999 and is presently ongoing. A high concentration pulse of lactic acid was injected at the start of this second phase. Once the material reached MW-3 on day 64, recirculation was discontinued. The remainder of this phase consists of long-term monitoring of chloroethenes, dissolved gases (ethene, ethane, and methane), sulfate, propionate, and acetate in the groundwater.

RESULTS AND DISCUSSION

In less than 60 days after the start of nutrient injections, sulfate concentrations were reduced from about 700 mg/L to less than 25 mg/L in MW-1, MW-2, and MW-3, as shown in Figure 3. A spike of sulfate appears around day 65 because groundwater was inadvertently recirculated without addition of lactic acid when a piece of equipment failed. Subsequent sulfate concentrations have remained consistently less than 5 mg/L from day 80 onwards.

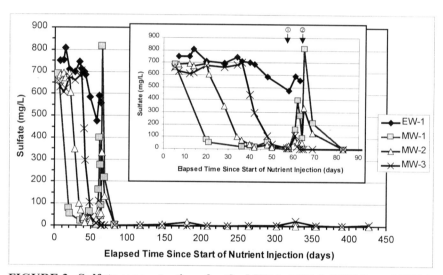

FIGURE 3. Sulfate concentrations for the MW-1, MW-2, MW-3, and EW-1. The arrow on the left (day 57) is when the high concentration pulse of lactic acid was injected. The arrow on the right (day 64) is when recirculation of groundwater was discontinued. Where error bars are not shown, they are within the size of the symbol.

As sulfate reduction occurred near the injection well and the zone of no sulfate proceeded to expand to MW-2 and MW-3, excess organic acids began to be detected. Figure 4 shows propionate and acetate concentrations at wells MW-2 and MW-3. By day 181, all organic acids were consumed. Lactate was rapidly metabolized, as evidenced by the fact that concentrations did not exceed 15 mg/L in any of the wells during the first 60 days.

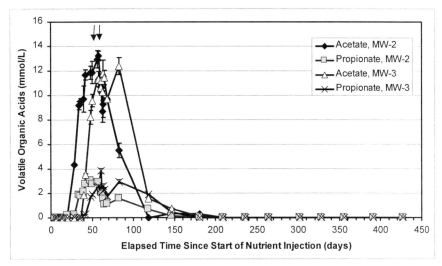

FIGURE 4. Propionate and acetate concentrations for wells MW-2 and MW-3. The arrow on the left (day 57) is when the high concentration pulse of lactic acid was injected. The arrow on the right (day 64) is when recirculation of groundwater was discontinued. Where error bars are not shown, they are within the size of the symbol.

Dissolved gas analysis results are shown in Figure 5 for methane and ethene plus ethane. Methane has maintained a fairly constant level from the time sulfate was reduced until present, supporting the notion that there is little groundwater flow at the site. The concentration of ethene (plus ethane) has risen steadily from the onset of dechlorination activity.

Upon removal of the sulfate, TCE was rapidly dechlorinated to DCE and subsequently to VC. A portion of the VC has been degraded to ethene. Figure 6 shows the concentration of chloroethenes and dissolved ethene at MW-3. The VC concentration profile is somewhat different than expected based on laboratory microcosm data (where site sediment and groundwater were used) [Keeling, 1999]. From the initial dissolved phase concentrations of TCE and DCE, it was expected that approximately 12 µmol/L of VC would be generated. The large increase in VC concentrations around day 180 and the subsequent decrease do not correspond to the ethene concentrations nor the laboratory microcosms. Aside from the peak at day 180, the VC concentrations have been relatively constant between 10 and 15 µmol/L. The chlorinated ethene concentrations at wells MW-1 and MW-2 (data not shown) mimic the concentrations at MW-3, but with concentrations approximately half of those found at MW-3.

The rate of VC dechlorination in the field is slower than the rate measured in the lactate-fed microcosm tests with site sediments and groundwater [Keeling, 1999]. The microcosms were reported to have a VC dechlorination rate in the range of 0.0125 µM/day to 0.034 µM/day (0.0025 µmol/day to 0.0068 µmol/day). Field data were analyzed using the change in ethene concentration as an indication of the VC transformation rate. From these calculations, the estimated VC transformation rate is $5.6 \cdot 10^{-4}$ µM/day in the field. This rate is two orders of magnitude lower than expected from the microcosms.

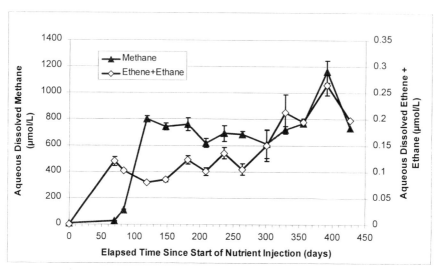

FIGURE 5. Concentrations of the dissolved gases methane and ethene plus ethane in groundwater at MW-3. Where error bars are not shown, they are within the size of the symbol.

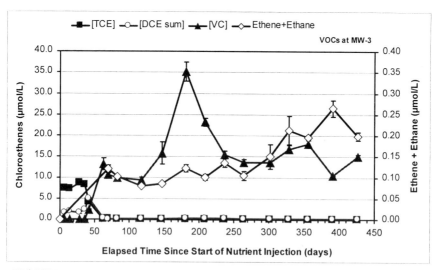

FIGURE 6. Concentrations of chlorinated ethenes and dissolved ethene in groundwater at MW-3. Where error bars are not shown, they are within the size of the symbol.

SUMMARY

The pilot test of accelerated ISB at NAS Point Mugu is currently underway. Injection of lactic acid into the aquifer has stimulated vigorous sulfate reduction throughout the treatment area. Rapid dechlorination of TCE to VC occurred upon reduction of the sulfate. Based on the ethene data, it appears that VC is being transformed at a rate of about $5.6 \cdot 10^{-4}$ µM/day, which is much less than expected based on laboratory microcosms. Other treatment alternatives, such as aerobic *in situ* bioremediation, are being considered to more quickly remediate the remaining VC.

REFERENCES

Clement, T.P. 1997. "RT3D - A Modular Computer Code for Simulating Reactive Multi-Species Transport in 3-Dimensional Groundwater Aquifers." Pacific Northwest National Laboratory, Richland, Washington. PNNL-11720. (http://bioprocess.pnl.gov/rt3d.htm)

Clement, T.P., Y. Sun, B.S. Hooker, and J.N. Petersen. 1998. "Modeling Multispecies Reactive Transport in Ground Water." *Ground Water Monitoring and Remediation*, 18(2):79-92.

Johnson, C.D., R.S. Skeen, D.P. Leigh, T. P. Clement, and Y. Sun. 1998. "Modeling Natural Attenuation of Chlorinated Ethenes Using the RT3D Code." In: *Proceedings of the Water Environment Federation 71st Annual Conference and Exposition, WEFTEC '98, Volume 3.* Water Environment Federation, Alexandria, Virginia. pp. 225-247.

Johnson, C.D., R.S. Skeen, M.G. Butcher, D.P. Leigh, L.A. Bienkowski, S. Granade, B. Harre, and T. Margrave. 1999. "Accelerated *In Situ* Bioremediation of Chlorinated Ethenes in Groundwater with High Sulfate Concentrations." In: *Engineered Approches for In Situ Bioremediation of Chlorinated Solvent Contamination*, A. Leeson and B.C. Alleman, eds. Battelle Press, Columbus, Ohio. pp. 165-170.

Keeling, M.T. 1999. *Bench-Scale Study for the Bioremediation of Chlorinated Ethylenes at Point Mugu Naval Air Weapons Station, Point Mugu California, IRP Site 24.* Oregon State University, Corvallis, Oregon. Thesis.

ENHANCED BIOREMEDIATION USING HYDROGEN RELEASE COMPOUND (HRC™) IN CLAY SOILS

Ms. Zahra M. Zahiraleslamzadeh (FMC Corporation, San Jose, CA)
Jeffrey C. Bensch (HSI GeoTrans, Rancho Cordova, CA)

ABSTRACT: This case study evaluates the pilot study application of Hydrogen Release Compound (HRC™) to enhance biodegradation of chlorinated solvents in groundwater beneath an active light-industrial property. The HRC application was successful by varying degrees in each test area as evidenced by the production of biodegradation daughter products and microbial end products. The success of the pilot test is being used to design and implement the full-scale in-situ remediation with minimal disturbances to ongoing business operations.

In this pilot study, three impacted areas of low, moderate, and high trichloroethylene (TCE) concentrations received HRC injections within the subsurface interval from 10- to 30-feet below ground surface. The HRC was applied in a silty clay soil matrix using direct push injection techniques. Site concerns for the HRC injection include: effectiveness enhancing the reductive dehalogenation process; horizontal distribution of HRC in the clay soils; and, vertical distribution of HRC from the injections. The area of moderate TCE concentration showed significant increased biodegradation due to the HRC application, while the areas of low and high TCE concentrations showed only some increased biodegradation. The variances are attributed to the heterogeneity of HRC migration and limited HRC available in specific areas. Areas of uncertainty under these site conditions include: understanding the duration of HRC effectiveness and the need for subsequent applications; and identifying an indicator parameter to monitor HRC migration and effectiveness.

INTRODUCTION

The biological reductive dehalogenation process of chlorinated solvents, such as tetrachloroethylene (PCE) and TCE, is an accepted viable groundwater remediation process. Various enhancements are available to stimulate biological activity and accelerate the dehalogenation process. Applying these enhancements to the subsurface for effective remediation can be difficult and uncertain. This paper presents the results of a pilot study that injected HRC into a silty clay soil to stimulate biodegradation of TCE in groundwater.

The Site is a 4.1-acre property with a 76,000 square-foot light-industrial retail building in a congested manufacturing neighborhood. The Site was used as agricultural land before 1960, then for various heavy manufacturing purposes through 1988. The building is currently leased to various light-industrial tenants.

Topography at the Site is essentially flat with existing grades designed to control storm water runoff. The Site was unpaved until approximately 1980. Currently, all unoccupied areas are paved.

Site soils are a relatively homogenous silty clay from ground surface to a depth of approximately 45 to 50 feet. A gravelly sand unit, approximately 30 to 35 feet thick, underlies the silty clay. Saturated soils are first encountered approximately 17 feet below ground surface in the silty clay; while the depth to groundwater is approximately 7 to 10 feet below ground surface in wells screened in the silty clay. The groundwater flow velocity in the silty clay is approximately 10 feet per year.

The Site is impacted with volatile organic compounds (VOCs) in soil and shallow groundwater in the northeastern corner of the property (Figure 1).

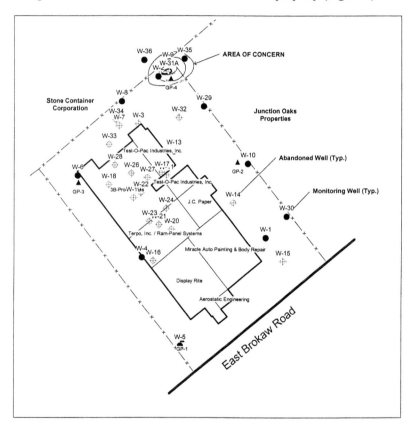

FIGURE 1. Site Features

Intrinsic biodegradation testing and evaluations were conducted in 1998. These tests indicated that biodegradation of VOCs occurred in the past, but was very slow, or in a dormant stage at the time of the evaluation. Various remediation alternatives were evaluated, and HRC injection was selected because the HRC application is a simple procedure with minimal impacts to the existing property and current business activities.

HRC is a food-grade polylactate ester that degrades to 92% lactic acid and 8% reduced sugars. The lactic acid becomes available to indigenous organisms that can metabolize it to produce hydrogen. This hydrogen serves as the electron donor for microbially-mediated reductive dehalogenation that can breakdown certain chlorinated hydrocarbons.

Anaerobic biodegradation of VOCs occurs in environments free of oxygen, where the parent chlorinated compound is progressively degraded into daughter products. The process at this Site is the degradation of TCE into subsequent daughter compounds: *cis*-1,2-dichloroethylene (*cis*-1,2-DCE), vinyl chloride (VC), and ethylene. Microorganisms mediate this process using chlorinated compounds as electron acceptors and a source of hydrogen as the electron donor.

In the anaerobic biodechlorination process, the *cis*-1,2-DCE isomer is produced preferentially to the *trans*-1,2, DCE isomer. As such, evidence for reductive dehalogenation of TCE can be obtained by observing the formation of *cis*-1,2-DCE in groundwater. Commercial formulations of TCE can have low levels of DCE with the ratio of the *cis/trans*- isomers being near one. Approximately 70% to 80% of the total DCE produced anaerobically is the *cis* isomer (Wiedemeier, et. al., 1996). *Cis/trans* ratios greater than one support the occurrence of anaerobic biodechlorination. The increasing concentrations of daughter product VC and microbial end product ethylene suggest the process is continuing toward completion.

The pilot study was designed to stimulate the reductive dechlorination process by introducing a critical mass of electron donor. The success of this process is evaluated by monitoring several key factors: the presence of dechlorinated daughter products; microbial end products; the microbial community structure; organic and inorganic parameters; dissolved oxygen; and oxidation reduction potential.

MATERIALS AND METHODS

The scope of pilot study activities included bench-scale testing using Site soils, and pilot-scale testing using field applications of HRC in three distinct areas.

Bench-Scale Testing. The bench-scale tests, performed by Applied Power Concepts, Inc., involved spiking Site soil samples with TCE, adding HRC, and monitoring chemical concentrations and microbial populations. The bench-scale test results indicated:

- TCE concentrations were reduced an order of magnitude from 10,000 microgram per liter (ug/L) to 1,100 ug/L and from 25,000 ug/L to 1,260 ug/L;
- 1,2-DCE and VC concentrations initially increased as TCE was dechlorinated;
- 1,2-DCE and VC concentrations declined as the reductive dechlorination process continued to completion;
- Lactic acid concentrations increased indicating breakdown of the HRC and available hydrogen to enhance the bioremediation processes; and

- Anaerobic microbial counts were high indicating a large native bacterial population capable of dechlorination.

Pilot-Scale Testing. The objectives of the pilot study were to evaluate:

1. The effectiveness of HRC to enhance the in-situ reductive dehalogenation biodegradation process;
2. The horizontal downgradient distribution of HRC in the silty clay soils; and
3. The vertical distribution of HRC using direct push injection.

Three adjacent areas were established to test the HRC application in areas of varying TCE concentrations. As shown in Figure 2, monitoring wells PW-2, W-2, and PW-3 are located in areas of approximately 1000, 3000, and 5000 ug/L TCE, respectively. Immediately upgradient of these wells are six HRC injection points: A-1 through A-3, and B-1 through B-3. Monitoring well PW-1 is the upgradient control well, and monitoring wells PW-4 and W-9 are downgradient control wells.

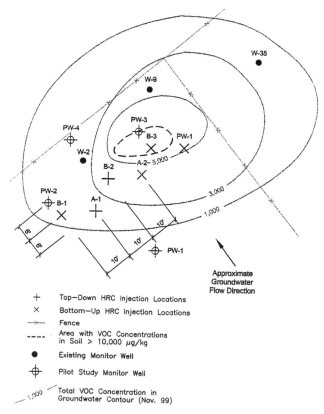

FIGURE 2. Pilot Study HRC Injection and Monitor Well Locations

HRC was injected into the subsurface on June 7, 1999. The direct push injection method involves hydraulically pushing or hammering a small diameter (1- to 2-inch) hollow steel probe into the subsurface. At designated depths the probe is withdrawn slightly to allow HRC injection using a high pressure pump attached to the direct push probe.

HRC was placed from 10 to 30 feet below ground surface at each of the six injection points. The injections were conducted using two methods: top-down injection and bottom-up injection. Top-down injection was used at points (A-1 and B-2). The direct push rod, with a 2-foot long perforated tip, was pushed to 12 feet below ground surface and approximately 1.6 gallons of HRC were injected. The rod was then pushed 2 feet farther and another 1.6 gallons of HRC were injected. This process continued to maximum depth of 30 feet below ground surface. Bottom-up injection was used at the remaining points (A-2, A-3, B-1, and B-3). The direct-push rod was pushed to the maximum depth and HRC was injected continuously while the rod was withdrawn. Approximately 0.8 gallons (9 pounds) of HRC were injected for every foot of the injection interval (from 30 feet to 10 feet below ground surface). Approximately 16 gallons of HRC were injected at each of the six points.

Groundwater samples for baseline testing were collected from the six monitoring wells prior to HRC injection. Monthly monitoring was conducted at W-2 for five months following HRC injection. Wells PW-1, PW-2, and PW-3 were monitored after the first quarter, then monthly thereafter. Wells PW-4 and W-9 were monitored after the first quarter and on the fifth month.

RESULTS AND DISCUSSION

Enhanced biodegradation occurred due to the HRC application. Monitoring wells W-2 and PW-4 demonstrated a substantial increase in the biodegradation of TCE to *cis*-1,2-DCE through VC to ethylene. PW-2 and PW-3 also showed some evidence of TCE biodegradation. Upgradient monitoring well PW-1 and downgradient monitoring well W-9 showed no signs of increased biodegradation activities. These wells provide further evidence that the increased biodegradation observed in other wells is due to the HRC application. The analytical results from the pilot study are summarized on Figure 3.

Monitoring Wells W-2 and PW-4 Indicate Biodegradation of TCE. Monitoring wells W-2 and PW-4 are located in an area with moderate TCE impacts and directly downgradient of top-down HRC injection points A-1 and B-2. TCE concentrations in this area historically ranged from 3,000 to 5,000 ug/L, although slightly lower and higher concentrations have been observed in W-2 during routine monitoring. The TCE daughter product concentrations detected in samples collected from W-2 in May 1998 and May 1999, prior to the pilot study, indicated little or no TCE breakdown. The *cis*-1,2-DCE to *trans*-1,2-DCE ratio (*cis/trans* ratio) prior to the pilot test was nearly equal to one, and the VC concentrations were less than 50 ug/L.

The organic acids and hydrogen produced by the HRC appear to have passed through W-2 during July and August 1999, approximately one to two months after the HRC injections. The organic acids historically and throughout the pilot tests were below detection limits, except during July 1999. The hydrogen concentrations generally increased from below 1 nano-molar (nM) to greater than 2 nM over the duration of the pilot test, with a peak concentration of 11 nM in August 1999.

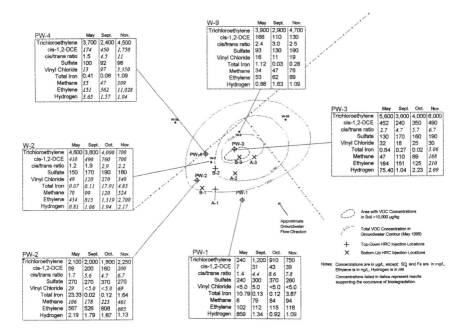

FIGURE 3. Pilot Study Monitoring Results

The TCE daughter product concentrations began increasing in September 1999. By November 1999, the *cis/trans* ratio was greater than 2 in W-2 and the VC concentration was 340 ug/L. In addition, the TCE concentration in W-2 was 700 ug/L, nearly an order of magnitude below historical levels. The November 1999 VOC concentrations from PW-4 provide additional evidence of significant biodegradation: the *cis/trans* ratio exceeded 10 and the VC concentration was over 3,000 ug/L.

The concentrations of microbial end products methane and ethylene also increased in wells W-2 and PW-4 during the pilot study. Ethylene concentrations increased from approximately 434 ug/L to 2,700 ug/L in W-2, and from 150 ug/L to over 11,000 ug/L in PW-4. These data indicate that the TCE biodegradation process is continuing through completion. Methane concentrations also increased, from approximately 70 ug/L to over 500 ug/L in W-2, and from 35 ug/L to over 300 ug/L in PW-4. The methane data indicate that organic compounds other than TCE are being degraded.

Other biodegradation parameters also indicated that the microbial activity increased due to the HRC application. Iron concentrations increased dramatically in W-2 during October and November 1999, indicating an increase in iron reducing bacteria activity. The sulfate concentrations, however, remained relatively high and consistent throughout the pilot study. Although the literature indicates this may hinder the biodegradation of VOCs, the data indicate the VOC reduction is occurring prior to substantial sulfate reduction.

Monitoring Wells PW-2 and PW-3 Show Some Indications of TCE Biodegradation. PW-2 and PW-3 exhibit signs of increased biodegradation, but have not reached the level of microbial activity present in wells W-2 and PW-4. This could be due to a variety of factors, including delayed growth of the microbial populations in these areas or limited elctron donors (HRC) available for the reductive dechlorination process.

PW-2 and PW-3 in the low and high TCE concentration areas, respectively, exhibited little or no TCE breakdown. The *cis/trans* ratios in each location, however, increased from near 1 to over 6. In addition, the VC concentration in PW-2 increased from non-detect to 69 ug/L, and the methane concentration in PW-3 increased from 47 ug/L to 167 ug/L during the period of the pilot test.

These encouraging results could be indications of increasing microbial activity and TCE reduction may be observed with continued monitoring. It is also possible that the HRC quantity was insufficient to further stimulate the reductive dehalogenation process. In either case, the results indicate that the HRC application had positive effects and a full-scale implementation is warranted.

Top-Down vs. Bottom-Up HRC Injection Methods. The top-down HRC injection method was used upgradient of monitoring well W-2, while the bottom-up injection method was used upgradient of monitoring wells PW-2 and PW-3. The bottom-up method is easier to implement than the top-down method because it is relatively straightforward to direct push the steel rods to the bottom of the injection interval without stopping. The rate of HRC injection can then be easily monitored while the direct push rods are removed.

The top-down method is more difficult than the bottom-up method because stopping the direct push process and attaching the HRC injection hose is time consuming. In addition, the injection pressures developed using the top-down approach caused breakthrough of the HRC up the sidewalls of the direct push rods. HRC was observed at the surface with each top-down attempt. After HRC breakthrough, the rods were pushed to the bottom of the injection interval and a bottom-up approach was used to complete the injection point.

Depth discrete sampling and field measurements were performed to evaluate the HRC vertical distribution differences between the top-down and bottom-up injection methods. The results indicated that there were no differences between the two methods. Dissolved oxygen and Eh measurements were essentially equal between depths in the three monitoring wells.

The top-down HRC injection method is intuitively preferred to the bottom-up method because it should create an even vertical distribution of HRC throughout the injection interval. The bottom-up method has the potential to apply a majority of the HRC at the bottom of the injection interval. Although there were field difficulties with the top-down approach and the depth discrete sampling did not demonstrate vertical distribution differences, the area of greatest biological activity, near monitoring well W-2, is downgradient of top-down injection points A-1 and B-2.

This suggests that continued improvements to the top-down method are warranted and this method should be used if it can be implemented economically and successfully.

CONCLUSIONS

The HRC application enhanced the biological reductive dehalogenation of TCE in the pilot test area. Of the three areas tested, one showed a substantial increase in biologic activity with a corresponding reduction in TCE concentration, while the other two showed only indications that reductive dehalogenation was occurring. The design of the full-scale implementation is under review by the lead regulatory agency. Full-scale implementation will provide passive in-situ remediation at an active light-industrial facility.

The pilot study also revealed areas of uncertainty in the application of HRC. The HRC effectiveness was documented; however, the duration of the HRC's influence has not been established. It is possible that subsequent HRC injections will be required to achieve Site remediation goals. Monitoring a full suite of analyses is currently necessary to understand the extent and rate of biological activity. An easily tested indicator parameter has not been established that will identify when the enhanced biological activity has slowed down to the point where another application of HRC is required. This type of indicator parameter would also facilitate identifying the extent of HRC migration and effectiveness.

REFERENCES

Wiedemeier, Todd H., M. Swanson, and D. Moutoux, E. K. Gordon; Drs. J. Wilson, B. Wilson, and D. Kampbell; J. Hansen and P. Haas; Dr. F. Chapelle, 1996. *Technical Protocol for Evaluating Natural Attenuation of Chlorinated Solvents in Groundwater.* Air Force Center for Environmental Excellence, November 1996.

A REDUCTIVE DECHLORINATION TREATABILITY STUDY OF A SHALLOW ALLUVIAL AQUIFER

Victoria Murt (Nebraska Department of Environmental Quality, Lincoln, Nebraska)
Theodore Huscher (Nebraska Department of Environmental Quality, Lincoln, Nebraska)
Diane Easley (U.S. EPA, Kansas City, Kansas)

Abstract: A treatability study was performed at the Ogallala Ground Water Contamination Site to determine if PCE could be reductively dechlorination in the shallow alluvial aquifer. An in-situ treatment cell was developed in a portion of the PCE plume downgradient of a dry cleaners. A substrate solution consisting of 60 % food grade sodium lactate was injected at intervals of 3 to 4 weeks to allow for adequate mixing and dispersion of the solution. Geochemical changes in the ground water were monitored to assess the REDOX conditions and to ascertain the best injection schedule. Reducing conditions developed in a step-wise fashion; denitrification followed by iron and manganese reduction, and finally sulfate reduction. Fermentation of the lactate was confirmed by the presence of volatile acids under methanogenic conditions, and an increase in bicarbonate. Although reducing conditions and microbial fermentation of the lactate indicate that conditions are appropriate, confirmation of reductive dechlorination cannot be confirmed at this time due to the initially low concentrations of PCE within the test cell and the lack of degradation product data.

INTRODUCTION

It is now recognized that standard pump-and-treat methods often require extensive cleanup time for chlorinated solvents and may not achieve water quality standards. Recently, several new innovative remediation technologies have been developed and are currently being tested or used at several sites in order to reduce the amount of time required for cleanup and the overall costs of the remedial action (ITRC, 1998). Other pilot studies have shown that enhanced biodegradation accelerates the cleanup process for chlorinated solvents such as perchloroethene (PCE) and trichloroethene (TCE). Degradation of PCE preferentially occurs under reducing conditions.

A pump and treat alternative was proposed in the 1996 Feasibility Study for the main PCE plume that is part of the Ogallala Ground Water Contamination Site. The costs were estimated at $1.5M and the system was expected to operate for a period of about 10 to 15 years. Based on information obtained on emerging technologies the Nebraska Department of Environmental Quality (NDEQ) and the Environmental Protection Agency (EPA) elected to perform a treatability study that would evaluate the reductive dechlorination process. A $100K grant was given to the NDEQ by EPA to perform the treatability study; the limited amount of funds available for the study resulted in a scaled down version of the treatment cell size, number of monitoring points and associated sampling activities.

The objective of the study was to assess enhanced reductive dechlorination and determine if it would be an appropriate remediation and cost effective alternative for the shallow alluvial aquifer. The first goal of the study was to determine if anaerobic conditions could be achieved within the cell, and if so, what substrate/nutrient requirements were necessary to maintain the indigenous anaerobes. The second goal of the study was to determine if appropriate electron donors were available in sufficient quantity and whether PCE was being degraded.

SITE BACKGROUND

Contamination of an Ogallala public water supply well was discovered during a 1989 quarterly sampling event conducted by the Nebraska Department of Health. Subsequent sampling indicated that five of the nine public water supply wells had been impacted by chlorinated volatile organic compounds (VOCs). NDEQ and EPA investigated several potential source areas through soil and soil-vapor sampling activities, and ground water monitoring studies. The Ogallala Ground Water Contamination Site was placed on the NPL in 1994. One of the identified potential source areas was the Tip Top Dry Cleaners located near Spruce and Fifth Street. PCE was detected in soil and soil-vapor samples. PCE has also been detected in the shallow unconfined alluvial aquifer downgradient of Tip-Top Dry Cleaners at concentrations that ranged from 0.78 ug/l to 1400 ug/l.

Site Geology and Hydrogeology

The City of Ogallala is underlain by a relatively thin layer of unconsolidated alluvial material which overlies the Ogallala Formation. Approximately 4 to 5 feet of top soil and fill material is present. The alluvial material generally consists of a silt to silty clay layer to a depth of about ten feet, followed by sand of variable grain size with occasional interbedded layers of silt or gravel to a depth of about 26 to 30 feet (7.9 m to 9.1 m) below ground surface (bgs). The top of the Ogallala Formation is delineated by the Ash Hollow Member which consists of a calcareous silty clay.

The alluvial aquifer in the treatment cell area consists of fine to medium grained sand overlying channel deposits that consist of coarse sands and gravels with a basal layer of cobbles, as shown in Figure 1. The Ash Hollow Member is encountered at 27 (8.2m) feet bgs. Depth to ground water is typically 11 feet (3.4m), with seasonal fluctuations of up to 2 feet (0.6 m). Ground water in the alluvial aquifer flows in a southeasterly direction with a relatively low horizontal gradient of about 0.002 ft/ft. The average linear ground water velocity is approximately 3 ft/day based on potassium bromide tracer data that was collected at the start of the study. Assuming a porosity of about 35 percent based on the alluvial materials present, the average hydraulic conductivity is about 0.42 ft/day.

Figure 1. Cross Section Of Test Cell Area

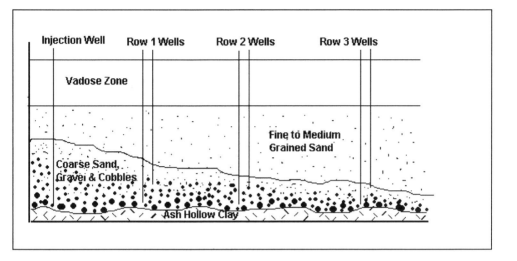

MATERIALS AND METHODS

Based on previous studies performed by Devlin (1994) a semipassive extraction - injection system was designed for the Ogallala site. The injection and extraction wells are aligned in a row perpendicular to the direction of ground water flow. This design allows for the formation of a recirculation cell. After the substrate has been injected the system is allowed to run for several hours during which the ground water/substrate solution is recirculated. After equilibration has been achieved the system is turned off. The substrate solution is then transported away from the injection system under natural ground water gradient conditions. This type of system promotes passive dispersive mixing which enhances the delivery of the substrate to the microbes in a continuous supply as it flows downgradient along with the contaminated water.

The location of the treatment cell was selected due to the presence of several commercial buildings, private residences, busy thoroughfares, and ease of accessibility. Production of hydrogen sulfide gas and seepage into basement structures was also a concern. The study area is located approximately 800 feet downgradient of the dry cleaners in St. Luke's Park. The cell location within the park was selected due to the use of the baseball diamond by St. Luke's Elementary School and summer tee-ball games, and the presence of an underground sprinkler system. The cell is slightly off-set from the longitudinal axis of the plume. The size of the treatment cell is 12 feet wide by 35 feet long (3.6m by 10.6m), and consists of one extraction well, two injection wells, and 6 sets of nested monitoring micro-wells (Figure 2). Nebraska well MW-3A, located approximately 70 feet upgradient of the test cell area, was used to collect upgradient geochemical and PCE data.

FIGURE 2. Schematic Diagram of Test Cell Configuration

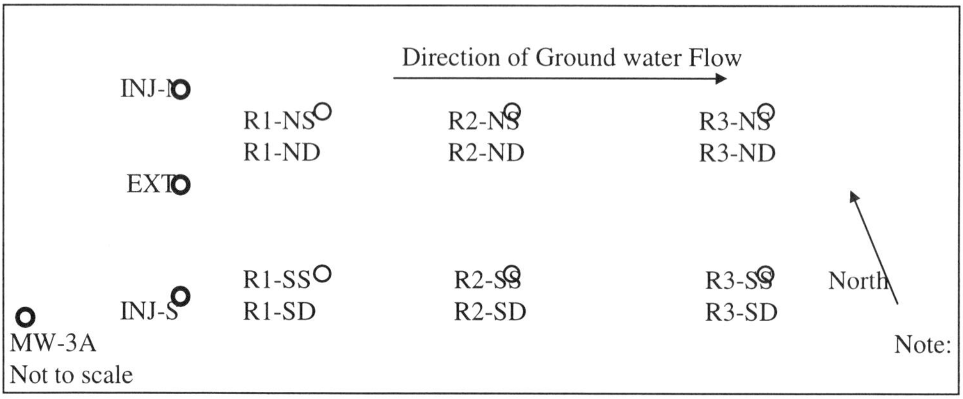

```
                        Direction of Ground water Flow
        INJ-N○                      ──────────▶
                  R1-NS○         R2-NS○         R3-NS○
                  R1-ND          R2-ND          R3-ND
        EXT○
                                                           ↗
                  R1-SS○         R2-SS○         R3-SS○   North
  ○    INJ-S○     R1-SD          R2-SD          R3-SD
MW-3A                                                      Note:
Not to scale
```

Treatment Cell Design

The extraction and injection wells are aligned in a row perpendicular to the direction of ground water flow. The wells are labeled as EXT, INJ-N and INJ-S. The distance between the extraction and injection wells, and pumping rates was based on two-dimensional simulations using Visual MODFLOW and MODPATH (Waterloo Hydrogeologic, 1999). A pumping rate of 10 to 12 U.S. gallons per minute (gpm) was determined to be sufficient for the development of a recirculation cell where the injection wells are located 6 feet (1.8m) from the extraction well. Large diameter boreholes were drilled using 14 inch hollow stem augers. The borehole size and a highly permeable filterpack material were selected to facilitate injection and circulation/mixing of the substrate solution in the recirculation cell. The wells were constructed of 4 inch (10.2 cm) ID Schedule 40 polyvinyl chloride (PVC) pipe. Stainless steel wire wrapped screens were used for these wells; the slot size for both injection wells was 0.030 inch (#30 slot), and the extraction well was 0.020 inch (#20 slot). The wells were screened between 16 and 26 feet (4.9m to 7.9m) bgs. Flush mount covers were used for each well.

Six sets of nested micro-wells were installed by direct push, and are located downgradient in three rows aligned approximately perpendicular to the direction of ground water flow. The wells were constructed of 3/4-inch (1.9 cm) ID Schedule 80 PVC, with screens that were 0.010-inch (#10 slot) and 3 feet (0.91m) in length. Flush mount covers were used for each micro-well. The shallow wells were screened between 18 and 21 feet bgs (5.5m and 6.4m), and the deeper wells were screened between 22 and 25 feet (6.7m and 7.6m) bgs. These intervals were selected on the basis that anaerobic conditions would be more readily achieved in the deeper portion of the aquifer and be less susceptible to chemical fluctuations during the irrigation season. A naming convention for each micro-well was adopted based on the row number, if it is located on the north or south side of the cell, and whether it is shallow or deep. For example, the designation R1-SD indicates that the well is located in the first row, on the south side and is deep.

Injection of the Primary Substrate and Geochemical Tracer

Regular injections of the substrate solution, which consisted of 60% food grade sodium lactate and subsequently referred to as the "lactate solution", were conducted to supply the microorganisms with a fermentable organic food source. Potassium bromide was also injected with the lactate solution during the first three injections. A schematic diagram of the system is shown on Figure 3. During operation of the system, extracted ground water was pumped through a closed-loop system to minimize aeration of the water. An in-line filter was used to remove sediments that potentially could cause clogging of the injection wells. The filtered ground water was then amended in-line with the lactate solution and geochemical tracer prior to injection. The extracted ground water was monitored for bromide and conductivity as an indication of the amount of lactate saturation and degree of equilibration during the recirculation process.

FIGURE 3. Schematic Diagram Of Injection System

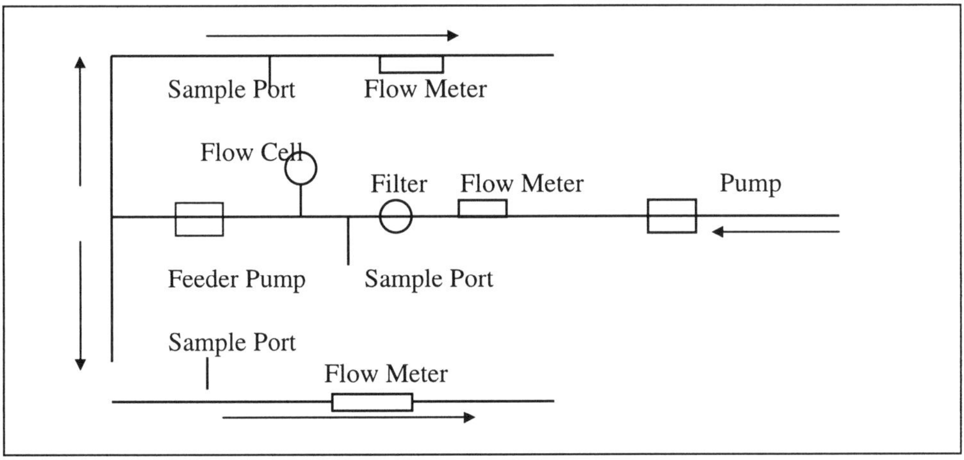

A utility pole with an electrical outlet was available for use at the site. The recirculation system was driven by a Grundfos JetStar JP-10 pump with foot valve to prevent back flow. Flow rates were measured with Headland EZ-View in-line flow meters with a maximum capacity of 10 and 16 gpm. The lactate solution and tracer were added in-line using a Stenner Dual Head Chemical Feeder Pump. The solution was pumped into a 2-inch ID clear acrylic pipe to monitor the injection and mixing process. Black PE plumbing pipe was attached to the JetStar pump and foot valve, and clear braided reinforced hose was used throughout the rest of the system. Hose connections and sampling port fittings were constructed of Cresline plastic pipe, and all valves were brass. Hose clamps were used for all connections of parts; no glue was used on any part of the system. A five foot long, 0.010 inch slot (#10), PVC screen with was attached to the end of the two injection well hose lines to promote dispersion and mixing of the lactate solution within the injection wells. The flexibility of the system allowed for the pump in-take and injection screens to be raised or lowered within the wells. The pump and filter were mounted on a low platform table constructed of plywood for stability. The system was designed to be portable, and allow for quick setup and breakdown of the hoses and other components.

Injection of the lactate solution began in April 1999. During the first three injections 2,500 grams of potassium bromide were injected along with the lactate. Initially, 25 kg of sodium lactate solution was injected once a month. Due to concerns about carbon limitation, the amount and frequency were increased to 75 kg every three weeks at the end of June. The injection and recirculation system was operated for about 7 to 8 hours, and then shut off. The study is expected to last through March 2000.

RESULTS AND DISCUSSION

Chemical Analysis

Field chemistry measurements were collected during the study using various portable instruments. The QED MicroPurge Flow Cell provided real-time data collection for temperature, pH, conductivity, dissolved oxygen (DO), and the oxidation-reduction potential (ORP). These parameters were monitored during the injection process, and when the monitoring wells were purged prior to sample collection. A select group of inorganic parameters were monitored periodically in the field; alkalinity was measured using the HACH Digital Titrator, bromide concentrations were monitored with an ORION Portable pH/ISE Meter and Bromide Half-Cell, and a HACH DR-700 Colorimeter was used to measure Fe^{2+} and nitrate/nitrite. Verification of the alkalinity, bromide and nitrate concentrations was performed in the laboratory.

Prior to the start of the study ground water samples were analyzed to establish baseline conditions at the test cell. Inorganic sampling parameters included HCO_3, Br^+, Cl^-, NO_3/NO_2, PO_4 and SO_4. Dissolved metals analyses included Ca^{2+}, Mg^{2+}, K^+, Na^+, $Fe^{3+/2+}$, and $Mn^{3+/2+}$. Analyses were also performed for total organic carbon (TOC), and a select group of VOCs including PCE, TCE, cis- and trans-1,2 DCE, and VC.

Weekly samples were collected for several months and analyzed for bromide and TOC. The data were used to estimate the ground water velocity, monitor dispersion and consumption of the lactate, and to make adjustments in the frequency and amount of injected lactate. Samples were collected periodically from different wells within the test cell to monitor HCO_3, NO_3/NO_2, SO_4, and VOCs. Additional samples were collected on occasion and analyzed for the suite of metals, ortho-phosphate, and chemical oxygen demand. Samples were collected later in the study and analyzed for ethene, ethane, methane, and volatile acids including acetate, butyrate and propionate.

Baseline measurements indicated elevated levels of nitrate and sulfate at 11 mg/l and 135 mg/l, respectively. Alkalinity was measured at 278 mg/l as HCO_3 and TOC was 1.80 mg/l. Flow-through cell measurements for DO ranged from 1.8 mg/l to 3.9 mg/l, and ORP from 149 to 254. Periodic measurements for total nitrate indicated that levels declined within the test cell through November (Table 1) when 1.8 mg/l was detected at INJ-S and less than the RL of 0.05 mg/l was reported for R3-SD. Nitrite had been detected in several wells starting in July and decreased to a concentration of 0.03 mg/l in R3-SD during August. Iron and manganese were not detected in the baseline samples. Field chemistry results for Fe^{2+} in August ranged from <0.5 mg/l in the row 1 north wells to 3.0 mg/l at R1-SD. Laboratory analyses for dissolved metals in December indicated that Fe^{2+} ranged from 0.28 mg/l (INJ-S) to 3.48 mg/l (R3-SD), and Mn^{2+} from 0.92 mg/l (INJ-S) to 2.75 mg/l (R3-SD). Beginning with the August 4 sampling event a discernible odor of hydrogen sulfide was noted in most of the wells. A stable decline of sulfate within the test cell was not evident (Table. 2), although a significant decrease occurred between INJ-S (106 mg/l) and R3-SD (20.2 mg/l) in the December samples. Alkalinity increased from the baseline level of 278 mg/l HCO_3 in most of the monitoring well locations throughout the study to a high of 906 mg/l observed at R3-SD in December.

Table 1. Analytical Results For Nitrate and Nitrite

Well	4/13/99	7/14/99	8/4/99	8/24/99	9/15/99	10/8/99	11/8/99
INJ-S	11.0		10.0				1.8
R2-SD Total Nitrate (mg/l)		7.9	6.1	3.7	2.1		
R3-SD Total Nitrate (mg/l)		4.7	0.93	<0.05	<0.05	<0.05	<0.05
R3-SD Nitrite (mg/l)		1.54	0.63	0.03			

Table 2. Analytical Results For Sulfate

Well	4/13/99	8/4/99	8/24/99	9/15/99	10/14/99	11/8/99	12/7/99
INJ-S (mg/l)	134			114		108	106
R1-SD (mg/l)		132	137				57.8
R3-SD (mg/l)		131	105	97.3	155	120	20.2

Table 3. Analytical Results For Alkalinity As HCO_3

Well	4/13/99	5/11/99	6/2/99	8/4/99	8/24/99	11/8/99	12/7/99
INJ-S (mg/l)	278			278	310	558	519
R2-SD		278	286	297	315		
R3-SD				322	373	352	906

The April baseline level of PCE within the test cell was reported at 20 ug/l, which was considerably lower than previous results reported (400 ug/l to 1,400 ug/l) for the upgradient monitoring wells MW-3A and MW-4. Subsequent samples collected from MW-3A indicated that the concentration of PCE had increased to 60 ug/l in November and 100 ug/l in December (Table 4). The fluctuation in the concentration of PCE at this location suggests that the plume is not homogenous and that a slug of more concentrated PCE was moving downgradient through the test cell area. Degradation products including TCE, DCE and VC were not detected in any of the samples collected from the wells that were sampled through December.

Samples were collected for the analysis of methane, ethene and ethane in November and December to evaluate whether methanogenic conditions were present within the test cell and complete degradation of the PCE had occurred. Samples were also analyzed for volatile organic acids including acetate, propionate and butyrate. The results of these analyses are presented in Table 5.

Table 4. Analytical Results For PCE

Well	4/13/99	5/11/99	8/24/99	9/15/99	11/8/99	12/7/99
MW-3A (ug/l)		30		20	60	100
INJ-S (ug/l)	20	20		<0.68	3.0	8.0
R1-SD (ug/l)		20	0.8		3.0	10.0
R2-SD (ug/l)		20	0.8	<0.68	2.0	
R3-SD (ug/l)			1	<0.68	0.90	3.0

Table 5. Analytical Results For Volatile Organic, Methane Gas and COD

Wells	Acetate (mg/l)	Propionate (mg/l)	Butyrate (mg/l)	Propionate/ Acetate	Methane (ug/l)	COD (mg/l)
INJ-S (Nov/Dec)	86.4/135	61.1/95.3	<5/<5	0.71/0.71	210/600	650/430
R1-SS (Dec)	30	16.9	ND	0.56	NA	NA
R1-SD (Nov/Dec)	107/158	61.4/120	ND/<5	0.57/0.76	73/140	1200/430
R2-SD (Nov)	73.6	25.4	ND	0.35	18	260
R3-SD (Nov/Dec)	28.1/311	<5/261	ND/<5	<0.18/0.84	12/180	65/910

a. NA indicates the sample was not analyzed. ND indicates the compound was not detected
b. <5 indicates the compound appears to be present but a less than the quantitation limit

The highest concentrations of acetate and propionate in samples collected in November were observed in R1-SD and decreased with distance downgradient. Chemical oxygen demand also followed this trend. Based on work performed at the INEEL, COD concentrations are typically close in value to the sum of lactate and its fermentation products (DOE-ID, 2000). The results indicate that the sums of lactate and the fermentation products are only a fraction of COD at the R1-SD location, but are higher at INJ-S and increase downgradient. This trend suggest that a more concentrated plug of lactate was in the vicinity of R1-SD and that fermentation was occurring downgradient. A decrease in the fermentation products downgradient of the injection system is to be expected due to the hydraulic conditions. The propionate/acetate ratios were less than 1 and decreased in the downgradient wells. Fermentation of lactate has been examined by Fennel and Gossett (1997). Two pathways were identified where in one pathway the lactate is fermented to acetate, bicarbonate, hydrogen ions and dissolved hydrogen. Fermentation in the second pathway leads to acetate, propionate, bicarbonate and hydrogen ions. The stoichiometric propionate:acetate ratio in the second pathway is 2:1, which suggests that the first pathway was more predominant during the November sampling event.

Results for the December sampling event indicate that the sums of the fermentation products account for over 50% of the COD. The increase in acetate and propionate with distance suggest that the lactate is being fermented as it moves downgradient towards R3-SD. At the same time, bicarbonate concentrations increase with distance away from INJ-S. The second pathway is more thermodynamically favored

(Fennel and Gossett, 1997), which suggests that a population of microbes that utilize it are becoming more efficient at lactate fermentation. The December results suggest that the bacteria are becoming more efficient at lactate fermentation within the test cell area (INEEL, personal communiqué). The increase in methane gas between the November and December samples suggest that micro environments with strongly reducing conditions are located in the test cell and may be increasing as more efficient fermentation of the lactate occurs.

CONCLUSIONS

The data suggest that acclimation of the microbes to the addition of the lactate during the first several months of operation was slow. Development of reducing conditions was first observed in July with the reduction of nitrate and the formation of nitrite. The appearance of Fe^{2+} was first noted in field chemistry samples collected in August. At the same time, the noticeable odor of hydrogen sulfide gas was detected in the monitoring wells and in the water. Sulfate reducing conditions were slow to develope. Confirmation of fermentation of the lactate solution was observed in samples collected in both November and December. The acetate/propionate ratios and an increase in HCO_3 suggest that the microbial community was still becoming acclimated to the lactate, but were successful in creating conditions that are appropriate for reductive dechlorination. Although data indicate that conditions within the cell are appropriate, without degradation data due to the initial concentrations of PCE, it cannot be confirmed at this time that reductive dechlorination is occurring.

ACKNOWLEDGMENTS

This treatability study was funded with a grant from EPA Region VII. Volatile acid analyses and additional assistance were provided by the Geo-Microbiology Group at Bechtel BWXT Idaho, LLC and the Idaho National Engineering and Environmental Laboratory (INEEL). Analysis of the samples for methane, ethane and ethene were provided by Quanterra Inc., Austin, Texas.

REFERENCES

Interstate Technology Work Group, 1998. Technical and Regulatory Requirements For Enhanced In Situ bioremediation Of Chlorinate Solvents In Groundwater.

Devlin, J.F., 1994. Enhanced in situ biodegradation of carbon tetrachloride and trichloroethene using a permeable wall injection system, Ph.D. Dissertation, 663 p. University of Waterloo, Waterloo, Ontario

Fennel, D.E. and J.M. Gossett, 1997. "Modeling the Production of and Competition for Hydrogen in a Dechlorinating Culture." *Environmental Science and Technology*, Vol. 32 pp. 2450-2460.

DOE-ID, 2000. *Field Demonstration Report, Test I Test Area North Final Groundwater Remediation, Operble Unit I-07B*, Revision C, U.S. Department of Energy Idaho Operations Office, DOE/ID-10178.

ENHANCED CLOSURE OF A TCE SITE USING INJECTABLE HRC™

Susan L. Boyle, Vincent B. Dick, Mark N. Ramsdell, P.E. (Haley & Aldrich, Inc., Rochester, New York)
Todd M. Caffoe, P.E. (New York State Department of Environmental Conservation, Avon, NY)

ABSTRACT: Clayey-silt soils were contaminated with trichloroethene (TCE) at a former industrial-filter-manufacturing site in Rochester, NY. 2-PHASE™ Extraction had reduced contaminant concentrations below clean-up criteria, except in the core source area. The regulatory agency agreed to testing of an innovative in-situ technology as an alternative to additional enhancements to the extraction system, and as a last step leading to site closure. Hydrogen Release Compound (HRC™) was injected into the subsurface utilizing Geoprobe® direct-push methods. Fifteen months of post-injection data indicate significant decreases in TCE concentrations with corresponding increases, then decreases in daughter product concentrations. Data has shown HRC to still be present fifteen months after injection. The regulatory agency has recommended site closure pending the results of the final sampling round.

INTRODUCTION

This paper describes current results from the first field-scale application of injectable Hydrogen Release Compound (HRC). HRC is a material formulated to promote the production of hydrogen by naturally occurring bacteria in an aquifer to enhance anaerobic biodegradation of chlorinated volatile organic compounds. Summarized herein are site setting, design criteria, monitoring methodology and results, interpretation of biodegradation processes, and regulatory issues for site closure.

SITE SETTING AND REMEDIATION PERFORMED

The site housed a former industrial filter manufacturer where chlorinated volatile organic compounds were utilized in degreasing operations. The predominate degreaser used was trichloroethene (TCE) which was determined to have affected the soil, groundwater, and sediments within the site. Site investigations and a feasibility study conducted under the guidance of the New York State Department of Environmental Conservation (NYSDEC) identified three areas of contamination: the former degreaser area, former drum storage area, and sediment in a stormwater retention pond. The site was listed on the New York State Registry of Inactive Hazardous Waste Sites.

Following performance of the RI/FS (Haley & Aldrich, 1991, 1992), a Record of Decision (ROD) was issued by the NYSDEC that selected removal and disposal of sediments with drainage controls; shallow soil removal; and an aggressive multi-phase high vacuum extraction for soil and groundwater remediation. Sediment and soil removal was successful upon completion and the 2-PHASE™ Extraction (2-PHASE) commenced in 1994.

Contaminants within the former drum storage and degreaser areas consist primarily of TCE with lesser concentrations of the breakdown compound 1,2-dichloroethene. Pre-remediation concentrations were >50 mg/kg in soil and >190 mg/L in groundwater. The soil concentration cleanup objectives identified in the ROD were 0.5 mg/kg for *cis*-1,2-DCE (DCE) and 1.0 mg/kg for TCE. The groundwater cleanup objective was to "design and operate to the extent practicable to mitigate and control shallow source area groundwater" (NYSDEC, 1993).

The 2-PHASE system operated for approximately 4 years and almost 30,000 operational hours. Although additional extraction wells had been installed to improve mass removal rates, periodic groundwater and soil sampling and analysis concluded that the system had reached asymptotic conditions and operation was discontinued. Groundwater concentrations remained elevated up to 28 mg/L in a "core" source area; still, it was determined these localized residual soil concentrations could fit a risk-based closure.

The localized residual groundwater concentrations provided the NYSDEC and the responsible party the opportunity to pilot test HRC and potentially improve the final remedial process for closure.

DESIGN CRITERIA

HRC promotes the production of hydrogen in an aquifer. Hydrogen is utilized by naturally occurring bacteria to anaerobically biodegrade chlorinated solvents. The compound is a proprietary food-grade polylactate ester. The formulation tested at this site was the consistency of viscous honey and injectable. Upon injection, HRC dissolves and degrades to lactic acid, which in turn is microbiologically metabolized to hydrogen. Hydrogen in turn is used to reductively dechlorinate the VOCs.

Many biodegradation processes, such as the breakdown of gasoline constituents, are limited by the availability of electron (e^-) acceptors such as oxygen and nitrate. During these processes, contaminants (i.e., BTEX) are oxidized as the e^- acceptors (i.e., oxygen) are reduced. Conversely, the processes associated with the biodegradation of many chlorinated solvents (particularly chlorinated ethenes) tend to be limited by the availability of e^- donors. Hydrogen, an excellent e^- donor, is released by indigenous microorganisms that metabolize the HRC, thereby enhancing the biodegradation of certain chlorinated solvents.

The project site was selected for the field pilot test due to the relatively small, controllable size of the VOC source core; the density of monitoring wells in and downgradient of the core area (Figure 1); and evidence of existing biodegradation. Subsurface materials consist of relatively low-permeability clayey silts, saturated to within 0.3 to 0.6m (1 to 2 ft.) of grade. Injection/treatment depth was up to 4.6± m (15± ft.) depth below grade.

Due to the need for anaerobic conditions for degradation, 2-PHASE was shut down to eliminate flow of air and relatively oxygenated surface water through the treatment zone. The NYSDEC allowed 2-PHASE shutdown prior to collection of pre-injection baseline samples.

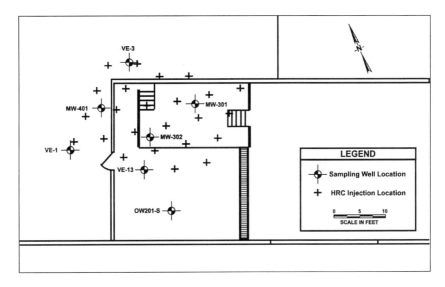

FIGURE 1: Site Plan

Site characteristics evaluated for injection design included: dissolved oxygen (DO) and other e^- acceptor concentrations, VOC concentrations, and hydrogeologic conditions. Hydrogeologic and VOC data existed prior to application. Based on this information, an initial grid pattern of injection locations was determined. DO and e^- acceptor data were collected for design purposes. Also, samples of relatively undisturbed soil were collected for a "microcosm study" of indigenous bacteria and laboratory inoculation with the HRC formulation. The initial injection grid was then fine-tuned and the amount of HRC to be injected, injection point spacing, and monitoring parameters were determined.

The grid was laid out in the field and the nodes were adjusted to permit rig access. HRC was injected with a Geoprobe rig fitted with a special pump to inhibit O_2 introduction. Final node spacing was targeted at 1.5±m (5+ ft.) intervals (see Figure 1) with 21 points injected with roughly 15.9 kg (35 lbs.) per hole.

MONITORING METHODS AND RESULTS

Groundwater Monitoring Program. Pre-injection groundwater samples were used to determine baseline conditions, the HRC was then injected, and subsequent progress of subsurface conditions was monitored on a quarterly basis. Fifteen months of monitoring data are included here (baseline event and five quarterly events). Monitoring parameters used for this site included: VOC concentrations; HRC component concentrations (metabolic acids); biodegradation indicators (e^-

acceptors, endpoint gases); microbial analyses; field parameters (DO, Eh). The monitoring program wells included: VE-1, VE-3, VE-13, MW-301, MW-302, OW-201-S, and MW-401 (see Figure 1).

Primary objectives of the study were to monitor and interpret decreases in parent (TCE) concentrations, coupled increases in daughter products, and changes in other indicator parameters. The expected primary degradation pathway is (Suthersan, 1996):

$$TCE \Rightarrow cis\text{-}1,2DCE \Rightarrow \text{Vinyl Chloride} \Rightarrow \text{Ethene}$$

Of particular interest was whether vinyl chloride would degrade completely to ethene under anaerobic conditions. Vinyl chloride is known to degrade more quickly under aerobic than anaerobic conditions (Lee, et. al., 1998). Production and potential accumulation of vinyl chloride is of interest relative to the reductive dechlorination process, as well as to site owners and regulators from an environmental risk standpoint.

Concentrations of HRC components and associated indicator parameters were also monitored to determine degradation impact and the longevity of HRC in the aquifer. Organic acids (lactic, pyruvic, and acetic) were analyzed in addition to total organic carbon (TOC). Because TOC should increase as HRC disperses in the aquifer and decrease as the HRC depletes, this parameter was used as an indirect indicator of HRC longevity. Field parameters (including DO and Eh/ORP) were also obtained to provide information regarding the development of the required anaerobic conditions.

INTERPRETATION OF RESULTS

The following sections provide an interpretive summary of the analytical results for the volatile organic compounds (VOC), HRC components, inorganic compounds, and field parameters, respectively.

Discussion of VOC Results. Overall, results of this field application indicate that the HRC is promoting complete and relatively rapid degradation of the source contaminant trichloroethene to ethene. Substantial decreases in the concentrations of the parent compound, TCE, were realized over the 15-month test period.

Table 1 summarizes the VOC data for the six quarterly events. The table also summarizes the percentage TCE loss for each well from the pre-injection event to the 15-month sampling event. Substantial decreases in TCE concentrations were observed in all of the seven wells in the test area. Five of the six wells located within the injection grid (VE-1, VE-3, VE-13, MW-302, and MW-401) experienced TCE reductions ranging from 82% to 100% over the 15 month test period.

Well VE-1, which had the highest pre-injection concentrations, exhibited steady decreases from 26 mg/L to 4.7 mg/L with corresponding increases in DCE. Well MW-301 decreased from the pre-injection level of 2.8 mg/L TCE to non-detect levels at the six month sampling event before increasing back to 4.3 mg/L.

This increase is attributed either to normal fluctuations of groundwater concentrations or to de-sorption of contaminants from the soils in this area.

TABLE 1: Summary of Analytical Results

Compound	Date	VE-1	VE-3	VE-13	MW-301	MW-302
TCE	Pre-HRC	26	3.5	0.68	2.8	1.1
	3 Month	14	NS	ND	3.1	0.34
	6 Month	11	0.03	0.007	ND	ND
	9 Month	8.6	3.5	0.095	0.61	0.17
	12 Month	13	1.6	0.022	1	0.32
	15 Month	4.7	0.82	0.0081	4.3	ND
	TCE Percent Loss	82%	94%	99%	-54%	100%
DCE	Pre-HRC	ND	0.34	0.25	0.26	0.18
	3 Month	1.3	NS	1.7	6.9	6.2
	6 Month	0.58	0.021	0.034	ND	2.4
	9 Month	0.61	1.4	0.4	8.9	11
	12 Month	0.72	0.640	0.063	1.5	12
	15 Month	0.26	0.22	0.028	1.8	18
Vinyl chloride	Pre-HRC	ND	ND	ND	ND	ND
	3 Month	ND	NS	0.84	ND	ND
	6 Month	ND	ND	0.024	0.14	0.21
	9 Month	ND	0.02	0.17	7.2	3.9
	12 Month	ND	0.0069	0.048	0.96	8.8
	15 Month	ND	ND	ND	0.19	ND
Ethene	Pre-HRC	ND	ND	ND	ND	ND
	3 Month	ND	ND	ND	ND	ND
	6 Month	ND	ND	0.02	ND	0.013
	9 Month	ND	ND	0.065	0.042	0.028
	12 Month	ND	ND	ND	ND	ND
	15 Month	ND	ND	0.0731	0.405	0.0505

VOC Mass Kriging Results. Figure 2 is a depiction of the mass of TCE, DCE and vinyl chloride present in the test area at the baseline event and at each of the subsequent monitoring events. Mass values were generated by kriging monitoring well concentrations over the area of treatment, and converting concentrations to total mass values for each compound. As seen on Figure 2, the TCE mass decreased by approximately 64% from approximately 625 grams to approximately 222 grams. Coupled with the decrease in TCE concentrations, 1,2-DCE concentrations increased from 43 grams prior to the HRC injection, peaked at 430 grams 89 days after injection, and were measured at 167 grams during the 461 day event. Vinyl chloride followed a similar pattern, increasing from non-detect levels prior to injection, to a peak at 159 grams at the 273 day event, then decreasing to 5 grams at the 461 day event. This result is of particular importance because of the particular physical properties of vinyl chloride and it's impact on successful closure at this site.

The reduced rate of degradation observed between six and nine months after installation appears to be a combination of the expected asymptotic degradation "slow down" combined with a depletion of the HRC.

HRC Component Analytical Results. As previously mentioned, as HRC dissolves and is microbiologically metabolized, hydrogen generation results. The

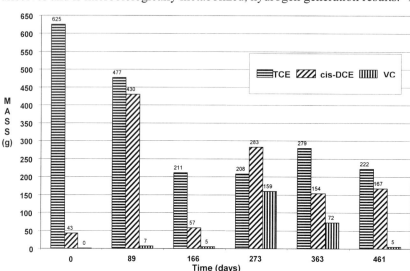

FIGURE 2: VOC Mass Graph

lactic acid is degraded in the following manner, with hydrogen being liberated at each step:

$$\text{lactic acid} \Rightarrow \text{pyruvic acid} \Rightarrow \text{acetic acid}$$

In general, detection of these acid breakdown products is evidence that hydrogen is being liberated into the aquifer. Note that at the time of this pilot test, direct measurement of hydrogen changes in groundwater was determined to be prohibitively difficult and was not attempted.

Lactic acid, in general, was seen at only low to non-detect values in the program wells. It is believed the lactic acid was rapidly converted to other volatile acids. Acetic and propionic acids were detected in the following wells: VE-13, MW-301, MW-302, OW-201-S, and MW-401. In particular, high concentrations of the various acids were detected at MW-301. This well has had consistently high concentrations of volatile acids including detections at the 15-month sampling event indicating the long-term presence of the HRC.

Inorganic Compound Analytical Results. The results of the inorganic compound analyses corroborate the trends seen in the VOC results. In general, the inorganic analyses showed the aquifer geochemistry changing toward anaerobic conditions. As expected, sulfate concentrations decreased as the anaerobic conditions developed and sulfate is reduced to sulfide. The sulfate concentrations rebounded somewhat in the more recent analyses indicating depletion of HRC. In addition, Fe^{+2} (dissolved iron) was produced and its concentrations peaked in the three to nine month timeframe after injection. Space limitations do not allow these data to be discussed in detail here but additional data is available in previous publications (Boyle, et. al., 1999). These inorganic data assist with determining the timeframe of HRC depletion, and therefore could be used if additional injections are considered in a particular application.

Field Parameter Results. Various field parameters were measured to assist in the interpretation of the subsurface conditions, including pH, temperature, conductivity, DO, and Eh. Two of these parameters, DO and Eh, are particularly important when evaluating HRC injection because they indicate the status of the aquifer with respect to development of anaerobic conditions. Both of these parameters decreased rapidly after HRC injection (DO went to <0.6mg/l average, and Eh to
<-107*ev* average). The field parameter results corroborated well with the results of VOC, acids, and inorganic analyses previously discussed. A more complete discussion of these parameters can be found in previous publications (Boyle, et. al., 1999).

Microbial Testing Results. Microbial samples were obtained from the site prior to the HRC test. These samples were obtained to test for the presence/number of appropriate bacteria, and to evaluate the capability of these bacteria for degrading the site contaminants—with and without the addition of HRC. Interpretation of the results from the pre-installation sample indicated that a viable bacterial community exists in the subsurface. The degradation rates seen in the laboratory HRC-amended samples were greater than the rates in the non-amended samples.

Contrary to the belief that remediation of contaminants in tight clayey soils can be prohibitively difficult, results from the 15-month study described in this paper indicate the clayey soils present at the site may have actually enhanced and/or lengthened the activity of the HRC in the aquifer. Low groundwater velocity likely allowed a longer residence time for the HRC, increasing contact time with microbes, and the overall extent of reductive dechlorination. HRC components (volatile acids) were still present at relatively elevated concentrations in the test area during the most recent sampling event (15 months after injection); much longer than initially expected.

REGULATORY CRITERIA

NYSDEC Record of Decision (ROD) Requirements. At project start, the selected remedy, 2-PHASE, was recognized by NYSDEC as an innovative technology with considerable potential to achieve soil cleanup goals. Further, it was recognized that the aggressive nature of 2-PHASE would remediate groundwater contamination to the extent practical. The ROD required operation of 2-PHASE until the cleanup goals were achieved or the system reached the limit of its effectiveness for site conditions. If cleanup goals for soils could not be achieved, then further cleanup would be evaluated in a focused feasibility study.

HRC Role in Advancing Site Closure. 2-PHASE was successful at reducing soil and groundwater concentrations to levels consistent with the ROD over the majority of the site. In general, total VOCs in soils were reduced to below the 0.5 to 1.0 mg/kg range, and concentrations of VOCs in groundwater were reduced from levels of several mg/L to a range generally less that 0.3± mg/L. Cleanup goals were not achieved in the core source area and continued operation of 2-PHASE had reached the limit of its effectiveness for site conditions. Pursuant to the ROD, a focused feasibility study was required.

As an alternative to a focused feasibility study, it was proposed to shutdown the 2-PHASE system and attempt HRC as an emerging/innovative technology. It was agreed that a focused feasibility study would not provide for site closure within a suitable time frame for the responsible party. The NYSDEC agreed that attempting additional reduction via HRC provided a better basis for final closure of the site. In exchange for implementing the HRC pilot along with a public notice, NYSDEC agreed that demonstration of an emerging/innovative technology at this site would preclude the need for a focus feasibility study. NYSDEC facilitated the public notice and provided some analytical services for the HRC pilot.

Ultimately, NYSDEC required several quarterly sampling events in order to demonstrate the effectiveness of the technology. As stated in the ROD, one of the criteria for remedy selection is that the remedy must reduce the toxicity, mobility, or volume of contamination. NYSDEC and the NYSDOH were concerned when vinyl chloride levels increased from non-detect to 8.8 mg/L levels in groundwater during the HRC pilot; therefore, the agencies required an extended monitoring period. Subsequent monitoring has alleviated these concerns with the most recent data showing vinyl chloride concentrations ranging from non-detect to 0.19 mg/L. If vinyl chloride levels remain constant or decreases, the NYSDEC will not require further actions and the site can be closed.

Regulatory Conclusions. Based upon the data generated to date, the NYSDEC believes that HRC technology has effectively reduced the levels of contamination in the core area of the site. As this technology becomes commercially available and more field data is generated, the monitoring requirements could be decreased making the technology even more economical.

REFERENCES

1. Boyle, S.L., et al, 1999. "Enhanced In-Situ Bioremediation of a Chlorinated VOC Site Using Injectable HRC", Proceedings of the Thirty-First Mid-Atlantic Industrial and Hazardous Waste Conference", June 20-23, 1999.
2. Haley & Aldrich of New York, 1991. Remedial Investigation Report.
3. Haley & Aldrich of New York, 1992. Feasibility Study Report.
4. Lee, T.H., et al, 1998. "A Combined Anaerobic and Aerobic Microbial System for Complete Degradation of Tetrachloroethylene", presented at the First International Conference on Remediation of Chlorinated and Recalcitrant Compounds, May 18-21, 1998.
5. New York State Department of Environmental Conservation, 1993. Record of Decision.
6. Suthersan, S., 1996. *Remediation Engineering, Design Concepts*, CRC Lewis Publishers, page 133.

ENHANCED REDUCTIVE DECHLORINATION: LESSONS LEARNED AT OVER TWENTY SITES

Michael A. Hansen, P.E., Jeffrey Burdick, Frank C. Lenzo, P.E., Suthan Suthersan, Ph.D.
(ARCADIS Geraghty & Miller, Inc., Langhorne, Pennsylvania)

ABSTRACT: Enhanced reductive dechlorination (ERD) using common, natural reagents as an organic carbon source to provide an in-situ reactive zone (IRZ) provides an economical means of remediating groundwater impacted with chlorinated constituents. A strong knowledge base of "lessons learned" has been developed through applying the ERD technology at sites with varied geographical, chemical, and hydrogeological environments. These lessons highlight the practical issues that need to be considered when applying ERD and IRZ technologies in general.

INTRODUCTION

ERD is a remedial technology that provides enhancement of ongoing reductive dechlorination of chlorinated volatile organic compounds (CVOCs) in groundwater via the creation of an IRZ. IRZ is a general term applicable to many in-situ treatment technologies. However, in the case of ERD, an IRZ is defined as a highly reducing groundwater environment, rich in organic carbon. ARCADIS Geraghty & Miller first patented the use of molasses as an electron donor for the enhanced treatment of CVOCs in groundwater in 1996 (Patent #5,554,290). Since initial application of the technique in 1995, the technology has been applied in both the pilot- and full-scale at over 20 sites throughout the United States and abroad.

During this period of time, ERD has been applied at a broad range of sites. These sites have included a variety of constituents to be treated - including PCE, TCE, 1,1,1-TCA, Carbon Tetrachloride, Pentachlorophenol, and chlorinated pesticides; various groundwater concentration ranges; and numerous hydrogeologic settings (including shale and karstic limestone bedrock, low permeability glacial tills and saprolite, and high permeability alluvium and glacial outwash environments).

As with all groundwater remediation activities both in-situ and ex-situ, the successful application of ERD relies mainly on sufficient and accurate hydrogeological information for the given site. However, more specifically, the following topics have shown to be of critical importance in the application of ERD: reagent delivery, microbial conditioning, natural surfactant effects, and fermentation/byproduct formation. In addition, overall project planning, the application of ERD under oxidizing and/or high groundwater velocity conditions, and the application of ERD in both low concentration or high concentration/DNAPL settings have been issues in which special considerations must be made in order to allow for successful application. These topics and the practical lessons learned related to each are discussed in greater detail in the following sections.

LESSONS LEARNED

Project Planning/Technology Screening. One of the first lessons learned in initial applications of the ERD technology was the importance of project planning. ERD cannot be applied without a careful consideration of the site hydrogeology and fate and transport of the target CVOCs. Typically the planning and screening stage of the project involves collecting biogeochemical data and determining the amount (if any) of ongoing natural attenuation (NA) in the impacted aquifer.

If the identified NA rates are high and a sufficient source of electron donor (natural organic carbon, petroleum hydrocarbons or other anthropogenic chemicals) is available, ERD may not be necessary. In this case, monitored natural attenuation (MNA) alone may gain site closure. Similarly, if the initial geochemical review shows the predominance of aerobic and oxidizing environments with very little evidence of natural degradation, ERD application may be problematic. In these situations, extractive or in-situ oxidation technologies may prove more cost effective.

Typically a pilot test or smaller-scale field test follows the initial screening process. The two primary issues to be addressed during the field testing phase are the provision of properly placed observation wells and allowing for sufficient time to demonstrate the success of ERD. The amount of time it takes to see positive results of the IRZ implementation is related to many factors, including groundwater velocity, the time required for introduced reagents to overcome the ambient redox conditions in groundwater, and the locations of observation wells in relation to the injection area. It is prudent to evaluate geochemistry and achievable degradation rates from data collected from wells located at least several months travel time apart from each other. Consequently, the minimum duration of a typical pilot study is six months with the flexibility to extend the testing based on data collection and site specific costs. Field tests that are shorter in duration, or are applied in too small of an area, often do not provide information that is applicable to a successful or economical large-scale implementation. This minimum period of time should be sufficient to overcome the initial aquifer redox conditions and allow for the degradation of constituents to a degree that will be observed in the collected pilot data. Some observation wells should also be placed within one to two months groundwater travel time from the injection area. This timing/placement should allow for early observations of the IRZ development, and allow for modification of the reagent injection program (strength and frequency) early enough in the planned test duration.

Reagent Delivery. The successful application of ERD to remediate chlorinated constituents in groundwater first and foremost relies on the timely and consistent delivery of the organic carbon reagent to the treatment zone.

ARCADIS Geraghty & Miller typically applies a dissolvable sucrose solution (molasses-based) as a reducing reagent. This serves as a cost effective reagent ($0.30/pound) that can aggressively alter the redox state of groundwater (oxidative poise) in a short time period. Other reagents, or electron donor substrates, such as edible oils and semi-solid forms of lactate (such as Regenesis' HRC®) will rely more on dissolution and diffusion for delivery. On a unit cost basis, these donors are more expensive. However, the application of a slow diffusing reagent may be more efficient in a highly reducing, slower groundwater velocity environment. The proper reagents should be selected based on the site hydrogeology and desired treatment time frame.

Based on the application of ERD to date, reagent delivery becomes most complicated in low permeability geologic environments [10^{-5} centimeters per second (cm/sec) or less] or those with low groundwater flow velocities (less than 50 feet/year). These settings can limit the area of influence of individual reagent injection points due to the absence of sufficient reagent dispersion. Poor donor delivery can also result in other potential complications. These complications include:

- Uneven application of reagent and resulting treatment; not achieving the treatment goals,

- Lack of sufficient or timely demonstration of the technology during pilot phase, or

- Requirement of too many injection points for a full-scale application.

In low permeability and/or low groundwater velocity environments, the reagent can also accumulate in the vicinity of the injection point. Careful monitoring of the pH and total organic carbon (TOC) levels in the groundwater near the injection well is necessary to avoid deleterious side effects. These effects are related to fermentation and byproduct formation and are discussed later in this paper.

Given these potential difficulties, prior to applying the ERD technology – or for that matter ANY in-situ technology - in full scale at a given site, a detailed review of the hydrogeology must be performed in order to determine if the reagent can be delivered to the desired portions of the impacted area. The area of influence for injection points installed in lower permeability settings can often be overcome by injecting soluble liquid reagents under pressure.

For these reasons, in a low groundwater velocity environment, ERD may prove to be more cost effective as a means of providing constituent containment. A low groundwater flow velocity can be used to an advantage and ERD may also be applied concurrently with an MNA remedy for other portions of the site.

Microbial Conditioning. Research and bench scale studies have been conducted to identify specific microbes capable of directly or cometabolically degrading CVOCs. Microbes have been identified in most redox (respiration) regimes that are capable of degrading some or all of the target CVOCs. Some groups within the scientific community maintain that these microbes are not ubiquitous in the environment and the sole application of electron donor amendments will therefore not result in the complete degradation of the target CVOCs. These groups maintain that bioaugmentation is necessary to achieve complete treatment. Bioaugmentation is the delivery of isolated strains of microbes that are known to degrade CVOCs. However, it has been ARCADIS Geraghty & Miller's experience that bioaugmentation is not necessary to degrade CVOCs in-situ. Moreover, the degradability of CVOCs in-situ, via an electron donor amendment, is related to microbial conditioning and the subsequent lag time before degradation byproducts are observed. As shown in the figure below, bioaugmentation may help shorten this lag time and also prove necessary in sterile environments. The additional costs associated with bio-assays and bioaugmentation should be weighed against the potentially longer microbial conditioning and treatment periods using only simple and cost-effective electron donor amendments.

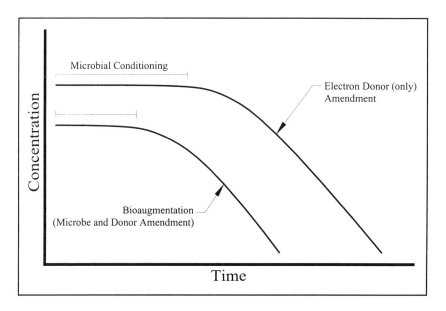

Natural Surfactant Effect. The presence of an abundant source of easily degradable organic carbon through the application of ERD typically results in a rapid and large increase in the population of microorganisms in the treatment zone. As in any microbiological system, this large population increase will also result in an increase in the production of natural biosurfactants by the microorganisms. Natural biosurfactants result in desorption of CVOCs adsorbed to the aquifer media. A high TOC gradient present between the groundwater and aquifer media will also result in desorption of organic chemicals with hydrophobic properties.

This desorption, or natural surfactant effect, is observed in many biological treatment processes as an increase in the constituent levels both in the treatment zone and in some cases downgradient of the treatment zone. In other cases, the constituent concentrations in the treatment zone remain constant even when biodegradation end-product data supports the conclusion that sufficient mass is being degraded by the ERD processes. The following figure is from an observation well from a successful full scale ERD remediation and illustrates a typical natural surfactant peak in target source CVOC concentrations followed by sequential formation and degradation of daughter products.

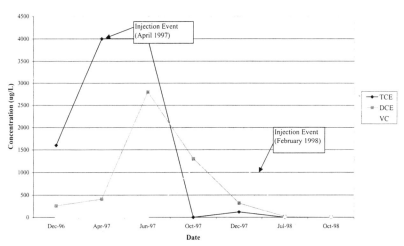

Figure 2 - Analytical Results for Well MW-4, Abandoned Manufacturing Facility, Emeryville, California

Intuitively, the increased desorption of target constituents allows for greater access to typically "untreatable" constituent mass. However, this natural surfactant effect must be expected and pilot or full-scale treatment should incorporate provisions to evaluate and account for it. For example:

- The potential for initial increases of stable parent constituent trends can be of concern to both clients and regulatory bodies as the data would tend to indicate the technology is not only not working, and in some cases could be considered as actually making conditions worse. Therefore, during the project or pilot test planning stages the possibility of this desorption effect must be evaluated in detail.

- The possibility for an increase of dissolved CVOCs to occur in areas downgradient of the treatment area must be addressed in regards to possible off-site migration and/or migration towards sensitive receptors. If these issues are of concern, the possibility for expansion of the ERD to treat these areas and/or the provision for additional downgradient monitoring and/or possible containment of the groundwater must be explored. Typically, an "outside-in" approach is applied, whereby ERD is established in a downgradient, lower concentration portion of the plume before applying ERD to a source area. Desorbed VOCs would then move into an area already amenable for treatment.

Fermentation & Byproduct Formation. During the application of ERD a highly reducing biogeochemical environment is generally created throughout the treatment zone. In addition, this zone will also contain a large excess of organic carbon. During the application of ERD, most commonly when the sites applied are of lower permeability (10^{-5} cm/sec or less), these conditions can result in the formation of organic acids in the groundwater as part of the degradation process. As a result of the formation of these acids, the ambient pH in the treatment zone can be lowered and in turn conditions conducive to fermentation-based reactions are then created.

This environment can create low pH conditions that are detrimental to methanogenic bacteria and also result in the formation of undesirable byproducts. The formation of undesirable byproducts, including acetone and 2-butanone has been observed at sites where reagent dosing has commenced without careful monitoring of groundwater conditions near the injection wells. The occurrences of these byproducts are generally limited in extent and often sporadic in nature. It is expected that these ketones are also utilized by microbes in the IRZ. However, the possibility of production of these byproducts needs to be accounted for in the project planning stage.

Therefore, the lessons learned regarding these potential occurrences are as follows:

- Careful and regular monitoring of groundwater within the treatment zone should be provided in order to ensure that pH levels are not depressed (pH < 5.0) and TOC levels are not excessive (site specific, but generally above 5,000 to 10,000 mg/L).

- The remedial plan for application of ERD should be flexible enough to allow for modification of both the delivery frequency and mass of organic carbon delivered to prevent the build-up of organic carbon and creation of conditions amenable to creation of these byproducts. Modifications in reagent delivery should be tied to the regular pH and TOC monitoring in the treatment zone.

Overcoming Oxidizing Conditions/High Groundwater Velocity. As discussed, the application of ERD relies on the creation of a highly reducing biogeochemical environment through the provision of excess organic carbon to the groundwater. These conditions must be present throughout the treatment zone. Achieving these conditions can be problematic in groundwater flow systems in which the ambient conditions are very oxidizing (due to shallow groundwater with abundant recharge) or the groundwater velocity is very high ($\geq 1,000$ ft/year). In both situations, the possibility exists for the amount of reagent needed to "overcome" the naturally oxidizing conditions to become either too great to easily deliver to the zone (narrow zones of influence from injection points) or in some cases, the reagent costs become too costly.

In high groundwater velocity settings the limited transverse dispersion in groundwater can limit the extent of the reactive zone created by an individual injection point. This is of particular importance in settings where drilling costs may be high, i.e., deep settings or complex geology. In such cases, these site specific considerations need to be weighed against other treatment alternatives.

Application in Areas of Low Constituent Concentration. The application of ERD to portions of an aquifer where the constituent concentrations are low (i.e., less than 100 ug/L) can pose additional challenges. A low concentration plume will impart less microbial conditioning, and it will therefore be even more difficult to enhance this microbial community. In these environments, a longer lag time for microbial growth and conditioning therefore should be expected. It is also more difficult to provide direct evidence of degradation through the monitoring program in a low concentration plume. The lessons learned in application of ERD in areas of low initial constituent concentration include the following:

- Data quality (i.e., detection limits) for groundwater analyses (both for the constituents and by- or end-products) must be sufficient to show remedial progress. This is especially true for measuring ethane and ethene.

- During the project or pilot test planning stages, the criteria by which successful application of the technology will be defined and determined should be evaluated and clearly defined to the clients and regulators.

Application in Areas of High Constituent Concentration/DNAPL. Given the inherent problems with the use of conventional remediation techniques in areas where the constituent concentrations are very high and/or where free phase constituent (DNAPL) may be present, ERD has been an attractive potential alternative for these settings. The benefits of applying ERD in high concentration regimes (>50 to 100 mg/L CVOCs) is related to the natural surfactant effect that usually accompanies this technique. When the groundwater equilibrium is altered, the transfer of more constituent mass from the free or adsorbed phase into the dissolved phase should be expected. An increase in the levels of dissolved constituents in groundwater results in an increase in a more treatable portion of the total CVOC mass. This effect can be used by itself or in conjunction with other ongoing technologies (such as pump and treat) to reduce treatment life span and costs. Care needs to be taken that desorption does not result in the vertical or horizontal migration of elevated dissolved concentrations away from the treatment area (i.e., expansion of the constituent plume).

The possibility of enhancing migration to off-site areas or sensitive receptors is even more pronounced when applying ERD in a potential DNAPL environment. Therefore, prior to ERD application in these settings a clear plan to address these possibilities must be available. This could include application of the technology on an "outside-in" type of approach in which the lower concentration areas downgradient of the areas of higher concentration are treated initially, or an approach in which the high concentration area is avoided altogether and the technology is applied in a containment role.

However, if properly accounted for, the possibility of concentration increases and/or migration of the impacts can be overcome and ERD can successfully be applied in these settings. The application of ERD in these areas will increase the levels of mass reduction in the subsurface, and once the initial disruption in phase equilibrium is overcome it can be expected that the technology will provide greater control of constituent migration from the source area.

SUMMARY & CONCLUSIONS

The application of ERD to treat CVOCs in groundwater at many sites located in varied in-situ hydrogeologic settings under different concentrations has provided a valuable knowledge base that has taught many lessons to be minded in future applications of the technology both at the pilot-and full-scale. These lessons learned are also applicable to applying other in-situ remedial techniques. Specific lessons learned have related to the changes in the ambient conditions in the groundwater system (such as production of natural surfactants and/or biodegradation byproducts), the need for collecting applicable baseline and ongoing groundwater monitoring data, and the need for flexibility in planning and application of the technology.

As expected, the possession of a strong understanding of the hydrogeologic conditions in the treatment area as well as the surrounding areas has been determined to be the most important factor for successful application of the technique. Without an understanding of these details it will likely be problematic or even impossible to apply the technology much less do so and achieve the remedial goals set out for the program.

COMPARISON OF NATURAL AND ENHANCED ATTENUATION RATES THROUGH SUBSTRATE AMENDMENTS

John F. Horst; Kurt A. Beil; Jeffrey S. Burdick; Suthan S. Suthersan, Ph.D. (ARCADIS Geraghty & Miller, Inc. Langhorne, Pennsylvania)

Abstract
The successful use of substrate amendments to enhance the anaerobic degradation of chlorinated volatile organic compounds (CVOCs) in groundwater has been documented by numerous published case studies. The level of success is typically based on a direct comparison of baseline (pre-treatment) constituent concentrations with post-treatment concentrations. Such direct comparison is sufficient to confirm the desired reduction of concentrations, but does so without consideration of the time intervals, or rates of attenuation involved. This study explores the extent to which first-order attenuation rates can be used to quantify the enhanced degradation of CVOCs achieved through the use of substrate amendments.

Four sites impacted with CVOCs were selected for this study. Groundwater analytical data collected from each of the four sites indicated conditions amenable to on-going natural transformation of CVOCs in groundwater. The subsequent use of substrate amendments at these sites resulted in a significant reduction of CVOC concentrations. For each site, first-order attenuation rates were calculated for CVOCs detected at individual wells over time using exponential regression methods. These "enhanced" rates were then compared to pre-treatment (natural) attenuation rates where sufficient data was available, as well as to published rates. The results of the study indicate that substrate amendments can speed up attenuation of CVOCs by reducing constituent half-lives observed under natural conditions by as much as 60%.

Both natural and enhanced attenuation rates can be affected by outside factors, most notably the initial constituent concentrations targeted for treatment and the site-specific hydrogeology. Additional data is needed to determine the effect of these factors on the ability to achieve enhanced attenuation rates.

Introduction
Anaerobic reductive dechlorination is a well-documented process whereby microbes can fortuitously degrade CVOCs in groundwater. The microbes utilize a primary substrate as a carbon and energy source, producing enzymes and co-factors that degrade other organic compounds present in the groundwater. The term reductive dechlorination stems from the method by which these reactions systematically strip chlorine atoms from CVOC molecules present in the groundwater. In order to facilitate their respiration while they metabolize available carbon-energy source material, subsurface microbes must utilize electron acceptors. As electron acceptors are depleted, the groundwater environment becomes increasingly reducing (lower oxidation-reduction potential)

and the microbes are forced to use successively alternative electron acceptors. The general sequence of the utilization of these alternate electron acceptors is as follows: oxygen, nitrate, ferric iron, manganese, sulfate, and carbon dioxide.

More highly chlorinated volatile organic compounds are typically the most susceptible to reductive dechlorination because of their higher state of oxidation (McCarty 1996). Consequently, dechlorination of CVOCs with fewer chlorine atoms in their molecular structures generally requires more strongly reducing conditions. Often a groundwater environment is not naturally reducing enough and the groundwater has been naturally depleted of an electron donor substrate (organic carbon). Both conditions can result in slowed CVOC attenuation rates and an accumulation of intermediate degradation compounds.

The introduction of substrate amendments into a groundwater environment to increase microbial activity and push the anaerobic dechlorination process to completion is an innovative technology known as enhanced reductive dechlorination (ERD). Case studies documenting the use of several types of electron donor reagents to promote ERD have been published over the last three years (i.e., USEPA 1999, *Engineered Approaches to In Situ Bioremediation of Chlorinated Solvents: Fundamentals and Field Applications*, Draft Report). Typically, the level of success at sites where ERD has been implemented is based on a direct comparison of pre- and post-treatment constituent concentrations. Such direct comparison is sufficient to confirm the desired reduction of concentrations, but does so without consideration of the time intervals, or rates of attenuation involved.

Four sites at which ERD has been successfully implemented to address CVOCs on either a pilot- or full-scale basis were selected for review and application of first order rate calculations. The selected sites are listed below with a brief summary of site conditions and relevant background information. Many of these sites also contained hexavalent chromium in groundwater, which was also targeted for remediation using the altered groundwater chemistry created by the ERD process.

Site No. 1: Former plating facility in California
Historic groundwater sampling data for Site No. 1 identified groundwater impacts including "source" CVOCs such as trichloroethene (TCE), as well as biotic degradation compounds such as cis-1,2-dicloroethene (cis-1,2-DCE) and vinyl chloride (VC). Total CVOC concentrations in the treatment area ranged up to approximately 10 parts per million (ppm). A six-month ERD pilot test was implemented in the spring of 1996 followed by a 24-month full-scale application initiated in the spring of 1997. The geology underlying the site is comprised of interbedded silts and clays. The ERD approach was successful in obtaining closure at this site after two years of application.

Site No. 2: Former plating facility in eastern Pennsylvania
Similar to Site No. 1, historic groundwater impacts at Site No. 2 included TCE and biotic degradation compounds such as cis-1,2-DCE and VC. Total CVOC concentrations in the treatment area ranged up to approximately 1.7 ppm. A six month ERD pilot test was initiated at the site in 1996 followed by a 16 month full-scale application. The implementation of ERD at this site involved treatment of shallow groundwater within glacially deposited silts and sands. The CVOC mass removed during the 16 month full-scale ERD application exceeded the mass previously removed via 10 years of groundwater extraction.

Site No. 3: Manufacturing facility in South Carolina
Historic groundwater sampling at Site No. 3 identified "source" CVOCs, primarily carbon tetrachloride (CT), with lesser concentrations of tetrachloroethene (PCE) and TCE. The presence of chloroform, dichloromethane, chloromethane, and cis-1,2-DCE suggested biotic degradation was on-going at the site. Total CVOC concentrations ranged up to 5 parts per million. A six month ERD pilot test was implemented at the site in 1998. Geology at the site is comprised of saprolite (weathered bedrock residuum consisting of sands, silts, and clay). This site is currently being evaluated for a full-scale ERD implementation.

Site No. 4: Former drycleaning facility in Wisconsin
Source CVOC impacts at Site No. 4 were limited to PCE. Biotic degradation of the PCE had produced significant concentrations of TCE, with lesser concentrations of cis-1,2-DCE and VC. Total CVOC concentrations in the treatment area ranged up to approximately 1.5 parts per million. A full-scale implementation of ERD was initiated in 1998. Geology at the site is comprised of glacially-deposited interbedded silts and clays, with a layer of higher permeability silty sand targeted for treatment. The full-scale ERD application is still underway, with PCE concentrations having declined to below laboratory detection limits.

The following section discusses the methodology used to evaluate the groundwater analytical data collected at each of the selected sites.

Methods
The rate of attenuation of a dissolved constituent through both destructive processes (i.e., biological and chemical) and non-destructive processes (i.e., dilution, dispersion, volatilization, and sorption) can often be estimated using first-order kinetics. Using data from individual monitoring wells at each of the four sites, both historical (pre-treatment) attenuation rates and attenuation rates achieved through the use of ERD were calculated for select CVOCs detected over time.

In each case, the rates were determined using exponential regression techniques at individual wells (Buscheck and Alcantar, 1995). For each well selected, constituent concentrations were plotted versus time (in days). An exponential regression was then fit to the resulting data set. On each plot, the following were posted to describe the exponential regression:

- The equation describing the exponential regression in the form $y_f = y_o e^{kx}$, where: y_o represents the initial constituent concentration; y_f represents the final constituent concentration; x represents the time element in days; and k represents the first order attenuation rate constant.

- The correlation coefficient (R^2) measures how well the regression equation represents the trend in the data. Regressions with R^2 values at and above 0.8 are generally considered to be useful.

Once the first-order attenuation rate constant (k) was determined, the attenuation half-life was calculated directly.

For the purposes of this study, only data obtained from observation wells were used to evaluate enhanced attenuation rates. CVOC data collected from injection wells are biased by the effects of dilution. Observation wells installed as part of an ERD well network are typically positioned sufficiently downgradient of the injection wells to mitigate the effects of dilution from the low volumes of substrate amendment injected. Increases in degradation product concentrations observed in conjunction with simultaneous decreases in "source" CVOC concentrations further support the enhancement of destructive processes (e.g., biotic degradation) rather than physical processes (e.g., dilution) at the observation wells. Thus, it is reasonable to conclude that groundwater analytical data obtained from observation wells are most representative of the attenuation processes occurring in the reactive zone established by the ERD injections.

As could be expected, within the data sets evaluated for each of the four sites, there were instances where constituent concentrations were below the limits of laboratory detection. Depending on the number of "non-detect" analytical results in a single data series and the magnitude of the reported laboratory detection limit, these points represented the potential for significant bias of the estimated attenuation rates. To minimize this potential, USEPA guidance (USEPA, 1998) was followed with respect to the treatment of these data points, as follows:

- Only data sets with less than 15 percent non-detects were utilized.

- Where detection limits were at or below the regulatory standard for the compound being analyzed, one half of the detection limit was plotted as the analytical result.

- Where detection limits were greater than the regulatory standard, but less than the previously detected concentrations of the compound being analyzed, one half of the detection limit was plotted as the analytical result.

Where detection limits were greater than both the regulatory standard and the previously detected concentrations of the compound being analyzed, the corresponding data point represented the greatest potential to skew the attenuation rate estimations. These data points were therefore considered to be invalid for this analysis and were omitted (high method detection limits are sometimes observed in laboratory analyses associated with ERD due to interference from dissolved gasses).

In addition to estimating enhanced attenuation rates, trend plots of the ERD monitoring data were generated for the selected observation wells. The trend plots included available analytical results for each compound in the complete reductive dechlorination sequence for the "source" CVOC involved. For example, the complete reductive dechlorination sequence for carbon tetrachloride yields chloroform, then dichloromethane, then chloromethane, then methane, and finally carbon dioxide and water.

The success of the ERD process in driving the reductive dechlorination of the source compounds to completion could be confirmed by examination of the trend plots (Wiedemeier, et al, 1998). This line of evidence was used as a final gauge to determine if biotic degradation was actually being enhanced, as opposed to physical processes (dilution, dispersion, etc.).

Results and Discussion
Using the methodology above, specific observation wells were targeted for evaluation at each site. Examples from Site No. 2 are provided in order to illustrate the evaluation that was completed for all four sites.

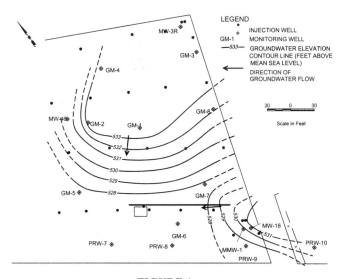

FIGURE 1

Figure 1 depicts the ERD injection and monitoring well network at Site No. 2, with groundwater elevation contours depicting flow direction. Based on the layout of the well network, Observation Well GM-1 was selected for trend evaluation. This well is located several months downgradient (based on groundwater flow rate) of the closest injection wells.

Trend plots of CVOC concentrations in groundwater samples collected during the implementation of ERD were then generated for Observation Well GM-1 and also for the wells selected at the other three sites. In each case, the trend plots confirmed the evidence of enhanced biotic degradation.

FIGURE 2

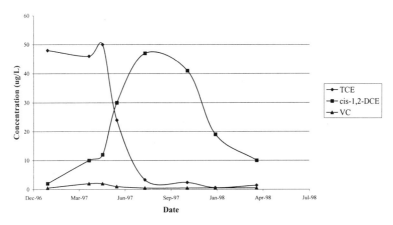

CVOC Concentration Trends During ERD Implementation, Observation Well GM-1, Site No. 2

Figure 2 depicts the trends observed at Site No. 2 during full-scale implementation of ERD. As Figure 2 illustrates, initial concentrations of the "source" CVOC at the site (TCE) increased slightly and then began a rapid decline. This initial increase in concentrations is typical of ERD implementation and is caused by microbial metabolic enzymes that desorb constituents from the soil matrix. This process is known as the "surfactant effect". As TCE was degraded, concentrations of it's primary degradation product (cis-1,2-DCE) increased, eventually reaching a peak before beginning a similar rapid decline. Concentrations of VC fluctuated around the laboratory detection limits.

Following the well selection and line of evidence evaluation, attenuation rates were determined. This portion of the study focused primarily on the "source" CVOCs at each site, namely PCE, TCE, and CT. Establishing reliable rates for the degradation products of these "source" CVOCs was difficult due to the fluctuation in concentrations illustrated by the trend plots.

Attenuation rates observed following implementation of ERD were calculated for each of the four sites. Historical groundwater analytical data sufficient for estimating pre-treatment (natural) attenuation rates were also available for Site No's. 1, 2, and 3. Using exponential regression techniques, Figure 3 depicts the estimation of TCE attenuation at Site No. 2 following implementation of ERD.

FIGURE 3

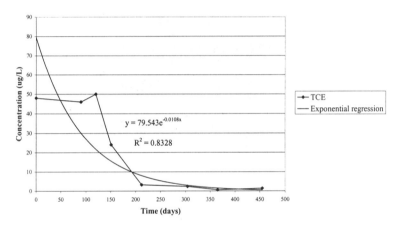

The results of both the historical and post-treatment attenuation rate calculations are summarized in Table 1. As shown in Table 1, the enhanced attenuation rates were significantly faster than the historical rates. The enhanced half-lives were, on average, approximately 60 percent less than the historical half-lives. As a conservative check, the enhanced half-lives were also compared to published half-life values for anaerobic degradation. The results were roughly the same, with the enhanced half-lives, on average, approximately 67 percent less than the published rates.

This evidence indicates that the ERD process is in fact effective in significantly accelerating the natural degradation process at sites where it is appropriate to implement.

At one of the four sites (Site No. 3), the published anaerobic degradation half-life used for comparison was lower than the observed enhanced attenuation rate. Review of the data for this site indicates that the magnitude of both natural and enhanced attenuation rates were affected by outside factors. Although not addressed by this study, it appears that the primary factors involved would be the initial constituent concentrations targeted for treatment and the low permeability site hydrogeology. Additional data is being gathered to quantify the effect of

CVOC concentrations and hydrogeology on the ability to achieve enhanced attenuation rates.

TABLE 1

Attenuation Rate Summary

Site	Published Half Life (days)	Historical k	R^2	Half Life (days)	Post-Treatment k	R^2	Half Life (days)	Pub. Rate Enhancement (%)	Hist. Rate Enhancement (%)
Site No. 1									
TCE	876	-0.0016	0.995	433	-0.0027	0.864	257	71	41
Site No. 2									
TCE	876	-0.0036	0.883	193	-0.0108	0.833	64	93	67
cis-1,2-DCE	416	-0.0041	0.967	169	-0.0067	0.913	103	75	39
Site No. 3									
CT	18	-0.0006	0.626	1155	-0.0098	0.857	71	--	94
Site No. 4									
PCE	98	--	--	--	-0.0104	0.831	67	32	--

Notes:
Published half-life values are averages of anaerobic half-lives
(Howard, Boethling, et al, 1991)

References

Buscheck and Alcantar, 1995. Regression Techniques and Analytical Solutions to Demonstrate Intrinsic Bioremediation. Intrinsic Bioremediation, Battelle Press.

Howard, P., Boethling, R., et al, 1991. Handbook of Environmental Degradation Rates. Lewis Publishers, Inc.

McCarty, P., 1996. Biotic and Abiotic Transformations of Chlorinated Solvents in Groundwater, Symposium on Natural Attenuation of Chlorinated Organics in Groundwater, EPA, September 11-13, Dallas, Texas.

USEPA, 1998. Guidance for Data Quality Assessment, Practical Methods for Data Analysis. EPA/600/R-96/084.

Wiedemeier, Todd H., et al., USEPA, 1998. Technical Protocol for Evaluating Attenuation of Chlorinated Solvents in Groundwater. EPA/600/R-98/128.

OPTIMIZATION STRATEGY FOR ENHANCING BIODEGRADATION IN AN UPLAND-WETLAND PLUME

W. Andrew Jackson (Texas Tech University, Lubbock TX)
John Pardue **(Louisiana State University, Baton Rouge LA)**
G. Nemeth and T. DeReamer (General Physics, Edgewood, MD)
Dean McInnis (CFR Technical Services, Baton Rouge, LA)
Don Green (U.S. Army, Aberdeen MD)

ABSTRACT: Treatability studies were conducted on aquifer material and groundwater slurries at seven locations encompassing a range of contaminant concentrations and terminal electron accepting conditions at the Lauderick Creek Area of Concern of Aberdeen Proving Ground, Maryland. Four generalized locations were selected: the primary plume (hot spot), and wells downgradient in the south, southeast, and northeast directions. The objectives of the study were two-fold: to determine the factors controlling biodegradation of chlorinated solvents and to assess whether addition of organic acids, yeast extract and vitamin B_{12} in various combinations has the potential to accelerate the rate or extent of biodegradation. Amendments were successful at accelerating the biodegradation of 1,1,2,2-tetrachloroethane (the primary parent compounds observed at all locations tested) in most locations. The two effective treatments were those in which butyrate and lactate were added. At PLC-6 in the main source area, concentrations of 1,1,2,2-TeCA were reduced from 17,000 µg/L to below 2,000 µg/L with the subsequent formation of *trans*-1,2-DCE and ethene. The two effective treatments were those in which butyrate and lactate were added in conjunction with yeast extract. At WLC-80A (immediately adjacent to the marsh) and PLC-01B (within the marsh) amendments resulted in complete degradation of parent 1,1,2,2-TeCA to ethene and ethane. Some intrinsic degradation of 1,1,2,2-TeCA was observed in the marsh controls. Results indicated that a population of halorespiring organisms exists naturally at the site. Creating a more reducing environment with ample H_2 concentrations can stimulate these organisms.

INTRODUCTION

In aquifers that are aerobic, or in anaerobic aquifers with low hydrogen concentrations, addition of readily degradable organic compounds can be performed to create better conditions for biodegradation. These compounds are usually volatile acids such as benzoate, acetate, formate, propionate or alcohols such as methanol. Addition of these compounds provides two functions: first, consuming the remaining electron acceptors that remain in the aquifer such as oxygen, nitrate or iron, and secondly, acting as a direct source of substrate for the organisms or undergoing fermentation to produce the hydrogen needed by the dechlorinating organisms (Fennell et al., 1997; Gerritse et al., 1999, Kengen et al., 1999, Smatlak et al., 1996; DiStefano et al., 1992). Treatability studies are required to identify the compound or mixtures of compounds that will optimize the biodegradation rate of chlorinated aliphatics. Recent evidence suggests that compounds such as propionate that release hydrogen slowly are preferred (Yang and McCarty, 1998). However, in other aquifers, compounds that generate hydrogen rapidly, i.e., benzoate, have been found to be effective (Beeman et al., 1994).

Objective. The objective of the treatability study is to identify the treatment(s) that increases the rate of anaerobic biodegradation and drives the dechlorination reaction to completion. Studies were conducted on aquifer material from the Lauderick Creek area of concern at Aberdeen Proving Ground (APG). This area historically served as a training facility for chemical ordinance. The surficial aquifer discharges into freshwater marsh areas fringing Lauderick Creek, a tidally influenced tributary of the Bush River. Four generalized locations were selected for the treatability study: the primary plume (hot spot), and wells downgradient in the south, southeast, and northeast directions. These locations include a range of geochemical groundwater types and VOC concentrations.

MATERIALS AND METHODS

Groundwater (4 L) was sampled from the selected wells into glove bags held under a N_2 atmosphere after well stabilization. Cores were obtained by direct push, sealed and transported to the field lab at APG. Groundwater and associated cores were homogenized under a nitrogen blanket and the resulting slurry from each location transferred to 125 ml serum bottles. GC-MS analysis was performed to determine the type and concentrations of chlorinated aliphatics. In the laboratory, slurries were amended with organic acids, yeast extract and vitamin B_{12} as described below. Each combination of organic compound (electron donor) and groundwater location was performed in triplicate.

The treatments used in the biostimulation study are described in Table 1. In general, the study design is based on a proposed treatability protocol developed by Battelle (1998). There were 7 treatments used in the study. Each of these treatments represents some combination of yeast extract, vitamin B_{12}, and electron donors (lactate, benzoate and butyrate). Groundwater samples contained measurable concentrations of 1,1,2,2-tetrachloroethane (1,1,2,2-TeCA), trichloroethene (TCE), 1,2-dichloroethane (1,2-DCA), chloroethane (CA), cis-1,2-dichloroethene (cis-DCE), trans-1,2-dichloroethene (trans-DCE), vinyl chloride (VC), ethene and ethane.

Three aspects of the microcosms were monitored initially, after 6 weeks and after 14 weeks: VOCs (using EPA Method 8260B), light gases (methane, ethane, ethene, CO_2, and H_2) using gas chromatography (GC-FID for hydrocarbons and a reduction gas detector for H_2), and organic acid concentrations (lactate, succinate, formate, butyrate, propionate, acetate and benzoate).

TABLE 1. Treatments used in the present study

Treatment	Individual Donor	Yeast extract	Vitamin B_{12}
A	None (autoclaved, abiotic "killed" control)	--	--
B	None (biotic control)	--	--
C	None	20 mg/L	0.05 mg/L
D	Lactate (3 mM)	--	--
E	Lactate (3 mM)	20 mg/L	0.05 mg/L
F	Butyrate (3 mM)	20 mg/L	0.05 mg/L
G	Lactate/benzoate (1.5 mM each)	20 mg/L	0.05 mg/L

RESULTS AND DISCUSSION

Main Plume (Source Area). Treatability studies were conducted in piezometer PLC-6 as an area representative of the source area. This site appears to be the source of VOCs for most of the Lauderick Creek Area Of Concern in addition to possibly containing some free phase DNAPL material. A focused feasibility study performed in 1998 and early 1999 reported high concentrations of VOC's in this area with 1,1,2,2-TeCA (~5,000-20,000 µg/l) and TCE (~1,000 µg/l). Natural attenuation parameters indicated that the biogeochemistry of the subsurface in this area was sulfate-reducing.

Two treatments (E and F) stimulated reductive dechlorination of 1,1,2,2-tetrachloroethene by conversion to *trans*-1,2-DCE over the first 14 weeks of the study. Transformation only occurred after a lag time that lasted between 6 and 14 weeks. No conversion was detected in the controls or killed controls. Treatments that enhanced biodegradation of the parent 1,1,2,2-TeCA at this location were those in which an electron donor (lactate or butyrate) was added in conjunction with the yeast extract. Treatment F (butyrate) was the most promising treatment, stimulating reduction of 1,1,2,2-TeCA from 17,000 to less than 2,000 µg/L over the first 14 weeks of the study (Figure 1).

In treatments A, B, and C with no added electron donors, concentrations of organic acids were low throughout the study, which is consistent with an aquifer location with relatively low carbon content. Treatments D, E, and G all involved the addition of lactate, which fermented to propionate and acetate within several weeks. In Treatment F, the butyrate added was fermented to acetate. In Treatment G, the lactate added was fermented to acetate and the benzoate also was degraded. Measured H_2 concentrations (>20 nM) were very high in the amended samples. Interestingly, no methane was detected in these samples until 1,1,2,2-TeCA was degraded. It is likely that these very high concentrations of parent VOC are inhibitory to methanogens.

Based on these results, PLC-6 is a candidate for a field application of this technology. Successful biostimulation will not only decrease the time for natural attenuation to reduce chlorinated solvent concentrations but also reduce contaminant loads on downgradient portions of the site which are undergoing biological natural attenuation presently. There is also the potential for increasing the rate of dissolution of any free phase product.

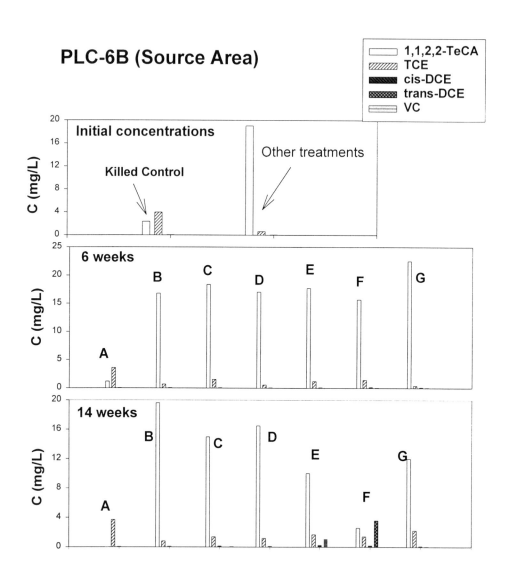

Figure 1. Chlorinated VOCs in PLC-6 microcosms over time. Treatments A (Killed control), B (Control), C (Yeast extract (YE)), D (Lactate alone), E (Lactate + YE), F (Butyrate + YE), G (Lactate + Benzoate + YE)

Southern Plume Area (WLC-80A). WLC-80A is on the edge of the Lauderick Creek marsh downgradient of the primary source area. VOC's in this are primarily 1,1,2,2-TeCA (6,283 μg/l). WLC-80A is located within an area classified as methanogenic. This location is important because high concentrations of 1,1,2,2-TeCA are found in groundwater, which is near a discharge area into the surface marsh. This area represents a potentially successful location for biostimulation given the high concentrations of 1,1,2,2-TeCA, ideal biogeochemical conditions, and possible ongoing reductive dechlorination. The location represents an opportunity to degrade the compounds before reaching the more sensitive marsh environment.

Reductive dechlorination of 1,1,2,2-TeCA was stimulated in treatments D, E, F, and G. Degradation was observed in all treatments where organic acids were added in conjunction with yeast extract. The most effective treatment was treatment E, however; essentially complete biodegradation to ethene was observed in all treatments where organic acids were added. No biodegradation has been observed in the control, killed control or when yeast extract was added alone. Despite the presence of methane production, no direct evidence of dechlorination was detected over 14 weeks in the controls. This likely indicates that concentrations of donors are not sufficient naturally at this location to drive fast dechlorination. Copious amounts of methane were produced in the amended samples. Methane was also detected in the control samples indicating the presence of methanogenic conditions in the microcosms. In treatments A, B, and C with no added electron donors, concentrations of organic acids were depleted rapidly during the study. Lactate quickly fermented to propionate and acetate within several weeks. The butyrate added was fermented to acetate and propionate, and the benzoate also was degraded. These transformations occurred rapidly and by 14 weeks, organic acids were essentially depleted. These fermentation reactions are important because they result in the production and release of H_2 that serves as the electron donor for the halorespiring organisms.

PLC-01B. This piezometer in the southern plume area is located within the marsh of the southern branch tributary of Lauderick Creek. Previous sampling in this location indicated 1,1,2,2-TeCA concentrations of 6,300 μg/L and non detect (<500 μg/L) concentrations of all other chlorinated compounds; groundwater is classified as methanogenic. This location is important because 1,1,2,2-TeCA may be discharging into the surface marsh.

As expected, degradation of the target compounds was very rapid in the marsh microcosms. Reductive dechlorination was observed in all amended treatments (C-G). In all of these treatments, degradation was complete to ethene and ethane by 14 weeks of incubation. Some evidence of biodegradation was also observed in the control but high concentrations of *cis* and *trans*-1,2-DCE remained after 14 weeks. No ethene was observed in the control samples. High concentrations of methane were observed in all samples indicating methanogenic conditions. Still, the dechlorination appeared "carbon-limited" as addition of the organic acids successfully stimulated biodegradation. Unlike the previous location, WLC-80A, concentrations of organic acids were not fully depleted during the study in treatments A, B, and C. This is attributed to the continued degradation of the marsh organic matter by the fermenting microorganisms. The availability of these donors and their subsequent generation of H_2 is responsible for the degradation observed in these unamended samples. Lactate quickly fermented to propionate and acetate within several weeks, butyrate added was fermented to acetate, and benzoate also was degraded. However, H_2 concentrations did not increase as high as in the other locations. This is likely due to the presence of a large population of methanogens (high CH_4 was observed) that will utilize the H_2.

Stimulation of biodegradation in the marsh environment may be necessary if concentrations of parent VOCs entering the marsh exceed the assimilative capacity. At this location, it appears that degradation of parent VOCs is occurring but could be stimulated to a faster rate. WLC-80A, which was adjacent to the marsh, may be a particularly good location to biostimulate, reducing some of the loading into the marsh itself.

Southeast Plume (WLC-84B). WLC-84B is a piezometer located downgradient and southeast of the primary hot spot at a site near (~200ft) the south branch of Lauderick Creek. This well is located within a smaller local hot spot. VOC's in this area are predominately 1,1,2,2-TeCA (7,470 µg/l). Natural attenuation parameters indicated that the biogeochemistry of the subsurface in this area was sulfate-reducing on the border of a downgradient methanogenic area. This area is an important location given its close proximity to Lauderick creek, its high concentration of 1,1,2,2-TeCA, and the lack of significant evidence of biological reductive dechlorination. Since additional hot spots may exist in the area, biostimulation in the main plume (PLC-6B) may have no effect on the plume at this location. At WLC-84B there was inconclusive evidence of whether biodegradation was enhanced by the addition of organic acids. Initial concentrations of VOCs were remarkably similar to concentrations determined in previous sampling events (i.e., 1,1,2,2-TeCA concentrations of 7,000-9,000 µg/L versus 7,470 µg/L measured previously). Over the course of the study, decreases in concentrations of 1,1,2,2-TeCA were observed in treatments E, F and G. These treatments were shown to stimulate biodegradation at the other locations in the study. Decreases in concentrations were approximately 30% over the first 14 weeks of the study. However, the loss of 1,1,2,2-TeCA was not balanced by the appearance of degradation products that we were monitoring. No ethene or ethane was observed. Concentrations were relatively stable in treatments A-D. Methane was not produced in the samples that create further uncertainty about whether degradation has occurred here. Decreases in concentration can be caused by a number of factors including experimental artifacts. However, it is likely that concentrations in the bottle decrease due to sorption over time.

WLC-22. WLC-22 is a piezometer located downgradient and just east of well WLC-84B on the edge of a sulfate reducing zone and a methanogenic zone. 1,1,2,2-TeCA is the main chlorinated solvent present (3,700 µg/L). Little or no evidence is present for intrinsic 1,1,2,2-TeCA reductive dechlorination. This area is an important location given its close proximity to Lauderick creek, its high concentrations of 1,1,2,2-TeCA, and the lack of significant evidence of biological reductive dechlorination. Biodegradation was stimulated in treatments E-G. Degradation was observed in all treatments where organic acids were added in conjunction with yeast extract. The most effective treatment was treatment G, however, substantial biodegradation was observed in all treatments where organic acids were added. No biodegradation has been observed in the control, killed control or when yeast extract was added alone. The primary intermediate observed was *trans*-1,2-DCE indicating biodegradation was through a dihaloelimination mechanism (1,1,2,2-TeCA to *trans*-1,2-DCE) followed by hydrogenolysis (to VC). VC was detected in treatments F and G. Copious amounts of methane were produced in the amended samples. Methane was also detected in the control samples indicating the presence of methanogenic conditions in the microcosms.

As in the other locations, low concentrations of organic acids were observed in treatments A, B, and C. These were rapidly depleted by approximately 6 weeks. In the lactate amended treatments (D, E, G), fermentation to proprionate and acetate were observed but much more slowly than in water from the wetland or near wetland locations (WLC-80A). Treatments F (butyrate added) also demonstrated fermentation of butyrate acid to acetate and propionate but with much slower degradation and larger residuals that was observed in the wetland piezometers. Benzoate was also degraded but at a comparatively slower rate to the other locations tested. High concentrations of H_2 were generated in only a few microcosms (>10 nM in treatments D, E, F and G).

Northern Plume. WLC-45 is a piezometer located northeast of the main plume hot spot on the northern side of the peninsula divide. Groundwater movement is in the direction of the north tributary of Lauderick creek. WLC-45 is located within a large area of the saturated zone classified as iron reducing and no specific VOC information is available.

Biodegradation was stimulated in treatments E-G. As in WLC-22, degradation was observed in all treatments where organic acids were added in conjunction with yeast extract. All three treatments (E-G) gave similar results. Biodegradation reduced concentrations of 1,1,2,2-TeCA by approximately 75% by week 14. No biodegradation has been observed in the control, killed control or when yeast extract was added alone. Copious amounts of methane were produced in the amended samples but not in the control samples. As in the other locations, low concentrations of organic acids were observed in treatments A, B, and C. These were rapidly depleted by approximately 6 weeks. Conversion of organic acids followed a similar trend as seen elsewhere but at a much slower rate.

CONCLUSIONS

Results of this study support the importance of conducting lab studies to determine the optimum amendment strategy at various areas of a contaminated system. Amendments were successful in promoting degradation of 1,1,2,2-TeCA in a number of locations. These locations had various initial biogeochemical conditions (Oxic-methanogenic). One site did not exhibit conclusive stimulation of reductive dechlorination further emphasizing the role of lab studies in the application of this technology. The results of this study fully support the initiation of a field trial at various locations within the site using optimized amendment recipes.

REFERENCES

Battelle, 1998. *A Treatability Test for Evaluating the Potential Applicability of Reductive Anaerobic Biological In Situ Treatment Technology (RABITT) to Remediate Chloroethenes*, Draft Technical Protocol. Prepared for US Air Force by Battelle Memorial Institute, Columbus, Ohio. July 1998.

Beeman, R.E., J.E. Howell, S.H. Shoemaker, E.A. Salazar, and J.R. Buttram. 1994. A field evaluation of *in situ* microbial reductive dehalogenation by the biotransformation of chlorinated ethenes. In, Bioremediation of Chlorinated and Polycyclic Aromatic Hydrocarbon Compounds, Hinchee/Leeson/Semprini/Ong (eds.), Lewis Publishers, pp. 14-27.

DiStefano, T.D., J.M. Gossett, and S.H. Zinder. 1992. Hydrogen as an electron donor for dechlorination of tetrachloroethene by an anaerobic mixed culture. *Applied and Environmental Microbiology* 57(8): 2287-2292.

Fennell, D.E., J.M. Gossett, and S.H. Zinder. 1997. Comparison of butyric acid, ethanol, lactic acid and propionic acid as hydrogen donors for the reductive dechlorination of tetrachloroethene. *Environmental Science and Technology* 31:918-926.

Lorah, M.M., L.D. Olsen, B.L Smith, M.A. Johnson, and W.B. Fleck. 1997. Natural attenuation of chlorinated volatile organic compounds in a freshwater tidal wetland, Aberdeen Proving Ground, Maryland. USGS Water-Resources Investigations Report 97-4171, 95 pp.

Smatlak, C.R., J.M. Gossett and S.H. Zinder. 1996. Comparative kinetics of hydrogen utilization for reductive dechlorination of tetrachloroethene and methanogenesis in an anaerobic enrichment culture. *Environmental Science and Technology* 30:2850-2858.

Yang, Y. and P.L. McCarty. 1998. Competition for hydrogen within a chlorinated solvent dehalogenating anaerobic mixed culture. *Environmental Science and Technology* 32:3591-3597.

HRC ENHANCED BIOREMEDIATION OF CHLORINATED SOLVENTS

Willard Murray and Maureen Dooley (Harding Lawson, Wakefield, MA)
Stephen Koenigsberg (Regenesis, San Clemente, CA)

ABSTRACT: Enhanced in situ bioremediation of chlorinated solvents in groundwater has been successfully demonstrated at several sites by supplying lactic acid as an electron donor. The source of lactate for these successful pilot tests is Hydrogen Release Compound (HRC™), a polylactate ester specially formulated for slow release of lactic acid upon hydration.

HRC was delivered to a chlorinated solvent groundwater plume by being contained in perforated canisters hung in wells, it can also be placed directly in a boring or injected into the contaminated aquifer using a direct push technology such as Geoprobe®.

This paper presents the results from three pilot tests where HRC has been used as the electron donor to enhance natural biological destruction of chlorinated solvents. Results show that HRC can effectively enhance the natural attenuation of chlorinated solvents with very efficient degradation half-lives.

INTRODUCTION:

Harding Lawson Associates (HLA) has conducted several field demonstrations to evaluate in-situ enhanced anaerobic biodegradation of chlorinated solvents using Hydrogen Release Compound (HRC). This paper presents the results from three field pilot tests that have been conducted recently.

Anaerobic biodegradation of chlorinated solvents such as perchloroethene (PCE) and trichloroethene (TCE) require highly reducing conditions to allow anaerobic bacteria to reductively dechlorinate the solvents. The technical approach involves provision of a carbon source that ultimately provides electrons used in the reductive dechlorination. Hydrogen actually serves as the ultimate electron donor and it is produced by the fermentation of lactic and other organic acids.

The technology is designed to provide these electron donors and facilitate a more rapid and complete degradation of the organic contaminants to non-toxic compounds such as ethylene.

MATERIALS AND METHODS:

Hydrogen Release Compound (HRC) is a lactic acid ester. When delivered to the subsurface, lactic acid, which has been shown to be an effective electron donor, is released continuously into groundwater. Since the material can be injected directly into the subsurface it can be used to engineer passive systems that enhance active biodegradation of chlorinated volatile organic compounds (cVOCs).

RESULTS:

HLA conducted several field demonstrations to evaluate the performance of HRC. The main objective of these tests was to evaluate the rate and extent of chlorinated solvent biodegradation using HRC as the electron donor source. In addition we also investigated design parameters for future work such as product longevity.

Pilot Test #1 in Massachusetts:

Site Description: The site is situated in a historically industrial section of Watertown, MA. The general soil profile consists of approximately 13 feet of sand and gravel over approximately 7 feet of silty sand; then glacial till (an aquitard) is encountered. Groundwater occurs at approximately 8 feet below land surface and is contaminated with chlorinated solvents, including PCE, TCE and degradation products characteristic of natural biological reductive dechlorination.

Design, Construction and Operation: In this field demonstration, which was the first field-scale test of the HRC product, groundwater is extracted from three downgradient wells, and injected into three wells 17 feet upgradient. Five 2-inch PVC monitoring wells are positioned between the injection and extraction wells to monitor the progress of the biodegradation process: a groundwater recirculation rate of 0.25 gallons per minute (gpm) establishes a single recirculation cell of about 30 feet in diameter.

Operation: The system was in full operation from February 1998 to November 1998 with a relatively constant recirculating flow rate of 0.25 gpm. HRC canisters were placed into the three injection wells beginning in February 1998 to passively deliver the electron donor. VOC and other data, that measured redox conditions and the concentration of electron acceptors (dissolved oxygen, ferrous iron, sulfate, nitrate) and electron donors (total organic carbon (TOC) and volatile organic acids), were collected over the time course of the study.

The pump was shut down in November and HRC remained in the injection wells to act as a barrier. Data were collected in January 1999 , approximately 2 months after the pump was shut down, to evaluate rebound within the treatment cell and to evaluate the effectiveness of HRC as a barrier.

Results: Results from the field demonstration indicate that under anaerobic conditions significant reductions in the concentration of all cVOCs was observed. The average TCE concentration at the beginning of the study was 9900 ug/L and after 206 days levels were reduced to <10ug/L. Initial PCE concentrations of approximately 740 ug/L were reduced to <1 ug/L. Dichloroethene (DCE) levels, which started at an average value of 2500ug/L, remained relatively constant for the first six months then, decreased to <100ug/L. Vinyl chloride levels rose from an average 250 ug/L to 3000 mg/L after the first six months, then decreased to less than 100 ug/L. Ethene was detected in IN-2 and the downgradient monitoring wells indicating complete biodegradation of the cVOCs. The reduction in total mass was calculated to be 97% based on the average concentration of cVOCs in the treatment cell.

Figure 1 shows concentrations of VOCs versus time in well EPA-2. It can be seen that reductive dechlorination is not apparent until 50 days after the testing had started. After 50 days redox levels had dropped from +100mv to less than – 50mv across the entire cell corresponding to the time when reductive dechlorination was first observed. Redox levels were maintained within a range of –150 to –50mv over the remaining duration of the study.

Volatile acid data were collected throughout the study from the injection well. Lactic acid levels were first detected in the range of 70 mg/L and increased to approximately 800 mg/L. The concentration of volatile acids dropped off dramatically by the time groundwater reached downgradient monitoring well EPA-3 with little to no lactic acid detected. Other volatile acids such as propionic acid were sporadically detected at levels between 1-15 mg/L at downgradient monitoring wells. Although volatile acid levels were significantly lower downgradient of the injection well, it was apparent that cVOC biodegradation was continuing based on the continued cVOC reduction observed along the flowpath.

Elevated sulfide levels were detected and dissolved iron levels were elevated across the cell, but there was no observed increase in methane. These results suggest that sulfate reduction and iron reduction were the predominant conditions favored across the treatment cell.

In November 1998 the recirculating pump was shut down and the HRC containing canisters remained in the injection wells. cVOC data were collected approximately 2 months after shut down. These analytical results showed that there was only a limited rebound in the concentration of cVOCs within the treatment cell (Figure 1). Under the conditions of this field demonstration, HRC was still present in the canisters after one year.

Biodegradation rates estimated for the individual cVOCs based on results from the field demonstration are: PCE 0.021 day^{-1}, TCE 0.018 day^{-1}, DCE 0.042 day^{-1}, and VC 0.044 day^{-1}; or half-lives of 33 days (PCE), 39 days (TCE), 17 days (DCE), and 16 days (VC).

Pilot Test #2 in Michigan:

Site Description: The site is located at a former manufacturing plant. The soil profile generally consists of sand and silty sands to a depth of about 25 feet where a low permeability clay is encountered. The water table is approximately 5 feet below ground surface. Hydraulic conductivity of the aquifer varies from 25 to 50 fpd, and hydraulic gradients vary from 0.0002 to 0.0008. Contaminants in the groundwater consist primarily of 1,2 dichloroethene (DCE) and vinyl chloride (VC).

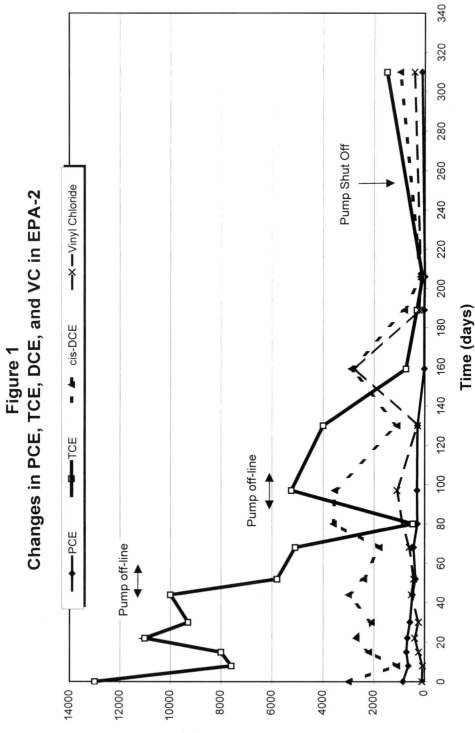

Figure 1
Changes in PCE, TCE, DCE, and VC in EPA-2

Design, Construction and Operation: At this site, the application of HRC is totally passive. HRC was injected into the aquifer at six injection points using a Geoprobe rig. The injection points are staggered in a "W" pattern over an area of approximately 15 feet by 5 feet, with the 10-foot side perpendicular to the groundwater flow. The contaminated groundwater simply flows past the "fence" of HRC which slowly dissolves into the naturally flowing groundwater to provide biological stimulation.

Results: After 60 days, the redox levels were decreased to less than -100 in the monitoring well (MW-26) located 15 feet downgradient from the HRC fence, and was maintained at this level through day 250 of the test. Figure 2 shows concentrations of VOCs versus time in MW-26. Upgradient concentrations remained relatively constant with DCE at approximately 40 mg/L and VC at approximately 10 mg/L. It can be seen that after initial in both DCE and VC, a function of parent product degradation, the concentrations decrease.

For DCE there is a continuous decrease from day 60 through day 301, and that a degradation half life of 80 days is indicated. The decline in VC is much less pronounced, but if the production of VC from the reductive dechlorination of the DCE is accounted for, the biodegradation half-life of VC is similar to that for DCE. Furthermore, the ratio of DCE to VC continually decreases from approximately 5.2 at day 0 to approximately 0.4 on day 301, which is a good indicator that reductive dechlorination is occurring. Upgradient the ratio remained constant at approximately 4. Detection of ethene in the range of 100 to 500 ug/L is further proof of total dechlorination.

Large concentrations of lactic acid fermentation products, acetic acid and propionic acid, were detected in MW-26 after day 30. Propionic acid is also an electron donor. However, after day 255, no organic acids were detected. Also, after day 255 redox levels begin to rise above –100. This would appear to indicate that depletion of HRC is occurring such that insufficient lactic acid or propionic acid is available to drive the reductive dechlorination process to completion. It is therefore concluded that additional HRC should have been added at about day 255 in order to maintain the consistent destruction of the DCE and VC.

Pilot Test #3 in Kansas:
    Site Conditions: The lithology at the site is composed of interbedded clays, silts and silty clays overlying weathered claystone. The depth to groundwater ranges from 5 to 9 feet, and the hydraulic conductivity is between 4.4 and 7.4 fpd. Groundwater velocity has been calculated to be approximately 0.03 fpd with a porosity of 0.40. The groundwater is contaminated with cVOCs including PCE with concentrations as high as 6400 ug/L.

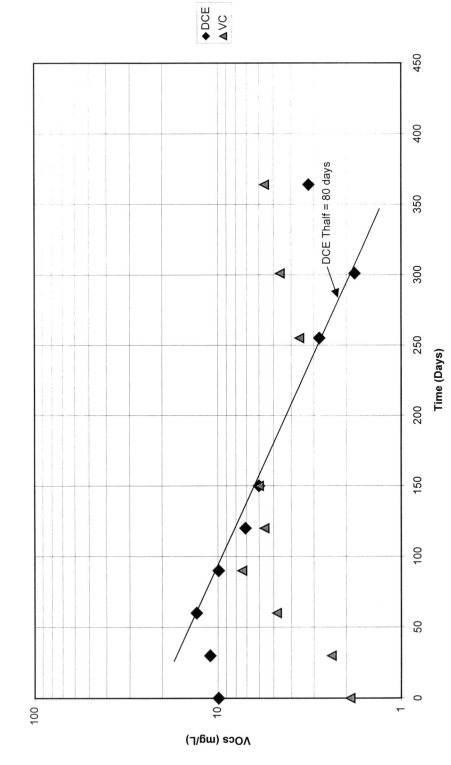

**Figure 2
MW-26 VOC Concentrations**

Design, Construction and Operation: HRC was injected into aquifer system using a direct push technology. Two rows of injection points with a point-to-point spacing of 5 feet were installed perpendicular to the groundwater flow direction approximately 7 feet apart. The upgradient row has 8 injection points and the downgradient row has 7. Each injection point was filled with HRC from 10 to 20 feet below ground surface. The natural groundwater flow simply passes through the "fence" of HRC to form a passive enhanced biodegradation treatment zone.

Results: The system has been operating for approximately 120 days. Upgradient concentrations of cVOCs are approximately 15 to 20 mg/L of PCE, 2 to 5 mg/L each of TCE and 1,2 DCE, and less than 0.3 mg/l of VC. At a monitoring well P-07 located approximately 5 feet downgradient from the HRC fence, the cVOCs as a function of time are shown in Figure 3.

It can be seen that the PCE was degraded from approximately 6 mg/L to 0.2 mg/L within 30 days. TCE was initially approximately 0.8 mg/L and has decreases in magnitude somewhat; VC has maintained its initial concentrations as well, with a slight increase from day 85 to day 120. The behavior of DCE is such that it increases steadily from approximately 0.6 mg/L on day 0 to 15 mg/L on day 120.

Very little ethene has been detected in downgradient monitoring wells, however levels of methane in the range of 0.2 to 0.8 have begun to appear after day 85. Ethene appears to be starting to increase in some downgradient wells at day 120. Levels of sulfate are decreasing in the downgradient wells from approximately 65 mg/L on day 0 to less than 10 mg/L after day 60. Large concentrations (1000 mg/L) of propionic acid, a lactic acid fermentation product that is still an electron donor, has developed within the HRC fence and immediately downgradient.

It appears that the reductive dechlorination process is just getting started and that the acclimation period for total destruction of the cVOCs is not yet over. However, it is apparent that successful reductive dechlorination has begun with almost complete destruction of PCE and the generation of significantly increased levels of DCE. Future sampling episodes will look for DCE to decrease and the complete dechlorination of the cVOCs to occur.

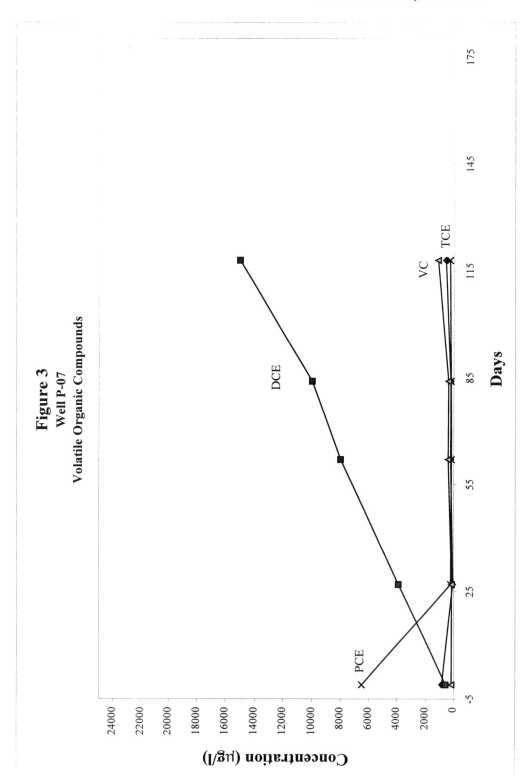

HRC-ENHANCED REDUCTIVE DECHLORINATION OF SOURCE TRICHLOROETHENE IN AN UNCONFINED AQUIFER

Willard D. Harms, Jr. (URS Greiner Woodward Clyde, Tallahassee, Florida)
Kristin A. Taylor, P.E. (URS Greiner Woodward Clyde, Tallahassee, Florida)
Brian S. Taylor (URS Greiner Woodward Clyde, Tallahassee, Florida)

Abstract: An interim measures (IM) effort to accelerate the reductive dechlorination of subsurface trichloroethene (TCE) at the source in a sandy unconfined aquifer in Florida injected 6000 pounds (2721 kg) of Hydrogen Release Compound (HRC™) at an interval from 35 to 65 feet (10.7 to 19.8 m) below land surface in a total of 24 direct-push injection points. The IM objective is to reduce numerically modeled natural attenuation timeframe predictions to less than five years for all ground water contaminants. Performance monitoring results through one year after HRC injection are presented. Significant reductions in TCE concentrations were observed immediately downgradient of the HRC injection area. Increases in ethene concentrations demonstrate that the dechlorination was being driven to completion. The data suggest that the effective life of the HRC amendments within the study area was at least eight months. It is concluded that HRC significantly accelerated the reductive dechlorination of TCE to completion until the HRC was depleted. Post-IM natural attenuation modeling, pending such time as plume equilibrium is re-established, will be used to assess whether the IM objective was met. It appears that application of HRC within a TCE source zone can effectively deplete the source if properly placed in sufficient quantity.

INTRODUCTION

Cost effective active remediation of ground water contaminated with chlorinated solvents to levels that meet ground water standards has proven to be a challenge with few efforts reporting complete success. Particularly prevalent and problematic are tetrachloroethene (PCE) and trichloroethene (TCE).

Understanding of the theory behind biological mechanisms governing transformation of chlorinated solvents in the environment has been an important goal for the environmental restoration industry in recent years. United States Environmental Protection Agency (USEPA) Publication EPA/600/R-98/128 entitled *Technical Protocol for Evaluating Natural Attenuation of Chlorinated Solvents in Ground Water* (Wiedemeier et al., 1998) presents a particularly thorough overview of advances in this direction.

Relatively recent recognition of monitored natural attenuation (MNA) by USEPA as a viable remedial technology (USEPA, 1999), at sites where its efficacy can be demonstrated, has increased activity in this area. In addition to being protective of human health and the environment, regulatory approval of MNA as part of a remedy is likely to be based on demonstrable source control measures and long-term performance monitoring. An additional component for regulatory approval of MNA is the expectation (and demonstration) that natural attenuation will occur in a reasonable time frame. A "reasonable" MNA timeframe is defined by USEPA in its MNA policy directive (USEPA, 1999) as that comparable to other alternatives. In contrast, the Florida Department of Environmental Protection (FDEP) appears to define a reasonable MNA timeframe to be not more than five years.

A fate and transport model conducted for the petroleum and chlorinated solvent plumes at the subject site suggested that PCE and TCE are the only ground water contaminants that would not attenuate naturally within five years (URS Consultants, 1998). It is on this basis that an interim measure (IM) was implemented in an effort to eliminate the remaining source of PCE and TCE dissolution to the intermediate surficial aquifer.

Objective. The IM objective is to reduce the occurrence of PCE and TCE from the intermediate zone of the surficial aquifer to such levels that natural attenuation is accelerated to a time frame not to exceed five years from implementation. The interim measure selected for implementation was the introduction of HRC into the intermediate zone of the surficial aquifer within the area of greatest historical occurrence of chlorinated solvent concentrations (URS Greiner Woodward Clyde, 1999).

Site Layout. Figure 1 shows the site layout, including monitoring well locations, local ground water flow direction, and the HRC™ injection/treatment area.

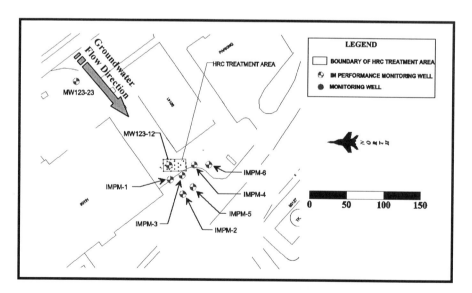

FIGURE 1. Site layout showing HRC™ injection/treatment area, performance monitoring wells, and ground water flow direction.

MATERIALS AND METHODS

Supplemental dissolved oxygen, oxidation-reduction potential, and iron measurements were obtained from the intermediate zone of the surficial aquifer. Ground water level measurements were collected to verify groundwater flow direction in the intermediate aquifer. Soil and groundwater samples suitable for bench-scale treatability testing were collected and shipped to Regenesis Bioremediation Products, the developer and supplier of HRC. Regenesis performed a microcosm treatability test and an Aquifer Simulation Vessel (ASV) treatability test. The findings of the microcosm test and ASV testing were used to prepare the HRC injection design. An IM performance monitoring plan was prepared. Six IM performance monitoring (IMPM) wells were installed in an array downgradient of the planned HRC injection/treatment area in an effort to achieve sufficient number and distribution of monitoring points to assess and demonstrate amended plume behavior. The spacing was selected based on predicted dissolved contaminant travel time. Request for approval of HRC injection was submitted and received from the FDEP.

The high-pressure HRC injection pump utilized a heated hopper to lower the viscosity of the HRC feed stock. A total of 6000 pounds (2721 kg) of HRC (less minor wastage) was injected into the subsurface at depth intervals of 35 to 65 feet (10.7 to 19.8 m) below land surface in 25 injection points between January 11, 1999 and January 17, 1999. Baseline and three post-injection ground water monitoring events were conducted over a period of 106 days.

Limited supplemental ground water monitoring was conducted by others, and the data was supplied to the authors by the Hurlburt Field Environmental Project Manager with permission to use and reproduce for the purpose of publication. These supplemental monitoring results reflect ground water samples collected on days 245 and 322 after HRC injection.

RESULTS AND DISCUSSION

Relatively low PCE concentrations were found in the baseline monitoring such that this study focuses only on the performance of the IM pursuant to TCE and its reductive dechlorination. The performance monitoring results are shown in Figures 2 through 5 as concentration isopleths generated by the computerized contouring program SURFER. These figures show changes in TCE, total dichloroethenes (DCE; mostly *cis*-1,2-DCE), vinyl chloride (VC), and ethene, respectively, over the initial 106-day study period.

Figure 6 graphically depicts all ground water monitoring results for TCE, DCE, VC, and ethene in all six IMPM wells, the source well (MW123-12), and the background well (MW123-23) through the 322-day data set. Note that these charts utilize a double Y-axis where ethene concentrations are correlated to the right-hand axis while the other compounds are represented on the more conventional left-hand axis.

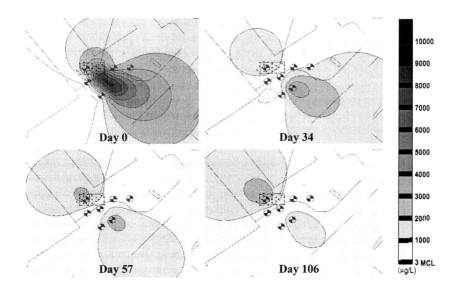

FIGURE 2. Trichloroethene (TCE) concentration isopleths at various times during the initial 106-day study period.

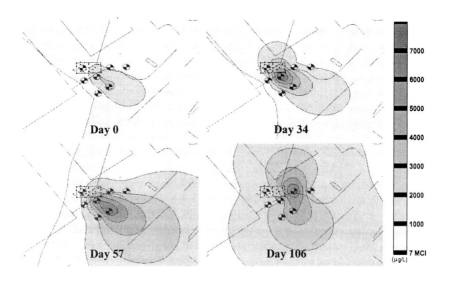

FIGURE 3. Total dichloroethenes (DCE; mostly *cis*-1,2-DCE) concentration isopleths at various times during the initial 106-day study period.

Field Applications of Enhanced Reductive Dechlorination of Chlorinated Solvents 299

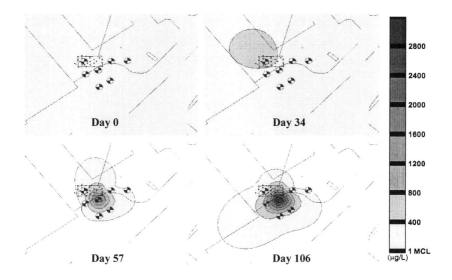

FIGURE 4. Vinyl chloride (VC) concentration isopleths at various times during the initial 106-day study period.

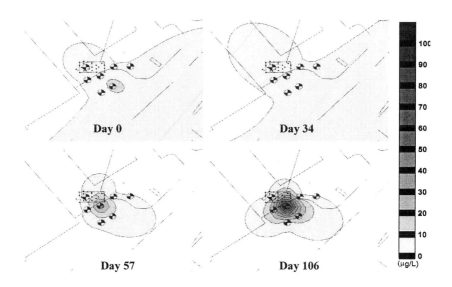

FIGURE 5. Ethene concentration isopleths at various times during the initial 106-day study period.

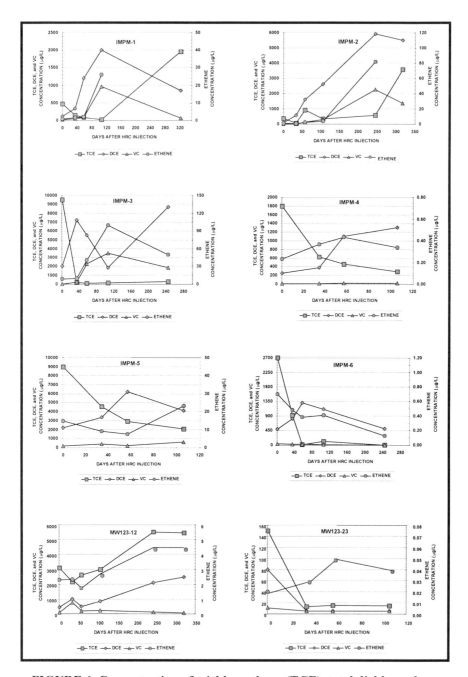

FIGURE 6. Concentration of trichloroethene (TCE), total dichloroethenes (DCE; mostly *cis*-1,2-DCE), vinyl chloride (VC), and ethene in interim measures performance monitoring wells as a function of elapsed time since HRC™ injection. Note that concentration scales differ from chart to chart.

It can be seen from Figure 2 that TCE concentrations downgradient of the HRC injection/treatment area diminished significantly over the initial 106-day study period. There is an insufficient number and distribution of supplemental monitoring data collected after the initial 106-day study period to accurately contour plumes beyond the initial 106-day study period (Figure 6).

The observed increases in TCE concentration as shown in Figure 6 for MW123-12 (the "source" well) are contrary to expectations. It is reasonable to expect that this well, being in such close proximity to multiple HRC injection points, would have exhibited a substantial depletion of TCE after HRC injection. The authors hypothesize that the observed behavior is related to enhanced desorption phenomena, but do not offer further explanation at this time.

As shown in Figure 6, IMPM-2 and MW123-12 (the "source" well located directly within the HRC treatment zone) are the only wells with results from all six sampling events (i.e., two supplemental data points). IMPM-1, IMPM-3, and IMPM-6 have one supplemental data point. Inspection of Figure 6 shows that appreciable TCE rebound occurred at IMPM-1 at some time prior to the 322-day monitoring event. TCE rebound can be seen to occur in IMPM-2 sometime between 245 days and 322 days after HRC injection. No other TCE rebound is noted, even in IMPM-3 and IMPM-6 which were monitored after 245 days. This evidence suggests that the effective life of the HRC amendments within the study area was at least 245 days (8 months).

Changes in total DCE concentration within the study area appear to follow expectations for the reductive dechlorination pathway. The predominant DCE species was the *cis*-1,2-DCE isomer. Formation of DCE is readily correlated to the depletion of TCE. Subsequent depletion of DCE is less readily correlated to the formation of vinyl chloride; however, IMPM-3 (the closest well downgradient of the HRC™ treatment zone) shows the DCE depleting as a function of vinyl chloride formation after an initial depletion of TCE (Figure 6).

The most prevalent production of ethene is shown to occur in IMPM-1, IMPM-2, IMPM-3, and IMPM-5. These wells are located directly down gradient of the HRC injection/treatment zone. The rather significant increases in ethene suggest that the dechlorination was stimulated to completion.

Initial concerns that vinyl chloride would accumulate and become recalcitrant were not realized as evidenced by the decline of vinyl chloride concentrations in IMPM-1 and IMPM-2 after the apparent depletion of HRC (Figure 6). Although data beyond the initial 106-day study period is sparse, it is sufficient to conclude that changes in ethene concentration are proportional to changes in vinyl chloride concentration. It is hypothesized that the reason ethene concentrations did not appear to increase during periods of vinyl chloride depletion is that the lag time between the formation and attenuation of ethene is relatively short.

CONCLUSIONS

It is concluded that HRC significantly accelerated the reductive dechlorination of TCE to completion until the HRC was depleted. It remains uncertain whether the IM objective has been met until such time as equilibrium is re-established and modeled post-IM natural attenuation predictions are performed. The data relieves initial concerns that accumulation of vinyl chloride might occur as a result of the IM. The HRC injection/treatment zone appears to have been well placed; however, rebound of TCE in IMPM-1 and IMPM-2 suggests that TCE remains in the subsurface as a source to dissolution in ground water. The data is insufficient in number and duration to conclude whether the remaining source exists within the HRC injection/treatment zone, beyond the HRC injection/treatment zone, or both. It is apparent that application of HRC within a TCE source zone can effectively deplete the source if properly placed in sufficient quantity.

ACKNOWLEDGEMENT AND DISCLAIMER

The authors are deeply grateful to our colleagues at URS Greiner Woodward Clyde in Mobile, Alabama (Rod Hames, P.E., Paul David, and Allison Davis) for their professional, efficient, and effective contribution to the management and execution of the IM. Special thanks is extended to Regenesis for their technical support and field assistance. The authors appreciate the leadership and vision extended by Ben Coulter, Hurlburt Field's Environmental Project Manager. The supplemental data (i.e. the 245- and 322-day results) provided by Hurlburt Field have not been validated by the authors. Although the information presented in this paper, with the exception of the noted "supplemental" data, has been previously peer reviewed by Hurlburt Field, this paper has not been subject to peer review by Hurlburt Field and therefore does not necessarily reflect the views of the U. S. Air Force.

REFERENCES

Wiedemeier, T. H., M. A. Swanson, D. E. Moutoux, E. K. Gordon, J. T. Wilson, B. H. Wilson, D. H. Kampbell, P. E. Haas, R. N. Miller, J. E. Hansen, and F. H. Chapelle. 1998. *Technical Protocol for Evaluating Natural Attenuation of Chlorinated Solvents in Ground Water*. USEPA Technical Report, EPA/600/R-98/128, USEPA Office of Research and Development, Washington, D.C.

USEPA. 1999. *Use of Monitored Natural Attenuation at Superfund, RCRA Corrective Action, and Underground Storage Sites*. USEPA Office of Solid Waste and Emergency Response Policy Directive, 9200.4-17P.

URS Consultants, Inc. 1998. *RCRA Facility Investigation/Baseline Risk Assessment; POL Fuel Yard, IRP Site ST-123, Hurlburt Field, Florida*. Prepared for U.S. Air Force Special Operations Command; Hurlburt Field, Florida.

URS Greiner Woodward Clyde. 1999. *Corrective Measures Study Report; POL Fuel Yard, IRP Site ST-123, Hurlburt Field, Florida*. Prepared for Air Force Center for Environmental Excellence, Brooks Air Force Base, Texas and U.S. Air Force Special Operations Command; Hurlburt Field, Florida.

EVALUATION OF IN-SITU BIOLOGICAL DEHALOGENATION ACTIVITY USING A REACTIVE TRACER

Sanjay Vancheeswaran (CH2M HILL, Santa Ana, California)
Gary Hickman, Randy Pratt and Scott McKinley (CH2M HILL, Corvallis, Oregon)
Matt Germon (CH2M HILL, Spokane, Washington)
John Gross (Weyerhaeuser Company, Tacoma, Washington)
Lewis Semprini and Jonathan Istok (Oregon State University, Corvallis, Oregon)

ABSTRACT: A reactive tracer was used in a field pilot study to evaluate the enhancement of in-situ biological dehalogenation activity, by the addition of lactate. Trichlorofluoroethene (TCFE), a compound considered to have comparable physical, chemical, and biotransformation characteristics, as that of the chlorinated ethenes and ethanes present at the study site, was the reactive tracer used for this study. The *Push-Pull* technique was employed to field test the reactive tracer. The *push-pull* test results indicated that the addition of lactate significantly improved in-situ reductive dechlorination activity. The biologically mediated dechlorination of the reactive tracer TCFE, evidenced by the disappearance of TCFE and the formation of the reduction daughter products (cis-dichlorofluoroethene (c-DCFE), chlorofluoroethenes (CFE), and fluoroethene (FE)), clearly demonstrated in-situ biological reduction activity. The degree and extent of dehalogenation was indicated by the production and proportions of c-DCFE, CFE and FE, which are progressively dechlorinated products of TCFE. These results suggest that TCFE analogues, such as tetrachloroethene (PCE) and trichloroethene (TCE), would also undergo a similar degree of dehalogenation to cis-dichloroethene (cis-DCE), vinyl chloride (VC), and finally to ethene. This hypothesis was supported by the observance of PCE and TCE reduction, and the transient accumulation of cis-DCE and VC at the site.

INTRODUCTION

Chlorinated ethenes such as PCE and TCE are common groundwater contaminants in aquifers throughout the United States. In view of their widespread distribution, there is considerable interest in the use of biologically mediated reductive dechlorination for remediation purposes. Evaluation of in-situ biological reduction activity is, therefore, essential for determining the potential for natural and enhanced attenuation of these compounds. Typically, in-situ transformation rates are indirectly evaluated from geochemical data and modeling approaches, or by studying the kinetics of contaminant disappearance and daughter product accumulation. The latter approach, however, can be misleading if there is a DNAPL present. In addition, the inability to discriminate between pre-existing and newly generated daughter products also makes it impracticable under certain situations.

An alternative approach that would still allow a near direct evaluation of the biological reduction of compounds such as PCE and TCE is the use of non-regulated analogues to these compounds. Ideally, a chlorinated ethene analogue should have 1) similar physical, structural and chemical properties as that of the compound of interest, 2) existing analytical tools for quantitation, 3) undergo the same biological transformation (at equivalent rates) as the compound of interest, and 4) retain a distinctive chemical feature or signature throughout the reduction process. TCFE, a fluorinated analogue of TCE and PCE was examined to contain all the above mentioned properties (Vancheeswaran et al., 1998). The field approach would involve the application of the single-well, "push-pull" test (Istok et al., 1997), which consists of controlled injection of a prepared test solution with TCFE into a monitor well and allowing sufficient reaction time for transformation processes to occur, followed by the extraction of the test solution/groundwater mixture from the same well. By measuring the concentrations of TCFE and its transformation products during the extraction phase, the reaction rates are evaluated.

Objective. The primary objective of the pilot test was to evaluate the enhancement of in-situ reductive dechlorination of PCE and TCE by the addition of lactate as an electron donor. The application of a fluorinated analogue, TCFE, for evaluating enhancement of in-situ biological reduction processes, and the degree and extent of dehalogenation was tested.

Site Description. The Weyerhaeuser Sycan Maintenance Shop site is a 160-acre parcel of property in Beatty, Oregon, which was used as a maintenance facility for railcars and locomotive engines until 1990. Historically, chlorinated solvents such as PCE and 1,1,1-trichloroethane (TCA) were used for locomotive cleaning at the site. Analytical sampling of soil and groundwater at the site revealed the presence of PCE, TCE and TCA. Detection of the reduction products of these compounds suggested that reductive dehalogenation was occurring naturally at the site. However, the prolonged observance of these chlorinated solvents suggested that these reductive dehalogenation reactions were relatively slow and probably limited by the availability of suitable electron donors.

The hydrogeologic features of the site are characterized by a perched zone of temporal groundwater occurrence that extends to about 9 feet below ground surface (bgs), an aquitard zone between 9-20 feet bgs, and an aquifer zone between 20 to 35 feet bgs. The upper aquitard is a fine-grained, dense, semi-consolidated silt and siltstone that represents a low permeability barrier between the ground surface and the underlying strata. Based on the remedial investigation results, it was determined that the hot spot of PCE and TCA contamination exists predominantly in the upper aquitard.

MATERIALS AND METHODS

The pilot test was conducted in a monitor well screened in the upper aquitard zone and consisted of three phases: test setup and injection, long-term monitoring, and final extraction/decommission phase. The methods adopted for pilot testing during each of these phases and the data collected are described below.

Test Setup and Injection Phase. Initially, a pre-test was conducted to determine the injection and extraction rates at the monitor well to ensure that the injection and extraction during push-pull testing was carried out at similar rates to promote similar zones of influence. After the pre-test, approximately 500 gallons of water was extracted from the well into a 500 gal polyethylene tank using a peristaltic pump. The extraction was carried out at an average extraction rate of 0.45 gpm. Five samples were collected from the discharge of the groundwater pump at a frequency of approximately one sample every 50 to 100 gallons. These samples were analyzed to establish baseline groundwater characteristics for field parameters (dissolved oxygen, pH, oxidation-reduction potential (ORP), conductivity), volatile organic compounds (VOCs), ethene, ethane, methane, and geochemical parameters (total organic carbon [TOC], nitrate, ferrous iron, sulfate, manganese, alkalinity).

The extracted water was then injected back into the well along with a stock solution containing lactate (substrate), potassium bromide (KBr) (conservative tracer) and TCFE (reactive tracer). The stock solution was made up in the CH2M HILL Applied Sciences Laboratory in Corvallis, Oregon in four 5-gal cubitainers. The solution was prepared so that when injected with the 500 gallons of extracted water, the concentrations of lactate, bromide, and TCFE achieved the target concentrations of 1500 mg/L, 100 mg/L and 2.1 mg/L, respectively. The extracted water and the stock solution were pumped at specified flow rates, which were matched to maintain a constant flow ratio. The groundwater and stock solutions were combined by means of a "tee" linking the two individual lines into a single injection line at the well head prior to injection, to achieve the target concentrations. The total mass of lactate, KBr, and TCFE added to the aquitard in 500 gallons of water at these concentrations was approximately 2.84 kg, 189 g, and 4.0 g, respectively.

The solution was pushed into the monitor well over a period of four 6- to 8-hr days at an average rate of 0.28 gpm. The slow injection allowed maximization of the radial zone of influence. The theoretical zone of influence for this 500-gallon push, assuming a porosity of 0.25, would be a cylindrical section 5.5 feet in diameter.

Long-Term Monitoring. Long-term monitoring was conducted to track the progress of biodegradation enhancement and to determine if it was appropriate to stop the testing and perform the final pull. Groundwater samples were collected from the well at the end of 5 and 10 weeks after injection of the substrate solution and analyzed for lactate, bromide, TCFE and its biotransformation products, and the baseline parameters. Samples were collected after purging approximately 10 gallons of groundwater to allow definite sampling of the injected fluid. The presence of tracers in the samples provided confirmation that the injected fluid was sampled.

Final Extraction/Decommission. The final pull/decommission phase (3months after injection) involved the extraction of 500 gallons of groundwater from the same monitor well. Extraction and sample collection procedures were similar to the initial extraction phase. Five samples were collected from the discharge of the groundwater pump at a frequency of approximately one sample every 50 to 100 gallons and analyzed for lactate, bromide, TCFE and its biotransformation products, and the baseline parameters.

Geoprobe groundwater samples were collected in two areas down gradient of the monitor well after the decommissioning to determine if residual concentrations of substrate, conservative tracer, or reactive tracer were present.

Analytical Methodology. Standard EPA approved analytical techniques were utilized for analyzing VOCs and all standard baseline parameters. For the quantitative analysis of TCFE and its reduction products, a GC/MS fitted with a DB-624 column and operated in the selective ion monitoring (SIM) mode was used. The ions monitored were m/z 46 for FE, m/z 79.9 for CFEs, m/z 96 for fluorobenzene (internal standard), m/z 113.9 for DCFEs and m/z 147.9 for TCFE.

RESULTS AND DISCUSSION

The observations and analysis of the results during the initial pull, long-term monitoring, and final pull are presented below grouped by the individual group of chemicals analyzed to provide clarity.

Substrate and Metabolite (Lactate and Acetate). A 50% reduction in lactate concentration was observed in the first month after injection, and then a slow decrease to about one-third its initial concentration was observed at the end of the pilot test. Acetate, a metabolic breakdown product of lactate appeared in the groundwater after lactate injection. Lactate utilization in conjunction with acetate formation confirms that indigenous microorganisms metabolized the added substrate.

Chlorinated Ethenes (PCE, TCE, and its reduction products). The effect of bio-enhancement (by the addition of lactate) on chlorinated ethenes is illustrated in Figure 1.

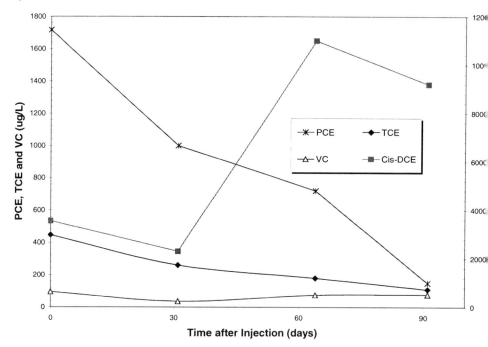

FIGURE 1. Concentration trends for PCE, TCE, cis-DCE, and VC observed during the pilot testing

The initial (or baseline) concentrations of PCE, TCE, cis-DCE, and VC prior to pilot testing were determined to be approximately 1716, 446, 3560 and 96 ug/L, respectively. These concentrations were determined by averaging the concentrations measured in five samples collected during the initial extraction phase. A decrease in concentration of all the chlorinated ethenes was observed in the first sample collected one month after injection. This decrease was likely a result of volatilization losses during storage, extraction and re-injection. However, the concentrations of PCE and TCE continued to decrease quite substantially until the final extraction conducted after 3 months. PCE concentrations decreased by an order of magnitude compared to baseline concentration, while TCE concentration decreased by a factor of three. A corresponding increase in cis-DCE, to concentrations about a factor of three higher than the baseline value was also observed. An increase in VC was also observed. Ethene was not detected in these samples. At the end of final extraction of 500 gallons of groundwater, the concentrations of PCE and TCE increased back to baseline concentrations, and the concentration of cis-DCE decreased down to the baseline concentration.

Significant reductions in PCE and TCE concentrations in conjunction with a corresponding increase in cis-DCE concentration provide compelling evidence of enhanced in-situ dehalogenation activity. The fact that rates were enhanced or accelerated is evidenced by a) comparing the considerable changes in chlorinated ethene concentrations measured in the pilot test to the relatively consistent historical chlorinated ethene levels, and b) the tendency of chlorinated ethene concentrations to return to baseline levels when unamended groundwater from outside the test area was sampled and analyzed at the end of the final extraction phase.

Reactive Tracer (TCFE and its reduction products). The rationale for using TCFE during the pilot test was to ascertain and evaluate biologically mediated dehalogenation. TCFE was injected along with the substrate and the tracer to achieve a target concentration of 2.1 mg/L (~ 14 µM) in 500 gallons of injectate. The pilot testing results for TCFE and its reduction products are illustrated in Figure 2. After about 1 month, significant concentrations of the reduction products c-DCFE and FE were measured, providing conclusive evidence of biological dehalogenation occurring in-situ. The identity of the trans- and cis- isomers for the dichlorofluoroethenes and the chlorofluoroethenes were inferred from analysis of literature (Vancheeswaran et al, 1998).

The large initial reduction in TCFE concentration between the calculated injection concentration and the 1-month measured value cannot be readily attributed to biodegradation based on the daughter products formed. It is more likely that sorption of TCFE to the soil organic carbon accounted for this initial decrease. However, it should also be noted that the measured aqueous concentrations of daughter products may also under represent their true production, since they would also undergo sorption to the solid organic phase. By the end of 2 months, the c-DCFE concentration had increased substantially and the TCFE concentration had decreased correspondingly. The general extent of TCFE biotransformation appears to have been considerable based on the observation that the aqueous concentration of break down products exceeded that of the parent compound. Between 2 and 3 months the concentrations of TCFE and c-DCFE remained relatively constant, and the concentration of FE was fairly constant from 1 to 3 months after the initial increase.

The formation of the reduction daughter products, c-DCFE and FE clearly demonstrates the in-situ biological reduction of TCFE. The degree of dehalogenation is indicated by the observation of CFE and FE, which are progressively dechlorinated products of TCFE (formed by the loss of 2 and 3 chlorine atoms from TCFE, respectively). This suggests that analogs PCE and TCE, would also undergo a similar degree of dehalogenation to VC and finally to ethene. A detailed analysis of TCFE transformation kinetics was not performed because the conservative tracer could not be analyzed and quantified due to co-elution of KBr and lactate in the ion chromatograph column that was employed.

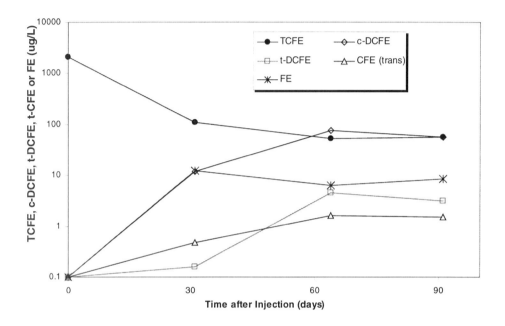

FIGURE 2. Concentration trends for TCFE and its reduction products

Chlorinated Ethanes (1,1,1-TCA, its reduction products and 1,1-DCE). A modest decrease in 1,1,1-TCA and 1,1-DCE and a corresponding increase in 1,1-DCA was observed by comparing the initial baseline and final concentration. This suggests that these compounds may have undergone biological degradation, but to a lesser extent. The apparently low biological reduction of 1,1,1-TCA maybe a result of competitive inhibition from PCE, which maybe a more favorable electron acceptor. (i.e. PCE in the system may preferentially scavenge the electrons donated by the substrate)

Field Parameters. The results of the groundwater pH analyses indicated a greater than one unit decrease in pH from the start to the conclusion of the test (from pH 6.9 to 5.4). This suggests that the H^+ ions released during the biological dehalogenation reaction may be lowering the pH. Alkalinity measurements also support the above observation.

The results of the oxidation-reduction potential measurements indicate a sharp reduction in the redox potential in the subsurface. The ORP decreased from +25 mV (1 month after injection) to highly reducing conditions of –400 to –520 mV (3 months after injection). The resulting redox conditions are in the optimum range for biological reductive dehalogenation of chlorinated solvents.

Recovery Evaluation. Groundwater sampling and analysis was conducted at three locations down gradient of the monitor well following the pilot test, to determine if any of the injected chemicals had migrated away from the test area in measurable levels. No lactate, bromide, TCFE or its reduction products were detected in any of these samples. Most of the injected mass was either re-captured during final injection, utilized or biodegraded in-situ, or otherwise attenuated.

CONCLUSIONS

The enhanced biodegradation pilot test provided encouraging results to support the enhancement of naturally occurring biological reduction processes. The degree and extent of dehalogenation indicated by the production and proportions of c-DCFE, CFE and FE, provide qualitative and quantitative evidence that PCE and TCE would also undergo a similar degree of dehalogenation to cis-DCE, VC, and finally to ethene. Potentially, with the availability of conservative tracer data, it would also be possible to estimate in-situ transformation rates.

Several laboratory scale experiments and pilot scale studies are currently being performed to evaluate the application of reactive tracers at other chlorinated ethene contaminated sites. Further research and field tests are required to fully develop an in-situ methodology using reactive tracers for evaluating in-situ dehalogenation activity for natural attenuation and enhanced biological reduction processes.

Disclaimer: It should be noted that some reactive tracers and their reduction products may be potentially hazardous, and adequate review and precaution should be taken prior to their selection and application.

ACKNOWLEDGEMENT

The authors would like to thank the Weyerhaeuser Company for their technical and financial support for this study. We would also like to thank Randy Eatherton, Weyerhaeuser Technology Center (Analysis and Testing), for his assistance in developing analytical methods for the analysis of TCFE and its reduction products.

REFERENCES

Istok, J.D.; Humphrey, M.D.; Schroth, M.H.; Hyman, M.R.; O'Reilly, K.T. 1997. "Single-Well, "Push-Pull" Test for In situ Determination of Microbial Activities", *Ground Water*, 35(4): 619-631

Vancheeswaran, S.; Semprini, L. and Hyman, M. 1999. "Anaerobic Biotransformation of Trichlorofluoroethene (TCFE) in Groundwater Microcosms", *Environmental Science & Technology,* 33: 2040-2045.

ENGINEERED PCE DECHLORINATION INCORPORATING COMPETITIVE BIOKINETICS: OPTIMIZATION AND TRANSPORT MODELING

Matthew Willis, *Christine Shoemaker*
(School of Civil and Environmental Engineering, Cornell University
Ithaca NY, USA 14853)

ABSTRACT : Halogenated organics (including chloroethenes) can be remediated with anaerobic biological methods in which hydrogen (produced from fermentation of a supplied organic electron donor) is used by dehalorespiring organisms to reductively dehalogenate the contaminants. Unfortunately, competing bacteria (e.g. methanogens and sulfate-reducers) can divert much of the hydrogen away from dehalogenation. Fennell and Gossett (1998) have recently proposed and verified a batch competitive biokinetic model to describe reductive dechlorination of tetrachloroethene (PCE) and lesser-chlorinated ethenes given bacterial competition. This batch model was tested in laboratory mixed cultures, but does not incorporate transport of chemical and biological species, which depends upon groundwater flow and sorptive/retardation aspects. We have extended this batch model to incorporate transport in the CORDITE (Cornell Competitive Reductive Dechlorination Incorporating Transport Equations) model. In this paper, optimization algorithms are applied to the CORDITE simulation model to identify cost-effective engineering design. Decisions determined by the optimization include location of injection and extraction wells, pumping rates, and concentrations of hydrogen donors. The goal is to identify design and operating plans that will minimize cost while meeting clean-up goals within the allowable clean-up time period

INTRODUCTION

Halogenated organics are ubiquitous contaminants of groundwaters throughout the U.S. Increasing attention is being devoted to anaerobic biological methods for remediation in which hydrogen, produced from fermentation of a supplied organic electron donor, is used by dehalorespiring organisms to reductively dehalogenate the contaminants. The Fennell and Gossett (1998) biokinetic model describes reductive dechlorination of tetrachloroethene (PCE) and lesser-chlorinated ethenes in the presence of bacterial competition. We have amended the batch kinetics for use as a subsurface transport model (Willis et al., 1999).

Recently the transport model has been expanded and improved numerically, and is now known as CORDITE. The CORDITE model is built upon the public model RT3D (Clement et al., 1998) and uses the ADIFOR automatic differentiation package (Bischof et al., 1996). Numerical results have been obtained with CORDITE for a range of field conditions.

Previous studies focusing on non-biological pump-and-treat systems have demonstrated the efficiency of optimization, for example, Ahlfeld (1990), Culver and Shoemaker (1992), Gorelick et al. (1984) and Lee and Kitiandis (1991). Subsequent studies examined the use of optimization techniques and in-situ bioremediation, such as Minsker and Shoemaker (1998a, 1998b), Yoon and Shoemaker (1999).

Derivative based methods (e.g. Minsker and Shoemaker, Culver and Shoemaker) offer the greatest numerical performance at the expense of implementational complexity. Non derivative methods (e.g. Yoon and Shoemaker) are applicable to a larger class of problems, parallelize easily, and are easily implemented. For this study, non-derivative based evolutionary strategies are used for optimization of the CORDITE model.

Evolutionary strategies can be effective for optimization of groundwater remediation problems (Yoon and Shoemaker, 1999). A large variety of evolutionary strategies exist. For purposes of this research, the GAFORTRAN code by D. L. Carroll (e.g. 1996) was used to examine remediation optimization.

Evolutionary Strategies: Modern Genetic Algorithms were described by Holland (1975) as an abstraction of biological evolution. Classical elements of GAs include chromosomes as a representation of strings of bits, coding an individual. The fitness of an individual controls its likelihood to pass on its genetic information. Processes of crossover and mutation work to ensure the GA adequately explores the search space. The GAFORTRAN code allows many variations on the archetypal GA problem, of which several are employed, an offspring selection process where the best child from the previous generation is kept.

A common improvement on GA, known as elitism, is for the replacement step to keep the best parent from the previous generation. Another variation is to use more or less than 2 parents. GAFORTRAN allows these and many other options which are not discussed here.

Optimization Trial: The Optimization Trial describes a relatively simple example with one variable rate injection well coupled to a downstream variable rate extraction well. It uses the geology and plume from the Base Case. As in the Base Case, butyrate is mixed with the extracted water to provide a hydrogen donor for the subsurface organisms. In the trial, this concentration of donor is variable. After a period of twelve years, the efficacy of a particular policy is assessed based on the concentrations of pollutant found in several observation wells. The objective is based on variable injection costs and violations of the MCL levels for a variety of pollutants. The field configuration used is shown in Figure 1.

The independent variables used in the optimization trial were the time-varying concentrations and injection/extraction rates used. Optimizations used 3 time steps of approximately 4 year duration, and donor injection concentrations were limited to less than 600 mg/L of butyrate. The quantity of water injected equaled the quantity of water extracted, and injection rates were limited to less than 3 m^3/h. A do-nothing policy results in practically no dechlorination, and a small displacement of the plume along the direction of groundwater

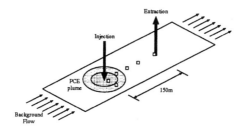

Figure 1: Treatment configuration for the CORDITE Optimization Trial. Open squares denote observation well locations.

flow. The variables used are shown in Table 1.

Table 1: Independent variables, CORDITE Optimization Trial

Decision Variable	Symbol	Range
Concentration	q_1, q_2, q_3	0-600mg/L
Injection Rate	c_1, c_2, c_3	0-3m^3/h

The cost formulation included both pumping costs and penalty costs. The pumping costs were calculated based on decision variables c_i and q_i, the donor concentrations and injection rates for each time period i:

$$f(c,q) = \sum_{i=1}^{N} \alpha_i c_i q_i + \beta_i q_i \quad (1)$$

where N, the number of time periods is 3 and α_i and β_i represent cost coefficients. (The values used were 2.33 and .167, respectively.) Penalty costs were based on violations of a maximum concentration limits at $M = 5$ observation wells:

$$\begin{aligned} f_p(c,q) &= \sum_{k=1}^{M} \gamma_{PCE} \cdot \max\left(0, c_k^{PCE} - MCL^{PCE}\right)^2 + \\ & \gamma_{TCE} \cdot \max\left(0, c_k^{TCE} - MCL^{TCE}\right)^2 + \\ & \gamma_{DCE} \cdot \max\left(0, c_k^{DCE} - MCL^{DCE}\right)^2 + \\ & \gamma_{VC} \cdot \max\left(0, c_k^{VC} - MCL^{VC}\right)^2 \end{aligned} \quad (2)$$

and observed concentrations for PCE, TCE, DCE and VC. The MCL levels are based on the USEPA Office of Ground Water and Drinking Water Guidelines, shown in Table 2. The weighting factors γ_{PCE} etc. were selected to penalize more greatly for higher chlorinated species, and are shown in Table 3. Vinyl chloride was penalized less so as to promote dechlorination, even if it results in large plumes of vinyl chloride. (Vinyl chloride is an order of magnitude more soluble than PCE, but may be more susceptible to traditional pump and treat extraction).

Table 2: Maximum Contaminant Levels (MCL's) for chlorinated ethenes. Ethene has no chlorine atoms and no MCL.

Species	MCL (mg/L)
PCE	0.005
TCE	0.005
DCE	0.07
Vinyl chloride	0.002
Ethene	no MCL

Table 3: Penalty weights for violation of MCL penalty function (Equation 2).

Weight Coefficient	Relative Weight
γ_{PCE}	4000
γ_{TCE}	3000
γ_{DCE}	2000
γ_{VC}	20

It is clear that the objective function is somewhat flat. The best policy discovered resulted in a large fraction of mass of pollutant being converted from PCE to other forms.

Table 4 gives the five best policies as found by the three optimization methods examined. GA with elitism found the best policy, with an objective function value of 2469.8. (Numbers closer to zero are better). GA with elitism outperformed GA without elitism, as the best objective under GA without elitism (2599.0) shows.

DISCUSSION

The optimization of the injection and extraction rates leads to considerable (but incomplete) dechlorination (Figure 2). Incomplete dechlorination is typical, since the vinyl chloride dechlorination step occurs catabolically in the CORDITE model. The best examined policy suggests that relatively high concentrations of donor are beneficial early in the simulation, to stimulate conversion of PCE to less chlorinated forms, but that high rate pumping becomes more effective as the plume moves. Indeed, by the third time step, considerable extraction occurs, with over half of the original chlorinated ethene mass

Table 4: Summary of best results found for five optimizations for GA and GA with elitism.

Method	c_1 mg/L	c_2 mg/L	c_3 mg/L	q_1 m^3/h	q_2 m^3/h	q_3 m^3/h	Objective
GA-e	409.41	91.76	35.29	0.36	1.65	2.86	2469.8
GA-e	520.00	235.29	56.47	0.40	0.79	2.98	2555.1
GA-e	454.12	89.41	42.35	0.34	2.04	2.72	2558.0
GA	298.82	23.53	37.65	1.36	0.27	2.99	2599.0
GA	263.53	451.76	21.18	1.14	0.39	2.92	2628.3
GA	334.12	75.29	28.24	1.27	0.94	2.93	2659.6
GA-e	564.71	171.76	28.24	0.48	1.15	2.36	2675.9
GA-e	214.12	167.06	7.06	2.13	0.47	2.84	2706.4
GA	228.24	148.24	2.35	2.15	0.00	2.55	2813.0
GA	197.65	174.12	32.94	0.59	2.24	2.45	2930.4

Table 5: Distribution of objective function evaluations for optimization methods GA and GA with elitism. Percentile refers to the number of examined solutions that were inferior to the given value.

Percentile	GA-e	GA
50%	4207.9	4846.5
80%	3212.0	3945.5
90%	2898.8	3515.3
95%	2706.4	3171.0
Best	2469.8	2599.0

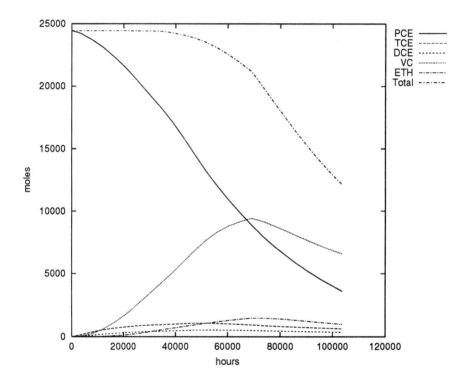

Figure 2: Total ethenes in the aquifer as a function of time for the optimal policy from Table 4. Time step 1 runs from 0 to 34560h, time step 2 runs from 34560h to 69120h, and time step 3 runs from 69120h to 103680h.

being removed. Interestingly, the species being extracted is primarily vinyl chloride, which has a higher solubility than PCE and is more mobile. This finding suggests that the incomplete biological conversion of PCE to ethene and vinyl chloride should be followed by aggressive extraction. Further studies will examine this approach.

Genetic algorithms with elitism outperformed genetic algorithms without elitism. This may be seen in both the final objective (Table 4) and the distribution of the trial objectives for each of five optimizations (Table 5). Not only did GA with elitism find a better solution, the top 20% of simulations for GA with elitism were better than the top 5% of simulations for GA without elitism. In the context of problems with relatively computationally expensive simulations, the consequence of improved convergence is a quicker result. The use of elitism is recommended.

ACKNOWLEDGEMENTS
This research has been funded by a grant to Christine Shoemaker and James Gossett from the New York Center for Advanced Technology (Biotechnology Program). The authors are grateful to T. P. Clement for access to the source code for the RT3D program, and to Christian Bischof and Alan Carle for the use of the ADIFOR automatic differentiation package. The program GAFORTRAN by D. Carroll was used for optimization.

REFERENCES
Ahlfeld, D. P., "Two Stage Ground Water Remediation Design", *ASCE Journal of Water Resources Planning and Management*, 116(4) 517-529 (1990)

Bischof, C., P. Khademi, A. Mauer and A. Carle, "Adifor 2.0: Automatic differentiation of Fortran 77 programs" *IEEE Computational Science & Engineering*, 3, 18-32 (1996)

Carroll, D. L., "Genetic Algorithms and Optimizing Oxygen-Iodine Lasers" *Developments in Theoretical and Applied Mechanics*, 18, 411-424 (1996)

Clement T. P., Y. Sun, B. S. Hooker and J. N. Petersen, "Modeling Multi-species Reactive Transport in Ground Water" *Ground Water Monitoring and Remediation*, 18, 79-92 (1998)

Culver, T. B. and C. A. Shoemaker, Dynamic Optimal Control for Groundwater Remediation with Flexible Management Periods, *Water Resources Research*, 28(3), 629-641, (1992)

Fennell, D. E. and J. M. Gossett. "Modeling the Production of and Competition for Hydrogen in a Dechlorinating Culture" *Environmental Science and Technology*, 32, 2450-2460 (1998)

Gorelick, S. M., C. I. Voss, P. E. Gill, W. Murray, M. A. Saunders and M. H. Wright, "Aquifer Reclamation Design: The Use of Contaminant Transport Simulation Coupled with Nonlinear Programming", *Water Resources Research*, 20(4), 415-427 (1984)

Holland, J. H. *Adaptation in Natural and Artificial Systems* (1975) MIT Press

Lee, S.-I. and P. K. Kitanidis, Optimal Estimation and "Scheduling in Aquifer Remediation with Incomplete Information", *Water Resources Research*, 27(9), 2203-2217 (1991)

Minsker, B.S. and C.A. Shoemaker, "Dynamic Optimal Control of In Situ Bioremediation of Groundwater" *ASCE Journal of Water Resources Planning and Management*, 124(3) 149-161 (1998)

Minsker, B.S. and C.A. Shoemaker, "Quantifying the Effects of Uncertainty on Optimal Groundwater Bioremediation Policies", *Water Resources Research*m 34, 3615-3625, (1998)

Willis, M., C. Shoemaker, J. Gossett, D. Fennell, "Applications of a Competitive Hydrogenotrophic Biological Dechlorination Transport Model for Groundwater Remediation" *Engineered Approaches for In Situ Bioremediation of Chlorinated Solvent Contamination*, A. Leeson and B.C. Alleman (eds.), 27-34, Vol.2, Battelle Press (1999)

Yoon, J. H. and C. A. Shoemaker, "Comparison of Optimization Metho ds for Groundwater Bioremediation" *ASCE Journal of Water Resources Planning and Management* 125: (1) 54-63 (1999)

PERFORMANCE OF FIELD-SCALE SEQUENTIAL ANAEROBIC/AEROBIC *IN SITU* BIOREMEDIATION DEMONSTRATION

Andrea Turpie and *Craig Lizotte* (Envirogen Inc., Canton MA, USA)
Mary F. DeFlaun and Joseph Quinnan (Envirogen Inc., Lawrenceville NJ, USA)
Michael Marley (XDD, LLC, Portsmouth, NH, USA)

ABSTRACT: Chlorinated solvents, petroleum hydrocarbons, and arsenic emanating from a landfill at a Superfund Site in New England are being treated in a field-scale pilot study with a biostimulation approach that includes sequential anaerobic and aerobic treatment zones. The test plot is 60 meters x 90 meters and is situated at the toe of the landfill. The Site groundwater in the vicinity of the toe of the landfill is anaerobic. Electron donor injection in the anaerobic zone is designed to enhance the biodegradation of the chlorinated solvents and tetrahydrofuran (THF). Oxygen injection is used to create an aerobic treatment zone to enhance biodegradation of petroleum hydrocarbons and non-chlorinated solvents and to stimulate the precipitation of arsenic. The system was turned on in January 1998. In the anaerobic zone, the complete dehalogenation of *cis* 1,2-dichloroethene (cDCE) to ethene and ethane was not detected prior to electron donor addition. Data collected through July 1999 in the anaerobic zone show that following an acclimation period, cDCE is now biodegrading to completion. Preliminary biodegradation rates indicate that the conversion of vinyl chloride (VC) to ethene and ethane is faster than the degradation of cDCE to VC. Electron donor is being distributed throughout the anaerobic zone. Concentrations of arsenic at the toe of the landfill compared to those within the anaerobic zone indicate that the arsenic may be precipitating with sulfate by a mechanism that occurs at low redox and oxygen concentration conditions, obviating the need for the aerobic zone for arsenic immobilization. The successful demonstration of this innovative technology provides a cost-effective alternative to the existing record of decision (ROD)-specified remedy of a landfill cap and pump and treat.

INTRODUCTION

This paper describes work conducted to date at a Superfund landfill site in New England related to the demonstration of sequential *in situ* anaerobic/aerobic bioremediation of Constituents of Concern (COCs) in groundwater. The objectives of this project include: 1) demonstrating *in situ* treatment of chlorinated and non-chlorinated solvents, petroleum hydrocarbons, and arsenic in groundwater emanating from the landfill; 2) calculating compound specific biodegradation rates for the target COCs; and 3) collecting sufficient information to develop a full scale design.

To collect sufficient information to achieve the project objectives, a large scale Treatment Zone Demonstration (TZD) was designed and implemented. The TZD was constructed in 1996 and 1997 and became operational in 1998. Details of the TZD design, construction and early operational data are in Lizotte et al. (1999). The TZD covers a 60 by 90 meter area of the Site at the toe of the landfill slope. The TZD includes two amendment injection systems, one for electron donor addition to the anaerobic treatment zone, and one for oxygen injection in the aerobic treatment zone. The performance of the anaerobic zone for treating chlorinated VOCs is the primary subject of this paper.

MATERIALS AND METHODS

The anaerobic portion of the TZD is located approximately 10 meters hydraulically downgradient from the toe of the landfill and extends approximately 50 meters into the TZD. The beginning of the anaerobic zone is marked by a row of small diameter groundwater monitoring points (MP Row 1) and a row of 5 centimeter diameter groundwater piezometers (PZ Row 1). Approximately 3 meters downgradient of MP Row 1 are 30 electron donor injection points (10 locations at three depth intervals). Two additional rows of monitoring points, (MP Row 2 and MP Row 3) are located approximately 7 and 33 meters, respectively, downgradient of the electron donor injection points. A second piezometer row (PZ Row 2) is located between MP Row 2 and MP Row 3. In addition, three clusters of monitoring points (1 cluster for each zone) known as the discrete monitoring point networks (DMPN) are located between MP Row 1 and MP Row 2. In each zone, shallow, intermediate and deep, the DMPN consists of three lines of tightly spaced points oriented parallel with groundwater flow direction.

Anaerobic TZD Operation. The electron donor injection system was designed and installed to distribute sodium benzoate solution into the Site groundwater. The injection system was designed to distribute an equivalent sodium benzoate concentration of approximately 4 mg/l in groundwater at MP Row 2. This electron donor equivalent concentration was shown by a treatability study (Envirogen, 1995) to provide a significant excess of donor to promote the anaerobic degradation of the target COCs. Approximately 5 liters/day of 1% sodium benzoate is injected into each sodium benzoate injection point (SIP). To facilitate laboratory analyses, groundwater samples collected to evaluate sodium benzoate distributions are acidified, forming benzoic acid. Approximately 0.85 grams of benzoic acid are produced for every gram of sodium benzoate acidified. Therefore, the sodium benzoate distribution target concentration of approximately 4 mg/l at MP Row 2 equates to a laboratory measured concentration of approximately 3.4 mg/l of benzoic acid. In addition to analyzing for sodium benzoate, Envirogen recently began monitoring for other volatile fatty acids (VFAs) to better evaluate the distribution of electron donor in the anaerobic zone. Through the end of 1999 approximately 974 kilograms of sodium benzoate have been injected into the anaerobic zone.

Anaerobic TZD Monitoring: Monitoring in the anaerobic zone focused on three general parameters: 1) groundwater flow direction; 2) target COC concentrations; and 3) electron donor distribution. Groundwater flow monitoring was essential for determining that the constructed monitoring point network orientation was consistent with the local and TZD-wide average groundwater flow directions and suitable to collect the data necessary to meet project goals.

RESULTS AND DISCUSSIONS

Target COC Concentrations. Baseline monitoring of the anaerobic portion of the TZD indicated that, in general, target COCs were detected only in the deep portion of the anaerobic zone in the vicinity of DMPN monitoring points MP-30 through MP-40. In addition, the presence of cDCE and the absence of daughter products, vinyl chloride (VC) and ethene and ethane (EE/EA) indicated that dehalogenation of cDCE was not occurring in this portion of the Site before electron donor addition was initiated. Monitoring in a portion of the deep DMPN wells for COCs and electron donor was conducted on a bi-weekly basis for 1998 and monthly basis in 1999. Current data show that cDCE concentrations are decreasing relative to 1998 data, while the concentrations of the dissolved gases VC and EE/EA were initially increasing with subsequent decreasing concentrations. These observations are consistent with enhanced anaerobic dehalogenation of cDCE.

Electron Donor Monitoring. As of July 31, 1999, benzoic acid and/or other VFAs, have been detected in samples collected from all of the fifteen points in MP Row 2. Some of the points in MP Row 2, where benzoic acid has not been detected, appear to be located outside of the zone specific groundwater flow tubes and, accordingly, are not located hydraulically downgradient of a SIP.

Samples collected at some of the points in the contaminated area of the deep DMPN have had inconsistent or only low level detections of benzoic acid. The distribution and temporal variability of benzoic acid may be attributable to the following:

- consumption of the electron donor by acclimated bacterial populations;
- degradation of benzoic acid to other VFAs, such as acetic acid; and
- possibly, localized soil heterogeneities (i.e., localized zones of low-transmissivity soil), which may retard benzoate migration.

Basis for Anaerobic Rate Estimation. Monitoring data collected during the initial 18 months of the project indicate that cDCE is degrading, resulting in the production of dissolved gases including vinyl chloride, ethene and ethane. At the end of the second reporting period (through December 1998), two points in the DMPN, MP-35-D and MP-39-D, exhibited discernible degradation trends (as confirmed by Mann-Whitney analyses) to allow derivation of initial biodegradation rate estimates. Through the end of July 1999, monitoring results from seven additional DMPN points (MP-32-D, MP-33-D, MP-34-D, MP-36-D, MP-37-D, MP-38-D and MP-40-D) exhibited discernible trends, suggesting that initial estimates of biodegradation rates were warranted for these points.

To calculate anaerobic biodegradation rates an iterative curve fitting procedure was developed that allowed the biodegradation rate constants for cDCE, vinyl chloride, and combined ethene and ethane to be estimated for individual monitoring points, based on a sequential, first-order degradation rate assumption. In general, the method back-calculates the biodegradation rates of VC and then cDCE from the rate of formation of ethene and ethane. This method ensures that the estimated biodegradation rate accounts only for biodegradation and not dilution or dispersion. Using this approach, groundwater quality at each monitoring point is treated as an individual microcosm or a "closed system," i.e., constituents do not enter or leave the monitoring point. This approach, therefore, assumes that transport of constituents does not significantly affect the concentrations at the individual points; i.e., there is not a loss of constituent mass due to migration away from the monitoring point. This assumption is conservative given the COC distribution in the deep DMPN, the relatively low groundwater flow velocities monitored in the area of the deep DMPN (0.09 feet/day), and the associated low constituent transport rates.

Assuming an initial concentration of cDCE, C_{cDCEo}, the resulting concentration at some later time, C_{cDCEt}, is a function of the first order biodegradation rate, k_{cDCE}, and the elapsed time, t:

$$C_{cDCEt} = C_{cDCEo}\, e^{-k_{cDCE}\, t} \qquad \text{[Equation 1]}$$

The net decrease in cDCE concentration was assumed to be instantaneously and completely converted to vinyl chloride on a molar basis. At 100 percent conversion efficiency, one mole of cDCE is instantaneously and completely converted to one mole of vinyl chloride. In concentration terms, the net decrease in cDCE is converted to vinyl chloride based on the ratio of the molecular weights. The concentration of vinyl chloride at the end of the same time interval, C_{VCt}, is calculated by adding the contribution from cDCE degradation, ΔVC_{cDCE}, to the initial vinyl chloride concentration, C_{VCo}, and applying the first-order decay as follows:

$$C_{VCt} = (C_{VCo} + \Delta VC_{cDCE})\, e^{-k_{vc}\, t} \qquad \text{[Equation 2]}$$

The degradation of vinyl chloride to produce ethene and ethane (Δee-ea) is similarly represented by:

$$C\text{ ee-ea}_t = (C\text{ee-ea}_o + \Delta\text{ee-ea}_{vc})\, e^{-k_{ee\text{-}ea}\, t} \qquad \text{[Equation 3]}$$

In cases where ethene and ethane concentrations accumulated without evidence of VC degradation, the degradation rate, $k_{ee\text{-}ea}$, was set to zero.

Because of the specific site conditions (distribution of cDCE, ground water flow rates and velocities) the closed system assumption is a conservative assumption for these calculations. If advection and dispersion were occurring to a significant degree, these processes would tend to reduce the observed COC concentrations in the deep DMPNs. As a result, concentrations of biodegradation daughter products would be measured at lower concentrations than actually produced. For example, assuming that VC is degrading, transport effects would result in lower measurements of ethene and ethane concentrations at the sampling locations. The rate of VC biodegradation would be underestimated in this case because it was treated as directly proportional to the observed accumulation of ethene and ethane.

The degradation rate constant estimates were derived by iteratively adjusting the cDCE and VC rate constants to match the modeled dissolved gas concentration trends with those exhibited for VC and combined ethene and ethane in the field data. Combined ethene and ethane degradation rate constants were also adjusted as necessary to achieve the best match on the combined ethene and ethane trends, based on the assumption that the area surrounding each well acts as a closed system within the time scale considered.

Anaerobic Biodegradation Rate Estimates. The time versus COC concentration trends at each well (see example Figure 1), were evaluated to identify regions on the decay curves that were indicative of enhanced biodegradation according to the following criteria: 1) increased cDCE degradation as indicated by an increased slope of the decay curve; and 2) increased production of vinyl chloride, ethene and ethane as indicated by increasing constituent concentrations.

Table 1 summarizes the results for each transformation step and provides Site-wide averages for comparison. The average degradation rate constant for cDCE was calculated to be 3.1×10^{-3} day^{-1} (255 day half-life), ranging from a minimum 1.6×10^{-3} day^{-1} of at MP-38-D to a maximum of 4.5×10^{-3} day^{-1} at MP-32-D. Vinyl chloride rate constants ranged between a minimum of 5.0×10^{-3} day^{-1} at MP-38-D to 4.0×10^{-2} day^{-1} at MP-36-D, with a DMPN-wide average of 2.2×10^{-2} day^{-1} or a half-life of 48 days. The estimated rate constants for combined ethene and ethane degradation fall between 0 day^{-1} at MP-34-D and MP-38-D and 1.2×10^{-2} day^{-1} at MP-37-D, yielding an average rate constant of 5.4×10^{-3} day^{-1} or 161 days in terms of half-life.

Evidence of the degradation of cDCE and VC and production of ethene and ethane was observed at all points in the deep DMPN, including points where benzoic acid concentrations were relatively low or not consistently detected (MP-33-D, MP-37-D, MP-38-D, and MP-40-D). Volatile fatty acid analyses were performed at three of these four points. Acetate, a fermentation product of sodium benzoate and another potential electron donor for dechlorination, was detected at relatively high concentrations (measured as acetic acid at 24 to 43 mg/l) at the three points (MP-33-D, MP-38-D and MP-40-D), as well as elsewhere in the deep DMPN. The acetic acid concentrations in the deep DMPN were higher than those in MP Row 1 just downgradient of the landfill, indicating that the concentrations in the DMPN were the result of benzoate degradation rather than being a component of the landfill leachate. These data, together with the observed COC data indicating active biodegradation, indicate that electron donor was being adequately distributed throughout the deep DMPN and was promoting degradation of the target chlorinated COCs.

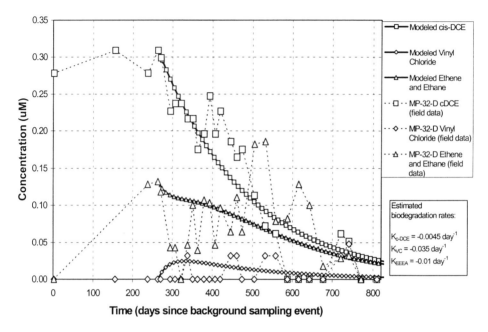

Figure 1. Estimated minimum biodegradation rate curves to fit MP32-D data.

Dissolved Arsenic Removal. The design for the TZD included the aerobic zone where less chlorinated solvents and petroleum hydrocarbons would degrade, and arsenic would co-precipitate with iron under elevated dissolved oxygen (DO) and redox conditions. During the evaluation of arsenic immobilization, it was observed that dissolved arsenic concentrations generally decreased with distance through the anaerobic treatment area. In all three strata, the dissolved arsenic concentrations are below the reporting limit (10 ug/l) at the beginning of the aerobic treatment area. Although the mechanism for the concentration decrease is not established at this Site, anaerobic arsenic immobilization has been documented at other sites (Masacheleyn et al., 1991; Pichler et al., 1999). Arsenic immobilization under anaerobic conditions was documented in an environment where dissolved sulfate and sulfate-reducing bacteria were present with arsenic precipitating as an iron-arsenic-sulfide solid phase using sulfide produced by microbial sulfate reduction (Rittle et al., 1995). Suitable conditions exist in the anaerobic treatment area of the TZD for this process to occur. Geochemical data in the anaerobic treatment area indicate that the appropriate redox conditions and concentrations of sulfate are present and generally decrease in concentration between MP Row 1 and MP Row 2. In addition, phospholipid fatty acid (PLFA) analyses on Site soil samples demonstrated the presence of sulfate-reducing bacteria in this area of the Site.

Table 1. Rates of anaerobic degradation of chlorinated compounds in the deep DMPN.

DMP	cis-1,2-DCE	Vinyl Chloride	Ethene & Ethane
	(day^{-1})	(day^{-1})	(day^{-1})
MP-32-D	4.5E-03	3.5E-02	1.0E-02
MP-33-D	3.5E-03	2.0E-02	9.0E-03
MP-34-D	1.7E-03	9.0E-03	0.0E+00
MP-35-D	2.7E-03	1.1E-02	2.0E-03
MP-36-D	2.5E-03	4.0E-02	4.5E-03
MP-37-D	4.5E-03	3.5E-02	1.2E-02
MP-38-D	1.6E-03	5.0E-03	0.0E+00
MP-39-D	4.0E-03	2.0E-02	2.0E-03
MP-40-D	3.0E-03	2.5E-02	9.0E-03
Average	3.1E-03	2.2E-02	5.4E-03
Coefficient of Variation	3.6E-01	5.6E-01	8.6E-01
	Half-Life	Half-Life	Half-Life
	(days)	(days)	(days)
MP-32-D	154	20	69
MP-33-D	198	35	77
MP-34-D	420	77	-
MP-35-D	257	63	347
MP-36-D	277	17	154
MP-37-D	154	20	58
MP-38-D	433	139	-
MP-39-D	173	35	347
MP-40-D	231	28	77
Average	255	48	161
Coefficient of Variation	4.2E-01	8.2E-01	

CONCLUSIONS

Data collected in the anaerobic portion of the TZD show that all deep DMPN points exhibit trends indicative of enhanced anaerobic biodegradation. The occurrence of increased dissolved gas production, particularly at points where benzoic acid and acetate were consistently detected, indicates that electron donor addition is enhancing the biodegradation of the target compounds. Based upon data available through this reporting period, estimated average enhanced anaerobic biodegradation rate constants are 3.1×10^{-3} day^{-1} for cDCE, 2.2×10^{-2} day^{-1} for vinyl chloride, and 5.4×10^{-3} day^{-1} for combined ethene and ethane. These rate constants are higher than the initial natural attenuation rate constant estimates at the Site.

Data collected to evaluate the biodegradation of COCs in the anaerobic zone indicate that the addition of an electron donor (sodium benzoate) will enhance the biodegradation rates of target COCs. As more data are collected, we will be able to more accurately calculate degradation rates at the Site, thereby allowing more accurate predictions of cost effectiveness and efficiency of a full scale remediation system for the Site.

The treatment system demonstrated during this project provides many advantages over remedial measures currently proposed for this Site and other similar sites (i.e., impermeable cap with pump and treat leachate collection). Preliminary results indicate that the dual treatment zone approach provides a mechanism for treating both compounds that are best degraded under anaerobic conditions (e.g., chlorinated solvents) and compounds that are best degraded aerobically (e.g., BTEX). It also provides a mechanism for immobilizing metals like arsenic (via oxidation in the aerobic zone) or others. The ability to treat the contaminants *in situ* eliminates the need to pump the contaminated groundwater to the ground surface, thereby limiting the need to handle or dispose of the contaminants. It also allows large volumes of groundwater to be treated simultaneously, without the flow limitations imposed by engineered aboveground systems like air strippers, carbon canisters, and/or bioreactors. Another advantage of the system is that it allows precipitation to continue to infiltrate and percolate through the landfill, maintaining conditions that support the existing microbial populations that have to date degraded contaminants within the landfill mass. Percolating precipitation will also flush residual contaminants to the treatment zone system for completion of degradation processes. The eventual installation of a permeable cap on the landfill also will preserve the bioreactor function of the landfill, while preventing potential direct contact with landfilled wastes and thus allowing beneficial re-use of the Site.

REFERENCES

Lizotte, C.C., M.C. Marley, S.C. Crawford, A.M. Lee, and R.J. Steffan. 1999. Demonstration of a sequential anaerobic/aerobic in-situ treatment system at a Superfund site. In: Phytoremediation and Innovative Strategies for Specialized Remedial Applications. A. Leeson and B.C. Alleman eds. Battelle Press, Columbus, Richland.

Masscheleyn, P.H., R.D. Delaune, and W.H. Patrick, Jr. 1991. Effect of redox potential and pH on arsenic speciation and solubility in a contaminated soil. Environ. Sci. Technol. 25: 1414-1419.

Pichler, T., J. Veizer, and G.E.M. Hall. 1999. Natural input of arsenic into a coral-reef ecosystem by hydrothermal fluids and its removal by Fe(III) oxyhydroxides. Environ. Sci. Technol. 33:1373-1378.

Rittle, K.A., J.I. Drever, and P.J.S. Colberg. 1995. Precipitation of aresenic during bacterial sulfate reduction. Geomicrobiol. J. 13:1-11.

ANAEROBIC/AEROBIC TREATMENT OF PCE USING A SINGLE MICROBIAL CONSORTIA

Scott W. Hoxworth, BSEnv, E.I. (Parsons Engineering Science, Winter Park, FL)
Andrew A. Randall, Ph.D, P.E. (University of Central Florida, Orlando, FL)
Terrence M. McCue, M.S., E.I. (University of Central Florida, Orlando, FL)

ABSTRACT: An experimental microcosm study was performed to investigate the feasibility of an alternating anaerobic/aerobic sequence to biologically transform PCE to non-hazardous end products such as ethylene, CO_2 and H_2 using a single microbial consortia. Analytical data from the experiments indicated that anaerobic reductive dechlorination resumed after aerobic sequences utilizing hydrogen peroxide and air.

INTRODUCTION

Halogenated aliphatic compounds (HACs) are among the most frequently encountered groundwater contaminants (Westrick et al., 1984). Tetrachloroethylene (PCE) and trichloroethylene (TCE) are HACs that are commonly used as industrial solvents, as degreasers, and in the dry cleaning industry (Fetter, 1993). The widespead occurrence of PCE and TCE in the groundwater is of concern because these compounds, and their lesser chlorinated degradation products cis-1,2 dichloroethylene (cDCE), trans-1,2 dichloroethylene (tDCE), 1,1 dichloroethylene (1,1DCE), and vinyl chloride (VC), can be both toxic and carcinogenic to humans (Ensley, 1991).

Anaerobic aquifers initially contaminated with PCE, TCE, and other chlorinated solvents were found to have accumulated dechlorination products over time (Beeman et al., 1994). Biotransformation of PCE and TCE to less halogenated daughter products under anaerobic conditions has subsequently been observed in field studies, fixed-film reactors, microcosm studies, and pure cultures. Methanogenic bacteria are the most documented bacteria capable of reductive dechlorination of PCE, but sulfidogenic (Bagley and Gossett, 1990) and homoacetogenic (DiStefano et al., 1991) bacteria have also been shown capable of reductive dechlorination. These studies have shown that the predominant end product of PCE dechlorination in an anaerobic environment is cDCE and VC, even though some methanogenic bacteria are capable of complete reductive dechlorination of PCE to ethylene (Freedman and Gossett, 1989). In general, the speed of transformation of PCE to less halogenated compounds decreases as the number of chlorines decrease. Unfortunately, the degradation products of PCE, in particular VC, are more hazardous to human health than PCE (Infante and Tsongas, 1982).

Various researchers have attempted to combine the benefits of anaerobic reductive dechlorination for highly chlorinated compounds such as PCE with aerobic co-metabolic oxidation of less chlorinated species (McCarty, 1991; Kastner, 1991; Zitomer and Speece, 1993). The problem with such a strategy lies in the fact that anaerobic bacteria, particularly methanogens, are well known for their intolerance of oxygen (Madigan et al., 1997). Thus most researchers propose ex-situ reactors in series or separate anaerobic and aerobic zones in-situ. Other species of anaerobic bacteria, however, are more aero-tolerant than methanogens (e.g., sulfate-reducing bacteria: Madigan et al, 1997). This suggests the possibility of using a single treatment zone or reactor with alternating anaerobic/aerobic conditions. Furthermore, diffusion limitations might even allow methanogens to survive aerobic phases of treatment in microanaerobic zones.

The objective of this laboratory research was to utilize an alternating anaerobic/aerobic sequence to biologically transform PCE, TCE, cDCE, and VC to non-hazardous end products such as ethylene, ethane, CO_2 and H_2 using a single microbial consortia. Both a methanogenic and a sulfidogenic series of experiments were conducted as anaerobic phase strategies during this research. Due to space limitations, the sulfidogenic experiments will not be discussed in great detail.

METHODS AND MATERIALS

The sequential anaerobic/aerobic environment microcosm experiment consisted of 46 biotic microcosms, 4 abiotic autoclaved and $HgCl_2$-dosed control microcosms, and 6 biotic control microcosms dosed with 50mM of BES to suppress methanogenic activity (Pavlostathis and Zhuang, 1991). The 46 biotic microcosms consisted of 16 microcosms to be used as anaerobic environment controls and 30 microcosms to be used with a sequenced anaerobic environment and aerobic environment. The 30 microcosms consisted of 3 sets of 10 microcosms using different aerobic environment strategies. Two of the aerobic environment strategies utilized hydrogen peroxide addition into the microcosm liquid phase to initiate the aerobic environment The third aerobic environment strategy utilized the injection of atmospheric air into the microcosm headspace to initiate the aerobic environment.

Microcosms were constructed in 120 mL amber glass bottles crimp sealed with PFTE butyl-lined septum tops. Initial liquid volume for the microcosms was 100 mL with 20 mL headspace. The biomass inocula for the microcosms came from a methanogenic environment reactor that was originally seeded with biomass from a wastewater treatment plant anaerobic digester (South Regional WWTP, Orlando, FL). Initial substrate concentrations for all of the sequential environment microcosms and controls were 500 mg/L acetic acid and 400 mg/L proprionic acid. Each microcosm included 1.0 mL of Wolfe's vitamin solution and 1.0 mL of Wolfe's mineral solution (Atlas, 1993). 50 mg/L of $Na_2S \cdot 9H_2O$ (reducing agent) was also added. Initial pH and ORP values measured after the addition of the reducing agent were 7.4 and –6 mV, respectively. All microcosms were maintained at 30°C for the duration of the experiment.

Abiotic control microcosms were autoclaved and dosed with 20 mg/L $HgCl_2$. Reductive dechlorination was not observed in three of four abiotic controls. Reductive dechlorination in the one abiotic control bottle was attributed to ineffective autoclaving. Further investigation has determined that in sulfate-reducing environments, mercuric ions (Hg^{+2}), which are typically converted into highly toxic methylmercury and dimethylmercury by bacteria in most anaerobic environments, are instead precipitated as HgS when in the presence of H_2S (Madigan, et al., 1997). Due to the use of sodium sulfide as a reducing agent during microcosm creation, H_2S is present in all of the microcosms. The low solubility of HgS apparently prevented $HgCl_2$ from being an effective biocide.

Liquid phase chlorinated solvents were measured using a Tekmar Co. purge-and-trap fitted to a Hewlett Packard gas chromatograph (GC) Model 5890 equipped with a flame ionization detector (FID) and a Vocol capillary column. Chlorinated solvents in the gas phase were calculated using Henry's Law. Volatile fatty acid (VFA) concentrations were measured using Shimadzu 14-A GC fitted with a FID and a packed column. Biogenic gases were measured using a Shimadzu 14-A GC fitted with a thermal conductivity detector and a packed column.

RESULTS AND DISCUSSION

The results of the methanogenic/aerobic environment experiments are summarized in Table 1.

TABLE 1. Summary of Results from Microcosm Studies

Microcosm Series	Experimental Time Frame	Complete Transformation of PCE to cDCE	CH_4 Production (1)	CO_2 Production (1)
Anaerobic	0 to 132	35 to 63 days	100 %	100 %
Anaerobic w/ BES	0 to 132	<81 days	4 %	70 to 78 %
Anaerobic/ Aerobic w/ 15 ppm H_2O_2	35 to 132	42 to 81 days	0 to 220 %	0 to 370 %
Anaerobic/ Aerobic w/ 75 ppm H_2O_2	35 to 132	Maximum converted to cDCE = 60% (Bottle #1 only)	0 to 47 %	0 to 1200 %
Anaerobic/ Aerobic w/ Air	84 to 132	Same as anaerobic series	Not Observed	Not Observed

Note: (1) CH_4/CO_2 production relative to non-suppressed anaerobic microcosms.

Anaerobic Microcosms. The results for the non-BES suppressed microcosm series are shown in Figure 1. Reductive dechlorination daughter products of PCE appeared in the anaerobic microcosms by Day 30 of the experiment. At this point, the data indicated that transformation of PCE to TCE was occurring. Chlorinated solvent measurements performed on Day 35 indicated that partial transformation of TCE to cDCE had also occurred. Complete transformation of PCE to cDCE was observed by Day 63 for microcosms which remained anaerobic. Quantitatively, an overall PCE to cDCE transformation rate of 9.3×10^{-2} µM/day from day 9 to day 63 was observed in the non-BES suppressed anaerobic microcosms. An overall average chloride release rate of 1.8×10^{-1} µM/day was observed in the anaerobic microcosms.

Methane gas production was detected within a few days of the start of the experiment. Methane gas production averaged 32 μmol/day during the first 63 days of reductive dechlorination and net CO_2 gas production averaged 5.5×10^{-1} μmol/day. After 63 days, the increasing pressure in the headspace made it difficult to obtain quantitative information on the headspace gas.

From day 63 on, VFA analysis performed on the anaerobic microcosms consistently showed over 98% of proprionate had been consumed from the start of the experiment. Acetate reduction varied between 34% and 99% for microcosms sampled on either day 63 or day 81 of the experiment. The VFA and biogenic gas data supports the occurrence of both proprionate fermentation and acetoclastic methanogenesis in the anaerobic phase of the experiment. H_2 utilizing methanogenesis and homoacetoclastic activity probably occurred also but their relative contribution to the overall metabolism could not be assessed.

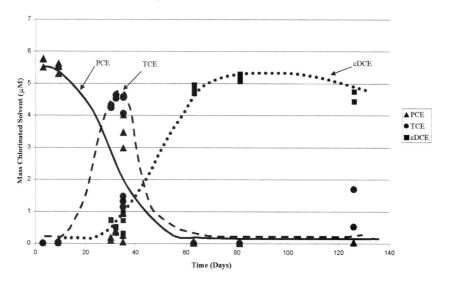

Figure 1: Biotic Anaerobic Microcosms

All 6 BES suppressed microcosms reductively dechlorinated the PCE to cDCE by day 81. The transformation was accomplished prior to day 81 of the experiment. The normalized chloride release rate was at least 1.2×10^{-2} μM/day for the BES suppressed versus 1.8×10^{-2} μM/day for the non-BES suppressed microcosms. Methane production was 3% to 4% and CO2 was two-thirds of that produced in the non-BES suppressed anaerobic controls. Reduced proprionate and acetate utilization compared to the non-BES suppressed microcosms was observed also. This data suggested that methanogenic microorganisms may not have been the primary metabolism responsible for reductive dechlorination in this experiment or certainly that other metabolisms (e.g., homoacetogenic, fermentative) can take over that function. For these microcosms it would be worthwhile to try anaerobic/aerobic experiments and identify the microbes responsible.

Anaerobic/Aerobic Microcosms (H_2O_2). On day 35 of the experiment, a 3 percent H_2O_2 solution was injected into 2 sets of 10 non-BES suppressed microcosms. At this point, parallel and identical microcosms not receiving H_2O_2 showed that 50% to 95% of the PCE had been transformed to mostly TCE with some cDCE. Ten microcosms were injected to produce an overall H_2O_2 concentration of 15 mg/L. The other ten microcosms were injected to produce an overall H_2O_2 concentration of 75 mg/L. 4 of the 10 bottles in each series was also injected with $HgCl_2$ to establish an abiotic control but $HgCl_2$ was not an effective biocide.

Following the H_2O_2 injection on day 35 of the experiment, the microcosms were then sampled over the following 12 weeks to check for indications of aerobic biodegradation (e.g., methanotrophic cometabolism) of the cDCE or continued reductive dechlorination of PCE or TCE. On day 116, selected bottles (4 of 6) were injected with additional PCE to determine if further reductive dechlorination was possible at the end of the experiment.

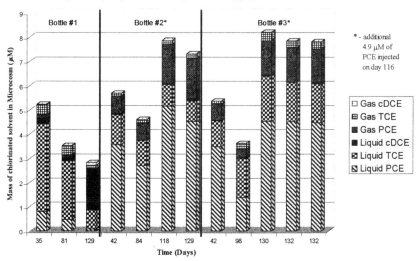

Figure 2: Hydrogen Peroxide Experiment (75 ppm), Selected Data

Some of the chlorinated solvent data for the H_2O_2 experiments are shown in Figures 2 and 3. The 75 ppm injection substantially inhibited the reductive dechlorination process while the 15 ppm injection bottles continued to reductively dechlorinate PCE and TCE to cDCE. However, no conclusive evidence of mineralization of the PCE daughter products was found during the experiment to show successful aerobic biodegradation had occurred. This was because the inocula probably did not contain methanotrophs. However, the experiments did show that anaerobic reductive dechlorination could be sustained through periodic aerobic conditions. These experiments should be repeated with inocula containing both anaerobes and methanotrophs or in-situ (anaerobes and methanotrophs are common in most soils).

Chlorinated solvent data for the high concentration (75 ppm) hydrogen peroxide injection is shown in Figure 2. Bottle #1 just prior to the H_2O_2 injection on day 35 shows only 17% of the HAC mass remains as PCE. The PCE has been predominantly converted to TCE with less than 5% converted to cDCE. Then on day 81 (46 days later), bottle #1 was resampled and showed little change. During this same time frame, identical anaerobic microcosms that didn't receive H_2O_2 converted all the PCE to cDCE. Reductive dechlorination resumed between day 81 and 129 and by day 129 about 65% of the PCE mass had been converted to cDCE.

Inhibition of reductive dechlorination was also apparent in bottle #2. However, unlike bottle #1, reductive dechlorination did not resume before the end of the experiment. Both bottle #2 and #3 received a second 0.5 µL (4.9 µM) PCE injection on day 116 of the experiment. Bottle #3 showed some additional reductive dechlorination after the 75 ppm H_2O_2 addition, although not as dramatic as that seen in Bottle #1. In bottle #3, overall HAC mass drops by approximately 33% while the TCE mass increased by 52% between day 42 and day 98 of the experiment, but further reductive dechlorination is not evident between day 98 and day 130.

PCE transformation to cDCE was less than 20% in bottles #4 through #6, supporting the inhibition of the reductive dechlorination mechanism after H_2O_2 injection. However on day 130, Bottle #6 showed about 70% of the total PCE mass was transformed to TCE and another 8% to cDCE after a second 0.5 µL PCE injection on day 116. Bottle #6 supported the presence of continued reductive dechlorination after the 75 ppm H_2O_2 injection.

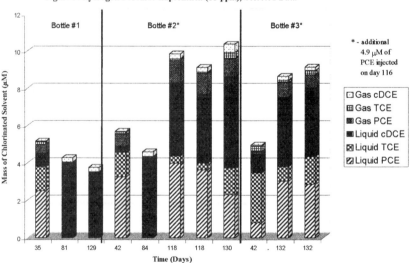

Figure 3: Hydrogen Peroxide Experiment (15 ppm), Selected Data

Chlorinated solvent data for the low concentration (15 ppm) hydrogen peroxide injection is shown in Figure 3. Unlike the high H_2O_2 concentration experiment, reductive dechlorination recovered in all 6 of the microcosm bottles.

Bottle #1 was sampled immediately prior to the H_2O_2 injection on day 35 of the experiment. Similar to the anaerobic control bottles, Bottle #1 showed partial transformation of the PCE with 45% of the PCE transformed to TCE or cDCE. By day 81 of the experiment, cDCE was the only detectable chlorinated solvent remaining in Bottle #1. This result was confirmed by Bottle #2 which showed partial PCE transformation (33% of PCE mass transformed to TCE or cDCE) 7 days after the H_2O_2 injection (day 42) and only cDCE detectable on day 84. Bottle #3 was sampled 7 days after the injection and showed 82% of PCE mass transformed to TCE or cDCE. Resampling Bottle #3 at the end of the experiment (two weeks after a PCE injection on day 116) showed a cDCE concentration increase of 270%. Bottle #1, #2, and #3 provided strong evidence that reductive dechlorination continued after a hydrogen peroxide injection of 15 ppm. In addition, Bottle #4 and Bottle #5 were sampled on days 84 and 89 respectively with the only detectable chlorinated solvent species being cDCE (results not shown). With complete transformation of PCE to cDCE, the Bottles #4 and #5 provide corroborating evidence for resumed reductive dechlorination following H_2O_2 injection. Bottle #6 was only sampled at the end of the experiment (two weeks after a second 0.5 µL PCE injection on day 116) and showed 57% of the injected PCE mass had transformed to cDCE.

Higher CH_4 gas production was observed in the 15 ppm H_2O_2 experiment as compared to the 75 ppm H_2O_2 experiment. After H_2O_2 addition on day 35 of the experiment, CH_4 generation continued in the 15 ppm microcosms with CH_4 mass increasing by 150% in Bottle #1 and 89% in Bottle #2. There appeared to be an overall increase in CH_4 mass in the 75 ppm microcosms of 20 to 40%. There was a correlation between continued CH_4 production and the resumption of reductive dechlorination when the CH_4 data was compared with the chlorinated solvent data. The 15 ppm experiments showed higher CH_4 production and better resumption of reductive dechlorination following H_2O_2 injection. This indicated methanogens and presumably other anaerobes tolerated 15 ppm better than 75 ppm.

The shutdown in reductive dechlorination at 75 ppm corresponded to a shutdown of CH_4 production and an increase in CO_2 production over that of identical anaerobic microcosms. Continued reductive dechlorination at 15 ppm corresponded to continued CH_4 production at perhaps a slightly reduced rate and steady CO_2 production similar to identical anaerobic microcosms. VFA data from samples drawn from the H_2O_2 experiments on days 84 and 89 indicated that reductive dechlorination was inhibited and proprionate fermentation was shut down or inhibited by the 75 ppm H_2O_2 addition. Proprionate fermentation appeared to have continued in the 15 ppm H_2O_2 microcosms along with reductive dechlorination after H_2O_2 injection.

Air Injection Experiments and Sulfidogenic Experiments. Injection of air into the microcosm headspace was utilized as a second strategy to induce an aerobic phase of the anaerobic/aerobic sequential environment and thereby stimulate methanotrophic bacteria in the microbial consortia to cometabolize the reductive dechlorination daughter products. Each microcosm was allowed to remain

anaerobic until day 84 to ensure PCE in the microcosm had been transformed to cDCE. To induce methanotrophic activity, the headspace of each microcosm was vented and flushed with 100 ml of atmospheric air. The vent was removed and each microcosm headspace was injected with 20 ml of 99.99% methane gas as a substrate for the methanotrophic bacteria.
The air injection was performed on days 84 and 102 of the experiment to induce the aerobic treatment (methanotrophic) phase of the experiment. The microcosms were then sampled over the one month period between the first air injection on day 84 and the end of the experiment on day 132.

The air experiments were started 7 weeks after the H_2O_2 experiments such that PCE had been completely transformed to cDCE. In this condition, further reductive dechlorination of cDCE to VC or ethene is significantly more difficult and unlikely to be observed. As VC and ethene were not observed, reductive dechlorination was not observed after injection of ambient air on day 84 and day 102 of the experiment. However, additional PCE was injected on day 116 of the experiment to test the microbial consortia for the capability of reductive dechlorination. On day 126 of the experiment (10 days after the PCE injection), samples from two of six microcosms showed detectable concentrations of TCE present. Previous samples from the two bottles on days 97 and 108 of the experiment prior to the PCE addition showed cDCE as the only detectable analyte. The detection of TCE in these microcosms demonstrates that further reductive dechlorination may be possible after ambient air injections to induce aerobic degradation. However, no conclusive evidence of mineralization of the PCE daughter products was found during the experiment to show successful aerobic biodegradation had occurred, probably due to the absence of methanotrophs in the inocula.

Results from sulfidogenic microcosm experiments utilizing a similar anaerobic/aerobic treatment strategy also indicated that anaerobic reductive dechlorination can continue after initiation of aerobic treatments (McCue, 1999). Partial reductive dechlorination from PCE to TCE was observed after 78 days at 30°C during the initial anaerobic stage. Experimental sulfidogenic microcosms were able to continue reductive dechlorination after three separate aerobic cycles, provided at least 10 days elapsed between aerobic treatments, thereby showing the potential of sulfidogenic bacteria to survive aerobic treatment cycles. Sulfidogenic microcosms with at least 7 days between aerobic treatments, however, showed no further reductive dechlorination after initiation of aerobic treatments. Furthermore, sulfidogenic microcosms that had aerobic treatments with 100 ppm H_2O_2 and with air stimulated PCE degradation to rates more rapid than purely anaerobic sulfidogenic microcosms. Paradoxically, the lowest oxygen tension aerobic phase treatments (10 ppm H_2O_2) suppressed further reductive dechlorination. No evidence of aerobic co-metabolic oxidation of PCE degradation products by methanotrophic bacteria was found in any of these experiments. Again repetition of the experiment with an inocula containing methanotrophs would be worthwhile.

CONCLUSIONS

- Reductive dechlorination of PCE was observed under methanogenic conditions in all unsuppressed anaerobic environment microcosms. PCE was observed to be sequentially reduce to TCE and to cDCE.
- Reductive declorination of PCE to cDCE was also observed in the non-methanogenic, BES suppressed, anaerobic microcosms at comparable rates to the methanogenic anaerobic microcosms.
- Results indicated reductive dechlorination of PCE and TCE resumed and/or continued after injection at the low H_2O_2 concentrations (15 ppm).
- Less conclusive results indicated reductive dechlorination resumed after a long reacclimation period for the injection of the higher H_2O_2 concentration (75 ppm).
- Limited reductive dechlorination was observed following exposure to air.
- Under sulfidogenic conditions, reductive dechlorination was resumed after both H_2O_2 and air exposure.
- No conclusive evidence was observed to indicate that aerobic degradation of cDCE during any of the aerobic phase techniques. This was probably due to the microcosm inocula not containing methanotrophs.

REFERENCES

Atlas, R. M. (1993). *Handbook of Microbiological Media*. Boca Raton: CRC/Lewis Press.

Bagley, David M., and James M. Gossett. (1990). "Tetrachloroethene Transformation to Trichloroethene and cis-1,2-Dichloroethene by Sulfate-Reducing Enrichment Cultures." *Applied and Environmental Microbiology, 56*, 2511-2516.

Beeman, R. E., J. E. Howell, S. H. Shoemaker, E. A. Salazar, and J. R. Buttram. (1994). "A Field Evaluation of in situ Microbial Reductive Dehalogenation by the Biotransformation of Chlorinated Ethenes." In Hinchee, R. E., L. Semprini, and S. K. Ong (Ed.), *Bioremediation of Chlorinated and Polycyclic Aromatic Hydrocarbon Compounds* (pp. 14-27). Ann Arbor, MI: Lewis Publishers.

DiStefano, T. D., J. M. Gossett, S. H. Zinder. (1991). "Reductive Dechlorination of High Concentrations of Tetrachloroethene to Ethene by an Anaerobic Enrichment Culture in the Absence of Methanogenesis." *Applied and Environmental Microbiology, 57*, 2287.

Ensley, B.D. (1991). "Biochemical Diversity of Trichloroethylene Metabolism." *Annual Review Microbiology, 45*, 283 – 299.

Fetter, C.W. (1993). *Contaminant Hydrogeology.* New York:Macmillan Publishing Company.

Freedman, David L., and James M. Gossett. (1989). "Biological Reductive Dechlorination of Tetrachloroethylene and Trichloroethylene to Ethylene under Methanogenic Conditions." *Applied and Environmental Microbiology. 55*, 2144-2151.

Infante, P. F., T. A. Tsongas. (1982). "Mutagenic and Oncogenic Effects of Chloromethane, Chloroethenes and Halogenated Analogues of Vinyl Chloride." *Environmental Science Research. 25*, 301-327.

Kastner, M. (1991). "Reductive Dechlorination of Tri- and Tetrachloroethylene Depends on Transition from Aerobic to Anaerobic Conditions." *Applied and Environmental Microbiology. 57*, 2039-2046.

Madigan, Michael T., John M. Martinko, Jack Parker. (1997). *Brock Biology of Microorganisms, Eighth Ed.* Upper Saddle River, NJ:Simon & Schuster.

McCarty, P.L. (1991). "Engineering Concepts for In-Situ Bioremediation." *Journal of Hazardous Materials. 28*, 1-11.

McCue, T. M. (1999). *Degradation of Halogenated Aliphatic Compounds in Sequential Anaerobic/Aerobic Sulfate-Reducing Environments.* M.S. Thesis, University of Central Florida, Orlando, FL.

Pavlostathis, Spyros G., and Ping Zhuang. (1991). "Transformation of Trichloroethylene by Sulfate-Reducing Cultures Enriched From a Contaminated Subsurface Soil." *Applied Microbiology and Biotechnology. 36*, 416-420.

Westrick, J. J., Mello, J. W., Thomas, R.F. (1984). "The Groundwater Supply Survey." *Journal of the American Water Works Association, 5*, 52-59.

Zitomer, D. H. and R. E. Speece. (1993). "Sequential Environments for Enhanced Biotransformation of Aqueous Contaminants." *Environmental Science and Technology*, 27(2), 227.

ENHANCED BIODEGRADATION OF ORGANIC COMPOUNDS IN GROUNDWATER VIA NUTRIENT INJECTION

Eric J. Raes, P.E. (Environmental Liability Management, Inc., Princeton, NJ)
Mary Ann Baviello, CHMM (ELM, Inc., Holicong, PA)
James Cook, P.E. (Beazer East, Inc., Pittsburgh, PA)
Steven Radel (Renewable Resources Co., LLC, Milford, MA; formerly with Beazer East, Inc.)

ABSTRACT: The natural attenuation processes that enhance degradation of organic constituents dissolved in groundwater were augmented via a nutrient addition field application to reduce the time and associated costs of a monitored natural attenuation program. The depletion of essential nutrients, nitrogen and phosphorus, in the site aquifer, was a natural degradation rate-limiting factor. Commercially available ammonium sulfate and hydrated sodium phosphate were injected into the affected groundwater at an empirically derived, desired carbon:nitrogen:phosphorous mass ratio of 100:10:1. The concentrations of the targeted organic constituents, benzene, chlorobenzene, cis-1,2-dichloroethene, ethylbenzene, toluene, naphthalene, vinyl chloride, and xylene, were reduced by 25% to 60% in the test wells over a six-week period, whereas historic site degradation rates were less than 4% annually. Pre-and post-injection microbial plate counts and phospholipid fatty acid (PLFA) analyses were also conducted. After a two-week lag period, plate counts increased by one to three orders of magnitude, followed by a slight decrease corresponding with the consumption/depletion of the injected nutrients. In addition, the evaluation of the PLFA analysis results confirmed that the biomass associated with the degradation of the organics, which was quantified as 60% of the total population, increased without adverse affects to the remainder of the population.

INTRODUCTION

A nitrogen (N) and phosphorous (P) based, nutrient addition field application was used at a former marine-based paint manufacturing facility in New Jersey to enhance the biodegradation processes of organic constituents in groundwater to expedite closure of the site regulatory process.

To facilitate the biodegradation of organic constituents in groundwater, additional nutrients are often needed to increase microbial activity. The stimulation of microbial populations by the addition/replenishment of depleted nitrogen and phosphorous has been shown to be effective in restoring organic-contaminated aquifers (American Petroleum Institute, 1980).

To allow survival, activity, and growth of its population, a microbial habitat must provide nutrients for the synthesis of cellular constituents and the generation of energy. Thus, sources of carbon, nitrogen, phosphorus, sulfur and low levels of various elements (Fe, Mg, Mn, etc.) must be available (Butler and Barker, 1996; Pankow and Cherry, 1996). Typically, nitrogen is the most limiting nutrient for degradation and is routinely added at rates to provide a carbon to nitrogen (C:N) ratio of 9:1 to 120:1 (Chang, et al., 1996; Dibble and Bartha, 1979). Phosphorous is typically the second most limiting nutrient and is often added along with N (Chang, et al., 1996). Recent studies have documented that a synergistic interaction between these two nutrients occurs (Walworth and Reynolds, 1995).

In most of the research reviewed, C:P mass ratios vary within the range presented by Walworth and Reynolds of 33:1 to 278:1. Based on our experience with nutrient addition and the methodology used for this field application (Kuo, 1999), a C:N:P mass ratio of 100:10:1 was used. This ratio is on a molar basis; thus for every 100 moles of carbon within the organic constituents to be degraded, 10 moles of nitrogen and 1 mole of phosphorous are required.

MATERIALS AND METHODS

The components of the biodegradation enhancement program consisted of the determination of the desired, optimum nutrient concentrations, injection program, and subsequent pre- and post-effectiveness-monitoring program, the specific details of which are discussed below.

Nutrients Calculations. Nitrogen and phosphorous were injected, in a diluted liquid form, into the former soil vapor extraction well points. N and P were added at concentrations ranging from 0.05% to 0.03% by weight, respectively (Kuo. 1999). Commercially available ammonium sulfate {$(NH_4)_2SO_4$} and hydrated sodium phosphate ($Na_3PO_4 \cdot 12H_2O$), in solid form, were mixed with tap water to produce the desired mass of dissolved nitrogen and phosphorus for injection. As a conservative measure, only 25% of the total calculated nutrient requirement was used.

Available groundwater data were used to calculate the total contaminant mass per cubic meter of aquifer material. For the purposes of the nutrient addition calculation, the maximum on-site contaminant concentration was used to determine the total contaminant mass in the aquifer material. This concentration was assumed to be present throughout the portion of the aquifer to be treated. The methodology for calculating the total mass of each organic contaminant per cubic meter of aquifer is summarized below.

- For each organic compound present, the organic carbon partition coefficient (K_{oc} in L/kg) was calculated using the chemical-specific octanol-water partition coefficient (K_{ow}) and equation 1:

$$K_{oc} \text{ (in L/kg)} = 0.63 K_{ow} \quad (1)$$

- The partition coefficient (K_p) was calculated using the organic carbon partition coefficient (K_{oc}) from the equation 1, along with the fraction organic content (f_{oc}), within the aquifer matrix:

$$K_p \text{ (in L/kg)} = f_{oc} K_{oc} \quad (2)$$

- The contaminant concentration (in mg/kg) adsorbed on the solid aquifer matrix (X) was determined with the equation 3:

$$X \text{ (in mg/kg)} = K_p \text{ (in L/kg)} \times C \text{ (in mg/L)} \qquad (3)$$

Where C = organic contaminant concentration in groundwater (mg/L).

- The bulk density of the aquifer matrix (in kg/m^3) was multiplied by 1 m^3 of aquifer volume to determine the mass (in kg) of a cubic meter of the aquifer matrix.
- The mass (in grams) of contaminant adsorbed onto the solid surface (S) was calculated as follows:

$$S = (X)(M_s) \times \frac{1 \text{ gram}}{1000 \text{ mg}} \qquad (4)$$

Where M_s = mass of the aquifer matrix (kg)

- The void space (Vϕ) per m^3 of aquifer was determined by multiplying the porosity (n$_e$) by the aquifer volume (1 m^3) and then converting to liters as follows:

$$V\phi \text{ (in L)} = (n_e)(1 \text{ m}^3)\frac{(1000 \text{ L})}{\text{m}^3} \qquad (5)$$

- The mass of contaminant dissolved in the groundwater in grams (W) was calculated with the equation 6:

$$W = (C)(V\phi) \times \frac{1 \text{ gram}}{1000 \text{ mg}} \qquad (6)$$

- The amount of dissolved (W) and adsorbed (S) contaminants were added to calculate the total mass of contaminant in grams per cubic meter of aquifer material.

Using this methodology, 41.1 grams of contaminants per cubic meter were estimated to be present within the affected groundwater area, for a total estimated mass of 357,570 grams. Using the calculated contaminant mass in one cubic meter of aquifer material, the nutrient requirements for optimal bioremediation were determined as follows.

- The total mass of each contaminant in the aquifer was divided by its molecular weight to determine the number of moles of the organic constituent per cubic meter of aquifer material.
- The number of moles of the organic constituent per cubic meter of aquifer material (from the previous calculation) was multiplied by the number carbon atoms per molecule to determine the number of moles of C.

- *The amount of nitrogen required for bioremediation* was determined using a C:N ratio of 100:10.
  1. Moles of nitrogen needed = (10/100) x (the number of moles of C).
  2. Amount (in grams) of nitrogen needed:

     (moles of N) x (molecular weight of N, which is 14)
  3. Moles of $(NH_4)_2SO_4$ needed:

     (moles of N) ÷ 2 (each mole of ammonium sulfate contains 2 moles of N)
  4. Amount (in grams) of $(NH_4)_2SO_4$ needed per cubic meter of aquifer:

     (moles of $(NH_4)_2SO_4$ needed) x (molecular weight of $(NH_4)_2SO_4$, 132)
  5. For the significant organic constituents of concern in this aquifer, the amount of nitrogen needed for optimal bioremediation is 4.8 g/m^3 as elemental nitrogen or 23 g/m^3 as $(NH_4)_2SO_4$.

The amount of phosphorous required for bioremediation was determined using a similar calculation (not shown) using the C:P ratio of 100:1.

The total nutrient requirement per cubic meter of aquifer was then multiplied by the total aquifer volume (8,700 m^3) to be treated via *in-situ* bioremediation. This calculation results in an estimation of the total nutrient requirement for the entire treatment area. The amount of nitrogen needed for optimal bioremediation of the entire treatment zone is 42,000 g (42 kg) as elemental nitrogen or 200,000 g (200 kg) as $(NH_4)_2SO_4$. The amount of phosphorous needed for optimal bioremediation of the entire treatment zone is 9,600 g (9.6 kg) as elemental phosphorous or 110,000 g (110 kg) as $Na_3PO_4 \cdot 12H_2O$. Equal volumes of the nutrient solutions were added to each injection of the 10 injection points.

Laboratory Procedures. In order to assess and quantify the effectiveness of the biodegradation enhancement program, chemical, geo-chemical and microbial analyses were conducted 14, 24 and 44 days subsequent to the nutrient injection from the four monitoring points (MW-7, -8R, -9, and -10). Baseline conditions were established prior to the injection. Concentration trends were evaluated based on the analysis of organic constituent and nutrient concentrations. In addition, geo-chemical and natural degradation indicator data were also collected, such as dissolved oxygen, redox potential, pH, temperature, etc. Finally, the effects of the nutrient addition on the existing microbial community were quantified through microbial plate counts, both total culturable heterotrophs (TCH) and total xylene degraders (TXD), and phospholipid fatty acid (PLFA) analyses.

The TCH is a standard plate count of the entire microbial population present; TXD is a subset of the TCH specifically cultured on an ethylbenzene and xylene carbon source. The ethylbenzene and xylene specific degraders were chosen since these constituents represent the majority of organic mass present at the site. These analyses were used to measure biomass growth in response to the injection of nutrients.

It is important to understand the indigenous microbial community structure and how this structure is affected by changes to the microbial environment, either through the release of chemicals or augmentation of natural degradation processes. Microbial communities have often been characterized by the analysis of signature lipid biomarkers (Brandt, et al., 1999). The phospholipid fatty acid (PLFA) analysis, which is a signature lipid biomarker analysis, was used to quantify: (1) viable biomass, (2) community composition/diversity, and (3) metabolic status. The viable biomass was used to enumerate the total abundance of the microbial population, including aerobic, anaerobic, culturable and non-culturable bacteria, for pre- and post-injection comparison purposes. The community composition provided a breakdown, by percent of the general classes of bacteria present, in each sample. The metabolic status can be determined by comparison of specific lipid biomarkers. The ratios generated from these comparisons were used to quantify the relative health of the bacteria by determining the growth stage and turnover rate.

RESULTS AND DISCUSSION

As a result of the nutrient injection enhancement program:

- Total organic concentrations in the two monitoring wells (MW-7 and MW-9) that contained the highest levels of targeted constituents were reduced by 25% to 60% over a 44-day period, whereas the pre-injection rate of attenuation was less than 4% annually;
- After a two-week lag period, TCH and TXD plate counts increased by one to three orders of magnitude followed by a slight decrease corresponding with the depletion of the injected nutrients;
- Biomass continued to increase in the monitoring point (MW-8R), but organic concentrations remained constant;
- There were no adverse affects on the microbial community, based on review of the PLFA analysis.

Decreasing Concentrations Trends. Although natural attenuation was occurring at the site, as evidenced from an overall decreasing trend in total organic constituents present, seasonal variations did occur for individual constituents from time to time. The annualized, percent reduction in total organics in the test wells over a six-year period was less than 4%, whereas the post-injection, percent reductions ranged from 25% to 60%. See Table 1 and Figure 1.

TABLE 1. Percent Reduction of Organic Constituents via Nutrient Injection.

Constituents	MW-7	MW-9	MW-8R	MW-10
Benzene	60.5	16.1	10.8	-5.3
Ethylbenzene	18.6	41.7	-9.1	-131.6
Toluene	ND	60.4	ND	-20.0
Xylenes (Total)	45.5	67.7	-7.5	0.0
Chlorobenzene	5.0	23.1	5.1	-47.1
cis-1,2-Dichloroethene	31.8	70.5	ND	-57.1
Naphthalene	15.7	67.3	38.6	ND
Vinyl Chloride	ND	100.0	ND	ND
Total Organics	**24.8**	**58.9**	**-3.2**	**-98.4**

ND – Constituent was not detected.

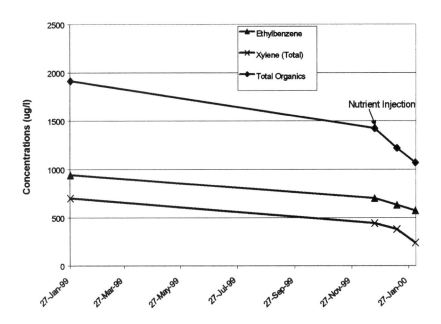

FIGURE 1. Total Organic Concentrations (µg/l) in MW-7 vs. Time.

Although biomass increased in monitoring points MW-8R and MW-10, organic concentrations remained constant or, in the case of MW-10, increased.

Nutrients were directly injected into well MW-8R. Although TCH and TXD results indicate significant growth in biomass, the PLFA data documents that the majority of the growth is of Eukaryote organisms (Figure 2).

Eukaryotes, such as fungi, protozoa, and algae, typically feed on aerobic bacteria. The surface area of the well may support colonization of Eukaroyotes, and this effect may be limited to the immediate vicinity of the well.

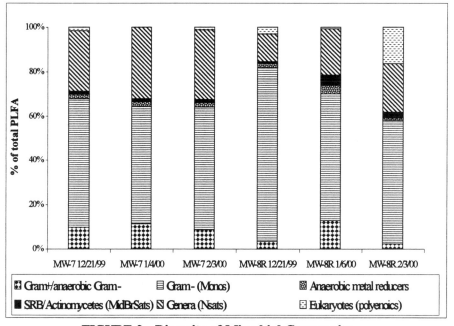

FIGURE 2. Diversity of Microbial Community.

The groundwater geo-chemical data of well MW-10 indicate strong reducing conditions, therefore indicating that dissolved oxygen is the controlling, rate-limiting degradation process in this area.

Increase in Biomass. After a two-week lag period, culturable-bacteria biomass increased from one to three orders of magnitude in the monitoring wells. The increase in biomass in all wells corresponds with the decrease in nutrients and organic constituent concentrations over time (Figure 3).

In addition, P in groundwater was an order of magnitude less than the injected concentrations, indicating that a significant portion was likely absorbed onto soil (Figure 3). This behavior is typical of P applied to a soil system. (Walworth and Reynolds, 1995).

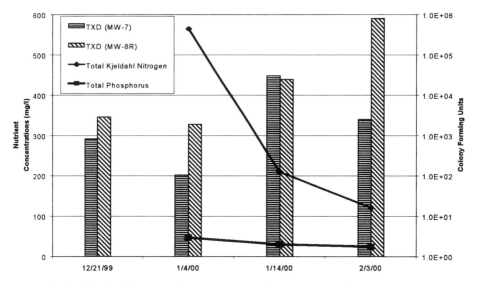

FIGURE 3. Biomass Growth vs. Nutrient Attenuation over Time.

CONCLUSIONS

Essential nutrients, specifically nitrogen and phosphorus, have increased the microbial population and biodegradation rates at this site. Therefore, these results support consideration of this remedial approach even at sites where natural attenuation is sufficient as a stand-alone remedy. The cost of this remedial approach is relatively inexpensive and the documented increase in degradation rates observed is predicted to reduce the length of time necessary to achieve applicable remediation goals at this site, and others. The overall cost of this field application, including nutrients, contractor injection, groundwater sampling and laboratory analysis (chemical and microbial), was less than $20,000.

Of paramount concern with the injection of nutrients or any action that alters the subsurface environment is to ensure that the remainder of the microbial community is not adversely affected by this change in conditions. The PLFA analysis served as an excellent tool to quantify the general categories of microbes present and monitor their condition over time. The review of the results of the initial, pre-injection PLFA analysis indicated a diverse and chemically stressed microbial community. The injection of nutrients did not adversely affect the microbial population.

Finally, as is common with nutrient addition programs, distribution limitations occurred. Direct injection into a monitoring point (MW-8R) increased growth of potentially competitive organisms, whereas the nutrients directly injected into the aquifer dissipated within six weeks of injection. A continuous feed trench system would alleviate this limitation, but would increase project costs and potentially create biofouling limitations.

REFERENCES

American Petroleum Institute. 1980. *Underground Spill Cleanup Manual.* Publication No. 1628. Washington, DC. (Cited in Ward and Lee. 1985.)

Brandt, C.C., J.C. Schryver, S.M. Pfiffner, A.V. Palumbo and S. Macnaughton. 1999. "Using Artificial Neural Networks to Assess Changes in Microbial Communities." *Bioremediation of Metals and Inorganic Compounds.* Battelle Press. Columbus, OH.

Butler and Barker. 1996. "Chemical and Microbial Transformations and Degradation of Chlorinated Solvent Compounds." (Cited in Pankow and Cherry. 1996).

Chang, Z.Z., R.W. Weaver, and R.L. Rhykerd. 1996. "Oil Bioremediation in a High and a Low Phosphorus Soil." *Journal of Soil Contamination.* 5(3):215-224.

Dibble, J.T., and R. Bartha. 1979. "Effect of Environmental Parameters on the Bioremediation of Oil Sludge." *Applied Environmental Microbiology.*

Kuo, J. 1999. *Practical Design Calculations for Groundwater and Soil Remediation.* Lewis Publishers. Boca Raton, FL.

Pankow, J.F. and J.A. Cherry (Eds). 1996. Dense Chlorinated Solvents and other DNAPLs in Groundwater: History, Behavior, and Remediation. Waterloo Press. Portland, Oregon.

USEPA. 1996. *Soil Screening Guidance: Technical Background Document.* Office of Solid Waste and Emergency Response. EPA/540/R-95/128.

Walworth, J.L. and C.M. Reynolds. 1995. "Bioremediation of a Petroleum-Contaminated Cryic Soil: Effects of Phosphorous, Nitrogen, and Temperature." *Journal of Soil Contamination.* 4(3):299-310.

Ward, C.H. and M.D. Lee. 1985. "In Situ Technologies." In Canter, L.W. and R.C. Knox (Eds.), *Ground Water Pollution Control.* Lewis Publishers, Chelsea, MI.

TREATMENT OF ORGANIC-CONTAMINATED WATER IN MICROBIAL MAT BIOREACTORS

Walter L. O'Niell and Valentine A. Nzengung
University of Georgia, Athens, Georgia, USA

Abstract: Bench scale experiments were conducted using sealed bioreactors to evaluate the degradation of tetrachloroethylene (PCE) and trichloroethylene (TCE) in a closed system. The sealed bioreactor represents a more practical design for a waste effluent treatment system because PCE and TCE were degraded without vapor loss, and mats in the reactor remained healthy throughout the experiments. PCE and TCE were degraded in a sealed microbial mat bioreactor, demonstrating half-lives as short as 9 days for PCE and 19 days for TCE. The addition of nutrients enhanced degradation rates, but the difference was not considered significant. Analyses of transformation products and toxicity tests using grass shrimp *(Palaemonetes pugio)* embryos indicated that aqueous samples dosed with PCE and treated in the bioreactor were not hazardous. This biotreatment system offers promise as an effective low-cost remediation technique for chlorinated solvent plumes upwelling into wetlands and other surface water bodies.

INTRODUCTION

Tetrachloroethylene (PCE) and trichloroethylene (TCE) are among the most common commercial and industrial contaminants. Their widespread use has led to extensive pollution of soils, groundwater and surface waters throughout the U.S. (Westrick et al., 1984). A bioreactor that uses microbial mat to sorb and degrade PCE and TCE may offer a lower cost alternative approach to the remediation of water affected by these compounds.

Microbial mats are photoautotrophs, which are composed of cyanobacteria (primarily *Oscillatoria* spp.) and related heterotrophic bacteria (primarily *Rhodopseudomonas* spp.) which stratify into upper aerobic and lower anaerobic zones within the mat. The heterotrophic anaerobes in the lower zones of the mat can biodegrade PCE and TCE through reductive dehalogenation, while the near-surface layers of the mat are photosynthetic and provide oxygen to degrade these compounds and their metabolites through oxidative pathways (Nzengung et al., 1999; O'Niell et al., 1998; Bender et al., 1995). Because of their self-sustaining nature, limited growth requirements, and exceptional tolerance to harsh conditions, microbial mats can be easily grown in a wide variety of environments.

Experiments were conducted to test the use of microbial mats in a sealed bioreactor system to degrade PCE under varying shades of natural light and to degrade TCE using mat supplemented with different nutrient supplements. Aqueous, solid, and vapor phase samples were periodically collected from the bioreactors and analyzed for parent compounds and transformation products. Controls were used in all experiments, and toxicity tests were conducted to verify that water dosed PCE and treated in the bioreactor did not contain toxic compounds.

PCE Transformation Experiments in Sealed Bioreactors. Previous research indicated that artificial and natural light conditions could be varied to enhance or decrease PCE degradation by constructed microbial mats (O'Niell and Nzengung, in press). The PCE transformation experiments were conducted using sealed bioreactors to investigate how varying natural light intensity affected mat health and PCE/TCE degradation rates.

A non-jacketed, glass fermentation vessel (Applikon Analyzers, Inc., 29200-10, Malverne, NY) was used as the bioreactor (Figure 1). The bioreactors were equipped with pH, Eh (pHoenix Electrode, Co., Houston, TX), and dissolved oxygen (DO) probes (Cole-Parmer Instrument Co., Chicago, IL). One port was fitted with a 1 cm thick Teflon® septum which allowed sampling of the aqueous phase using a syringe. A motorized impeller was used to mix the reactor liquids prior to sampling. After adding the Allen-Arnon liquid growth media and mat, the bioreactor was sealed and maintained in a greenhouse at $25 \pm 2°C$ while the mat was acclimated to these conditions. A healthy mat generally formed a thick, cohesive layer floating near the surface of the bioreactor and produced measurable amounts of oxygen when exposed to sufficient sunlight. At peak sunlight, water in a bioreactor under a 50% shade cloth was saturated with more than 15 mg/L dissolved oxygen. Three reactors were prepared and each was exposed to different light intensities. Each bioreactor was filled with 2.5 L of growth media and 200 g of mat, leaving a headspace volume of 0.4 L. One reactor (designated LM for "light mat") was exposed to unfiltered natural light, and a second reactor (HM for 50% natural light or "half" light) was placed under a 50% shade cloth. Opaque foil was wrapped around the third reactor (designated DM for "dark mat") from the base to the liquid-mat interface to prevent most light from reaching the aqueous phase, while still allowing light to reach the mat. After 72 hours, the bioreactors were dosed with PCE at initial solution concentrations of 10 mg/L. Aqueous phase samples were collected at 24 to 72 hour intervals, and were analyzed for PCE and its metabolites.

The initial PCE aqueous phase concentrations detected in the bioreactors varied from 13 to 15 mg/L due to incomplete mixing of the PCE stock and bioreactor liquids (Figure 2). After 24 hours, the aqueous phase concentrations decreased to approximately 6 mg/L, primarily due to more complete mixing and sorption of PCE to the mat. The aqueous phase concentrations then decreased as PCE was degraded (Figure 2). The half-life of PCE in bioreactor LM and DM were similar at 28 and 21 days, but the transformation rate in reactor HM was faster with a half-life of only 9 days. Also, TCE, a metabolite of PCE, was detected in reactor HM after 27 days and was completely degraded (Figure 2). Note on Figure 2 that TCE was not detected in bioreactor experiments LM and DM, and both experiments were discontinued after 41 days.

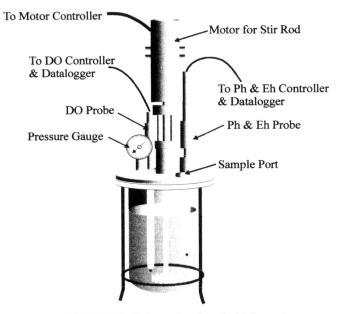

FIGURE 1. Schematic of Sealed Bioreactor

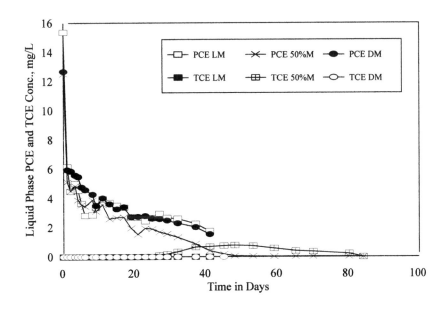

FIGURE 2. Liquid Phase PCE and TCE Concentrations in Bioreactors Dosed with PCE and Maintained Under Different Light Conditions

TCE Transformation Experiments in Sealed Bioreactors. TCE transformation experiments were conducted as described above, and all bioreactors were maintained under a 50% shade cloth. Four different treatments were tested. One sealed bioreactor (SBNO) did not receive any nutrient supplements other than the initial 2.5 L of AA media added at the start of the experiment (this experiment was replicated). The other three reactors received different nutrient supplements each week to stimulate mat and/or bacterial growth. A different nutrient supplement was prepared for each reactor. One reactor (SBAA) was supplemented with five mL of AA media concentrate sufficient to provide the aqueous phase with a full strength concentration of AA media every two weeks. A second reactor (SBAL) was supplemented with five mL of AA concentrate plus acetate and lactate sufficient to raise the aqueous phase concentration of these two compounds to 1 mM (0.5 mM/each) every 2 weeks. A third reactor (SBP) was supplemented with five mL of AA concentrate plus sodium pyruvate sufficient to raise the aqueous phase concentration of this compound to 0.5 mM every 2 weeks. The reactors were dosed with TCE to obtain an aqueous phase concentration of 10 mg/L, and samples were collected every 24 to 72 hours.

Transformation Kinetics of TCE in Sealed Bioreactors
The sorption rate constants, K_s, were 0.20 hr^{-1} ($t_{1/2}$ = 3.43 hr^{-1}) for bioreactor SBNO, 0.45 hr^{-1} ($t_{1/2}$ = 1.55 hr^{-1}) for bioreactor SBAL, 0.43 hr^{-1} ($t_{1/2}$ = 1.63 hr^{-1}) for bioreactor SBP, and 0.74 hr^{-1} ($t_{1/2}$ = 0.94 hr^{-1}) for bioreactor SBAA.

The degradation rates of TCE were compared for the reactors receiving nutrient supplements and the reactors that did not receive supplements. The degradation kinetics were modeled as first-order reactions and these data were used to calculate TCE half-lives in the bioreactors. The TCE half-lives in bioreactors SBNO, SBP and SBAA varied from 21 to 25 days, and the half-life in reactor SBAL was 19 days (Figure 3).

Toxicology tests. A toxicology test was conducted by Dr. Gi Kim and Dr. Richard Lee at the Skidaway Institute of Oceanography to determine the effects of toxicity of aqueous samples dosed with PCE after treatment in the microbial mat bioreactor on hatching rates of grass shrimp (*Palaemonetes pugio*) embryos. No hazardous compounds were detected by GC/ECD and GC/FID analysis in aqueous phase samples collected from the reactor after treatment. The toxicology tests indicated that the treated liquids did not affect the hatching rates of the shrimp embryos.

CONCLUSIONS
A sealed system is a favorable design for a microbial mat bioreactor. The mat is able to survive and increase its mass in the sealed chamber, and losses to volatilization are minimized. PCE half-lives of 9 days and TCE half-lives of 19 days were calculated. The addition of nutrient supplements seemed to stimulate bacterial activity in the bioreactors dosed with TCE, but transformation rates were

not significantly affected. Analytical results and toxicology tested indicated that hazardous metabolites (e.g., dichloroethylene and vinyl chloride) were not formed in the bioreactors.

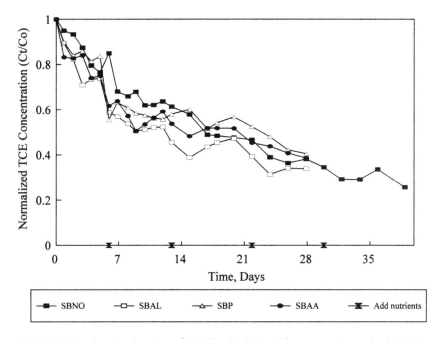

FIGURE 3. Time Series Data for Microbial Mat Bioreactors Dosed with TCE and Amended with Different Nutrient Supplements

Microbial mats were able to survive in sealed bioreactors, and mat health was not affected by dosing with PCE or TCE. The ability of mats to degrade PCE and TCE in a sealed system, without producing hazardous metabolites or releasing toxic compounds to the atmosphere, in a system powered only by solar energy make them an excellent candidate for use in remediation of contaminated compounds. These mats also sequester and hyperaccumulate metals. Potential applications of microbial mats to remediate contaminated media include: (1) decontamination of industrial wastewater containing chlorinated organic solvents prior to disposal; (2) treatment of shipboard wastewater at sea; (3) as a component of a constructed wetland system for treatment of wastewater; and (4) low-cost treatment of contaminated potable water from a contaminated well or stream in remote locations of less developed nations.

ACKNOWLEDGMENTS

Funding was provided by the U.S. Department of Energy through a subcontract between Florida A&M University and the University of Georgia - Grant/Contract #C-9594/10-21-RR176-234. The mats used in these experiments were grown in-house using mat inocula provided by Dr. Judith Bender and Dr. Peter Phillips of Clark-Atlanta University.

REFERENCES

Bender, J., P. Phillips, R. Lee, S. Rodrigues-Eaton, G. Saha, B Longanathan and L. Sonnenberg, 1995. "Degradation of Chlorinated Organic Compounds By Microbial Mats." Biological Unit Processes for Hazardous Waste Treatment. *Third International In Situ and On-Site Bioreclamation Symposium, v. 9. Battelle et al.*, pp. 299-310.

Nzengung V.A., N.L. Wolfe, D. Rennels. and S.C. McCutcheon. 1999. "Use of Aquatic Plants and Algae For Decontamination of Waters Polluted With Chlorinated Alkanes." *International Journal of Phytoremediation*, 1(3): 203-226.

O'Niell, W.L. and V.A. Nzengung. 2000. "Biosorption and Transformation of Tetrachloroethylene and Trichloroethylene Using Mixed Species Microbial Mats." *Journal of Hazardous Substance Research* (in press).

O'Niell, W.L., V.A. Nzengung, J.E. Noakes, J. Bender and P. Phillips, 1998. "Biodegradation of Tetrachloroethylene and Trichloroethylene Using Mixed Species Microbial Mats." *Proceedings of the First International Conference on Remediation of Chlorinated and Recalcitrant Compounds, May 18-21, 1998, Monterey, California.*, pp. 233-237.

Westrick, J.J., J.W. Mello and R.F. Thomas, 1980. "The Groundwater Supply Survey." *Journal of the American Water Works Association* ,76: 52-59.

TCE DEGRADATION IN ANAEROBIC/AEROBIC CIRCULATING COLUMN

A. Narjoux[1,2], Y. Comeau[2], J.-C. Frigon[1], **S. R. Guiot**[1,2,3]
([1] Biotechnology Research Institute, NRC, Montréal, Québec, Canada)
([2] École Polytechnique, Montréal, Québec, Canada)
([3] Université de Sherbrooke, Sherbrooke, Québec, Canada)

ABSTRACT: The potential for anaerobic/aerobic degradation of trichloroethylene (TCE) was studied in a coupled anaerobic/aerobic integrated column, using ethanol as primary substrate. Air was directly sparged into the liquid recycled from the top to the bottom of the reactive column, providing oxygen at different heights in an otherwise anaerobic granular bed. Even under these conditions, methane was still produced and, the concomitant presence of CH_4 and O_2 promoted growth of aerobic methanotrophic bacteria. TCE load was increased progressively to a maximum of 20 mg TCE/$L_{reactor}$.day. At TCE loads below 15 mg TCE/$L_{reactor}$.day, no chlorinated metabolites were detected in the effluent. At TCE loads above 17.5 mg TCE/$L_{reactor}$.day, effluent analysis showed cis-dichloroethylene (cis-DCE) to be the dominant metabolite, while vinyl chloride (VC) was not at all detected. At a load of 20 mgTCE/$L_{reactor}$.day, aeration rate was increased. Concurrently, cis-DCE proportion in effluent increased gradually. At the same time, biomass activities had decreased, thus prompting a cessation of TCE feeding for a 2 months period. During these 2 months, organic loading rate was decreased to modify the coupling conditions and increase the aerobic-to-anaerobic metabolism ratio. For the next 5 months, TCE loading was resumed however at 2 mg TCE/$L_{reactor}$.day. During this period, cis-DCE was the terminal product of TCE dechlorination. The above TCE-acclimated anaerobic/aerobic granules were shown to be able to mineralize over 90% of the TCE (2 mg TCE/g biomass dry weight) in the presence of CH_4 and O_2, eventhough the biomass in the system was only performing the first step of TCE reductive dechlorination. This indicates that a potential for mineralization existed, but could not manifest within the actual operational conditions of the system.

INTRODUCTION

Trichloroethylene (TCE) is a widespread groundwater contaminant due to its extensive industrial use as a solvent and degreaser. The anaerobic degradation of TCE proceeds by a sequential reductive dechlorination to dichloroethylene (DCE), vinyl chloride (VC), and even to ethylene and ethane. However, in many cases, this anaerobic dechlorination is incomplete. The rate of transformation decreases with the decrease of chlorine substitution (Vogel et al., 1987). In contrast to the anaerobic processes, aerobic microorganisms are efficient degraders of less substituted compounds up to complete mineralization. Aerobic and anaerobic environments each have limitations in their biodegrading abilities that often complement each other when they are combined (Zitomer and Speece, 1993). Therefore, coupling of reductive and oxidative functions in complex microbial ecosystems might allow for complete TCE biodegradation. The anaerobic/aerobic granular biofilms formed in the CANOXIS single bioreactor technology (Guiot, 1997) should be an appropriate matrix for integrating reductive and oxidative catabolisms.

TCE degradation in a coupled anaerobic/aerobic circulating column was studied during a step by step increase of TCE loading to a maximum of 20 mgTCE/$L_{reactor}$·day. The system was operated for 16 months, with a 4 month startup period, a 5 month TCE acclimation period, a 2 month maintenance period on ethanol alone, and a 5 month period at low TCE loading. TCE degradation performance was evaluated in the system. Concurrently specific activity tests were performed at different TCE loading conditions to follow the evolution of the biomass' trophic potential. Microcosm tests with [1,2-^{14}C]TCE were performed at the end of the experiment, to appraise whether the biomass presented a potential for TCE mineralization under aerobic conditions.

MATERIALS AND METHODS

Experimental System Setup and Operation. The treatment system was based on the CANOXIS technology (Guiot, 1997). It consisted of a reactive column (10 cm inside diameter (I.D.), effective liquid volume 15 L) connected to a recirculation line with two aeration columns (5 cm I.D., effective liquid volume of 0.5 L each) (Figure 1). The reactive column along with both aeration columns were glass-made, and all connecting lines and pieces were either glass, teflon or viton. Liquid was recycled in an upflow manner at a rate of 400 L/d. Air was sparged in the aeration columns to maintain dissolved oxygen at around 2 mg/L in the influent of the reactive column. Three oxygenated liquid lines were connected to the reactive column, one to provide oxygen with ethanol and TCE at the bottom, and two to provide oxygen alone to the granular bed. A stock solution of TCE in ethanol was prepared when TCE was fed. Measurements and calculations were done on a daily basis and then averaged. The system was operated at room temperature (27°C), at a hydraulic retention time (HRT) of 1 day, and a liquid upflow velocity of 2 m/h.

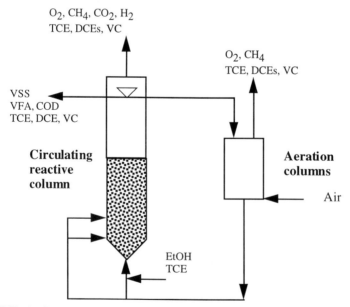

FIGURE 1. Schematic view of coupled anaerobic/aerobic integrated system

The inoculum consisted of a mixture of anaerobic granular sludge from a food industry wastewater treating reactor (Champlain Industries, Cornwall,

Ontario), activated sludge from a municipale treatment plant (Sainte-Catherine, Quebec) and an anaerobic PCE-dechlorinating sludge from a reactor previously operated in the laboratory (Guiot et al., 1995). Before inoculation, the sludge was homogenized for 2 min using a Kinematica CH-600 homogenizer (Kinematica GmbH, Luzern, Switzerland), and to accelerate granulation, a chitosan solution at 1% in acetic acid was added, to provide 10 mg chitosan/g SS (El-Mamouni et al., 1998).

The feed solution contained (in mg/L) : KH_2PO_4, 22.0; K_2HPO_4, 28.1; NH_4HCO_3 339; $KHCO_3$ 867; $NaHCO_3$ 667; $FeSO_4.7H_2O$, 5.2; $MgSO_4.7H_2O$, 8.2; EDTA, 1.19; $MnSO_4.H_2O$, 1.04; $(NH_4)_6Mo_7O_{24}.4H_2O$, 0.33; $Co(NO_3)_2.6H_2O$, 0.41; $ZnSO_4.7H_2O$, 0.33; $NiSO_4.6H_2O$, 0.19; $CuSO_4$, 0.12; H_3BO_3, 0.09; $AlK(SO_4).12H_2O$, 0.04; Na_2SeO_4, 0.03. This formula was modified when the OLR was changed to achieve a COD:N:P:S ratio of 100:12:1:0.2 required for optimal aerobic growth.

During the initial 4 month startup period, ethanol was fed to reach a specific organic loading rate (SOLR) of 0.2 g COD/g VSS.day in order to promote granulation. For the next 5 months, TCE was fed at a stepwise increasing rate until a maximum of 20 mg TCE/$L_{reactor}$.day. Throughout this period, TCE and chlorinated metabolites were analysed in the gaseous and liquid effluents. Also monitored were process parameters such as volatile fatty acids (VFA) and ethanol concentrations, chemical oxygen demand (COD), biogas production and composition, dissolved oxygen concentration (DO), system content in biomass based on volatile suspended solids (VSS), pH and temperature, as described in details elsewhere (El-Mamouni et al., 1998; Stephenson et al., 1999).

Specific Metabolic Activities. The biomass specific activities were determined by separately measuring the rate of uptake of defined substrates (acetate, H_2/CO_2 and ethanol) under non-limiting conditions in strictly anaerobic batch tests. Thus, the results of these tests were indicative of the relative amount of the corresponding bacterial trophic group. Each treatment was prepared in triplicates. The activity tests were conducted at 35°C in 120 mL serum bottles shaken at 100 rpm for liquid substrates and 400 rpm for gaseous substrates. Each bottle contained 10 mL of phosphate buffer and biomass, at 5 g VSS/L for liquid substrate and 2 g VSS/L for gaseous substrate. The specific activity of biomass was calculated by dividing the rate of substrate consumption by the VSS content of the bottle.

Microcosms Experiments: Mineralization of [1,2-^{14}C]TCE. The mineralization potential of the biomass at the end of the experiment was assessed under aerobic and methanogenic conditions. Aerobic microcosms were carried out in the presence of either O_2/CH_4 (a mixture of air and CO_2 plus CH_4 giving : N_2=57%, O_2=15%, CO_2=18% and CH_4=10%), or O_2/ethanol/BES (O_2=100%; ethanol initial concentration of 3 g/L; BES (bromoethanesulfonate), an inhibitor of methanogens at 20 mM). Methanogenic microcosms were carried out under anaerobic conditions in the presence of ethanol (initial concentration at 3 g/L). In aerobic microcosms, headspace oxygen content was calculated to allow for complete mineralization of the carbon source provided. Concurrently, three identical aerobic activity tests were run without [1,2-^{14}C]TCE to monitor the carbon source and oxygen. In aerobic microcosm tests and control activity tests, CH_4 and O_2 were respiked when depleted. Serum bottles were prepared with a minimum salt medium (Boyd et al., 1983) and biomass concentration was around 1-2 g VSS/L in all bottles. Experimental series were prepared in triplicate. Duplicate killed control microcosms were prepared as described and autoclaved three times for 20 min at 15 psi and 121°C. Microcosms were amended with approximately 0.023 µCi of [1,2-

[^{14}C]TCE. Initial dissolved TCE concentration in equilibrium with the headspace was estimated neglecting adsorption and using Henry's Law (Heron et al., 1998) to be about 0.354 mg TCE/L. $^{14}CO_2$ was collected in 0.5 mL of KOH (1 M) and quantified by liquid scintillation counting (Liquid Scintillation Analyser Tri-Carb 2100TR, Packard).

RESULTS AND DISCUSSION

Column Startup and Operation. The column has been inoculated at 15 gVSS/L$_{reactor}$. Before adding TCE it was operated for 4 months at an approximative organic loading rate (OLR) of 2-3 g COD/L$_{reactor}$.day. OLR was continuously adjusted to take into account important biomass washout during this period. At the end, biomass in the system was around 5 gVSS/L$_{reactor}$, and was well granulated, with smaller and darker granules at the top of the bed.

During 40 days, system was then operated at a TCE load of 2 mg TCE/L$_{reactor}$.day. Once stable operation was achieved, TCE load was gradually increased for 50 days to end up at 20 mg TCE/L$_{reactor}$.day load. The aeration rate was increased as well and the system was operated under these conditions for 60 days. During this 5 month run, performance of the coupled system was evaluated based on gas balances, batch activity tests, and efficiency of TCE dechlorination was evaluated on the results of TCE, DCEs and VC analysis in liquid effluent and off-gas. At the end of this 5 month period, biomass in the system was around 10 g VSS/L$_{reactor}$. Afterwards TCE feeding was ceased and the system was maintained on ethanol alone, with enhanced aeration, for 2 months. Biomass content decreased to around 8 g VSS/L$_{reactor}$. For the next 5 months, low TCE feeding was resumed (2 mg TCE/L$_{reactor}$.day). Biomass density kept decreasing, and stabilized around 4 g VSS/L$_{reactor}$ at the end.

Anaerobic activity tests showed that at the end of the startup period (day 110), an enrichment in methanogens had been achieved (Table 1). At this time in the experiment, the coupled system was operated with a low oxygen supply, and less than 10% of the OLR was aerobically degraded.

TABLE 1. Specific substrate biomass activity as a function of TCE loading rate and experiment step in reactive column

Experimental stage [1]		Specific Substrate Biomass Activity (mg substrate/g VSS.d)		
TCE Loading Rate (mg TCE/L$_{reactor}$.d)	Operational Time (day)	EtOH	AcH	H$_2$
Inoculum	0	264 ± 27	116 ± 30	352 ± 46
0	110	1265 ± 122	652 ± 94	469 ± 106
20	270	825 ± 127	506 ± 178	339 ± 52
2	470	464 ± 112	385 ± 165	202 ± 24

[1] : batch tests conducted at the end of each period, when stable operation conditions achieved

Table 2 summarizes the effect of the change in the reactor coupling conditions on the methane productivity yield and the COD oxidation yield (O$_2$ consumed per unit overall COD removed). When the system COD oxidation yield increased, first due to an increasing aeration rate (days 250-320) and afterwards due to an OLR decrease (days 320-500), expectedly the methane productivity yield decreased. The theoretical methane yield is 0.35 L(STP) CH_4/g $COD_{removed}$, less the growth of biomass (10% of the COD removed). This results in an expected methane yield of 0.32 for a strictly anaerobic system. Oxygen addition could cause a decrease in methane production by directly inhibiting the methanogens (Patel et al., 1984). However, as it has already been shown previously that methanogens are not inhibited under aerated conditions similar to those reported in this experiment (Stephenson et al., 1999), a decrease in the CH_4 yield under the actual oxygenation conditions could be attributed largely to a diversion of substrate away from the methanogens to the aerobic bacteria. It is likely for the same reason that the anaerobic specific activities decreased compared to activities at the end of startup period (Table 1). From the end of startup period to the end of experiment, the relative decreases in ethanol and H_2 anaerobic activities were of the same order : from end of the startup period (day 120) to end of the acclimatation period (day 270), methane productivity decrease was 40% and decreases in anaerobic activities were 35-40%, and from the end of the acclimation period to the end of the experiment, a decrease of methane productivity was around 60% and decreases in activities were 57-63%. While not quantitatively similar, the acetate activity followed the same trends. The relative decrease in acetate activity from end of TCE acclimation (day 270) to the end of the experiment is 24%.

In addition, the growth of methanotrophic bacteria, due to the concomitant presence of O_2 and CH_4, would be another reason for the net methane production to decrease. The development of methane aerobic activity potential of the biomass showed some parallel with the O_2-to-COD consumption ratio increase, although this could have been limited by some inhibitory interference of the TCE concentration (results not shown).

TABLE 2. Impact of coupling conditions on system performance

Experimental stage			Coupling and performance parameters [1]			
Operational Time (day)	TCE (mg/L_{rx}.d)	Biomass Density ($gVSS/L_{rx}$)	ORR (%)	ORR ($gCOD_{rem}/gVSS.d$)	CH_4 yield ($LCH_4/gCOD_{rem}$)	O_2 consumption ($gO_2/gCOD_{rem}$)
100-120	0	5.0	97 ± 5	0.24 ± 0.04	0.30 ± 0.16	0.07 ± 0.03
192-202	10	7.6	97 ± 2	0.18 ± 0.03	0.32 ± 0.08	0.05 ± 0.02
203-221	15	8.2	100 ± 1	0.19 ± 0.02	0.27 ± 0.08	0.06 ± 0.02
222-287	20	9.8	94 ± 11	0.17 ± 0.03	0.19 ± 0.05	0.11 ± 0.06
287-336	0	9.0	91 ± 20	0.10 ± 0.05	0.14 ± 0.06	0.11 ± 0.05
337-530	2	3.6	94 ± 6	0.14 ± 0.04	0.12 ± 0.06	0.28 ± 0.15

[1] averages on daily measurements when stable operation conditions achieved
rx = reactor
rem = removed

TCE degradation in circulating column. Up to a load of 15 mg $TCE/L_{reactor}$.day, TCE and metabolites were not detected in biogas and liquid effluent. More than 95% of TCE was totally dechlorinated (Table 3). During this period, the rates of TCE removal and TCE complete dechlorination increased proportionally with the TCE loading rate. Above 15 mg $TCE/L_{reactor}$.day, cis-DCE and TCE started to be detected. At 20 mg TCE/L_{reator}.day, TCE and cis-DCE in effluent increased to reach 15-25% (mole to mole) of TCE in the influent. In the last 10 days of the highest TCE loading, cis-DCE in the effluent increased to be in the same order of magnitude as in the influent (mole to mole). Chlorinated ethenes detected were mainly in the liquid effluent. The 20 mg TCE/L$_{reactor}$.day load apparently impaired the TCE-dechlorinating capacity of the once acclimated biomass, although the first resductive dechlorination step was still performed. Even an increase in the aeration rate did not help the system to recover better efficiency of complete dechlorination. A 2 month period of feeding without TCE (ethanol alone) was then performed, to restore an active coupled biomass. TCE feeding was then resumed, but at a much lower rate (2 mg/L$_{reactor}$.day). TCE transformation rate was stable around 95%, but complete dechlorination rate stayed very low compared to the first TCE loading over the 5 months following TCE onset. During the same time, biomass density in the system kept slightly decreasing. It seems that a longer operating time would be needed for the biomass to recover its previous activity potential and/or the system operational and/or nutritional conditions were sub-optimal for the complete dechlorination to be achieved within the retention time imposed.

TABLE 3. Specific rates of TCE transformation and TCE complete dechlorination throughout the experiment

Experimental stage TCE loading rate		Specific TCE transformation rate	Specific TCE total dechlorination rate
(mg TCE/L$_{rx}$.d)	(μmol TCE/g VSS.d)	(μmol TCE/g VSS.d)	(μmol TCE/g VSS.d)
2	2.28	2.10 ± 0.46	2.10 ± 0.46
10	8.23	7.92 ± 1.22	7.92 ± 1.22
15	13.11	12.67 ± 8.81	12.70 ± 8.79
20	14.11	14.60 ± 0.77	10.39 ± 4.58
2 [1]	2.14	1.89 ± 0.43	0.29 ± 0.32
2 [2]	5.67	2.87 ± 1.90	0.90 ± 1.91

[1] : beginning of low TCE period
[2] : end of low TCE period

TCE mineralization potential of biomass. The results of mineralization tests were interpreted using an empirical method proposed by Millette et al. (1995) to calculate response parameters and a normalized biodegradation index (NBDI) (Table 4). This index NBDI is useful to compare different processing conditions. A value of 1 is indicative of no mineralization, and a value > 1 is indicative of a mineralization potential.

Mineralization tests showed that no mineralization could occur under strict anaerobic conditions (NBDI calculated was 1). The tests under aerobic conditions, using ethanol as primary substrate, exhibited some potential for mineralization, as NBDI index was 6.1. The addition of BES was to ensure that no methane could be produced given the possibility of O_2 transfer limitation and of some anoxic zones, which could otherwise interfer with ethanol as a primary substrate. Clearly the biomass expressed the highest mineralization potential under aerobic conditions in presence of CH_4. Such a high mineralization level (more than 90% in some cases) could be explained by a restricted assimilation ^{14}C-carbon rate in biomass, due to the cometabolic pathway of TCE degradation. However it should be noted that earlier primary substrate depletion could have limited the final mineralization level in the ethanol-spiked aerobic tests. The apparent mineralization detected in all the control tests was likely the result of any physico-chemical reaction and of the transfer of some volatile [1,2-^{14}C]TCE to the KOH trap in the microcosm bottles.

TABLE 4. Mineralization results of microcosm tests

Experimental conditions		LAG (day)	MAX% (%)	RMAX (%.d^{-1})	RAVG (%.d^{-1})	BDI (-)	NBDI (-)
O$_2$/CH$_4$	Control	3.8	16% ± 11%	0.8%	0.8%	0.15	
	Sample	8.3	91% ± 15%	2.3%	2.3%	2.45	15.82
O$_2$/EtOH/BES	Control	14.9	20% ±	0.7%	0.5%	0.11	
	Sample	4.6	34% ± 13%	1.7%	1.6%	0.68	6.10
N$_2$/CO$_2$ [1]	Control	1.9	4% ± 13%	0.5%	0.5%	0.02	
	Sample	2.1	5% ± 10%	0.5%	0.4%	0.02	0.99

[1] : sacrificed bottles
MAX% : final percentage of radioactivity accumulated as $^{14}CO_2$
LAG : time required to reach 1/5 MAX%
RMAX : rate of mineralization between LAG and 2/3 MAX%
RAVG : rate of mineralization between 0% and 2/3 MAX%
BDI : biodegradation index, defined as : $BDI = RAVG \times \dfrac{MAX\%}{100}$
NBDI : normalized biodegradation index, defined as : $NBDI = \dfrac{BDI_{sample}}{BDI_{control}}$

CONCLUSIONS

In conclusion, a coupled anaerobic/aerobic system may allow for multiple interactions between bacterial populations and complex biotransformations in a single reactor. The biomass consortium developed within such a system may be covering a broad spectrum of catabolic functions, as shown by the capacity of near-to-complete mineralization of TCE, when conditions are optimal as supposedly so in the microcosms test. However the fact that a significantly lower performance was achieved in the reactor, indicated that in such a continuous system, proper control might be difficult to effect, and optimal conditions, to achieve, as far as the TCE complete degradation is concerned.

ACKNOWLEDGMENTS

The authors are grateful for the technical expertise of X. Ducarre, C. Cantin, S. Deschamps, C. Beaulieu, and A. Corriveau, for the discussions with D. Millette, and for the english proofreading of M.F. Manuel.

REFERENCES

Boyd, S. A., D. R. Shelton, D. Berry, J. M. Tiedje. 1983. "Anaerobic Degradation of Phenolic Compounds in Digested Sludges." *Appl. Environ. Microbiol.* 46: 50-54.

El-Mamouni R., R. Leduc and S. R. Guiot. 1998. "Influence of Natural and Synthetic Polymers on the Anaerobic Granulation Process." *Water Sci. Technol.* 38(8-9): 349-357

Guiot, S.R. 1997. "Anaerobic and Aerobic Integrated System for Biotreatment of Toxic Wastes (CANOXIS)" - *US Patent* No. 5,599,451, Feb. 4, 1997.

Guiot, S. R., X. Kuang, C. Beaulieu, A. Corriveau, and J. A. Hawari. 1995. "Anaerobic and Aerobic/Anaerobic Treatment for Tetrachloroethylene (PCE)." Proc. Third Int. In-situ and On-site Bioreclamation Symp., San Diego, CA. In: *Bioremediation of Chorinated Solvents*, Hinchee, R., Leeson, A. and Semprini, L. (Eds.), Vol. 3(4), Battelle Press, Columbus, OH, pp. 191-198.

Heron G., T. H. Christensen, C. G. Enfield. 1998. "Henry's Law Constant for Trichloroethylene Between 10 and 95 °C." *Environ. Sci. Technol* 32(10): 1433-1437.

Millette, D., J. F. Barker, Y. Comeau, B. J. Butler, E. O. Frind, B. Clément and R. Samson. 1995. "Substrate Interaction During Aerobic Biodegradation of Creosote-related Compounds : a Factorial Batch Experiment." *Environ. Sci. Technol.* 29(8): 1944-1952.

Patel G. B., L. A. Roth and B. J. Agnew. 1984. "Death Rates of Obligate Anaerobes Exposed to Oxygen and the Effect of Media Prereduction on Cell Viability." *Can. J. Microbiol.*, 30, 228-235.

Stephenson, R., A. Patoine, and S. R. Guiot. 1999. "Effects of Oxygenation and Upflow Liquid Velocity on a Coupled Anaerobic/Aerobic Reactor System." *Water Res.*, 33, 2855-2863.

Vogel, T.M., C.S. Criddle and P.L. McCarty. 1987. "Transformations of Halogenated Aliphatic Compounds." *Environ. Sci. Technol.*, 21, 722-736.

Zitomer, D.H., and R.E. Speece. 1993. "Sequential Environments for Enhanced Biotransformation of Aqueous Contaminants." *Environ. Sci. Technol.* 27(2), 227-244.

ENHANCING *IN SITU* DEGRADATION OF RESIDUAL CHEMICALS IN GROUNDWATER NEAR A CLOSED WASTE DISPOSAL CELL

Scott D. Warner, James H. Honniball, and Thomas A. Delfino
(Geomatrix Consultants, Inc., Oakland, California)
Catherine E.B. Goering (Chemical Waste Management, Valley Center, Kansas)

ABSTRACT: A pilot program to evaluate enhancing the *in situ* degradation of residual chemicals in groundwater was performed at a closed landfill near Wichita, Kansas. Groundwater downgradient from closed disposal cells is affected by residual chlorinated volatile organic compounds (VOCs). Groundwater extraction controls migration of affected groundwater, but the cost of the remedy has increased as the concentration of chemicals in groundwater has decreased to a constant level.

Evaluation of groundwater chemistry at the site suggests that limited natural anaerobic degradation of target chemicals is occurring. Enhanced reductive dehalogenation of the groundwater system was recommended to reduce remedial costs. In this study, a sucrose/yeast extract solution and bromide tracer were injected into groundwater. One test was conducted in wells at the fringe of the known plume. A second test was made in an area immediately downgradient from the closed disposal cells, where VOC concentrations in groundwater were higher.

Field monitoring showed that soon after the injections, the redox potential, DO, specific conductance, pH, and bromide concentration in the injection well were strongly affected (e.g., redox potential decreased by 400 millivolts (mV) to less than approximately −300 mV as Eh). After several days, bromide concentration increased by 10 milligrams per liter above background, and redox potential decreased by 200 mV in nearby wells. Effects on target VOCs were not discernible during the first test even though the change in ambient chemistry indicated a more reducing environment. During the second test, both VOC concentrations and ambient chemical conditions were affected by the injections, changing concentration of parent compounds and daughter products.

INTRODUCTION

For more than a decade, research has shown that chlorinated ethenes and ethanes dissolved in an aqueous environment can be degraded through reductive dechlorination (e.g., Vogel and McCarthy, 1985; DiStefano et al., 1991). Recently biological and chemical methods have been developed that can enhance natural reductive processes to degrade these chemicals *in situ* within a groundwater system. One method, which utilizes an injection of disaccharide and yeast extract in water to promote an endogenous microbial population capable of reducing chlorinated organic compounds, has been successfully applied in laboratory (Bolesh et al., 1995) and small-scale field (Honniball et al., 1998) studies. This paper describes a more recent application of the method; a field test both of larger scale and within a more complex hydrogeologic environment.

A former hazardous waste disposal facility lies approximately 10 miles northeast of Wichita, Kansas. The site has undergone several successful activities to close the former disposal ponds and landfills under approval from state and federal regulatory agencies. However, residual concentrations of chlorinated ethenes and ethanes remain in groundwater (Figure 1). To contain affected groundwater and to remove chemical mass, a series of groundwater extraction trenches and wells have been operating to remove affected groundwater from both a perched and deeper groundwater zone. Groundwater in the perched (A-zone) and deeper (B-zone) groundwater zones is monitored in wells within the facility and at downgradient compliance locations.

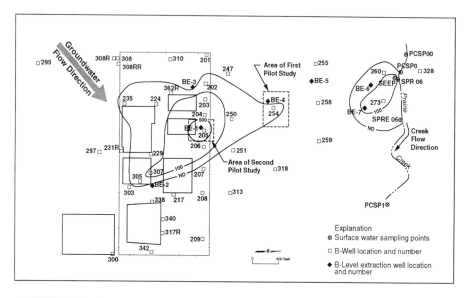

FIGURE 1: Site plan showing general distribution of total VOCs in B-zone and locations of Pilot Test 1 (square) and Pilot Test 2 (square)

The B-zone, which is dominated by clay, siltstone, and weathered shale, is considered the uppermost regional water-bearing zone. Fractures, some in-filled with calcite and gypsum, are present in the shale bedrock and may influence groundwater flow. The direction of groundwater flow within the B-zone generally is toward the north and a small creek approximately 1 mile from the former disposal cells.

The residual chemicals in groundwater within the B-zone include chlorinated ethenes (tetrachloroethylene [PCE], trichloroethylene [TCE], the 1,2-dichloroethylene [cis and trans 1,2-DCE] isomers, and vinyl chloride) and chlorinated ethenes (1,1,2-trichloroethane [TCA], 1,2-dichloroethane [DCA]) and 1,2-dichlorobenzene and chlorobenzene). Generally, concentrations of total VOCs in B-zone groundwater are less than 2 parts per million (ppm), and the presence of lower chlorinated daughter products (e.g., vinyl chloride) indicated that some natural degradation of VOCs was occurring. A long-term strategy for the site is to enhance the degradation of these chemicals *in situ* and reduce the emphasis on groundwater extraction.

MATERIALS AND METHODS

Two pilot tests of enhanced biodegradation were performed at the site (Figure 1). The first test was performed at the downgradient fringe of the plume as the proof of concept required by the regulatory agency in an area with low concentrations of VOCs. The goal was to assess the effect of the process on the ambient hydrochemistry before moving to an area affected by greater concentrations. The second test was performed following an assessment of the initial test. The following paragraphs describe the general methods used during the study.

Testing consisted of injecting the sucrose/yeast extract mixture into an existing well and monitoring the effects in that well and other nearby wells. Injection wells for each test consisted of former groundwater extraction wells (BE-4 for Test 1 and BE-1 for Test 2). The solution consisted of approximately 0.3 percent sucrose supplemented with 0.3 percent yeast extract by weight and mixed in municipal tap water. The tap water was aerated for at least 24 hours to remove trihalomethanes before being used to dissolve the sucrose and yeast extract. A bromide tracer was added to the solution to monitor its migration in the groundwater system. The rate of injection was the rate at which the aquifer accepted the solution, generally about 1 to 2 gallons per minute. In each pilot test, injections occurred over approximately 3 to 4 consecutive days, with a repeat injection approximately one month later. Total injection volumes were approximately 3425 gallons for Test 1 and 9000 gallons for Test 2. Daily injection volumes for each test are given in Table 1.

TABLE 1. Daily injection volumes (in gallons) of reducing solution for pilot tests 1 and 2.

No.	Test 1			Test 2		
Day	1	2	3	1	2	3
Month 1	50	435	220	1500	1500	1500
Month 2	1260	420	1040	1500	1500	1500

Monitoring was performed immediately before and after the injections. The nearest observation well for Test 1 was well B254 (approximately 80 feet downgradient from the injection well) and for Test 2 was well B205 (located approximately 50 feet downgradient from the injection well). No wells were installed for the tests. Field, water quality, and analytical parameters included: water level, redox potential, DO concentration, pH, specific conductance, chloride, bromide, and VOCs using EPA Test Method 8240. Parameters were monitored in the injection well, nearby observation wells, and background wells. Parameters were measured at approximately 48- to 72-hour intervals during the first 2 weeks following the injections, weekly for the next 2 weeks, then monthly. Groundwater samples for VOC analyses were collected immediately before the injections and monthly thereafter.

RESULTS AND DISCUSSION

Results from pilot test 1 in a fringe area of the plume having low VOC concentrations indicated a significant effect from the injected solution on ambient hydrochemistry (i.e., the redox potential and DO decreased significantly). However, the effect on VOCs was inconclusive because of low to negligible starting concentrations. Measurements of water levels did not indicate significant mounding or head fluctuation due to injection (Figure 2). Generally, regional water levels began to decrease through the monitoring period. The results for Test 2 are summarized in Figures 3 through 7.

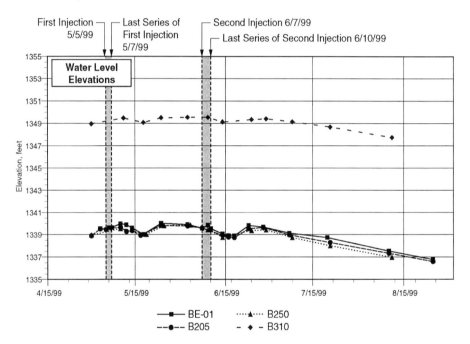

FIGURE 2. Water level measurements for second pilot test.

Redox Potential. Figure 3 presents measurements of redox potential concentration in the Pilot Test 2 injection well (BE-1), the nearest observation well (B205), and upgradient (B310) and downgradient (B250) background wells. Before injection of the carbohydrate yeast extract solution, background measurements of redox potential were between approximately 100 and 200 mV (300 to 400 mV as Eh). Within 48 hours of the initial injection, redox measurements decreased dramatically to greater than 500 mV less than background. One week following injection, the field redox potential in the nearest observation well had decreased by more than 600 mV. The resulting redox levels of between –200 and –300 mV as Eh approach the levels necessary for complete destruction of chlorinated ethene compounds. Approximately one month after the first series of injections, field redox values had recovered to between 50% and 70% of pre-injection values. Following a second series of injections with the carbohydrate yeast extract solution, field redox values were lowered again to 250 to 350mV less than background in both the injection well (BE-1) and nearest observation well (B205).

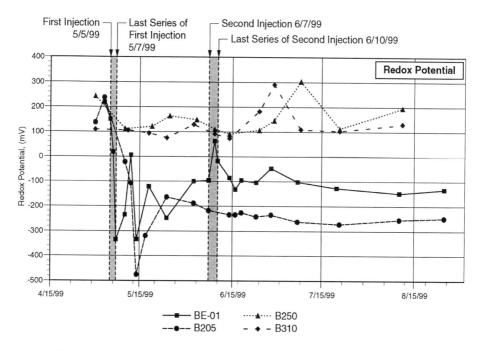

FIGURE 3. Measurements of Redox potential for second pilot test.

Dissolved Oxygen Concentration. DO concentrations (Figure 4) decreased quickly following the initial injection. At the injection well, DO in groundwater decreased from a background concentrations slightly greater than 1 milligram per liter (mg/L) to less than 0.2 mg/L within one week and to near non-detectable values within two weeks. At observation well B205, DO decreased more gradually from a pre-injection concentration of greater than 1 mg/L. DO decreased to only about 0.8 mg/L until after the second injection, after which it quickly decreased to approximately 0.6 mg/L and eventually to a sustained DO near 0.5 mg/L, or about one-half of background concentrations. Over the same period, DO concentrations in background well B250 appeared to decrease by about 0.5 mg/L (from approximately 2.5 to 2.0 mg/L).

pH and Specific Conductance. Values of pH and specific conductance in the injection and observations wells (Figures 5a, 5b) show strong effects from the carbohydrate yeast extract solution. As expected, pH dropped immediately from an ambient value near pH 7 to less than pH 4.5 in well BE-1, due in part to the production of hydrogen and consumption of carbon. As hydrogen is consumed, the pH returns to ambient values. This pattern was observed after each series of injections. For observation well B205, the pH dropped from approximately pH 7 to approximately 6.4 following the first injection, dropping another 0.2 pH units following the second injection before stabilizing at approximately pH 6.4 or about 0.6 pH units lower than pre-injection values (Figure 5a).

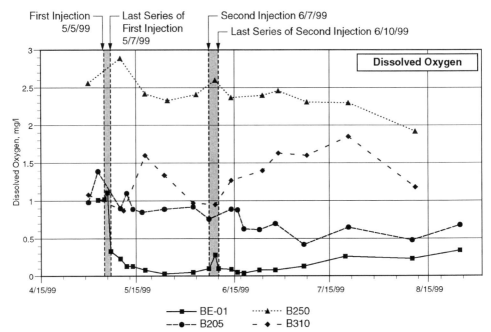

FIGURE 4. Measurements of dissolved oxygen for second pilot test.

Specific conductance showed a response similar to that of pH in both the injection well and nearest observation well. Specific conductance increased dramatically in well BE-1 following each injection, returning to near ambient levels following the first injection, decreasing only slightly after the second injection. The effects also were seen, although less dramatically, in observation well B205 (Figure 5b).

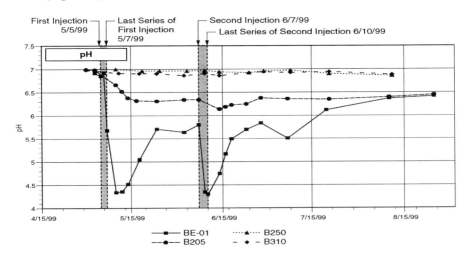

FIGURE 5a. Field measurements of pH for second pilot test.

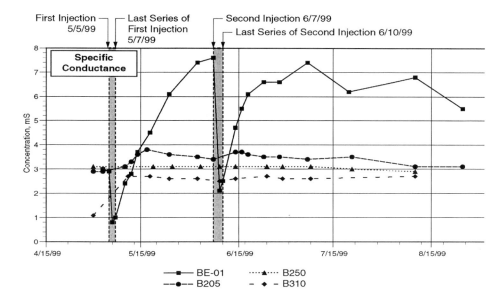

FIGURE 5b. Field measurements of specific conductance for second pilot test.

Bromide and Chloride. Bromide was added to the injection solution as a tracer for monitoring the migration of the solution through the aquifer. Bromide was added to the injection solution at a concentration of approximately 1000 mg/L. The effect was readily apparent in measurements made with an ion-specific electrode within injection well BE-1 following injection (Figure 6a). Approximately one week following the injection an increased concentration of bromide at well B205, located approximately 50 feet downgradient, indicated that physical effects of the solution were observed quickly at a distant location. This was due finding likely to advective transport through the aquifer and any fractures within it, as well as from the initial head put on the system during injections. The migration of bromide for at least 50 feet indicates that the injection solution penetrated a large volume of aquifer.

Chloride also was measured in the field using an ion-specific electrode to provide a gross assessment of mass balance during dechlorination. The concentration of chloride in groundwater at the injection well increased following the initial injection before decreasing to ambient levels. Following the second injection, the concentration steadily increased (Figure 6b). The field measurements likely were influenced by the bromide tracer, and the steady increase in chloride is not attributed solely to dechlorination, although some contribution is expected.

Analytical Results. For the second pilot test, groundwater samples were collected from the injection well, BE-1 (Figure 7), and the nearest observation well, B205 (Figure 8), before the first injection and approximately monthly following the injections. Samples were analyzed for VOCs using EPA Test Method 8240. Review of the data indicates the following.

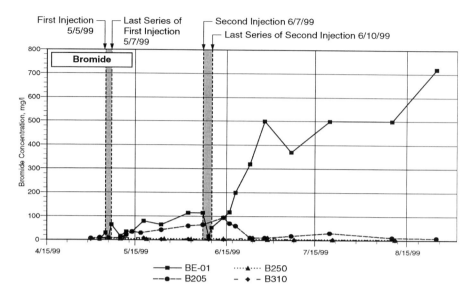

FIGURE 6a. Concentrations of bromide in injection well BE-1 for second pilot test.

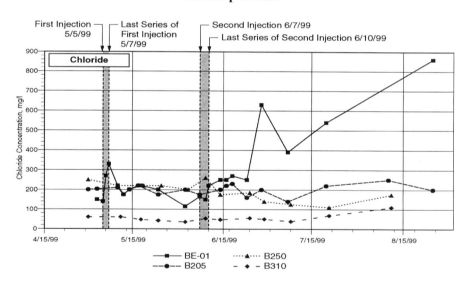

FIGURE 6b. Concentrations of chloride in injection well BE-1 for second pilot test.

Concentrations of several chlorinated ethenes and ethanes in groundwater from well BE-1 decreased by more than 50 percent following the series of injections. Concentrations of parent compounds (PCE and TCE) remained depressed following the injections, while concentrations of the lower chlorinated daughter products (cis-1,2-DCE and vinyl chloride) rebounded to about 30 to 40 percent below pre-injection concentrations.

After the injections, concentrations of chlorinated ethenes and ethanes in groundwater from well B205 appeared to increase before showing signs of reduction.

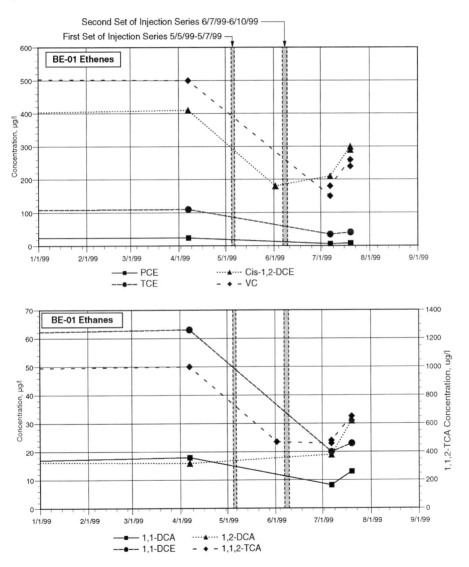

FIGURE 7. Concentrations of chlorinated ethenes and ethanes observation well BE-01 for second pilot test.

Concentrations of other VOCs including benzene and 1,2-dichlorobenzene, have showed little change to date. Concentrations of chloroform decreased significantly; however, the decreasing concentration appears to be due somewhat to ambient fluctuations in the groundwater system and not completely to injection of the sucrose yeast extract solution.

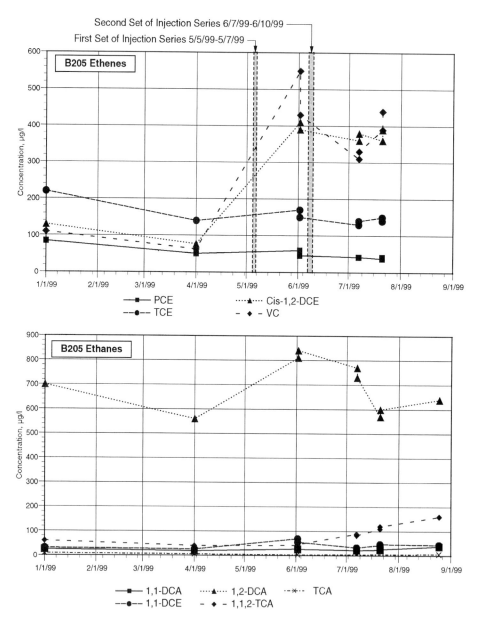

FIGURE 8. Concentrations of chlorinated ethenes and ethanes in observation well B205 for second pilot test.

CONCLUSIONS

Redox potential was lowered significantly by the reaction of the sugar yeast extract solution in the groundwater near the injection well. The decrease in redox potential in an observation well 50 feet distant indicates that a large volume of aquifer was influenced by the injected solution.

Changes in pH (decrease), DO (decrease), and specific conductance (increase) in the injection and observation wells confirm that the process affects the ambient geochemical environment in a manner consistent with anaerobic biological activity.

Concentrations of chlorinated ethenes and ethanes were reduced in groundwater samples from the injection well. The concentrations of daughter compounds began to rise following the initial decrease, likely due partly to the degradation of parent compounds and rebound from the injection process.

Initial increases in concentration of VOCs at observation well B205 likely were due to a modified hydraulic flow field and slight shift in the source area following shut-down of extraction well BE-1 prior to the pilot test. The beginning of a decreasing trend in concentration of VOCs at B205 following this initial increase may reflect enhanced reduction, although additional time is necessary to confirm this trend.

The pilot testing program demonstrated that a solution of sugar and yeast extract can dramatically lower the redox potential of groundwater to the point of enhancing reductive dechlorination of VOCs. It also is apparent that the strongest effects occur nearest the injection well, although diffuse effects were observed, at this site, at least 50 feet distant. Observations suggest that a strategy for this site, would be to inject the reducing solution into a series of wells located across the direction of plume migration and repeating the injection at some interval. The result would be a biologically active *in situ* "barrier" sufficient to reduce concentrations of VOCs in groundwater to acceptable levels and prevent downgradient migration of recalcitrant compounds.

ACKNOWLEDGEMENTS

The authors thank J. Baker and P. Craig of Waste Management, Inc. for supporting this project as well as the Kansas Department of Health and Environment for approving the pilot testing.

REFERENCES

Bolesch, D.G., Delfino, T.A., and Keasling, J.A., 1995, "Complete Anaerobic Dechlorination of Trichloroethene by a Groundwater Enrichment Culture," Department of Chemical Engineering, University of California, Berkeley, California.

DiStefano, T.D., Gossett, J.M., and Zinder, H., 1991, "Reductive Dechlorination of High Concentration of Tetrachloroethene to Ethene by an Anaerobic Enrichment Culture in the Absence of Methanogenesis," *Applied Environmental Microbiology*, Vol. 57, pp. 1187-2292.

Honniball, J.H., Delfino, T.A., and Gallinatti, J.D., 1998, "Removing Recalcitrant Volatile Organic Compounds Using Disaccharide and Yeast Extract," *Bioremediation and Phytoremediation, First International Conference, Chlorinated and Recalcitrant Compounds*, pp. 103-108, Battelle Memorial Institute.

Vogel, T.M., and McCarty, P.L., 1985, "Biotransformation of Tetrachloroethylene to Trichloroethylene, Dichloroethylene, Vinyl Chloride and Carbon Dioxide Under Methanogenic Conditions," *Applied Environmental Microbiology*, Vol. 49, pp. 1080-1083.

AEROBIC AND ANAEROBIC BIOREMEDIATION OF cis-1,2-DICHLOROETHENE AND VINYL CHLORIDE

Thomas S. Cornuet (WESTON, West Chester, PA)
Craig Sandefur (Regenesis, San Clemente, CA)
W. Mark Eliason, (WESTON, West Chester, PA)
Steven E. Johnson, (WESTON, West Chester, PA)
Carlos Serna, (WESTON, Vernon Hills, IL)

ABSTRACT: This paper presents a comparative study of aerobic and anaerobic bioremediation of cis-1,2-dichloroethene (DCE) and vinyl chloride (VC). The study includes laboratory bench scale testing of anaerobic biodegradation followed by a side by side field pilot test comparing the effectiveness of aerobic and anaerobic biodegradation processes. Prior to initiating the pilot test, DCE and VC concentrations ranged from 220 to 5700 µg/L in the test wells. A proprietary time release oxygen and hydrogen source was utilized for the enhancement of aerobic and anaerobic degradation processes, respectively. The side by side field pilot tests consist of two 4-well application arrays with monitor wells located 20 ft (6.1 m) upgradient, 5 ft (1.5 m) downgradient and 25 ft (7.6 m) downgradient at both test areas. After six months of monthly post-application sampling, the data indicate that DCE and VC concentrations have been reduced significantly in both the aerobic and anaerobic field pilot tests. DCE and VC concentrations have been reduced in the two 25 ft (7.6 m) downgradient wells by an average of 78% in the aerobic test and 96% in the anaerobic test. Full-scale groundwater remediation of the entire 400 ft (122 m) by 600 ft (183 m) sandstone bedrock groundwater plume is planned for later this year.

INTRODUCTION
Objective. The objective of this study is to determine the feasibility of using either enhanced aerobic or anaerobic biodegradation to remediate DCE and VC contamination in groundwater at a site in Ohio, and to evaluate which approach is most cost-effective to complete the onsite groundwater remediation.

Scope. In order to develop the data necessary to successfully remediate the constituents of concern this remedial effort is structured in the following three phases:
- Phase I—Anaerobic Bench Scale Test
- Phase II—Aerobic/Anaerobic Comparative Study Field Pilot Test
- Phase III—Full Scale Implementation

Phase I activities include conducting an anaerobic bench scale test to determine how well the proposed hydrogen enrichment source would enhance the natural biodegradation process at the site. The results of the bench scale testing indicate that the naturally occurring bacteria at the site can significantly lower site volatile organic compound (VOC) contamination in the groundwater, and the rate of contaminant reduction can be increased by the proposed hydrogen enrichment material.

Phase II consists of conducting a comparative study field pilot test to confirm laboratory bench-scale results and provide information required to complete the final design of the full-scale remediation system. The objectives of the field pilot test include determining if the groundwater contamination in the VOC plume can be reduced more effectively aerobically (using oxygen enrichment material) or anaerobically (using hydrogen enrichment material). The Phase II field pilot testing activities include well installation, application, and groundwater sampling. The results of the field pilot test indicate that the site VOC groundwater contamination can be very effectively remediated by both aerobic and anaerobic processes. The currently available data collected from the pilot test indicate that anaerobic biodegradation may be the most cost-effective approach for site groundwater remediation.

Phase III activities will include the full-scale implementation of the in situ remediation effort planned for the entire 400 ft (122 m) by 600 ft (183 m) VOC groundwater plume, later this year.

Site Description. The site is located in Ohio, and manufacturing activities have been conducted at the site for over 100 years. The site is underlain by approximately five to eight ft of overburden, which is underlain by sandstone and shale bedrock of Mississippian age. The 30 ft (9.1 m) thick sandstone is fine to medium-grained, with interbedded siltstone and shale layers in the lower part of the formation. The sandstone is underlain by low permeable shale with some thin siltstone layers in the upper part of the formation.

A groundwater elevation contour map of wells screened in the sandstone bedrock is shown in Figure 1. The depth to water at the site generally ranges from 9.5 to 11.5 ft (2.9 to 3.5 m) below ground surface (bgs) and the overburden is unsaturated across most of the site. The average hydraulic conductivity (K) calculated from slug and pump tests in site wells is 4 ft/day (1.2 m/day). The horizontal groundwater gradient is 0.008 ft/ft in the northeast direction. Based on these data and an assumed effective porosity of 15%, the groundwater velocity is approximately 0.2 ft/day (0.06 m/day) in the pilot test area.

Soil vapor extraction (SVE) and soil removal efforts were used to remediate TCE soil contamination, which was the source of the DCE and VC groundwater contamination. Groundwater contamination now consists almost entirely of DCE and VC. The general absence of elevated TCE concentrations in the groundwater demonstrates that significant biologic degradation processes are present in the sandstone water bearing unit and that the soil source reduction efforts were effective.

MATERIALS AND METHODS
Biochemistry. This study utilizes a proprietary time release oxygen or hydrogen source for the enhancement of aerobic or anaerobic degradation processes, respectively. The aerobic pilot test utilizes Oxygen Release Compound (ORC®) as an oxygen enrichment material to stimulate the aerobic degradation of the DCE

and VC contamination. The anaerobic pilot test utilizes Hydrogen Release Compound (HRC™) as a hydrogen enrichment material to stimulate anaerobic reductive dechlorination of the DCE and VC contamination. Both of these products are manufactured by Regenesis Bioremediation Products.

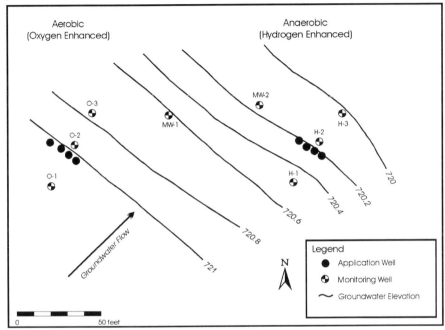

Figure 1. Site Map

The oxygen enhancement material used in the aerobic pilot test is a unique formulation of magnesium peroxide that releases oxygen slowly when hydrated. The compound releases oxygen while being converted to ordinary magnesium hydroxide in accordance with the following equation:

$$MgO_2 + H_2O \longrightarrow 1/2\ O_2 + Mg(OH)_2$$

Magnesium peroxide and magnesium hydroxide are both environmentally benign and actually safe enough to ingest. They are both recognized as medical antacids, magnesium hydroxide is known as milk of magnesia. Aerobic degradation of DCE and VC can occur by two processes including direct metabolism or cometabolism. Direct metabolism is an intracellular process in which DCE and VC serve as a primary substrate for oxygen dependent microbial growth. Cometabolism is an extracellular degradation process that requires an enzyme-inducing substrate (Koenigsberg and Sandefur, 1999).

The hydrogen enhancement material used in the anaerobic test is a proprietary, environmentally safe, food-quality polylactate ester specially formulated for the slow release of lactic acid upon hydration. The lactic acid is converted to several other acids and produces hydrogen along the way. The hydrogen produced by this process is used by reductive dechlorinators, which are capable of dechlorinating DCE and VC (Koenigsberg and Farone, 1999).

Anaerobic Bench Scale Test. The purpose of the anaerobic laboratory bench-scale test was to determine how well the proposed hydrogen enrichment source would enhance the natural biodegradation process at the site. Existing site data indicate that in situ bioremediation is already occurring naturally at the site. The focus of the bench scale test was to determine whether the bedrock groundwater contains a population of bacteria that is suitable to perform the remediation and respond to an increase in both the carbon compound biochemical energy and the hydrogen generated from the proposed hydrogen enrichment source material.

In situ rock and groundwater samples were collected at the site to conduct four laboratory bench-scale tests. One test was conducted using bedrock chips and groundwater. Three additional tests were conducted using groundwater samples collected from well MW-2, one additional upgradient well, and well O-3. The test using bedrock chips and site groundwater is considered to be the best representation of biodegradation occurring at the site. Varying concentrations of trichloroethene (TCE) were added to the bench-scale test tubes and subsequent samples were collected for analysis of target VOCs and lactic acid. At the conclusion of the test, the water was plated to determine the actual bacterial concentrations. The bench scale test was conducted by William Farone at Applied Power Concepts, Anaheim, California (as outlined in Farone and Koenigsberg, 1999).

TCE concentration reductions were observed during the first week of all four bench scale tests. TCE concentrations were significantly reduced in all four tests after four weeks, and the tests were terminated at that time. The bedrock chip test showed that the concentrations had decreased from initial concentrations of 25 and 10 mg/l to final concentrations of 4.41 and 1.63 mg/L, respectively. The test samples were also analyzed for DCE and VC. These data showed that DCE and VC were generated during the test from the breakdown of TCE. After four weeks, the bedrock chip bench-scale test indicated significant degradation of DCE and VC. The results of the bench scale testing indicate that the naturally occurring bacteria at the site can significantly lower site VOC contamination in the groundwater, and the rate of contaminant reduction can be increased by the proposed hydrogen enrichment material.

Pilot Test Design. This comparative study consists of two side by side field pilot tests. Both tests consist of two 4-well application arrays with monitor wells located 20 ft (6.1 m) upgradient, 5 ft (1.5 m) downgradient and 25 ft (7.6 m) downgradient. The application wells are spaced 5 ft (1.5 m) apart. The two test areas are located in the same 30 ft (9.1 m) thick sandstone bedrock water bearing unit and are separated by approximately 150 ft (46 m) in a cross gradient direction. The sandstone unit is underlain by very low permeability shale. Figure 1 shows a general layout of the two pilot tests. Wells O-3, H-3, MW-1, and MW-2 were installed 6 years before the pilot test and groundwater samples were collected from them 6 years, 3 years, and 1 year prior to the test. The additional pilot test wells O-1, O-2, H-1, H-2, and all of the application wells were installed 1 month before the pilot test. The test wells were installed to a depth of approximately 35 ft (10.7 m) bgs. The aerobic test application wells are 8-inch (20 cm) diameter wells and the anaerobic test application wells and all monitor wells are 5-inch (13 cm) diameter wells. Both sets of pilot test wells were drilled using water rotary drilling techniques in order to limit the amount of aeration during drilling.

Application of the oxygen and hydrogen enhancement material was conducted on the same day. The aerobic pilot test consists of stacking four PVC screened canisters containing five 'socks' of material per canister across the entire saturated thickness in each of the four aerobic test application wells. Each filled 'sock' contains 11 pounds of material amounting to a total of 220 pounds (100 kg) of applied material into each of the four aerobic test application wells. The rate of canister application was approximately one application well per hour.

The anaerobic pilot test consists of completely filling the saturated thickness of the open-hole bedrock application wells with the hydrogen enhancement material. Thirty gallons of the material were applied into each of the four anaerobic application wells. The application rate was approximately one to two gallons per minute. Both of these tests simply consisted of filling each application well up with one well volume of bioremediation enhancement material. This application had minimal affect on the surrounding water levels, which dissipated shortly after application, and was not enough material to either spread or block the natural movement of the plume.

Groundwater Sampling and Analysis. Baseline groundwater samples were collected from all upgradient and downgradient monitor wells prior to application. Groundwater samples were also collected from wells O-3 and H-3 prior to the test 6 years, 3 years, and 1 year before the application. Following application, groundwater samples were collected monthly (approximately every 30 days), to monitor the effectiveness of the aerobic and anaerobic pilot tests. Groundwater samples were collected using low flow sampling techniques. Laboratory samples collected during the pilot test included VOCs, inorganics, total organic carbon (TOC), and acids. Field parameters collected throughout the sampling period included dissolved iron, pH, temperature, Eh, specific conductance, and dissolved oxygen.

RESULTS AND DISCUSSION

After six months of monthly post application sampling, the data indicate that DCE and VC concentrations have been reduced significantly in both the aerobic and anaerobic field pilot tests. The two downgradient wells in the anaerobic test area (H-2 and H-3) were sampled for acids every month following the application. Sampling results for well H-2 are presented in Table 1 and indicate that acids were detected in that well during every monthly post-application sampling event. Lactic acid was detected in well H-2 five months after the application indicating that the lactic acid lasted at least five months in the treatment area. The data indicate that propionic acid was still present at a concentration of 273 mg/L in well H-2 six months after application, indicating that hydrogen will continue to be produced beyond the six month pilot test sampling effort. Propionic and acetic acid were detected in well H-3 120 days after application at concentrations of 12 and 21 mg/L, respectively, indicating that acids migrated at least 25 ft (7.6 m) in the sandstone bedrock water bearing unit.

Sulfate, dissolved iron, TOC and pH data for well H-2 are also presented in Table 1. The general decrease in sulfate (with the exception of the 120 and 150 day samples) and increase in iron are geochemical indicators that anaerobic conditions favoring reductive dechlorination were developed. The elevated TOC and decreased pH indicate that the hydrogen enrichment material migrated to well H-2. These same trends were also observed in well H-3. Conversely, the downgradient aerobic test wells O-2 and O-3 exhibited increasing sulfate and decreasing iron and no significant changes in TOC and pH.

Table 1. Well H-2 Sampling Results (mg/L) – Anaerobic Application

Analyte	Baseline	30 days	60 days	90 days	120 days	150 days	180 days
Lactic Acid	<100	571	6900	3830	279	214	<1
Pyruvic Acid	<100	1.2	10.4	4.2	0.4	2.1	<0.1
Butyric Acid	<1	166	309	313	156	103	52
Propionic Acid	<1	208	434	146	142	211	273
Acetic Acid	<1	284	566	272	131	243.0	272
Sulfate	310	29	7	29	420	860	62
Iron, dissolved	24	44	311	486	142	183	53
TOC	9	880	300	670	520	570	310
pH	6.45	5.74	4.53	5.1	NA	5.15	5.62

Table 2: DCE and VC Concentration (ug/L) 25 ft (7.6 m) Downgradient

	Prior Historical Data			Baseline	Post Application					
	6 years	3 years	1 year	0 days	30 days	60 days	90 days	120 days	150 days	180 days
Aerobic: Well O-3										
DCE	2900	990	1900	2500	1800	1700	930	613	181	448
VC	1300	1300	1300	800	570	490	560	736	64	275
DCE + VC	4200	2290	3200	3300	2370	2190	1490	1349	245	723
Anaerobic: Well H-3										
DCE	2700	1100	1400	590	690	1600	120	88	23	20
VC	230	320	1300	210	160	420	220	248	50	12
DCE + VC	2930	1420	2700	800	850	2020	340	336	73	32

Historical VOC sampling data are available from wells O-3 and H-3. The DCE and VC data collected from these wells 6 years, 3 years, and 1 year before application is summarized in Table 2. These data show that some contaminant reduction occurred during that time period but the rate of reduction was relatively slow. Extrapolation of these data show that it would take several decades for the DCE and VC concentrations to be naturally lowered to the U.S. EPA Maximum Contaminant Levels (MCLs) of 70 and 2 ug/L, respectively. The pilot test was conducted to determine if adding oxygen and hydrogen enrichment products could increase these natural degradation rates.

Table 2 also shows the DCE, VC, and total (DCE + VC) concentrations for the baseline and the 6 monthly post-application sampling events. These data show that while some fluctuations occurred during the six-month sampling period, a significant trend of reduced DCE and VC concentrations was observed in wells H-3 and O-3. Most notably in the anaerobic test, the DCE concentration in well H-3 was reduced from 590 ug/L during the baseline sampling to 20 ug/L (significantly below the MCL of 70 ug/L) 180 days later. The VC concentration was also reduced in well H-3 from 210 ug/L during the baseline sampling to 12 ug/L (much closer to the MCL of 2 ug/L) 180 days later. The aerobic test well O-3 also exhibited significant DCE and VC concentration reductions during the 180 day pilot test, although slightly less on a percentage basis than the anaerobic test.

A graph of the DCE + VC concentrations in wells H-3 and O-3 is shown in Figure 2. The DCE and VC concentrations for the three historical sampling events are also listed on this figure for comparison. This figure shows that the rate of DCE and VC contaminant reduction was significantly increased after the application of the oxygen and hydrogen enrichment material. The figure also shows that the DCE and VC concentrations steadily declined in the aerobic test well O-3 during the first five months of the test but started to rebound during the sixth month. This rebound may have been due to the depletion of the oxygen enhancement material. Conversely, the DCE and VC concentrations continued to decrease in the sixth month and it is anticipated that this reduction will continue due to the remaining presence of butyric and propionic acids in well H-2.

The percent reduction in DCE, VC, and DCE + VC was calculated for the aerobic and anaerobic monitor wells and is summarized on Table 3. This table shows that the DCE and VC concentrations were reduced in the aerobic test well O-3 by 82% and 66%, respectively at the end of the 180 day pilot test. The table also shows that the DCE and VC concentrations were reduced in the anaerobic well H-3 by an even more significant amount 97% and 94%, respectively. The data summarized in Table 3 and Figure 2 show that the DCE and VC concentrations were significantly reduced in both pilot tests and that the reductions were higher and more sustained in the anaerobic pilot test.

Future Work. Additional data collection will be used to confirm the results of the 180 day aerobic and anaerobic comparative study pilot tests and to design a full-scale groundwater bioremediation program for the entire 400 ft (122 m) by 600 ft (183 m) groundwater plume.

Table 3: DCE and VC Sampling Results (ug/L) and Percent Reduction

Aerobic Test		DCE			VC		
Well	Location	Baseline	180 days	Percent Reduction	Baseline	180 days	Percent Reduction
O-1	20' upgradient	740	675	9%	1,100	553	50%
O-2	5' downgradient	420	339	19%	1,700	1,040	39%
O-3	25' downgradient	2,500	448	**82%**	800	275	**66%**
Anaerobic Test		DCE			VC		
Well	Location	Baseline	180 days	Percent Reduction	Baseline	180 days	Percent Reduction
H-1	20' upgradient	5,700	5,600	2%	450	290	36%
H-2	5' downgradient	2,600	1,640	37%	1,200	253	79%
H-3	25' downgradient	590	20	**97%**	210	12	**94%**

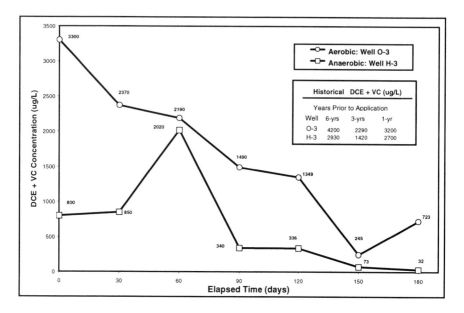

Figure 2. Summary of Results 25 ft (7.6 m) Downgradient of Application

REFERENCES

Farone, W.A., S.S. Koenigsberg and J. Hughes. 1999. "A Chemical Dynamics Model for CAH Remediation with Polylactate Esters". In: A. Leeson and B.C. Alleman (Eds.), *Engineered Approaches for In Situ Bioremediation of Chlorinated Solvent Contamination*, pp. 287-292. Battelle Press, Columbus, OH.

Koenigsberg, S.S. and W. Farone. 1999. "The Use of Hydrogen Release Compound (HRC™) for CAH Bioremediation". In: A. Leeson and B.C. Alleman (Eds.), *Engineered Approaches for In Situ Bioremediation of Chlorinated Solvent Contamination*, pp. 67-72. Battelle Press, Columbus, OH.

Koenigsberg, S.S. and C.A. Sandefur. 1999. "The Use of Oxygen Release Compound for the Accelerated Bioremediation of Aerobically Degradable Contaminants: The Advent of Time-Release Electron Acceptors." *Remediation Journal*. 10(1): 31-53.

DESIGN OF IN SITU MICROBIAL FILTER FOR THE REMEDIATION OF NAPHTHALENE

M. WARITH (Civil Eng., Ryerson Polytechnic University, Toronto, Canada)
L. FERNANDES (Civil Eng., University of Ottawa, Ottawa, Canada)

ABSTRACT: The in-situ microbial filter (ISMF) is a specific application of in-situ bioremediation (ISB) for the treatment of groundwater contaminated plumes. The ISMF filter consists of an in-situ reactor, composed of sand mixed with non-indigenous microorganisms, that is placed ahead of migrating contaminant plumes and removes contaminants by biological and abiological processes. This paper presents the results of an experimental study investigating the remediation of naphthalene contaminated water through a laboratory scale ISMF. Mathematical modeling was performed to determine the optimum dimensions and configuration of the ISMF, and to stimulate the migration of naphthalene through the ISMF.

INTRODUCTION

The purpose and underlying concept of in-situ bioremediation (ISB) is to attenuate hazardous compounds in the soil by bio-transforming these substances into innocuous forms. The first commercial ISB system was implemented in 1972 at a pipeline failure site in Pennsylvania (NRC, 1993). ISB system involves the addition of nutrients and suitable electron acceptors to the contaminated soil to promote the breakdown of the contaminants by microorganisms in place. ISB does not include any treatments that require excavation of contaminated soil or pumping of groundwater to a treatment system. Groundwater pumped for the purpose of hydrogeological control, nutrient addition or pretreatment prior to its re-injection into the subsurface such that biodegradation activity occurs in place, is considered part of ISB system (Flatham 1993; Major, 1992; Keitkamp *et al.*, 1989). The in-situ microbial filter (ISMF) represents a specific application of ISB. This application involves placing sand mixed with non-indigenous microorganisms into a trench in the subsurface ahead of the contaminant plumes (Talyer et al., 1993). The contaminants in the groundwater are metabolized by the microorganisms as the groundwater flows through the trench. This remediation method can reduce or eliminate groundwater contamination, thus reducing the need for extensive monitoring and treatment requirements. ISMFs have the following advantages:

- removal of contaminated soil is only limited to the trench excavation;
- contaminants in the subsurface are degraded in-situ instead of transferring them to another medium;

- a continuous input of energy for pumping groundwater is not required, thus the system will not be prone to failure due to mechanical breakdown or power outage;
- the filters will continue to operate with only minor inputs of energy to supply the required oxygen and nutrients, and maintenance is limited; and
- since water is not brought to the ground surface for treatment, technical and regulatory problems related to discharge of treated water is avoided and scarce groundwater resources are not wasted.

ISMFs are relatively inexpensive treatment method for the above reasons. Also, using an ISMF will often allow remediation to proceed without interrupting the normal site activities, and there are no additional costs due to temporary storage or transportation of contaminated soil. In addition, the sandy soil biofilter material will not need removal once remediation is complete.

The main objectives of this study are to: (1) investigate the possibility of using an ISMF to remediate naphthalene contaminated water through laboratory analysis and; (2) determine the ideal conditions for in-situ remediation of naphthalene contaminated water.

MATERIALS AND METHODS

The materials used in the laboratory experiments included: sand as the column media; microorganisms for the soil column tests; naphthalene as the selected hydrocarbon contaminant; and a glucose based co-substrate. High purity oxygen dissolved in water was used as the electron acceptor for the soil column studies.

Batch Adsorption:

Batch adsorption experiments were used in determining the adsorption kinetics and adsorption isotherms of naphthalene and two different adsorbents, sand and biomass. Naphthalene adsorption to the two adsorbents was tested independently. The sand tested was the sand used in the soil column study for naphthalene biodegradation. The source of biomass for this test was settled mixed liquor obtained from the bioreactors that were being used to cultivate microorganisms for the soil column study. Prior to conducting the test for naphthalene adsorption to biomass, the microorganisms were biologically inhibited by exposing them to 0.1% sodium azide for 12 hours.

To establish the time required for the adsorption reaction to reach equilibrium between the adsorbent and liquid phases, a kinetic plot of naphthalene remaining in solution versus time was examined. Once the equilibrium contact time was established, the degree of adsorption of different concentrations of naphthalene to the two adsorbents was tested using the Freundlich and Langmuir isotherm models.

Soil Column Test:

A total of five soil columns (200 mm in length) were used during this testing procedure. Four of the columns were filled with approximately 2670 g of sand that was thoroughly mixed with biomass such that the biomass/sand ratio was 1.3 mg VSS/g sand. The average density of the sand/biomass mixture packed in each of the columns was 1669 kg/m^3. Two of the test columns were composed of sand mixed with microorganisms previously acclimatized to naphthalene, and two columns consisted of sand mixed with non-acclimatized microoorganisms. In addition, a control column composed of only sand was run concurrently with the test columns to verify the loss of contaminant due to abiotic means such as volatilization, adsorption and abiotic degradation. Sodium azide (0.1%) was continuously added to the influent of the control column in order to inhibit any microbial activity within the column if it became contaminated with biomass throughout the course of the test. The columns were packed very carefully so that their inital hydraulic conductivities would be as close as possible, and to minimize the number of large voids that could promote preferential drainage. The coefficient of variation of the initial hydraulic conductivities of the five columns was 6%.

RESULTS AND DISCUSSION

Batch Adsorption:

Batch adsorption experiments were performed using both biomass and sand as adsorbates in order to determine the potential retardation of naphthalene in the soil/biomass columns due to the adsorption process. Kinetic studies were undertaken to establish the time required for the adsorption to reach equilibrium between the solid and liquid phases. It was found that 24-hour period is reasonable time to attain equilibrium with sand and biomass. There was an 18% decrease in the naphthalene concentration while in contact with 5 g of sand in the batch adsorption bottles, while there was only a 5% decrease in naphthalene concentration while in contact with 140 mg of microbial biomass in the batch adsorption bottles.

The Freundlich equations derived based on the batch adsorption analysis are presented in Table 1. Higher values of the Freundlich adsorption coefficient (K_f) are indicative of greater adsorptive capacity; sand (K_f = 0.00114) was found to have a greater adsorptive capacity for naphthalene than biomass (K_f = 0.00024).

Hydraulic Conductivity Results:

Preliminary soil column tests were undertaken to determine the variation in the sand medium hydraulic conductivity over time due to biomass growth, the hydraulic conductivity that results in the most effective biodegradation and the optimum initial biomass concentration to achieve this hydraulic conductivity. The following biomass/sand ratios were used to examine the change in soil hydraulic conductivity (k): 0, 0.34, 0.68 and 1.04 mg VSS/g sand. The soil columns containing biomass were run over a period of approximately 175 h until the columns became relatively impermeable. The k in the soil column containing the highest concentration of biomass, 1.04 mg VSS/g sand, decreased at a rate of 1.9 x 10^{-4} m/s/h and declined to the lowest k value achieved in this test, 4 x 10^{-7} m/s. The hydraulic conductivity in the column containing 0.68 mg VSS/g sand decreased very quickly at a rate of approximately 2.4 x 10^{-4} m/s/hour but only decreased to a value of 1.4 x 10^{-6} m/s during the test period. The k in the column containing the lowest concentration of biomass, 0.34 mg VSS/g sand, decreased to a minimum value of 1 x 10^{-5} m/s, and the hydraulic conductivity was maintained above 5 x 10^{-5} m/s for over 140 h of column operation.

TABLE 1. Freundlich Model Parameters for Naphthalene Adsorption to Sand and Biomass

Adsorbent	Isotherm Equation	Regression Equation	Correlation Factor
BIOMASS	$(X/M) = 0.00024 C^{1.377}$ [mg/mg]	Log $(X/M) = 1.4253$ log C_e -3.6352	0.98
SAND	$(X/M) = 0.00114 C^{2.03}$ [mg/g]	Log $(X/M) = 2.0351$ log C_e -2.9462	0.94

Soil Column Test:

Figure 1 presents the average naphthalene removal efficiency over time in soil columns with their corresponding standard deviations. The naphthalene removal efficiency increased at the beginning of the experiment in both the control and the biologically active columns. The initial increase in naphthalene removal in the control column stopped after a period of 48 h and became relatively stable at approximately 47%, while the naphthalene removal efficiency continued to increase in the biologically active columns. The naphthalene removal in the acclimatized and non-acclimatized columns increased steadily to eventually attain about 100% and 95%, respectively. The initial increase in naphthalene removal efficiency in the control column was considered to be due to solid phase sorption, while the continuing increase in the biologically active columns was due to a combination of both biodegradation and solid phase sorption.

The results in Figure 1 indicate that both the non-acclimatized and the acclimatized columns attain high naphthalene removal efficiencies. During the first 200 hours of the study, the naphthalene removal efficiency was almost equal in both the acclimatized and non-acclimatized columns. From 200 to 300 hours, the naphthalene removal efficiency in the acclimatized columns appears to increase at a faster rate and attains a maximum efficiency more quickly than the non-acclimatized columns. However, after about 330 hours both the acclimatized and the non-acclimatized columns achieved approximately equal naphthalene removal.

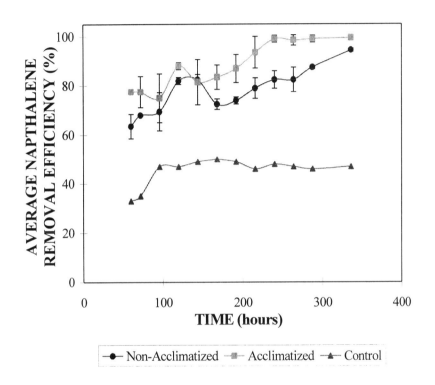

FIGURE 1. Removal Efficiency of Naphthalene

The naphthalene removal efficiency in the control column remained relatively constant and it can be assumed that any changes were unrelated to the number of pore volumes that passed through the column. However in the biologically active columns, the naphthalene removal efficiency increased as the number of pore volumes increased. The naphthalene removal efficiency increased more quickly in the acclimatized columns, 100% naphthalene removal after passing 180 pore volumes, while in the non-acclimatized columns, 74% naphthalene removal was obtained. These results indicate that using acclimatized biomass increases the rate of naphthalene attenuation in the column.

The average contaminant residence time in the biologically active columns changed throughout this study due to the changing porosity and hydraulic conductivity. Figure 2 shows that in the soil columns containing non-acclimatized microorganisms, the naphthalene removal efficiency increased from 64% to 95% as the residence time increased from 0.5 h to 15 hours. In the columns containing acclimatized microorganisms the naphthalene removal efficiency also increased from 77.5% to 100% as the residence time increased from 0.5 to 14 hours. The naphthalene removal efficiency increased as the hydraulic conductivity decreased and as the residence time consequently increased. In the control column, residence time did not change significantly during the test because the hydraulic conductivity and porosity in the column remained quite stable.

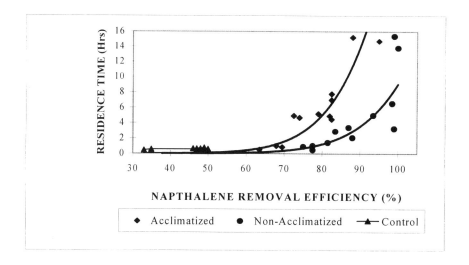

FIGURE 2. Naphthalene Removal Efficiency vs. Residence Time

Biodegradation Rate Constants in Column Test

In order to determine the biodegradation rate constants in the present study, it was necessary to isolate the attenuation of naphthalene due to biodegradation from the attenuation of naphthalene due to the combined effect of sorption and biodegradation. This was accomplished by subtracting the naphthalene removal that occurred in the biologically inhibited control column, which represents attenuation, by adsorption. The biodegradation rate constants were determined by substituting the concentration values that were achieved during the column study into a first-order biodegradation equation (MacIntyre *et al*. 1993). Table 2 lists the average naphthalene removal due to biodegradation, the average column residence time, and the first-order biodegradation rate constants for both the acclimatized and non-acclimatized columns.

TABLE 2. Biodegradation Rate Constants

Biomass	Naphthalene Removal due to Biodegradation	Residence Time (d)	First-Order Biodegradation Rate Constant (d^{-1})
Acclimatized	43%	0.15	3.75
Non-Acclimatized	32%	0.20	1.93

CONCLUSIONS

This study focused on the use of ISMFs for the remediation of naphthalene contaminated groundwater. The transport and ultimate fate of contaminants is greatly affected by the sorption phenomenon. The Freundlich isotherm model was determined to be appropriate to describe naphthalene adsorption to sand and biomass. A retardation factor of 205 was calculated for naphthalene adsorption to sand. Column tests performed to determine the optimum biomass concentration and hydraulic conductivity within an in situ microbial filter, indicated that a biomass concentration of 1.4 mg VSS/g sand and a hydraulic conductivity of 1 x 10^{-5} m/s resulted in the most effective biodegradation. Continuous flow column tests indicated that the naphthalene removal in the acclimatized columns increased steadily to attain 100%, while there was 95% naphthalene removal in the non-acclimatized columns. At the beginning of the study, the acclimatized columns achieved higher naphthalene removal efficiencies in shorter times than the non-acclimatized columns however, by the end of the study both the acclimatized and the non-acclimatized columns achieved very high naphthalene removal efficiencies. This indicated that the time and effort required to acclimatize microorganisms is not justified since it does not result in appreciable increases in naphthalene removal over time. The first-order linear biodegradation rate constant determined for the columns containing acclimatized biomass was 3.75 d^{-1} while the biodegradation rate constant determined from the columns containing non-acclimatized biomass was 1.93 d^{-1}.

REFERENCES

Bjerg, P.L., Hinsby, K., Christensen, T.H. and Gravesen, P. (1992), "Spatial Variability of Hydraulic Conductivity of an Unconfined Sandy Aquifer Determined by a Mini Slug Test", *Journal of Hydrology*, 136, pp.107-122.

Flatham, P. (1993). "Background on bioremediation", *Proceedings, EPA Seminar on Bioremediation of Hazardous Waste Sites: Practical Approaches to Implementation*. June 10-11, 1993. Chicago, Ill.

Heitkamp, M.A. and Cernigila, C.E. (1989). "Polycyclic Aromatic Hydrocarbon Degrdation by a *Microbacterium sp.* in Microcosms Containing Sediment and Water from a Pristine Ecosystem", *Applied and Environmental*

MacIntyre, W.G., Boggs, M., Antworth, C.P. and Stauffer, T.B. (1993). "Degradation Kinetics of Aromatic Organic Solutes Introduced into a Heterogeneous Aquifer", *Water Resources Research*, 29,12, pp.4045-4051.

Major, D.W. and Cox, E. (1992). "The Current State-of-the-Art of In Situ Bioremediation: Considerations, Limitations, Potential and Future Directions", MOEE, copy Queen's Printer ISPN 0-7778-0322-4.

National Research Council (NRC), (1993). *In Situ Bioremediation: When does it work?* National Academy Press, Washington D.C.

BIOAUGMENTATION POTENTIAL AT A CARBON TETRACHLORIDE CONTAMINATED SITE

S. M. Pfiffner (The University of Tennessee, Knoxville, TN)
T. J. Phelps, and A.V. Palumbo (Oak Ridge National Laboratory, Oak Ridge, TN)

ABSTRACT: A plume of carbon tetrachloride (CT) contaminated groundwater at the Y-12 site in Oak Ridge poses a difficult long-term remediation problem and several solutions including bioremediation are being considered. *Pseudomonas stutzeri* strain KC has the potential to degrade CT, without producing undesired degradation products (e.g., chloroform). However, specific environmental conditions, e.g., low iron - alkaline pH, appear necessary to achieve significant degradation with this strain. We designed a microcosm study to examine the feasibility of bioaugmentation with the KC strain at the Y-12 site. In this initial work, additions of nitrate, acetate, nitrilotriacetic acid [NTA], phosphate, and bacteria were tested, to determine if degradation was feasible. The microcosms contained site sediment (13 mL) and groundwater (17 mL) with an addition of 500 µg/L CT. With the addition of 10^7 cells, the CT was degraded rapidly. Generally complete degradation was achieved within 48 hours and nitrate was exhausted to undetectable levels. No pH adjustments were necessary. However, in the presence of NTA, pH adjustment inhibited degradation. Apparently, neither the pH nor iron concentration in the CT plume posed a problem for CT degradation. Thus, the geochemistry of the site appears suitable for bioaugmentation with the KC strain.

INTRODUCTION

A plume of carbon tetrachloride (CT) contaminated water at the Y-12 site in Oak Ridge is present in fractured bedrock and apparently originates from a DNAPL source (ORNL, 1997). Thus, the plume poses a difficult long-term remediation problem. Pump-and-treat and bioremediation are being considered for site remediation.

Various laboratory studies have demonstrated the potential for *Pseudomonas stutzeri* KC to degrade CT, without producing undesired degradation products (e.g., chloroform). Several investigators have demonstrated CT degradation in field and laboratory studies (Criddle et al., 1990; Witt et al., 1995; Mayotte et al., 1996). Degradation apparently proceeds with the involvement of an extracellular low molecular weight molecule, pyridine-2,6-bis(thiocarboxylate) [PBTC] and the production of PBTC is critical to the process (Dybas et al., 1995a; Lee et al., 1999).

Specific environmental conditions, low iron, appear necessary for PBTC production and CT degradation (Tatara et al., 1993; 1995). In the presence of PBTC, CT transformation takes place in the proximity of actively growing or respiring cells (e.g., KC or indigenous aquifer flora). Thus, it is critical that iron is either at a level low enough to stimulate production of the extracellular PBTC. To achieve this at some sites, it has been necessary to adjust the pH in the CT plume to reduce iron availability (Dybas et al., 1995b, 1998).

We conducted a feasibility study with the goals of assessing the potential for bioremediation using this bacterial strain. The feasibility study focused on these questions. 1) Is in situ iron low enough to permit significant degradative activity? 2) Can potential amendments (e.g., pH alteration, chelator addition) control iron to the degree that significant degradation activity takes place?

MATERIALS AND METHODS

Field Site and Sampling. In late February 1999, 8 liters of groundwater from well GW606 and 8 liters of groundwater-sediment slurry from well GW605 were recovered (Fig. 1). Contaminants present in well GW606 were carbon tetrachloride (~ 150 µg/L), chloroform (~ 150 µg/L), tetrachloroethene (~ 8 µg/L), sulfate (~ 48 mg/L), and nitrate (N ~ 4.7 mg/L). The oxidation-reduction potential was ~ 150 mV and the dissolved oxygen ~ 0.2 mg/L (ORNL, 1997) indicating a near anaerobic environment. Concentrations are not available for the groundwater-sediment slurry from GW605 but the well was contaminated and there were some anaerobic degradation products of CT present. Groundwater pH was consistently slightly alkaline but addition of sediments caused a shift to lower pH values.

Groundwater sampling was performed in accordance with USEPA Region I "Low Stress (low flow) Purging and Sampling Procedure for the Collection of Ground Water Samples from Monitoring Wells." A QED MicroPurge Flow Cell/Analyzer was used to analyze all intrinsic field parameters prior to sample collection. A dedicated submersible bladder pump, the associated pneumatic controller, and an oilless compressor were used to collect all the groundwater samples. Teflon-lined tubing was used for both the pneumatic and groundwater lines. The flow-through-cell was disconnected prior to groundwater sample collection. All groundwater samples were collected under a nitrogen blanket to minimize exposure to atmospheric air. In a field sampling bag flushed with 99% nitrogen, groundwater was collected in glass bottles, which had Teflon® caps. The bottles were completely filling the bottles to before capping, to maintain zero headspace.

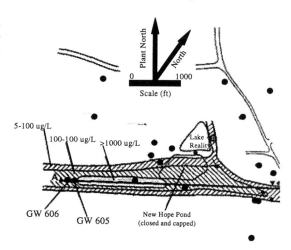

FIGURE 1. Map of Y-12 CT plume with sampling wells GW606 and GW605.

Because fresh core material was not available for the study, we used sediments recovered from the bottom of well GW605 (Fig. 1) in order to examine degradation in the presence of aquifer solids. Sediment sampling began by agitating sediment settled in the bottom of each well using a weighted tape (or equivalent). Sediment-laden groundwater was pumped from the bottom of the well to the ground surface using the same bladder pump system used for the groundwater sampling. With the bladder pump at a flow rate of approximately 1 L per minute, the sediment agitation continued throughout sampling of each well. Sediment-laden groundwater samples were collected in 10 L plastic bottles under a nitrogen blanket to minimize exposure to atmospheric air. The bottles were stored in an ice chest (4°C) for transportation under chain-of-custody procedures to our nearby laboratory.

Experimental Design. Triplicate microcosms were prepared in EPA vials with minnert caps for repeated head-space sampling for each treatment. The microcosms contained 13 mL of the groundwater-sediment slurry (2.5 g dry weight sediment) and 17 mL of groundwater, plus a combination of the following CT (500 µg/L), acetate (0 or 500 mg/L), nitrate (0 or 50 mg/L), other nutrients (phosphate 10 mg/L), NTA (0 or 100 mg/L), bacteria (10^7 cells/mL), and pH adjustment. Supplement concentrations were set to yield a 100:10:1 C:N:P. ratio. The KC strain was grown overnight, in the phosphate-buffered mineral salts media (PBBM) described by Palumbo et al., (1995) modified by removal of iron and copper from the trace minerals, centrifuged, and the cell pellet was washed with anaerobic sodium chloride solution to remove residual medium. The washed cell pellet was resuspended in site groundwater. The cell concentration was determined microscopically and the stock suspension was diluted with site groundwater to the appropriate concentration for addition to the microcosms. The final cell concentration in the microcosms varied in the experiments from 10^{6-7} cells/mL. Incubations were at room temperature (about 22°C). A number of different controls were included in the experiments. Controls included vials, which contained no bacteria or bacteria killed with formaldehyde (negative controls) or bacteria in medium D (positive controls). Subsequent experiments focused on the effect of bacterial, acetate, nitrate and CT concentrations. We measured degradation at several time points, usually at 0, 1, 3 and 7 days using GC analysis for CT and degradation products. A secondary study was initiated using the stored (4°C) groundwater and groundwater-sediment slurry with differing concentrations of acetate, and nitrate. CT concentration was 500 µg/L. Acetate concentrations of 500 or 1000 mg/L were paired in all combinations with nitrate concentrations of 50 and 100 mg/L. Bacteria were added to a final concentration of 10^7 cells/mL. Time point measurements were 0, 1, 3 and 7 days. In this experiment, cells were grown on PBBM before inoculation. Due to problems in achieving degradation with cells grown on PBBM, as discussed below, medium D (Criddle et al., 1990), with 2100 mg/L acetate and 1400 mg/L nitrate, was used in all subsequent experiments.

RESULTS AND DISCUSSION

In this microcosm study significant CT degradation was detected in the presence of nitrate and acetate (Fig. 2A). Planned sampling at 7, and 14 days was canceled due to the highly positive results from the first two time points in favor of additional studies on concentrations of nitrate and acetate. Results of this microcosm experiment indicated that acetate and nitrate stimulated the added bacteria to degrade 500 µg/L CT (Figure 2A and 2B). The rates of disappearance of CT were comparable to those seen in other microcosm studies using the KC strain (e.g., Tatara et al., 1993) in which the pH was adjusted to 8.2. The difference in our study was that groundwater pH was all ready in a favorable range and pH adjustment was not necessary.

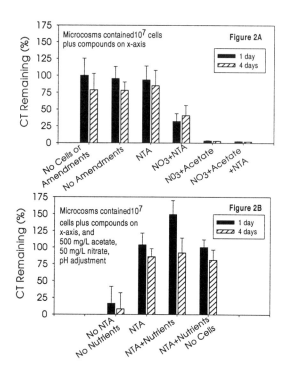

FIGURE 2. CT degradation in the first experiment.

Other adjustments, e.g., pH adjustment and addition of NTA did not have a positive effect. Adjusting the pH to 8.2 was not necessary (Fig 2B) and in the presence of NTA the pH manipulation appeared to inhibit degradation. This may have been an artifact of the manipulation required or due to geochemical interactions. The addition of NTA alone did not stimulate degradation (Fig. 2A). In the presence of nitrate, without pH adjustment or the addition of acetate, NTA did stimulate degradation (Fig. 2A). This may indicate the NTA was used as an electron donor.

Subsequent experiments (below) indicated that growth in medium D was necessary for reproducible degradation results. The bacteria for the first two experiments were grown in PBBM prior to inoculation rather than in medium D. The results of the second round of tests were negative due to the inability of cultures used to degrade CT (Table 1) even in the positive controls. Additional testing was performed to determine what growth conditions were required to ensure high degradation activity.

Up to this point we had been using PBBM to grow the cells for inoculation and upon switching to medium D reproducibility problems were eliminated and high rates of degradation were always achieved. With medium D the degradation was reproducible (e.g., Table 1). The additional investigations revealed our initial growth medium apparently often inhibited the ability of the bacteria to degrade CT, perhaps due to small amounts of iron and copper in the medium (Tatara et al., 1993). The degree of inhibition observed with our original growth medium may depend on the trace amount of iron in the batch of materials used to prepare the medium. The subsequent microcosm studies, using medium D to grown the bacteria, all exhibited CT degradation using acetate (1000 mg/L), nitrate (50, 100, or 200 mg/L), and KC at 10^{6-8} cells/mL (e.g., Table 1).

TABLE 1. Example results from experiments with cells grown on PBBM and medium D.

	Carbon Tetrachloride Remaining (%)			
Experiment	Day 1		Last Day*	
	0 cells	+ cells	0 cells	+ cells
PBBM	108	107	113	113
medium D	111	79	102	33

* day 5 for medium D and day 7 for PBBM.

CONCLUSIONS

It appears that the site conditions are very suitable to bioaugmentation approaches. The pH and iron concentrations appear to pose no problems for achieving high degradation rates. Our findings also were in agreement with studies by Tatara et al., (1993) that indicated that media had a significant effect on the ability of the *Pseudomonas stutzeri* KC strain to degrade CT.

ACKNOWLEDGMENTS

Special thanks go to Shirley Scarborough, Lisa Fagan and Susan Carroll, who assisted with the experiments. This research was supported by IRTD. We also thank Envirogen, who took the field samples with logistical support from Bechtel-Jacobs. Oak Ridge National Laboratory is managed by University of Tennessee-Battelle LLC for the U.S. Department of Energy under contract number DE-AC05-00OR22725.

REFERENCES

Criddle, C. S., J. T. Dewitt, D. Grbic-Galic, and P. L. McCarty 1990. Transformation of carbon-tetrachloride by *Pseudomonas sp* strain KC under denitrification conditions. Appl. Environ. Microbiol 56: 3240-3246.

Dybas, M. J.,G. M. Tatara, C. S. Criddle. 1995a. Localization and characterization of the carbon-tetrachloride transformation activity of *Pseudomonas sp* strain KC. Appl. Environ. Microbiol 61: 758-762.

Dybas, M. J., G. M. Tatara, W. H. Knoll, T. J. Mayotte, and C. S. Criddle. 1995b. Niche adjustment for bioaugmentation with *Pseudomonas sp.* strain KC. pp. 77-84 In Bioaugmentation for Site Remediation R E. Hinchee, J. Frederickson, and B. C. Allemen (eds). 3(3) Battelle Press, Columbus.

Dybas M. J., M. Barcelona, S. Bezborodnikov, S. Davies, L. Forney, H. Heuer, O. Kawka, T. Mayotte, L. Sepulveda-Torres, K. Smalla, M. Sneathen, J. Tiedje, T. Voice, D. C. Wiggert , M. E. Witt, C. S. Criddle. 1998 Pilot-scale evaluation of bioaugmentation for in-situ remediation of a carbon tetrachloride contaminated aquifer. Environ Sci Technol 32: 3598-3611.

Lee, C. H., T. A. Lewis, A. Paszczynski, and R. L. Crawford. 1999. Identification of an extracellular catalyst of carbon tetrachloride dehalogenation from *Pseudomonas stutzeri* strain KC as pyridine-2,6-bis(thiocarboxylate). Biochem. Bioph. Res. Co. 261: 562-566

Mayotte, T. J., M. J. Dybas, and C. S. Criddle. 1996. Bench-scale evaluation of bioaugmentation to remediate carbon tetrachloride-contaminated aquifer materials. Ground Water 34:358-367.

ORNL 1997. Evaluation of Calendar Year 1996 Groundwater and Surface Water Quality Data for the Upper Fork Poplar Creek Hydrogeologic Regime at the U.S. Department of Energy Y-12 Plant, Oak Ridge, Tennessee (Y/SUB/97-KDS15V/6).

Palumbo, A. V., S. P. Scarborough, S. M. Pfiffner, and T. J. Phelps. 1995. Influence of nitrogen and phosphorous on the in-situ bioremediation of trichloroethylene. Appl. Biochem. and Biotech. 55/56: 635-647.

Tatara, G. M., M. J. Dybas, C. S. Criddle.1993. Effects of medium and trace-metals on kinetics of carbon-tetrachloride transformation by Pseudomonas sp Strain-KC. Appl. Environ. Microbiol. 59: 2126-2131.

Tatara G. M., M. J. Dybas, C. S. Criddle 1995. Biofactor-mediated transformation of carbon tetrachloride by diverse cell types. pp 69-76. In: Bioremediation of Chlorinated Solvents. R.L. Hinchee, A. Lesson, and L. Semprini, (eds). 3(4) Battelle Press, Columbus.

Witt, M. E., M. J. Dybas, R. L. Heine, S. Nair, C. S. Criddle, and D. C. Wiggert. 1995. Bioaugmentation and transformation of carbon tetrachloride in a model aquifer. pp 221-228.In Bioaugmentation for Site Remediation R E. Hinchee, J. Frederickson, and B. C. Allemen (eds). 3(3) Battelle Press, Columbus.

THE EFFECT OF SOIL HETEROGENEITY ON THE VADOSE ZONE TRANSPORT OF BACTERIA FOR BIOAUGMENTATION

Barry L. Kinsall (Oak Ridge National Laboratory, Oak Ridge, Tennessee)
Glenn V. Wilson (Desert Research Institute, Las Vegas, Nevada)
Anthony V. Palumbo (Oak Ridge National Laboratory, Oak Ridge, Tennessee)

ABSTRACT: Heterogeneity in hydraulic, physical and chemical properties of porous media can not only limit microbial dispersion but may also complicate the quantification of microbial transport processes and resultant microbial activities. The objectives of this research were to examine the potential for bacterial transport through an unsaturated soil block under transient flow conditions and to determine the influences of soil properties and phosphate additions. Despite the block consisting of >99% sand and appearing to be completely homogeneous, (i.e. structure-less) flow was extremely heterogeneous, as only 6-16% of the cross-sectional area exhibited flow with 88% of this flow occurring through just 4% of the area. The preferential flow paths exhibiting high and moderate flow rates were spatially consistent among water additions suggesting that soil properties caused the heterogeneity in flow rather than unstable wetting fronts. Transport of the GFP bacteria was extremely rapid with breakthrough occurring at the initiation of flow (0.1 h) and bimodal in fast flow areas. The soil texture, rather than porosity, was the most significant property controlling the microbial transport as areas dominated by fine sand trapped the GFP bacteria. These findings demonstrate how apparent homogeneity in media properties does not equate with homogeneity in flow or transport of solutes and colloids. While bioremedial feasibility studies often center on soil chemical properties, this study indicates that consideration should be given to the physical and hydraulic properties of the soil as well.

INTRODUCTION

Most bioremedial practices involve the utilization of indigenous populations to degrade existing contaminants, however, recent studies have suggested that non-indigenous and genetically engineered microorganisms (GEM's) could increase the efficiency of bioremediation. Thus, questions related to bacterial transport may be critical to predicting the effectiveness of bioaugmentation. Sediment heterogeneity (i.e. physical, chemical, and hydraulic properties variation) plays a significant role in determining the fate of microbial populations during transport. Water movement through soils with preferential flow pathways or immobile-regions could significantly decrease bioremediation efficiency by preventing contaminant degrading bacteria from reaching the contaminants that are dispersed within the soil matrix.

Objective. The objectives of this research were to determine the potential for bacterial transport through an unsaturated soil block under transient conditions and to determine the influences of soil properties and phosphate additions on bacterial transport and retention.

Site Description. An undisturbed block was obtained from the 5.9 m depth, i.e. the top of the groundwater table, at a DOE research site (Oyster Borrow Pit) approximately 4 miles east of the township of Cheriton in Northhampton County, Virginia on the Chesapeake Bay peninsula. The soil at the sampling location was the Molena series (Sandy, mixed, thermic Psammentic Hapludults) and consists of coarse texture sediments. This soil is a strong brown loamy sand (0.71 m thick), with a substratum of strong brown sand to a depth of 1.8 m or more (USDA, 1989).

MATERIALS AND METHODS

A 35 cm x 35 cm x 70 cm pedon was shaped in the face of a borrow pit at the site. A stainless steel box (32 cm x 32 cm x 50 cm), with open ends, was constructed to encase the undisturbed block. The stainless steel box was hydraulically pressed into the exposed pedon with negligible soil disturbance (Kinsall et al., 1997). A stainless steel plate with a cutting edge was driven horizontally across the bottom to sever the bottom of the encased block from the soil pedestal.

An acrylic grid of 64 flow collection chambers (3.75 cm x 3.75 cm) inside a 1 cm wide border flow collection area was fitted and sealed to the bottom of the stainless steel block by insetting it 1.5 cm into the soil. Each collection chamber had been partially filled with sterilized coarse sand to establish a hydrologic continuity between the soil and collection chambers. Two holes were drilled within the 1.0 cm annulus between the steel box and the 30 x 30 cm flow collection chambers to allow for collection of drainage from the border of the encased block. A rain simulator positioned above the soil block consisted of 64 application drippers spaced in a grid pattern 3.75 cm. Each dripper was controlled by a valve, which allowed for rainfall application rates to be easily adjusted as needed.

Simulated Rain Events and Effluent Sampling. A 0.001 M $CaCl_2$ solution, similar to the ionic strength of rainfall was applied to the surface of the blocks at a continuous rate of 5400 cm^3 h^{-1} for a duration of one hour, thereby simulating a 60 mm h^{-1} rainfall. The soil block received four simulated rainfall applications, at least 2 days apart, prior to the start of the experiment to wet the soil to field capacity. Collection grids, containing Nalgene polypropylene (autoclavable) bottles, were used to collect effluent from the soil into each respective collection chamber. When the first bottle was filled to capacity, the entire collection grid of 64 bottles was removed and replaced with another grid containing 64 empty bottles, and the time recorded. Effluent was continuously collected over a period of two days after each simulated rainfall application until free drainage ceased. Each bottle containing effluent was weighed to obtain the flow rate.

Two water applications were made on the block. During the first simulated rainfall application, a solution containing a fluorescing bacteria, *Pseudomonas putida* (1 x 10^9 CFU L^{-1}), was applied to the surface of the block. This *P. putida* strain was modified by the addition of the Green Fluorescent Protein (GFP) gene, which allows detection by fluorescent signal (Burlage et al., 1996). The bacterial solution was followed by two additional applications of 0.001 M $CaCl_2$, which was used to determine the flushing of phosphate and *P. putida*. Effluent samples from a fast, moderate, and a slow collection cell were analyzed with respect to time for microbial and chemical constituents

Soil Sampling and Analysis. Following the completion of the simulated rainfall applications, the soil block was dissected into 5 cm depth increments with 64 individual samples (3.75 cm x 3.75 cm cell area) per depth increment. Horizontal slits, spaced at 5 cm increments, were made in the front face of the steel box at construction. These slits allowed the encased block to be sectioned into 5 cm depth increments through the use of a 3.2 mm thick stainless-steel plate. After the sheet had been pressed through the entire width of the block, the encased 5 cm layer was carefully lifted out of the steel box. The layer was then sectioned into 64, 3.75 cm x 3.75 cm area increments in a grid pattern that was aligned with the grid of 64 effluent collection cells. Samples from the 3-8 cm, 23-28 cm, and 43-48 cm depths were utilized for complete analysis. *P. putida* analysis for effluent and soil-extraction samples was conducted by the fluorescent signal given off by the GFP bacteria. A fluorescence spectrometer was used to determine the transmittance at a wavelength of 509 nm, with an excitable wavelength of 395 nm. The colony forming units (CFU) was determined by direct count of colonies growing on Luria-Bertani selective media which contained 50 mg tetracycline L^{-1}. The correlation between microbial concentration, i.e., CFU cm^{-3} of effluent, and the fluorescent signal was used to estimate the cells cm^{-3} of each sample. The lower limit detection in this analysis was approximately 10^1 cells cm^{-3}.

After subsamples of wet soil had been removed for microbial analysis, all samples were analyzed for bulk density(ρ_b) and gravimetric (θ_g) water content. To determine (θ_g), a 10g subsample was removed, oven dried at 105 °C and reweighed. The remaining sample was weighed wet, air-dried for two days and then reweighed. The (θ_g) of the air dried sample was also determined from a 10 g subsample. Bulk densities, volumetric and gravimetric water contents were calculated by using the samples cube volume of 3.75cm x 3.75cm x 5cm, and the wet and oven-dried weights as described by Gardner (1986). Percent porosity was calculated in each sample by using an expanded version of Danielson and Sutherland's method (1986), which involved the use bulk density and an estimated particle density of 2.65 g cm^{-3}. This method depends on the relationship:

$$\% \text{ Porosity} = (1 - \rho_b / 2.65 \text{ g cm}^{-3}) \times 100 \quad (1)$$

Particle size analysis of the sand block samples was conducted by using the Dry Sieving Method for fractionation of sand particles (Gee and Bauder,

1986), based on the USDA scheme for particle sizes. Specific Surface area's were calculated by using the summation equation of:

$$\text{Specific surface } (a_m) = (6/p_s) \sum(c_i/d_i) \quad (2)$$

where c_i is the mass fraction of particles of average diameter d_i, and p_s is a particle density of approximately 2.65 g cm^{-3} (Hillel, 1982).

Soil chemistry was determined on soil-water extracts. Three grams of soil were combined with 27 ml of deionized water and placed on a shaker for 20 minutes. Soil solutions were thoroughly mixed and then filtered through a 0.45 μm Acrodisc filter. Effluent samples (10 ml) were analyzed for anions on the Ion Chromatograph (IC), and cations on the ICAP. A 1:9 ratio of soil/water was used for soil extractions to determine soil pH. Both soil and effluent pH was determined on an expandable ion analyzer(McLean, 1982). Total phosphate concentrations were determined from soil extractions on the ICP and Ion Chromatograph (IC) respectively.

Pearson correlation analysis was conducted using SAS on all variables for each effluent samples during each rain event, and for all soil property variables in each layer. Cumulative flow and effluent bacteria concentrations were also analyzed for correlation, by layer and rain event, to all soil properties of their vertically aligned samples.

RESULTS AND DISCUSSION

Spatial Distribution of Flow. During the 4 simulated rain events, only 6-16% of the 64 collection cells exhibited outflow. The numbers of collection cells where flow was observed included 3 fast flowing cells(>400 cm/rain event), 1 medium flowing cell (200-400 cm/rain event) and 5-7 slow flowing cells (<200 cm/rain event). Spatial distributions of flow among the collection cells were consistent for all events, which indicates that the preferential flow paths were controlled by soil properties. The results were consistent with the funnel flow concept, the occurrence of preferential flow paths through structure-less media, appears to be supported by these findings (Selker et al., 1992). Flow volumes were highest during the first simulated rainfall (Figure 1). Approximately 98% of the applied solution was collected from the base of the block indicating that the block was at field capacity before the event. The second event, however, produced the lowest effluent volumes, with only 88% of the

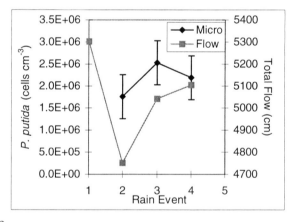

Figure1. Total Flow vs *P. putida* Concentrations

applied solution collected. The two remaining events produced effluent volumes similar to the first event with collection of 94-95% of the applied solution. The first occurrence of flow was approximately 12 minutes after initiation of rainfall for all of the rain events, and always occurred in the same collection cell.

***Pseudomonas putida (GFP)* Populations in Effluent.** GFP concentrations in collected effluent were found to vary both spatially and temporally within each rainfall event. Effluent samples revealed that only 4% of the applied *P. putida* populations were accounted for in the collection chambers. The second rainfall event produced the lowest GFP numbers (average 1.76×10^6 cells cm^{-3}) while the highest populations were detected during the third and fourth rainfall events (Figure 1). This increase was likely due to the flushing of GFP from the matrix into preferential flowpaths. A slight decrease in GFP numbers was noted during the fourth rainfall.

At any particular location, the indigenous microbial populations generally exhibited the highest concentration at the initiation of flow and as flow continued, indigenous populations decreased. Secondary peaks in GFP concentrations were seen during the recession limits of the hydrograph (Figure 2a). Such bimodal breakthrough curves are typically explained as proof of a dual pore system, in which rapid breakthrough occurs from the preferential flow paths followed by a delayed breakthrough from matrix pores. This was attributed to the flushing of previously immobile cells from low porosity areas into preferential flow pathways. However, flow occurring in areas where slower flow rates were observed produced the highest variability (Figure 2b). This fluctuation in effluent microbial concentrations was attributed to the hydrodynamic dispersion effects of a greater contribution of flow through the soil matrix regions. As flow continues, diffusion from the less mobile pore regions into the preferential flow paths is not able to maintain this initial high concentration so populations decrease.

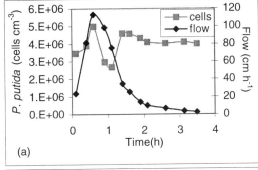

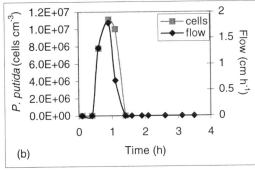

Figure 2. Effluent *P. putida* concentrations vs Flow in (a) Fast Flow Cells and (b) Medium Flow Cells.

Soil Physical Properties. In our studies, the soil particle sizes, rather than soil porosity, determined the mobility of the bacteria (Table 1). Soil porosity is based on the total porosity of a volume of soil, rather than individual pore size. Clay has a higher porosity than that of sand, however individual pore diameters are larger in sands due to the large size of sand particles. Thus, bacteria could move farther in sandy soils even with a lower porosity than that of clay. While there is no correlation between porosity and flow or porosity and microbial transport, an indirect relationship does exist between soil texture and flow due to the arrangement of soil particles (soil structure). Fine sand within the 3-8cm

Table 1: Mean Values and Standard Deviations for Soil Physical Properties and *Pseudomonas putida* Concentrations for the Soil Block (*Size fractions defined by the USDA scheme.)

Physical Property	3-8 cm	23-28 cm	43-48 cm
Porosity (%)	44.63 ± 5.38	44.88 ± 3.81	36.12 ± 3.75
Grav. Water (g g^{-1})	0.07 ± 0.08	0.20 ± 0.01	0.14 ± 0.03
Sand (%)	99.82 ± 0.40	99.84 ± 0.04	99.51 ± 0.002
* Very Coarse (%)	0.18 ± 0.19	0.52 ± 0.44	27.15 ± 7.30
* Coarse (%)	8.60 ±1.9	1.16 ± 0.57	35.4 ± 5.12
* Medium (%)	4.60 ± 1.1	2.34 ± 0.47	1.9 ± 0.70
* Fine (%)	85.1 ± 2.0	94.7 ± 0.61	34.03 ± 7.77
* Very Fine (%)	1.43 ± 0.09	1.12 ± 0.18	1.04 ± 0.33
* Residual Silt (%)	0.09 ± 0.03	0.16 ± 0.04	0.22 ± 0.13
Surface Area (cm^2 g^{-1})	57.70 ± 0.24	57.76 ± .003	57.60 ± 0.08

layer of the block produced a filtering effect, which significantly lowered the mobility of GFP bacteria. The small particle sizes, such as fine sands, trapped microbes while regions with coarse sand particles allowed the microbes to be readily flushed through the soil block. The highest gravimetric water contents were detected in the samples that had the highest percentages of fine sand particles. Gravimetric water contents in both blocks were found to be relatively low, which was attributed to evaporation losses. The highest microbial concentrations were observed in the top layer (3-8cm) and middle (23-28cm) layers of the block, which consisted primarily of smaller sand particles.

Soil Chemical Properties. Soil chemistry was found to be less influential on the distribution of microbial populations than the physical properties of the soil, however there were detectable influences. Phosphate concentrations were highly correlated with microbial distributions within the layers. It was expected that retention to iron and magnesium oxides would greatly inhibited phosphate transport. While phosphate distributions were essentially limited to the surface layers of both blocks, the highest concentrations were detected in the lower porosity areas. Limited phosphorous mobility was also reported by Jones and Lee (1977). They observed that phosphorus movement was greater in coarse sands, while effective P removal occurred in soils composed primarily of medium and fine sands. Rubaek et al., (1999) reported that total and organic phosphorus contents in soil samples increased as soil particle sizes decreased. Correlations between phosphate contents and soil pH were difficult to interpret in the soil layers, since most of the samples from both blocks had a slightly acidic to neutral pH, which is the conditions in which phosphate is most available to microbial populations. Changes in soil pH were associated with changes in microbial distributions, with the highest populations existing in the samples with a pH of 6.0-7.0, and the lowest numbers in the samples that were > 7.0. The limited correlation between microbial populations and chemical properties is likely due to the limited clay content of the soil block.

Distribution of *Pseudomonas putida* populations. In many laboratory transport studies injected bacterial populations decline steadily from the injection point (e.g. Toran and Palumbo, 1992), however in this study we saw a different pattern. *Pseudomonas putida* populations were approximately equal in the 3-8 cm and 23-28 cm layers of the soil block with average distributions of 1.10×10^{12} and 1.07×10^{12} cells kg^{-1} respectively. Standard errors for these layers were 7.26×10^9 and 1.90×10^{10} cells kg^{-1}. These were the layers that were predominately composed of fine sand particles. The lowest populations were seen in the 43-48 cm layer with average distributions of 5.26×10^{11} cells kg^{-1} and a standard error of 2.85×10^{10} cells kg^{-1}. In general, as the particle size decreases so does the average pore entrance size, thereby restricting microbial penetration. This lack of penetration inhibited the transportability of microorganisms through the block, and limited the movement of microbes to the 43-48 cm layer. Microbial filtration or adsorption to the fine sand particles could have affected flow movement by reducing the dimensions of the pores by aggregation of the filtered bacteria cells. This would decrease the transportability of the microbes with time and possibly reroute flow, as well as prevent further bacterial transport. Particle-size distributions also likely affected spatial heterogeneity of soil water and phosphates, which were found to be highly correlated to distributions of microbial populations.

CONCLUSION

While many manipulative practices can be utilized to increase the efficiency of bioremediation, i.e. nutrient additions and GEM's, bioremedial success will be predominately governed by the spatial variability of soil properties existing at the contaminated site. Variability in soil conditions can greatly vary within a small distance and must be taken into account when determining if bioremedial practices are practical for contaminate cleanup. While the general practice of most environmental companies is to limit the number of soil samples analyzed for bioremediation feasibility studies, decisions made from these samples may prove to lessen bioremedial efficiency, since the samples analyzed may not fully represent existing soil conditions. Increasing the number of samples for analysis will not only assist the researcher in determining the spatial variability of soil properties, but will also help determine what practices need to be applied for microbial proliferation, which in turn will increase the success of bioremediation. While bioremedial feasibility studies often center on soil chemistry and nutrient concentrations, our studies indicate that equal consideration should be given to the physical properties of the soil.

REFERENCES

Alexander, M. and Scow, K.M. 1989. Kinetics of Biodegradation in Soil. In Reactions and Movement of Organic Chemicals in Soils. *Soil Science Society of America Journal.* 22:243-269.

Burlage, R. S., Z. Yang, and T. Mehlhorn. 1996. "A Transposon for Green Fluorescent Protein Transcriptional Fusions: Application for Bacterial Transport Experiments." *Gene* 173:53-58.

Danielson, R.E. and Sutherland, P.L. 1986. "Porosity." In Klute, A., et al. (ed). Methods of Soil Analysis, Part 1. Physical and Mineralogical Methods. *Agronomy* 9:443-445.

Gee, G.W. and Bauder, J.W. 1986. "Particle Size Analysis." In Klute, A. et al. (ed) Methods of Soil Analysis, Part 1. Physical and Mineralogical Methods. *Agronomy* 9:383-409.

Hillel, D. 1982. "Texture, Particle Size Distribution, and Specific Surface." In *Introduction to Soil Physics,* pp. 21-34. Academic Press, Inc., San Diego, CA.

Jones, R.A. and Lee, G.F. 1977. *Septic Tank Disposal Systems as Phosphorous Sources for Surface Waters.* USEPA Rep. 6013-77-129. USEPA, Adam, OK.

Kinsall, B.L. 1987. "Correlation of Heterogeneity in Physical and Chemical Properties of the Vadose Zone with Microbial Populations and Migrations." M.S. Thesis, University of Tennessee, Knoxville, TN.

McLean, E.O. 1982. "Soil pH and Lime Requirement." In Methods of Soil Analysis, Microbial and Chemical Properties. *Soil Science Society of America Journal.* 9:199-209.

Rubaek, G.H., Guggenberger, G., Zech, W. and Christensen, B.T. 1999. "Organic Phosphorus in Soil Size Separates Characterized by Phosphorus-31 Nuclear Magnetic Resonance and Resin Extraction." *Soil Science Society of America Journal.* 63:1123-1132.

Selker, J.S., Steenhuis, T.S. and Parlange, J.Y. 1992. "Wetting Front Instability in Homogeneous Sandy Soils Under Continuous Infiltration." *Soil Science Society of America Journal.* 56:1346-1350.

Toran, L. and Palumbo, A.V. 1992. "Colloid Transport Through Fractured and Unfractured Laboratory Sand Columns." *Journal of Contaminant Hydrology.* 9:289-301.

United States Department of Agriculture. 1989. *Soil Survey of Northhampton County, Virginia.* United States Department of Agricriculture, Soil and Conservation.

*This research was funded by the Department of Energy Subsurface Science Program (Frank Wobber, program manager). Oak Ridge National Laboratory (ORNL) is managed by Lockheed Martin Energy Research for the U.S. Department of Energy (DOE) under contract DE-AC05-96OR22464.

ANAEROBIC BIOREMEDIATION OF CHLORINATED VOCS IN A FRACTURED BEDROCK AQUIFER – PREPARATIONS FOR AN IN SITU PILOT STUDY

R. Joseph Fiacco, Jr. (ERM, Boston, Massachusetts)
Matthew H. Daly, James D. Fitzgerald and Gregg Demers (ERM, Boston, Massachusetts)
Michael D. Lee (TerraSystems, Wilmington, Delaware)
Duane Wanty (The Gillette Company, Boston, Massachusetts)

ABSTRACT: A bench-scale pilot study demonstrated that a combination of enhanced bioremediation and bioaugmentation resulted in degradation of trichloroethene (TCE) to vinyl chloride (VC) and ethene over a 27 week period in bedrock aquifer samples and complete degradation of TCE to ethene over a three to seven week period in till aquifer samples. TCE concentrations as high as 252 mg/L were degraded with the bioaugmented treatments, demonstrating the potential application of anaerobic bioremediation to dense non-aqueous phase liquid (DNAPL) impacted zones. This study demonstrated that the optimal substrate additive is lactate and that the Pinellas dechlorinating culture was an effective exogenous culture to facilitate reductive dechlorination of chlorinated ethenes at the site. A forced gradient tracer study was conducted to determine hydraulic parameters for the shallow bedrock aquifer necessary to design an in situ bioremediation pilot study.

INTRODUCTION

Chlorinated volatile organic compounds (VOCs) were detected in overburden and bedrock aquifers beneath an active manufacturing facility. DNAPL was detected in the shallow overburden aquifer beneath the manufacturing building. However, migration of chlorinated VOCs within the shallow overburden aquifer is limited and does not extend off site. Significant downward migration of chlorinated VOCs and possibly DNAPL has occurred along structural supports beneath the manufacturing building to the deep overburden, shallow bedrock and deep bedrock aquifers.

Based on evidence of intrinsic bioremediation at the site, the property owner elected to evaluate implementation of enhanced bioremediation and/or bioaugmentation. This paper presents results of the feasibility study designed to evaluate the potential effectiveness of this remedial approach.

PREVIOUS STUDIES

Under anaerobic conditions, TCE can be biodegraded to cis-1,2-dichloroethene (DCE), VC, ethene, and ethane (Lee et al. 1998). At many sites, dechlorination stops at DCE or VC because there is not sufficient organic carbon to drive the groundwater anaerobic and to support reductive dechlorination or a microbial population capable of complete dechlorination has not developed. Bioaugmentation with a TCE-dechlorinating enrichment has been demonstrated in laboratory column studies (Harkness et al. 1999) and in a recirculating pilot study at Dover Air Force Base (Grindstaff 1998).

SITE HYDROGEOLOGY

Site geology consists of the following stratigraphic units, from top to bottom: urban fill, silty peat, clay, till and bedrock. Overburden units vary in thickness across the site, with some units absent beneath portions of the site. The silty peat, clay and lodgement till units are relatively impermeable and act as barriers to substantial vertical migration. However, numerous structural supports for the manufacturing building were installed to and into bedrock, and appear to act as preferred pathways for vertical migration. Bedrock consists of an argillite, which is highly fractured at the surface with generally increasing competency with depth. Depths to bedrock range from approximately 80 to 100 feet below ground surface.

The site is abutted to the north by a tidally influenced water body, which may act as a discharge boundary for the overburden aquifer, but not for the bedrock aquifer. Groundwater elevations at the site are tidally influenced and saltwater intrusions appear to exist. Under natural conditions, groundwater generally flows to the north at an average horizontal gradient of 0.0125 feet per foot (ft/ft). However, due to the proximity of a major construction project, which is conducting significant dewatering and recharging activities, site groundwater currently flows to the west-northwest at an average horizontal gradient of 0.062 ft/ft. Site hydrologic conditions have been significantly complicated due to these construction activities.

TCE was detected in ablation till and shallow bedrock at two locations downgradient of the source area and upgradient of the abutting channel (Figure 1). The TCE degradation product cis-1,2-dichloroethene (DCE) was detected in both of these aquifers, suggesting intrinsic bioremediation at the site. Groundwater samples were collected from these aquifers for analysis of intrinsic bioremediation indicator parameters, in general accordance with the protocol presented by Wiedemeier, et al. (1997). A summary of chlorinated ethene concentrations and bioremediation indicator parameter data are presented in Table 1.

The presence of generally reducing (i.e., nitrate- to iron-reducing) conditions in both the till and bedrock aquifers, combined with the presence of DCE, suggested that ambient conditions could support enhanced or augmented bioremediation. However, the general absence of vinyl chloride and ethene suggested either that there was not enough organic carbon (i.e., electron donor) to favor geochemical conditions necessary for a native microbial population to completely degrade TCE, or that the native microbial population was not capable of completely degrading TCE. Therefore, microcosm studies were completed to establish the rate limiting factor(s) and evaluate feasibility of enhanced or augmented bioremediation.

TABLE 1. Chlorinated ethene and biogeochemical indicator parameter data for June 1999.

Well I.D.	MW801T	MW801B	MW813T	MW813B
Depth (feet below grade)	69-89	121-131	77-97	120-130
VOCs (mg/L)				
Trichloroethene	45	479	23	171
Dichloroethene, (total)	59	12	5	ND
Vinyl Chloride	ND	ND	ND	ND
Biogeochemical Indicator Parameters				
Oxidation-Reduction Potential (mV)	-82.4	-54.2	-68.2	-25.1
Dissolved Oxygen (mg/L)	0.26	0.34	0.63	0.41
Nitrate Nitrogen as N (mg/L)	ND	ND	ND	ND
Nitrite Nitrogen as N (mg/L)	ND	ND	ND	ND
Ammonia Nitrogen as N (mg/L)	4	0.53	1.1	0.96
Dissolved Iron (mg/L)	ND	2.8	0.87	10
Dissolved Manganese (mg/L)	0.52	3.7	1.6	1.9
Sulfate (mg/L)	180	260	150	460
Sulfide (mg/L)	ND	ND	ND	ND
Alkalinity (mg/L as $CaCO_3$)	260	360	250	210
Chloride (mg/L)	4,400	5,300	3,800	6,700
Total Organic Carbon (mg/L)	4.5	18	1.5	1.9
Phosphate, total as P (mg/L)	ND	ND	ND	ND
pH	8.03	6.75	6.94	6.49
Temperature (C)	18.99	17.41	17.91	20.65
Conductivity (mS/cm)	11,963	13,742	8,355	17,294
Total Dissolved Solids (g/L)	6.87	7.99	4.67	10.18

LABORATORY MICROCOSM STUDY

Field Sample Collection Procedures. To accurately simulate in situ aquifer conditions, a new borehole (i.e., MW901) was advanced to collect representative samples of aquifer material (i.e., till and bedrock) under anaerobic conditions. Split spoon samples of till and core samples of bedrock were collected and immediately placed inside a glove bag filled with a nitrogen atmosphere. The samples were collected into bottles that were then topped off with groundwater collected under anaerobic conditions (i.e., a nitrogen atmosphere) from each of the respective aquifers at the upgradient MW801 couplet. Additional groundwater was collected from the MW801 couplet under anaerobic conditions for use in the microcosm studies.

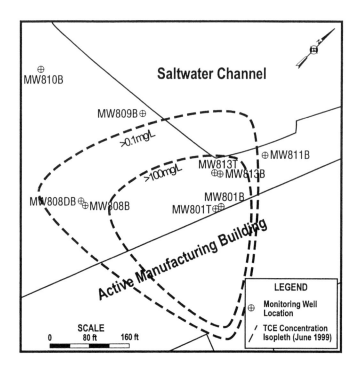

FIGURE 1. Site plan showing the locations of relevant site features, bedrock and till monitoring wells and generalized TCE concentration isopleths.

Microcosm Study Procedures. A total of 16 microcosms were prepared using till and bedrock samples and various combinations of substrates (i.e., sodium lactate, molasses and soybean oil), yeast extract and nutrients (i.e., nitrogen, phosphorous and trace elements and vitamins) and an exogenous dechlorinating culture (i.e., Pinellas dechlorinating enrichment (PDE)). The PDE was derived from the Department of Energy facility in Pinellas, Florida and has been shown to support complete dechlorination of TCE to ethene when bioaugmented into soil columns (Harkness, et al., 1999). The microcosms contained 10% (volume/volume) solids (i.e., till or bedrock) and 90% groundwater collected from the till or bedrock aquifers. Some of the bedrock microcosms were diluted to evaluate the potential toxicity of the elevated TCE concentration (i.e., 479 mg/L). Samples were collected from each microcosm on an incremental basis throughout the study and analyzed for VOCs and metabolic end product gases (i.e., ethene, ethane and methane). Microcosm studies were conducted for 12 to 27 weeks, depending on the rate of degradation. An anaerobic atmosphere was maintained in all microcosms throughout the duration of the study. The microcosm studies were completed at room temperature. Therefore, the reaction kinetics defined in the laboratory may not reflect reaction kinetics under ambient conditions.

Results. Till microcosms that were only amended with various substrates (i.e., no PDE) resulted in incomplete degradation of TCE, generally resulting in the production of DCE or VC. Complete reductive dechlorination of TCE (i.e., to an ethene endpoint) was observed in the till microcosms three to seven weeks after bioaugmentation with the PDE (Figure 2).

Bedrock microcosms that were only amended with various substrates (i.e., no PDE) resulted in incomplete degradation of TCE, generally resulting in the production of DCE or VC. The bedrock microcosm amended with PDE was able to generate both VC and ethene from an initial TCE concentration of 252 mg/L, although VC did not completely degrade to ethene by the end of the study (Figure 3). The diluted bedrock microcosm amended with lactate and bioaugmented with PDE was able to completely dechlorinate TCE to ethene from an initial concentration of 24.5 mg/L (Figure 4).

Results of the bioaugmentation microcosms indicate that lactate is the optimal substrate to facilitate complete dechlorination of TCE at the site. Microcosm results indicate that an in situ pilot study is warranted at the site.

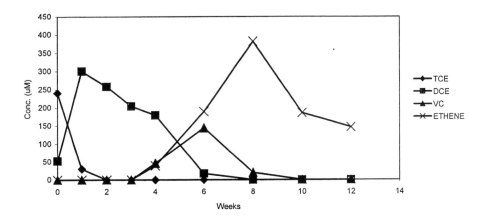

FIGURE 2. Plot showing VOC concentrations versus time for a till aquifer microcosm amended with lactate plus yeast extract and bioaugmented with the PDE. Demonstrates complete degradation of TCE to ethene.

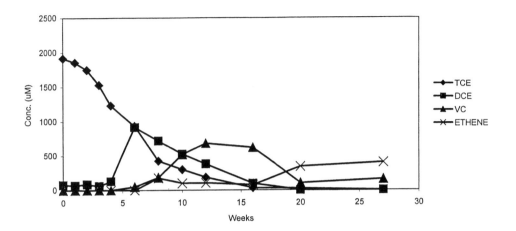

FIGURE 3. Plot showing VOC concentrations versus time for a bedrock aquifer microcosm amended with lactate plus yeast extract and bioaugmented with the PDE. Demonstrates near complete degradation of TCE to VC and ethene.

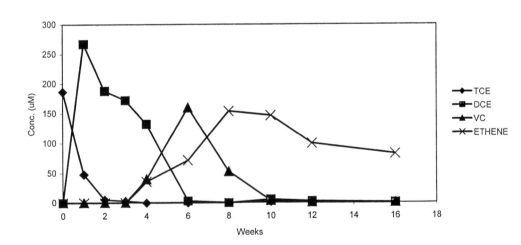

FIGURE 4. Plot showing VOC concentrations versus time for a diluted bedrock aquifer microcosm amended with lactate plus yeast extract and bioaugmented with the PDE. Demonstrates complete degradation of TCE to ethene.

TRACER STUDY

Field Methods. A forced gradient tracer study was conducted in the shallow bedrock aquifer within the TCE plume to characterize hydraulic parameters necessary to design an in situ bioremediation pilot study. The tracer study was conducted for a duration of 15 hours. The test was conducted under pumping conditions to accelerate its completion. Two conservative tracers (i.e., Rhodamine WT dye and potassium iodide) were used to provide confirmation of results and to evaluate the potential for retardation of one tracer relative to the other.

Results. Rhodamine WT concentrations began increasing at the extraction well approximately 1.75 hours after tracer injection Concentrations increased from baseline to 50% maximum breakthrough concentration at 2.3 hours after injection. Mass recovery for Rhodamine WT was 7%. Low recovery for tracer is likely due to dilution caused by radial flow to the pumping well. Iodide breakthrough is less defined than Rhodamine WT because laboratory analyses were performed at a lower frequency. The initial and 50% maximum breakthrough times for iodide were similar to those for Rhodamine WT. However, iodide mass recovery for iodide at MW81B was significantly higher (i.e., 27%). Increased mass recovery for iodide relative to Rhodamine WT is most likely due to sorption of Rhodamine onto both suspended and fixed particles in bedrock.

Using data obtained from the tracer study, effective porosity (n_e) for bedrock was calculated at 0.005. Hydraulic conductivity (K) for bedrock was calculated at 0.219 feet per day (ft/d). Using these data, the groundwater velocity within the anticipated in situ pilot study treatment zone was calculated to be 0.42 ft/d, using the current hydraulic gradient. These data will be used to calculate residence times under varying scenarios to design and effective in situ bioremediation pilot study.

CONCLUSIONS

Limited intrinsic bioremediation is occurring in the till and bedrock aquifers at the site. Nitrate- to iron-reducing conditions were documented in the till and bedrock aquifers, resulting in limited degradation of TCE to DCE. Microcosm studies indicated that the introduction of lactate and PDE to aquifer samples resulted in complete degradation of TCE to ethene in till and near complete degradation of TCE to VC and ethene in bedrock. Complete degradation of TCE to ethene was demonstrated for diluted bedrock aquifer samples. TCE concentrations as high as 252 mg/L were degraded with the bioaugmented treatments, demonstrating the potential application of anaerobic bioremediation to DNAPL-impacted zones. Hydraulic conductivity and effective porosity were determined for the bedrock aquifer using a tracer study.

REFERENCES

Grindstaff, M. 1998. "Bioremediation of Chlorinated Solvent Contaminated Groundwater." Prepared for USEPA Technology Innovation Office under a

National Network of Environmental Management Studies Fellowship. At http://clu-in.org.

Harkness, M.R., A.A. Bracco, M.J. Brennan, Jr., K.A. DeWeerd and J.L. Spicvak. 1999. "Use of bioaugmentation to stimulate complete reductive dechlorination of TCE in Dover soil columns, Environmental Science and Technology 33(7): 1100-1109.

Lee, M. D., J. M. Odom, and R. J. Buchanan, Jr. 1998. "New Perspectives on Microbial Dehalogenation of Chlorinated Solvents. Insights from the Field." *Annual Reviews in Microbiology* 52: 423-452.

Wiedemeier, T.H., M.A. Swanson, D.E. Moutoux, J.T. Wilson, D. Kampbell, P. Haas and J. Hansen. 1997. "Protocol for supporting natural attenuation of chlorinated solvents with examples." In: *In Situ and On-Site Bioremediation: Volume 3*, p. 147. Battelle Press, Columbus/Richland.

FUNGAL TREATMENT FOR WASTEWATER CONTAINING RECALCITRANT COMPOUNDS

S.V.Srinivasan[*] and D.V.S.Murthy,
Environmental Engineering Laboratory, Department of Chemical Engineering
Indian Institute of Technology Madras, 600 036 INDIA

ABSTRACT

Due to rapid industrialization and urbanization, large amounts of wastewater are discharged into the environment. Most of the municipal wastewaters contain biodegradable substances which can be treated by conventional biological treatment methods such as activated sludge process, aerated lagoon and aeration pond. Industrial wastewater generated from the pulp and paper industry, and from dyeing and dye manufacturing, cannot be treated by these conventional methods due to the presence of recalcitrant compounds. These compounds are found to be toxic, carcinogenic and mutagenic. White rot fungi, which have the ability to degrade lignin, can be used for the treatment of effluent generated from these industries. These fungi have a non-specific enzyme system, which oxidizes the recalcitrant compounds present in the wastewater.

In this paper, the application of fungal treatment for wastewater from pulp and paper plant and dye industries using the white rot fungus *Trametes versicolor* is reported. Batch experiments were carried out in shake flasks. The effect of carbon source (glucose or sucrose) concentration, pH and initial color concentration on decolorization efficiency was evaluated. It was found that the presence of easily biodegradable primary carbon source was required for the growth of the fungus and decolorization. From these investigations, it was concluded that the fungal treatment could be used as primary treatment before conventional secondary treatment.

INTRODUCTION

Due to rapid industrialization and urbanization, a large number and volume of chemicals are manufactured for use in day to day life. Dyes, pesticides, insecticides and other chemicals are found in the wastewater generated from industries where these compounds are manufactured and/or processed. Because of their recalcitrant nature, these compounds are not easily amenable to conventional biological treatment processes such as activated sludge process, aerated lagoon and oxidation ponds. Effluents discharged from pulp and paper mills and dyeing industries is highly colored due to the presence of lignin-related compounds and dyes, respectively.

---

[*] Present address : Assistant Environmental Engineer, Central Pollution Control Board, East Arjun Nagar, Delhi – 110 032 INDIA (Email : srinivasansv@yahoo.com)

Pulp and paper is another major polluting industry in India. The effluent from pulp and paper mills is dark brown in color and contains chlorinated organic compounds formed during the use of chlorine and its derivatives in the pulp bleaching process. The chlorinated organic compounds are identified as potentially hazardous, carcinogenic, mutagenic, persistent and bioaccumulative (Walden & Howard, 1977). Conventional biological treatment system currently used in the pulp and paper industry such as activated sludge process, aerated lagoon, anaerobic lagoon, stabilization ponds etc. are successful in reducing the biochemical oxygen demand (BOD) from the effluent whereas color and a large percentage of the chlorinated organic compounds are not effectively removed. Mixed cultures used in biological treatment systems are capable of metabolizing soluble sugars and some low molecular weight compounds, leading to reduction of BOD, but they lack in an enzyme system capable of oxidizing color-causing compounds (chloro-lignins).

The use of many synthetic dyes has been increasing in textile and dye industries because of their ease and cost effectiveness in synthesis, firmness, and variety in color, compared to that of natural dyes. About 100,000 commercial dyes are manufactured which include several varieties of dyes, such as acidic, reactive, basic, disperse, azo, diazo, anthraquinone-based and meta-complex dyes. The toxicological and ecological aspects of these dyes are complex to evaluate. Some of the dyes are known to be toxic, carcinogenic and mutagenic (Eastlander 1988; Khanna et al., 1991).

Water pollution and other environmental regulations in India are beginning to require substantial reductions in the color of effluents from these industries. In general, the colored effluents from these industries are aesthetically unacceptable and reduce the light transmission through the contaminated waterways. This reduces photosynthesis which leads to depletion of dissolved oxygen and poses a health hazard to the aquatic life in the receiving water bodies.

Methods such as chemical coagulation, chemical oxidation, electrochemical, UV irradiation, ozonation and adsorption are available for color removal from dye and textile industries but they are not implemented at industrial scale due to high cost or technical feasibility. The ability of white rot fungi to degrade a wide range of synthetic chemicals, many of which are recalcitrant to biodegradation, has been reported (Field et al., 1993). Baipai and Baipai (1993) and Banat et al (1996) have reviewed biological color removal of pulp and paper mill wastewater and textile dye containing wastewater, respectively. Treatment of hazardous waste (Haimann, 1995), xenobiotic compounds (Paszczynski and Crawford, 1995) organo-pollutants (Hammel, 1989) and the mechanism by which the white rot fungi degrade pollutants (Barr and Aust,1994) have also been reviewed.

The white rot fungus, *Trametes versicolor* has a non-specific enzyme system that oxidizes recalcitrant compounds. It has been used for the decolorization of pulp and paper mill effluents (Pallerla and Chambers, 1995; Manazanares et al., 1995), and for biodegradation of azo dyes (Yesilada, 1995; Heinfling et al., 1997) and polycyclic aromatic hydrocarbon compounds such as acenaphthene and acenaphthylene (Johannes et al., 1998). In the present investigation, the removal of color from pulp and paper mill effluent and dye from aqueous solution by the white rot fungus *Trametes versicolor* MTCC 138 grown under different culture conditions is reported.

MATERIALS AND METHODS

Culture: The white rot fungus culture, *Trametes versicolor* MTCC 138 was procured from the Institute of Microbial Technology, Chandigarh, India. The organism was maintained on agar slants containing yeast extract (5 g/L), glucose (10 g/L) and agar-agar (15 g/L) and maintained at 40°C. The pH of the medium before solidification was adjusted to 5.8.

Medium Composition: The basic growth medium composition consists of 10g/L Glucose; 1g/L KH_2PO_4; 1.75 g/L NH_4Cl; 0.5 g/L KCl; 0.5 g/L $MgSO_4 \cdot 7H_2O$ and 0.01 g/L $FeSO_4 \cdot 7H_2O$. The initial pH of the medium was adjusted to 4.5

Pulp mill Effluent: Effluent samples were obtained from Tamil Nadu Newsprint and Paper Limited, Kagithapuram (Tamil Nadu), India. The highly colored effluent was from a plant utilizing bagasse as the raw material and was collected in air-tight plastic cans (30 L) and stored at 4 ± .1°C. Chemical characteristics of the effluent, except color, were determined according to Standard Methods for Examination of Water and Wastewater (APHA,1989) and are reported in Table 1.

Color measurements: The color of effluent was measured according to the National Council of the Paper Industry for Air and Stream Improvement Standard Methods (NCASI). The pH of the original effluent sample was in the range of 8.5-9.5 and was adjusted to 7.6 followed by filtration through 0.45 µm filters. The absorbance of the filtrate was measured at 465 nm against distilled water using a 1-cm light path cuvette in an UV spectrophotometer (Shimadzu, Japan). Platinum Cobalt Color units were calculated as follows.

$$\text{Color units (CU)} = 500 \times A2 / A1$$

Where: A1 - Absorbance of sample at 465 nm
A2 - Absorbance of 500 CU Pt-Co standard at 465 nm

Dyes: Remazol orange and Remazol brown, which are commonly used dyes in industries, were selected for the study and were obtained from Color-chem, Mumbai, India.

Dye measurements: Dye removal/disappearance was monitored spectrophotometerically by measuring the absorbance at or near the wavelength of maximum absorbance for each dye (i.e., at 490nm and 480nm for Remazol orange and remazol brown, respectively).

EXPERIMENTAL

The effect of varying concentrations of glucose and sucrose on color and COD removal from pulp and paper mill effluent were carried out in shake flasks (250 ml). The basic nutrients were added to 100 ml of the effluent sample with varying concentrations of glucose (2.5 to 50 g/L) and sucrose (2.5 to 25 g/L). Similarly, batch experiments were carried out in shake flasks (250 ml) containing 100 ml of aqueous solution containing the medium mentioned above and dyes at 250 and 500 mg/L concentrations. The pH of the effluent was adjusted to 4.5 (optimum) and autoclaved at 120°C for 15 min. After autoclaving, the flasks were inoculated with the white rot fungus *Trametes versicolor* and placed on a rotary shaker (180 rpm). After 7 days, the effluent were withdrawn, filtered through 0.45 µm filters and analyzed for color and COD using the methods mentioned earlier.

RESULTS AND DISCUSSION

Treatment with pulp mill effluent

The characteristics of effluent from the pulp and paper mill utilizing bagasse as the raw material is shown in Table 1. The effluent characteristics show a large fraction of non-biodegradable substances. Based on BOD/COD ratio of the effluent, which is less than one, the effluent is not suitable for conventional biological treatment.

Table 1   Characteristics of pulp mill effluent

Parameter	Range
pH	8.5-9.5
Color concentration (Pt-Co units)	4500-4700
Total Solids (mg/L)	3200-3800
BOD (mg/L)	260-360
COD (mg/L)	4500-4800

Effect of Glucose Concentration

Color removal is a secondary metabolic process and the fungus requires glucose or cellulose for energy. The effect of initial glucose concentration on color removal efficiency is shown in Figure 1. In the present study, the maximum color removal of 92.3% was obtained at a glucose concentration of 50 g/L. It was observed that the color removal efficiency did not increase significantly at glucose concentrations greater than 15 g/L, where about an 85.2% reduction was obtained. Also, it was observed that the presence of easily metabolizable sugars was required for the growth of the fungus and color removal. This confirmed the observations of earlier investigations (Royer et al., 1985; Eaton et al., 1980). Incubation with this fungus has not only reduced the color but also reduced considerably the COD in the effluent. A maximum COD removal of 78% was observed at a glucose concentration of 10 g/L and further addition of glucose resulted in an increase in effluent COD due to the presence of unmetabolizable glucose.

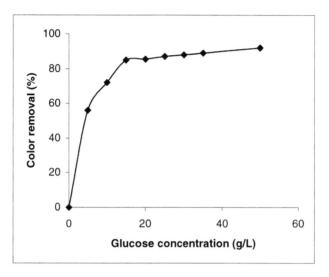

Figure 1. Effect of glucose concentration on Color removal efficiency (pH- 4.5; NH_4Cl -0.5 g/L; Initial Color- 4700 Pt-Co and basic nutrients)

Effect of Sucrose Concentration

The effect of initial sucrose concentration on color removal efficiency is shown in Figure 2. It was observed that the maximum color removal efficiency of 80.3% occurred at an initial sucrose concentration of 5 g/L. Further addition of sucrose had no significant effect on the color removal efficiency. In the control flask (i.e., the flask without the addition of sucrose) no growth of the fungus was observed and hence there was no color removal. This supported the hypotehesis that the fungus requires easily biodegradable sugars for growth. The maximum COD removal of 73.3% was obtained at an initial sucrose concentration of 5 g/L and as with glucose, the further addition of sucrose increased the COD of the treated effluent due to the presence of unused sucrose.

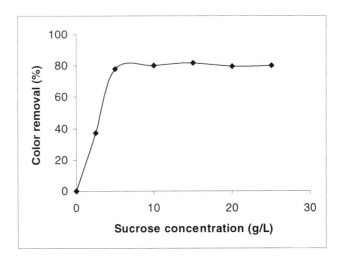

Figure 2. Effect of Sucrose Concentration on Color removal efficiency (pH-4.5; NH$_4$Cl -0.5 g/L; Initial Color- 4700 Pt-Co and basic nutrients)

Removal of dyes from aqueous solution

The results from the dye experiments are shown in Figure 3. The data show the ability of the white rot fungus *Trametes versicolor* to degrade remazol orange and remazol brown. It was observed that extensive removal of the two dyes by the fungus occurred based on the decrease in the absorbance of the culture medium. In the case of Remazol brown, 96.7% and 98.5% color removal efficiencies were attained at 250 mg/L and 500 mg/L dye concentration over a incubation period of 7 days, respectively. In the case of Remazol orange, decolorization achieved 70.4% and 81.3% over the same time period with 250 mg/L and 500 mg/L initial concentrations, respectively.

Control flasks incubated without additional carbon source (glucose) showed no decrease in absorbance or growth of the fungus (data not shown). It was evident that the addition of external carbon source was required for the growth of the fungus and the decolorization process. Also, it was found that the higher concentration of 500 mg/L has no adverse effect on the growth of the fungus and in turn the dye removal percent of dye was higher. For Ramazol organge, decolorization efficiency of 81.3% was obtained at an initial concentration of 500 mg/L compared to 70.4% at an initial concentration of 250 mg/L. There was not much difference in the decolorization efficiency when the fungus was grown in culture medium with 250 mg/L and 500-mg/L remazol brown dye concentration.

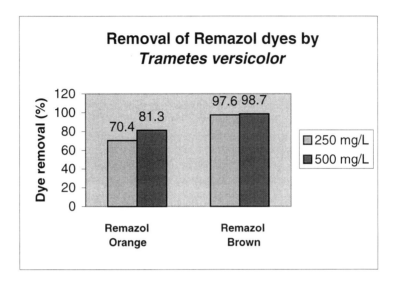

Figure 3. Percent Removals of Ramazol Orange and Ramazol Brown by Trametes versicolor in aqueous culture.

CONCULSIONS

Based on the results using the white rot fungus *Trametes versicolor* for the removal of color from pulp and paper mill wastewater and dye from aqueous solution, it was concluded that:

1. An external carbon source is required for the growth of the fungus and for decolorization to take place.

2. The fungus is effective in the removal of color and dye in the presence of additional carbon source and nutrients.

3. Further experiments are needed to determine the conditions at which the maximum decolorization of effluent and the optimal biodegradation of pollutants occur.

REFERENCES

Bajpai P. and Bajpai P.K. (1994). Biological color removal of pulp and paper mill wastewaters. J. of Biotechnology, 33, 211-220.

Banat I.M., Nigam P., Singh D. and Marchant R.(1996). Microbial decolorization of textile-dye-containing effluents:A review.Bioresource Technology, 58, 217-227.

Barr D.P. and Aust S.D. (1994). Mechanism white rot fungi use to degrade pollutants. Environ. Sci. Technology, 28(2), 79A-87A.

Eastlander, T. (1988) Allergic dermatoses and respiratory diseases from reactive dyes. Contact Dermatitis, Vol. 18, pp 290-197.
Eaton, D.; Chang, H.M.; and Kirk, T.K. (1980) Fungal decolorization of Kraft bleach effluents. TAPPI J 63, 103-106
Field J.A., De Jong E., Feijoo-costa and deBont J.A. (1993). Screening for ligninolytic fungi applicable to the biodegradation of xenobiotics. TIBTECH, 11, 44-49.
Haimann R.A. (1995). Fungal technologies for the treatment of hazardous waste. Environmental Progress, 14(3), 201-203.
Hammel K.E. (1989). Organopollutant degradation by ligninolytic fungi. Enzyme Microb. Technology, 11, 776-777.
Heinfling A., Bergbauer M. and Szewzyk U. (1997). Biodegradation of azo and phthalocyanine dyes by Trametes versicolor and Bjerkandera adusta. Appl. Microb. Technol., 48, 261-266.
Johannes C., Majecherczyk A. and Huttermann A. (1998). Oxidation of acenaphthene and acenaphthylene by laccase of Trametes versicolor in a laccase-mediator system. J. of Biotechnol., 61, 151-156.
Khanna, S.K. and Mulul das (1991) Toxicity, Carcinogenic potential and Clinico-epidemiological studies on dyes and dye intermediates. J Scientific Industrial Res., Vol.50, pp 965-974.
Manzanares P., Fajardo S. and Martin C. (1995). Production of ligninolytic activities when treating paper pulp effluents by Trametes versicolor. J. of Biotechnology, 43, 125-132.
Royer, G.; Desrochers, M.; Jurasek, L.; Rouleau, D ; Mayer, R. (1985) Batch and continuous decolorization of bleach kraft effluents by a white-rot fungus. J. Chem. Tech. Biotech. 35B, 14-22
Standard methods for Examination of Water and Wastewater (1989), 17th edition, APHA, Washington D.C.
Walden C.C.; Howard J.E.(1977) Toxicity of pulp and paper mill effluents . TAPPI J , 60, 122-125
Yesilada, O (1995) Decolorization of Crystal violet by fungi. World. J. of Microb. Biotechnol., 11, 601-602.

COMETABOLISM OF POORLY BIODEGRADABLE ETHERS IN ENGINEERED BIOREACTORS

Matthew J. Zenker, Robert C. Borden, and Morton A. Barlaz
(North Carolina State University, Raleigh, North Carolina)

ABSTRACT: A microbial consortium with the ability to biodegrade cyclic and alkyl ethers was studied to determine its ability to be used in an engineered process for the treatment of ether-contaminated groundwater. The consortium was isolated from an aquifer contaminated with 1,4-dioxane. Previous batch experiments suggested that the consortium was cometabolizing 1,4-dioxane using tetrahydrofuran (THF) as the growth substrate. A lab-scale trickling filter and rotating biological contactor (RBC) bioreactor were constructed with this consortium. Both reactors demonstrated an ability to effectively remove low levels of 1,4-dioxane when simultaneously fed THF. The trickling filter received different combinations of both chemicals to determine a minimum THF loading rate that yielded optimum 1,4-dioxane removal. At a flowrate of approximately 3.95×10^3 L/m^2-day, the trickling filter was capable of biodegrading influent concentrations of 20, 10, and 5 mg/L of THF to approximately 1-3 µg/L. 1,4-Dioxane at an influent concentration of either 1 or 0.2 mg/L was biodegraded to an average of below 65 µg/L at all loading rates. The RBC received a constant loading of approximately 20 mg/L of both THF and 1,4-dioxane at a flowrate of approximately 6 L/day. Both chemicals were degraded to below 0.8 mg/L. The reactors were also capable of biodegrading methyl *t*-butyl ether (MTBE) in the presence of THF.

INTRODUCTION

Synthetic ethers have many uses including gasoline additives, solvents, stabilizers, agrochemicals and detergents. Thus, ethers can appear as contaminants of air, surface water, groundwater, and even drinking water. 1,4-Dioxane is a cyclic ether that is formed as a byproduct during the formation of organic fibers using terephthalic acid and ethylene glycol (Popoola, 1991). It is also used as a stabilizer for chlorinated solvents and a wetting agent for textile processes. 1,4-Dioxane has become a contaminant in wastewater streams, ground and surface water due to improper treatment and/or storage. In 1997, approximately 324 tons of 1,4-dioxane was released into the environment because of industrial uses (TRI, 1997).

MTBE is an alkyl ether that is widely used as an octane enhancer in reformulated gasoline. A recent study listed MTBE as the second most common contaminant detected in shallow groundwater from eight urban areas (Squillace et al., 1996). Both 1,4-dioxane and MTBE are very soluble in water, and are difficult to remove from aqueous solution via physical separation processes.

Although recent research has established that both chemicals are biodegradable (Parales *et al.*, 1994; Hanson *et al.*, 1999), their rates of biodegradation are very slow (Sock, 1993; Salanitro *et al.*, 1994). Due to these low growth rates, the feasibility of using ether-degrading microorganisms in engineered bioreactors remains questionable. The slow biodegradation of these chemicals presents two possible disadvantages. (1) The biodegradation of these chemicals may require large reactors to achieve sufficient removal. (2) Low concentrations of the target contaminant may not provide sufficient carbon/energy to sustain growth of the 'active' biomass population. An engineered bioreactor treating contaminated groundwater, for example, may suffer from the latter disadvantage. Contaminated groundwater typically has relatively low concentrations of organic chemicals compared to industrial wastewater streams.

The objective of this study is to utilize an ether-degrading microbial consortium in engineered attached-growth bioreactors for the treatment of groundwater contaminated with either 1,4-dioxane or MTBE. The reactors will be fed low concentrations of 1,4-dioxane and MTBE to simulate concentrations frequently detected in groundwater contaminated with these chemicals. THF will be added to sustain growth and activity of the biomass for 1,4-dioxane and MTBE biodegradation.

MATERIALS AND METHODS

The consortium was isolated from an aquifer that had been contaminated with 1,4-dioxane. A sample of the aquifer sediment was retrieved and incubated in triplicate with native groundwater, 200 mg/L 1,4-dioxane, and 200 mg/L THF at 35 °C. Once there was evidence of 1,4-dioxane degradation, aliquots were taken from each soil microcosm and transferred to a flask containing mineral medium L (Leadbetter & Foster, 1958). Mineral medium L had the following composition: $NaNO_3$, 2.0 g; $MgSO_4$, 0.2 g; Na_2HPO_4, 0.1 g; NaH_2PO_4, 0.15 g; KCl, 0.04 g, $CaCl_2$, 0.015g; $FeSO_4\text{-}7H_2O$, 1 mg; $CuSO_4\text{-}5H_2O$, 20 mg; H_3BO_3, 57 mg; $MnCl_2\text{-}4H_2O$, 36 mg; $ZnSO_4\text{-}7H_2O$, 44 mg; $Na_2MoO_4\text{-}2H_2O$, 177 mg; and deionized water, 1L. THF and 1,4-dioxane were added at 200 mg/L and the flask was incubated at 35 °C. Disappearance of both chemicals was observed in two weeks, along with the appearance of white, filamentous colonies floating on the water surface.

RBC Construction and Operation. A plastic RBC reactor was initially constructed to determine if sustained biodegradation of THF and 1,4-dioxane could be accomplished. The reactor was subdivided into four chambers, each of which had two 7 in. (17.7 cm.) diameter by ¼ in. (0.63 cm.) thick circular disks. The disks were 33% submerged, and rotated at a speed of 2 rpm. The reactor received a 6 L/day flowrate of 25% mineral medium L. THF and 1,4-dioxane contained in a concentrated aqueous solution were fed into the influent stream via a syringe pump. Average influent concentrations of THF and 1,4-dioxane over the course of operation were approximately 20 and 30 mg/L, respectively. The RBC was initially inoculated with aliquots from flasks that exhibited THF and 1,4-dioxane biodegradation. The reactor was operated in a temperature-controlled room at 35 °C.

Trickling Filter Construction and Operation. The trickling filter was constructed of ½ in. (1.3 cm.) (I.D.) x 4 ft. (1.2 m) rigid plastic tubing. Filter packing material consisted of 6 mm. ceramic saddles. The filter received an average flowrate of 3.95 x 10^3 L/m^2-day of 25% mineral medium L and THF/1,4-dioxane was delivered by a syringe pump. The trickling filter would receive varying influent THF concentrations of 20, 10 and 5 mg/L. At each THF loading rate, influent 1,4-dioxane concentrations of 1 and 0.2 mg/L would be utilized. Influent concentrations of 1,4-dioxane were selected to mimic typical concentrations of 1,4-dioxane in contaminated groundwater. Field investigations of four sites contaminated with 1,4-dioxane revealed that 1,4-dioxane concentrations were usually less than 1 mg/L. These loading rates were tested after the trickling filter had been in operation for approximately 240 days. Prior to this testing, the filter influent contained approximately 10 mg/L of THF and 1,4-dioxane. The trickling filter was operated at each loading rate for approximately three weeks. The trickling filter was inoculated with biomass from the RBC and operated in a 35 °C room.

The trickling filter was backwashed daily by a water/air scour to prevent clogging of the filter media with biomass. Furthermore, this procedure allowed the filter biomass to more rapidly acclimate to the adjustments in influent ether concentrations. The water was pumped upwards through the column at approximately 1 L/min for one minute, while air was supplied at a pressure of 140 kpa. The backwash water was collected and analyzed for suspended solids to determine the biomass removed.

A peristaltic pump was used at the effluent end of the trickling filter to (1) draw effluent water from the column, and (2) allow airflow through the column. A vent opening was placed at the top of the column to achieve this airflow. The air flowrate through the column was measured by attaching the effluent tubing into a sealed glass serum bottle to collect the effluent water. The headspace of the serum bottle was then vented into a gasbag to collect air that had passed through the filter. Subsamples of the air captured in the gasbag were analyzed via gas chromatography (GC) to determine the amount of ethers that had volatilized. The remaining volume contained in the gasbag was quantified to determine the air flowrate.

Analytical Techniques. Aqueous samples from the trickling filter were analyzed by purge and trap gas chromatography to determine ether concentrations. Samples were analyzed using a Tekmar® LSC 3000 Purge and Trap with a 2016 autosampler. This was connected to a Perkin-Elmer Model 9000 Auto System Gas Chromatograph fitted with a J&W Scientific DB-VRX capillary column and flame ionization detector. Since the ethers tested are extremely soluble, the autosampler was outfitted with a Tekmar® AUTOmatic Sample Heater that heated samples to 80 °C for 15 minutes before analysis. 1,3-Dioxane was added to each sample to serve as an internal standard. The detection limits for each chemical were 1 µg/L.

Aqueous samples from the RBC were analyzed by direct aqueous injection of a 0.5 mL sample into a Hewlett-Packard 5890 Gas Chromatograph (GC) equipped with a flame ionization detector (FID) and a 75-meter DB-VRX column. The column was held at 70 °C for 2 min, raised to 160 °C at 15 °C/min, then raised to 220 °C at 70 °C/min and held at 220 °C for 2 min. The detection limits for all chemicals was 0.8 mg/L. Gaseous samples from the trickling filter were analyzed by direct injection of 2 mL samples into a Shimadzu 14A GC. This instrument was equipped with a Supelco Hayesep Q packed column operated isothermally at 220 °C. The gaseous detection limit of all chemicals was approximately 1 ppmv. Total suspended solids (TSS) were determined according to Standard Method 2540D (APHA *et al.*, 1995).

RESULTS AND DISCUSSION

At the flow and loading conditions tested, the RBC reactor continually biodegraded THF and 1,4-dioxane to below 0.8 mg/L. A biofilm of approximately 0.1 in. (0.2 cm.) formed on the RBC disks in the first basin after one month of operation. The remaining three basins only developed very thin biofilms throughout reactor operation. The purpose of the RBC was to determine if sustained biodegradation of 1,4-dioxane was possible, and to sustain the culture throughout experimentation. With these goals being achieved, the trickling filter was analyzed more thoroughly to produce a reasonable understanding of the factors governing its operation.

Airflow through the filter column was determined to be approximately 0.18 L/min. GC analysis of the air determined that less than 3% of the ethers were volatilized. Biomass removal due to filter backwashing was approximately 20 mg TSS/day. The influent and effluent concentrations of THF and 1,4-dioxane are illustrated in Figures 1 and 2, respectively.

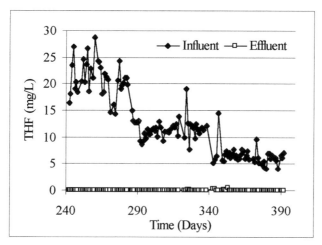

FIGURE 1. Influent and effluent concentrations of THF in the trickling filter.

THF loading began at the highest influent concentration (20 mg/L) at the beginning of experimentation. The average effluent THF concentration throughout reactor operation was 18 µg/L. This average value is deceptively high, however, due in large part to effluent concentration fluctuations resulting from frequent changes in loading rates. A more representative average concentration would be 1-3 µg/L, which is where effluent THF values stabilized 10-14 days after a loading change.

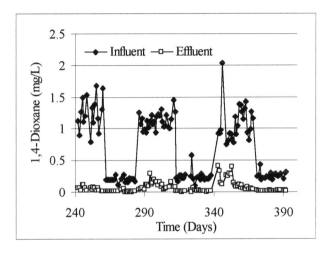

FIGURE 2. Influent and effluent concentrations of 1,4-dioxane in the trickling filter.

1,4-Dioxane loading began at an influent concentration of 1 mg/L, and was oscillated between this value and 0.2 mg/L. The trickling filter produced an average effluent 1,4-dioxane concentration of 65 µg/L for each loading rate. The extent of 1,4-dioxane removal, however, was dependent on the influent THF/1,4-dioxane ratio. Percent removals for 1,4-dioxane and THF at each loading rate are presented in Table 1.

It is noteworthy that optimum 1,4-dioxane removal occurred at the intermediate THF/1,4-dioxane ratios of 14-18. This suggests that a high influent concentration of THF may inhibit 1,4-dioxane biodegradation. Cometabolic substrates often competitively inhibit one another, which can explain the lower 1,4-dioxane removals at high ratios. Also, the decrease in 1,4-dioxane removal at lower ratios suggests that a THF:1,4-dioxane ratio of approximately 15 will be required to achieve high 1,4-dioxane removal.

TABLE 1. Percent removals of 1,4-dioxane and THF at various loading ratios.

THF/1,4-Dioxane Loading (mole/mole)	Percentage 1,4-Dioxane Removal	Percentage THF Removal
81.7	93.6%	99.96%
39.5	93.6%	99.96%
18.5	94.4%	99.96%
14.4	95.5%	99.86%
8.2	90.3%	99.12%
5.1	86.1%	99.95%

At the conclusion of the 1,4-dioxane testing, MTBE was added to the trickling filter influent at approximately 0.8 mg/L. Influent THF concentrations were raised back to 20 mg/L. The filter performed satisfactorily for the removal of MTBE as well. After three weeks of operation, effluent MTBE concentrations stabilized at approximately 27 µg/L. The formation of t-butyl alcohol (TBA), a biodegradation product of MTBE, was also monitored during this phase of research. TBA was present only in small concentrations (1-3 µg/L) in the effluent. Based on these results, it is reasonable to assume that the trickling filter can also be used to effectively biodegrade MTBE.

This research has shown that biological attached growth reactors can be effective tools to remediate ether-contaminated groundwater. The trickling filter sustained long-term biodegradation of low levels of 1,4-dioxane. Contaminated groundwater typically has low concentrations of organic chemicals, which could pose a problem for continuous-flow biological processes. Low concentrations of poor growth and energy substrates, such as 1,4-dioxane and MTBE, may not sustain growth in such a system. The addition of THF not only induced biodegradation activity of MTBE and 1,4-dioxane, but also continued the growth of the reactor biomass throughout operation.

REFERENCES

American Public Health Assocation, American Water Works Association, Water Environment Federation. 1995. "Standard Methods for the Examination of Water and Wastewater." 19th Edition.

Hanson, J.R., C.E. Ackerman, and K.M. Scow. 1999. "Biodegradation of Methyl *t*-butyl ether by a Bacterial Pure Culture." *Applied and Environmental Microbiology*. 65:4788-4792.

Leadbetter, E.R., and J.W. Foster. 1958. "Studies on Some Methane-Utilizing Bacteria." *Archiv fur Mikrobiologie*. 30:91-118.

Parales, R.E., J.E. Adamus, N. White, and H.D. May. 1994. "Degradation of 1,4-dioxane by an actinomycete in pure culture." *Applied and Environmental Microbiology*. 60:4527-4530.

Popoola, A.V. 1991. "Mechanism of the Reaction Involving the Formation of Dioxane Byproduct During the Production of Poly(ethylene terephthalate)." *Journal of Applied Polymer Science*. 43:1875-1877.

Salanitro, J.P., L.A. Diaz, M.P. Williams, and H.L. Wisniewski. 1994. "Isolation of a Bacterial Culture that Degrades Methyl *t*-butyl ether." *Applied and Environmental Microbiology*. 60: 2593-2596.

Sock, S.M. 1993. "A Comprehensive Evaluation of Biodegradation as a Treatment Alternative for the Removal of 1,4-Dioxane." M.S. Thesis, Clemson University, Clemson, SC.

Squillace, P.J., J.S. Zogorski, W.G. Wilber, and C.V. Price. 1996. "Preliminary Assessment of the Occurrence and Possible Sources of MTBE in Groundwater in the United States, 1993-1994." *Environmental Science and Technology*. 30: 1721-1730.

Toxics Release Inventory. 1997. Public Data Release. Office of Pollution Prevention and Toxics. United States Environmental Protection Agency, Washington, D.C.

PREDICTING ENHANCED BIOREMEDIATION PERFORMANCE AT THE SRS SANITARY LANDFILL

Neal D. Durant (HSI GeoTrans, Sterling, Virginia)
Philip A. Weeber (HSI GeoTrans, Roswell, Georgia)
Pete Andersen (HSI GeoTrans, Roswell, Georgia)
Dennis G. Jackson (Westinghouse Savannah River Co., Aiken, South Carolina)
Bryan J. Travis (Los Alamos National Laboratory, Los Alamos, New Mexico)

ABSTRACT: Horizontal well gaseous nutrient injection systems are being operated at the Savannah River Site Sanitary Landfill to promote the cometabolic (methanotrophic) biodegradation of trichloroethene (TCE) and the aerobic biodegradation of vinyl chloride (VC). Prior to system start-up, modeling was conducted with the numerical code TRAMPP to predict bioremediation performance and radius of influence, and to select monitoring point locations. TRAMPP has the capability to simulate a variety of processes in three dimensions, including solute air-water partitioning, Monod kinetics, cometabolism, rate-limited sorption, and nutrient limitation. Methanotrophic biodegradation of TCE is described by triple Monod kinetics, with TCE utilization depending on CH_4 and O_2 concentrations. TRAMPP also allows specification of pulsed injection schedules, a feature that allowed close approximation of the system at SRS. The model predicted that remediation of the TCE and VC plumes will require approximately 1 and 5 years, respectively. The model also predicted that the bioactive zones around each horizontal injection well have a 50 to 75 foot (15.2 to 22.9 m) radius. These results suggest that ground-water velocities will be the primary factor controlling plume remediation. Significant biotreatment should only occur when advection and dispersion bring the plume into contact with the bioactive zone surrounding the horizontal wells.

INTRODUCTION

Gaseous nutrient injection is being used to enhance the in situ biodegradation of trichloroethene (TCE) and vinyl chloride (VC) in groundwater at the Savannah River Site (SRS) Sanitary Landfill (SLF). Cometabolic (methanotrophic) biodegradation of a TCE plume is promoted by injecting air, CH_4, N_2O, and $TEPO_4$ (triethyl phosphate) via a horizontal well installed at the downgradient edge of the landfill. Enhanced aerobic biodegradation of VC in a separate plume is achieved by injecting air from a second horizontal well. Both injection systems are operated in a pulsed injection mode with the goal of maximizing biodegradation and minimizing volatilization.

These gaseous injection systems have been operating since August 1999. Prior to system start-up, numerical modeling was performed to predict the bioremediation performance efficiency and the radius of influence (ROI) for the given operating parameters. Monitoring well locations were selected based on the ROI predictions from the model. In this paper we describe the application of a numerical model for evaluating the performance of the gaseous nutrient injection systems at the SLF.

SITE DESCRIPTION

The SLF is a 70-acre (28.3 hectare) unlined landfill that received wastes from 1974 to 1994. Wastes disposed in the SLF include sanitary wastes, construction debris, paint cans, paint thinners, and rags soaked with TCE, tetrachloroethene, and other organic solvents. A RCRA cap installed on the landfill in 1997 is expected to reduce leachate by 99.9% (WSRC, 1992).

Groundwater Chemistry. Although a variety of chlorinated aliphatic compounds have been detected in the SLF ground water, TCE and VC are the most widespread and exhibit the greatest spatial continuity. Groundwater flow at the site is generally from north to south, and relatively distinct VC and TCE plumes have formed along the southwestern and southern edges of the SLF (Figure 1). These plumes developed prior to the installation of the cap. Several lines of evidence indicate that natural attenuation is occurring, including an apparent retardation of the plume, and the presence of TCE biodegradation products (dichloroethene, VC, and chloride) (Brigmon and Fliermans, 1997). Groundwater is generally aerobic in the vicinity of the TCE plume, and anaerobic in the vicinity of the VC plume.

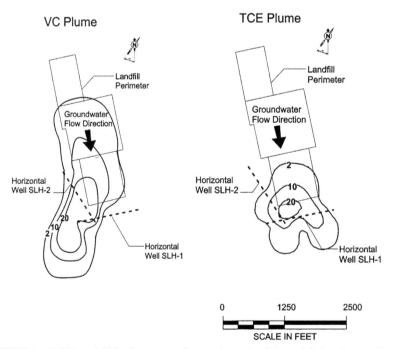

FIGURE 1. TCE and VC plume configurations (µg/L) and injection well locations.

Gaseous Nutrient Injection System. Pilot tests conducted previously at the SLF demonstrated the feasibility of this technology and allowed determination of design parameters (e.g., composition of bioremediation gases and injection rates) (WSRC, 1996; Altman et al. 1997). Horizontal wells for gaseous nutrient injection were installed after the completion of the pilot tests. Horizontal well SLH-1 is located within the TCE plume, and SLH-2 is located within the VC plume (see Figure 1). Both wells are installed approximately 60 feet (18.3 m) below ground surface in saturated aquifer material possessing a hydraulic conductivity of $> 10^{-5}$ m/s. The objective of the injection systems is to intercept and treat the plumes by enhancing the cometabolic biodegradation of TCE and the aerobic biooxidation of VC. The systems became operational in August 1999 (performance results are not yet available).

CONCEPTUAL MODEL

The conceptual model for each system consists of: 1) a ground-water flow field; 2) an injection schedule; and 3) a biodegradation kinetics component. Advection and dispersion transport the TCE and VC from underneath the landfill to the respective bioactive zones at each injection well. The rate of transport is a function of formation permeability and hydraulic gradient. Due to the low organic carbon content of the SLF aquifer sands (WSRC 1996), sorption of TCE and VC is assumed to be insignificant. Given the installation of a RCRA cap in 1997, it is assumed for the purposes of this modeling task that any TCE source has been eliminated. The initial TCE and VC mass in the plumes can be computed from the isopleths shown in Figure 1.

When the gas injection commences, bioremediation gases will gradually displace the water from the pore space in a zone overlying the injection wells. The migration of gas from the injection well can be described in terms of injection pressures (bubbling pressures and capillary pressures), densities of the injected gases, relative (air/water) permeabilities, formation porosity, and air/water partitioning. As injection proceeds, the sparged zone will grow until it reaches a dynamic steady state in which each subsequent pulse results in a sparged zone of the same size as the zone resulting from the previous pulse.

It is assumed that biodegradation of VC will proceed by double Monod kinetics, in which VC oxidation is coupled to biomass growth and O_2 consumption. Methanotrophic biodegradation of TCE is assumed to proceed by triple Monod kinetics, in which TCE degradation is coupled to biomass growth and O_2 and CH_4 consumption.

NUMERICAL MODEL

MODFLOW-96 simulations were used to establish the groundwater flow field in and around the SLF. The results of the calibrated flow model simulations were used to specify the flow field across the bioremediation system model domains. The simulated hydraulic gradient was 0.005.

Since the biochemistry of TCE and VC bioremediation systems is significantly different, and the TCE and VC plumes are spatially distinct, the two systems were modeled separately. Bioremediation modeling was accomplished using the Los Alamos National Laboratory (LANL) code TRAMPP. Travis and Rosenberg (1997) applied TRAMPP to model a TCE bioremediation demonstration at the SRS M Area. TRAMPP simulates 2-phase (air, water) flow and transport of contaminants in porous media, coupled with microbial growth and contaminant biodegradation. The governing equations are described in Travis and Rosenberg (1994; 1997). TRAMPP is also capable of simulating nutrient-limitation, product toxicity, and protozoa grazing, but these processes were not modeled at the SLF.

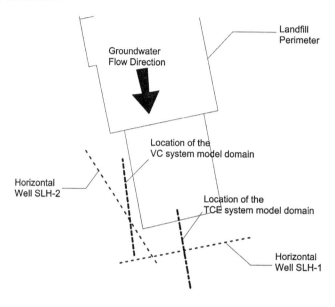

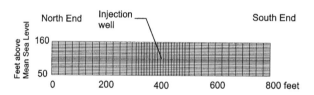

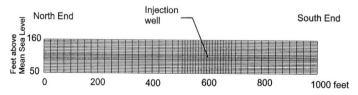

FIGURE 2. Location and dimensions of TCE and VC bioremediation system model domains.

The bioremediation model domains were constructed as 2-D cross-sectional planes that transect each of the horizontal injection wells (see Figure 2). Each model domain was oriented in-line with the hydraulic gradient. Vertically, each model domain included the entire sequence of saturated and unsaturated sediments overlying the injection well, and extended down to the top of an underlying confining unit. The top boundary of the domain was open to the atmosphere, except for the northern most 100 ft (30.5 m) of the TCE cross-sectional profile that was simulated as capped. The domain consisted of one cell in the direction parallel to the injection well. A cell width of 9 ft (2.7 m) was assigned in the model, representing 1% of the 900-foot (274 m) length of the actual injection well. As such, only 1% of the anticipated air flow was specified.

The initial O_2 concentration in the VC and TCE plumes was set at 0 and 2 mg/L, respectively. Initial methanotroph cell concentrations in groundwater at the SLF are on the order of 10^3 cells/mL (Altman et al., 1997). Injection, hydrogeologic, and biokinetic input parameters are summarized in Table 1.

RESULTS AND DISCUSSION

The results of the TRAMPP simulations of the gaseous nutrient injection systems are depicted in Figure 3. The model predicted that TCE upgradient of the injection well will be remediated within less than a year. This rapid cleanup time is a reflection of effective CH_4 and O_2 distribution from the injection well, efficient biodegradation performance by the indigenous bacteria, and the relatively small mass of TCE occurring upgradient of the injection well. After 300 days of operation, the simulated ROI ranged from 50 to 70 feet (15.2 to 22.9 m), with the higher ROI occurring at shallower depths. The model also simulated CH_4 and O_2 distributions around the injection well (results not shown). Sensitivity analysis indicated that order of magnitude variations in K_{CH4} and k_{CH4} each could affect predicted efficiency of the TCE bioremediation system by 15%.

The model predicted significantly slower remediation of the VC plume than the TCE plume. This result is not surprising given that the model specified a higher half saturation concentration for VC than for CH_4, and a lower maximum substrate utilization rate and growth yield for VC than for CH_4. In addition, the initial mass of VC upgradient of SLH-2 exceeded the initial specified mass of TCE upgradient of SLH-1. Modeling predicted that remediation of the VC plume will require approximately 5 years (data not shown).

For both TCE and VC bioremediation systems, the length of time predicted for remediation was proportional to the mass of contaminant occurring upgradient of the injection well at the time of system start-up. As such, the value of these predictions strongly depends on the spatial characterization of the plume and the initial contaminant mass specified in the model domain. A paucity of monitoring wells in the landfill limited our ability to clearly define the northern-most extent of the TCE plume. Model predictions may be improved by adjustment of the initial contaminant distribution in the model domain. In 2000, performance data at the injection systems will be collected and the model will be calibrated against TCE and VC measurements.

TABLE 1. Operating, hydrogeologic, and biokinetic parameter values used in bioremediation simulations.

Injection Schedules							
System	On Interval	Off Interval	Rate (scfm)	Air (%)	CH_4 (%)	N_2O (%)	$TEPO_4$ (%)
TCE	8 hr/day every other day for 7 days	7 days	240	98	1	0.07	0.007
VC	8 hr/day every other day for 7 days	7 days	270	100	-	-	-

Hydrogeologic parameters affecting air/water relative permeabilities[A]	
Initial Saturation (%)	27
Bubbling Pressure (kPa)	10
pore-size distribution index (0-1)	0.62
Irreducible saturation (%)	25

Chemical Parameters (19°C)[A]				
	VC	TCE	CH_4	O_2
Henry's Law coefficient (dimensionless)	0.87^B	0.28^B	27	29
Water Diffusivity (D_w) (cm²/s)	$8.5^{-6\ C}$	10^{-5}	10^{-5}	10^{-5}
Air Diffusivity (D_a) (cm²/s)	0.11^D	0.1	0.23	0.14

Biokinetic Parameters[A]		
Parameter	Symbol	Value
Initial bacteria concentration	M_{TCE} M_{VC}	1×10^{-12} g/mL 1×10^{-9} g/mL [E]
Background microbial growth rate	k_c	3×10^{-7} g cells/g substrate/s
Microbial half saturation Concentration	K_{TCE} K_{CH4} K_{VC} K_{O2}	0.1 mg/L 0.4 mg/L 0.79 mg/L [F] 1.0 mg/L
Max. rate of substrate Utilization	K_{TCE} K_{CH4} k_{VC}	5×10^{-5} g TCE/g cells/s 3×10^{-5} g CH_4/g cells/s 1.4×10^{-5} g VC/g cells/s [F, G]
Microbial growth yield	Y_{CH4} Y_{VC}	0.5 g cells/g CH_4 0.1 g cells/g VC [G]
Mass ratio of O_2 consumed per mass substrate consumed	F_{TCE} F_{CH4} F_{VC}	0.3 g O_2/g TCE 4.0 g O_2/g CH_4 1.3 g O_2/g VC
Microbial decay rate	k_d	1.5×10^{-6}/s

Notes: **A.** Values from Travis and Rosenberg (1997), unless noted otherwise. **B.** Gossett, 1987; **C.** Hayduk and Laudie, 1974; **D.** Fuller et al., 1966; **E.** Norland et al., 1987; **F.** Bradley and Chapelle, 1998; **G.** Clement et al., 1998.

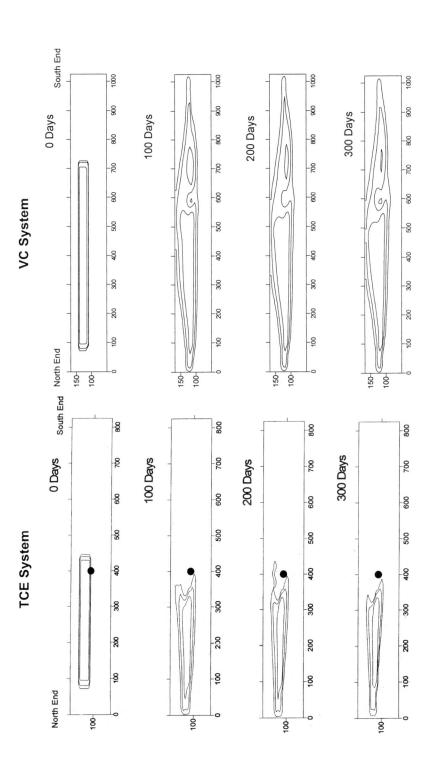

FIGURE 3. Predicted TCE and VC bioremediation performance in SLF injection systems. Contour intervals of 1, 5, and 25 µg/L. In VC domain, injection well is located at x=600, z=109. In TCE domain injection well is located at x=400, z=109.

REFERENCES

Altman, D.J., C.J. Berry, A. Bourquin, R. Brigmon, M.M. Franck, T.C. Hazen, D. Mosteller, and F.A. Washburn. 1997. *Sanitary Landfill Supplemental Test Final Report*. Westinghouse Savannah River Company Report, WSRC-RP-97-17.

Bradley, P.M., and F.H. Chapelle. 1998. "Effect of Contaminant Concentration on Aerobic Microbial Mineralization of DCE and VC in Stream-Bed Sediments." *Environmental Science and Technology*. *32*(5): 553-557.

Brigmon, R.L., and C.B. Fliermans. 1997. *Intrinsic Bioremediation of Landfills Interim Report*. Westinghouse Savannah River Company Report, WSRC-RP-97-323.

Clement, T.P., Y. Sun, B.S. Hooker, and J.N. Petersen. 1998. "Modeling Multispecies Reactive Transport in Ground Water." *Groundwater Monitoring Review and Remediation*. Spring:79-92.

Fuller, E.N., P.D. Schettler, and J.C. Giddings. 1966. "A New Method for Prediction of Binary Gas-Phase Diffusion Coefficients." *Industrial Engineering Chemistry*. *58*(5):18-27.

Gossett, J.M. 1987. "Measurement of Henry's Law Constants for C_1 and C_2 Chlorinated Hydrocarbons." *Environmental Science and Technology*. *21*(2): 202-208.

Hayduk, W., and H. Laudie. 1974. "Prediction of Diffusion Coefficients for Non-Electrolytes in Dilute Aqueous Solutions." *AIChE Journal*. *20*:611-615.

Norland, S., M. Heldal, and O. Tumyr. 1987. "On the Relation Between Dry Matter and Volume of Bacteria." *Microbial Ecology*. *13*:95-101.

Travis, B.J., and N.D. Rosenberg. 1994. *Numerical Simulations in Support of the In Situ Bioremediation Demonstration at Savannah River*. Los Alamos National Laboratory Report, LA-127890-MS, UC-940.

Travis, B.J., and N.D. Rosenberg. 1997. "Modeling In Situ Bioremediation of TCE at Savannah River: Effects of Product Toxicity and Microbial Interactions on TCE Degradation." *Environmental Science and Technology*. *31*:3093-3102.

WSRC. 1992. *HELP Modeling of the Sanitary Landfill Cap (U)*. Westinghouse Savannah River Company Inter-Office Memorandum, 350:EPD-SE-92-1282.

WSRC. 1996. *Sanitary Landfill In Situ Bioremediation Optimization Test Final Report*. Westinghouse Savannah River Company Report, WSRC-TR-96-0065.

ENHANCED BIOREMEDIATION OF SOLVENTS, ACETONE, AND ISOPROPANOL IN BEDROCK GROUNDWATER

Gregory L. Carter (Earth Tech, Roanoke, Virginia) Tiffany Dalton, Jennifer C. Vincent and Barbara B. Lemos (Earth Tech, Concord, Massachusetts) Rosann Kryczkowski (ITT Night Vision, Roanoke, Virginia)

ABSTRACT: Site investigations at a manufacturing facility in Virginia delineated soil and groundwater contamination by chlorinated solvents (trichloroethylene (TCE); 1,1,1-trichloroethane (TCA); and associated breakdown products) as well as acetone and isopropanol. Chemical and microbiological sampling over time showed that natural biodegradation of these compounds was occurring at a limited rate. In-situ enhancement of cometabolic bioremediation through the injection of air, gaseous-phase nutrients, and methane was undertaken to stimulate existing microbial populations to promote and accelerate contaminant breakdown.

A key factor in the selection of this remedy was the need to avoid interference with manufacturing operations. Natural subsurface challenges included groundwater flow in a clay overburden and underlying fractured interbedded shale and limestone. Man-made challenges included extensive subsurface utilities and above ground structures essential to plant operations.

Under the facility's RCRA Corrective Action program, an Interim Measure (IM) was initiated in one of three contaminant source areas as a full-scale enhanced bioremediation field test. The design of the injection system was performed in consultation with technology developers from the Department of Energy's Savannah River Site. EPA and state agency approval was secured in early 1997, and system start-up occurred in March 1998. This field test is also being reviewed by the EPA Superfund Innovative Technology Evaluation (SITE) program.

The system features injection of oxygen, gaseous-phase nutrients (nitrous oxide and triethyl phosphate), and carbon source (methane). Baseline and periodic monitoring has included testing of soil gas for contaminant concentrations and indicator gases such as methane, carbon dioxide, and oxygen. Groundwater samples have been analyzed for site contaminants (and associated daughter products), microbial populations, and groundwater quality parameters.

Monitoring of the injection zone of influence has shown that the enhanced bioremediation system has been successful in stimulating microbial growth based on increases in phospholipid fatty acid (PLFA) biomass measurements of several orders of magnitude within four months of system start-up. The results from groundwater monitoring indicate significant (~60 to 99.96%) total volatile organic compound (VOC) reductions in the pilot test area and down gradient monitoring locations since the initiation of the injection campaign. This indicates that the bioremediation process is very effective treating dissolved phase contaminants.

Based on these observations, the system has recently been expanded to full scale application to increase the delivery of the necessary amendments to complete the site restoration process in the source area.

INTRODUCTION

An in-situ groundwater bioremediation project was pilot tested at the ITT Industries Night Vision plant located in Roanoke, Virginia. The facility manufactures Night Vision devices for both government and commercial customers.

The pilot test area is located adjacent to one of the manufacturing facility buildings. When evaluating the technology options, particular emphasis was placed on treatment technologies that could be applied in-situ given the site restrictions. After review of several technologies, in-situ enhanced bioremediation was selected as the technology best suited to the contaminants, geology, and logistical factors present at this site. The basis of the chosen technology, developed at the Westinghouse Savannah River Plant site (Hazen, 1995) and licensed by the U. S. Department of Energy, is an injection system used to deliver a gaseous phase mixture of air, nutrients (nitrous oxide and triethyl phosphate), and methane to the targeted subsurface area to stimulate the growth of methanotrophs. These bacteria produce enzymes (methane monooxygenase) which degrade the chlorinated solvents and their daughter products to non-hazardous constituents.

PROJECT OBJECTIVE

The purpose of this pilot test, which was implemented as a RCRA Interim Measure (IM), was to observe the effectiveness of the system in reducing VOC concentrations in groundwater in the pilot test area. The degree of effectiveness of the pilot study would determine whether this technology would be expanded in this source area and potential application at numerous other sites with similar conditions.

SITE DESCRIPTION

This facility is a Night Vision devices manufacturing complex consisting of three major buildings and associated infrastructure located on 26.7 acres of land. The facility is situated on a small knoll, with the surrounding topography characterized by small hills and valleys and intermittent and continuous streams.

The contaminants of concern in groundwater at the Night Vision facility consist of chlorinated solvents including TCE, TCA and associated daughter products (cis 1,2 dichloroethene, vinyl chloride, 1,1 dichloroethane, and chloroethane), as well as acetone and isopropanol. Remediation is complicated by the target VOCs occurring in the clay and fractured bedrock. The VOC releases entered the very low hydraulic conductivity (10^{-4} to 10^{-7} cm/sec) clay overburden then migrated to the underlying bedrock. Groundwater flow and VOC migration primarily occurs via fractures in the bedrock, including high hydraulic conductivity (10^{-2} to 10^{-4} cm/sec) fractures. Groundwater is typically encountered at 5 to 15 ft (1.5 to 4.6 m) below ground surface (BGS) in this area.

METHODS

The primary components of the pilot test system consist of: an injection well; five monitoring wells; four soil gas monitoring points; and the air, nutrient and methane injection equipment (Figure 1). The monitoring wells were constructed to intercept the water-bearing fracture zone encountered between 40 and 50 ft (12.2 and 15.2 m) BGS and the shallower 15 to 30 ft (4.5 to 9.1 m) BGS water-bearing fracture zones. The injection well was constructed of 1-inch diameter steel to be compatible with the injection media. The soil gas points were installed with 2-inch diameter PVC intercepting the overburden and transition zones.

The injection system, comprised of the air, nutrient, and methane injection equipment, is housed in a temporary building located in the pilot test area utilizing an existing concrete pad. The air source is an existing facility air compressor which includes a condensate tank with a drain, an air line, coalescing filters and pressure regulator and valves. The methane and nitrous oxide are provided in standard cylinders and are piped into the main air line using regulators and flow meters. The methane is stored in an attached shed and is piped into the main line through appropriate meters and regulators.

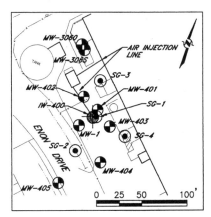

Figure 1 Site Map

The triethyl phosphate (TEP) is in liquid state stored in a steel tank. Air from the main line is diverted through the tank to volatilize the TEP for subsurface delivery. The air, nitrous oxide, and TEP are injected continuously while the methane is injected on a pulsed schedule of 8 hours per day, 5 days per week. The methane is closely monitored to ensure that the injection concentration does not exceed 4% by volume, thus avoiding the methane lower explosive limit of 5%.

The injection campaign was implemented in a phased approach to determine the optimum system operating conditions. Initially, only air injection was performed for 6 weeks until the groundwater data indicated limited nutrient availability. An air and nutrient injection phase was performed for 10 weeks, during which a depletion of methane was observed. When methane levels became limited, methane was added to the injection stream to maintain an adequate amount of cometabolic carbon source, air and nutrients.

Periodic headspace field screening was performed on the soil gas points and IM monitoring wells for the presence of methane, carbon dioxide, and oxygen. In addition, pressure readings at the monitoring points were made using magnehelic gauges. The soil gas was periodically sampled for laboratory analyses of VOCs, methane, and carbon dioxide by the USEPA as part of the Superfund Innovative Technology Evaluation (SITE) Program.

The treatment system's effectiveness was monitored throughout the pilot test through low-flow groundwater sampling of the IM wells. Groundwater sampling events were performed pre-injection, post air-only injection, post air/nutrient injection, and at different air/nutrient/methane injection periods to assess the performance of each of the injection phases and monitor VOC changes over time. Groundwater samples were analyzed for VOCs, methanotroph most probable number (MPN) counts, phospholipid fatty acids (PLFA) total biomass, soluble methane monooxygenase (sMMO) by DNA testing, and general water quality parameters such as nutrients and metals. Additionally, field measurements were made to monitor dissolved oxygen (DO), reduction/oxidation potential (redox), pH, conductivity, temperature, and depth to groundwater prior to sample collection. These field parameters were also monitored during injection periods in several key wells.

RESULTS AND DISCUSSION

Microbial Analysis. Baseline microbial sampling results indicated that a diverse relatively limited microbial community existed at the site. The target methanotrophic population was a minority of the total microbial community at the start of the injection process. Following the addition of nutrients and methane, microbial data indicated that the methanotrophs had increased by 1 to 4 orders of magnitude as shown on Table 1. The total biomass increased as well by 1 to 3 orders of magnitude. Significant increases in biomass and methanotrophs were not observed in the IM wells until the post-nutrient sampling event. The DNA analysis of methanotrophic populations targeted the identification of soluble methane monooxygenase (sMMO) which if present provides further support that the necessary cometabolic components are present for bioremediation to have occurred. The DNA results indicated that sMMO during the baseline sampling was either not detected or slightly above the detection limits at 10^2. Following the nutrient and methane injection, the sMMO relative abundance was reported at 10^5 to 10^7. This further indicates that the methanotrophic population was significantly increased during the injection campaign.

TABLE 1. Microbial Population Data for the baseline sampling events and injection campaign sampling events.

Well Number	Baseline Average PLFA Biomass (PLFA/ml filtered water)	Injection Average PLFA Biomass (PLFA/ml filtered water)	Baseline Average Methanotroph (MPN)	Injection Average Methanotroph (MPN)
MW-306S	5.1	413	26	171

MW-1	1.4	162	286	3,013
IW-400	0.12	62	26	10,430
MW-401	0.29	27	48	2,610
MW-402	2.7	7.8	1,124	7,533
MW-403	0.34	19	286	7,335,000

VOC Analysis. Throughout the pilot test, decreases in VOC concentrations, as compared with the lowest baseline data, have been observed in the majority of the area wells. VOC reductions have been observed in monitoring wells 75 feet (22.8 m) from the injection well. Specifically, MW-306S located in the source area and MW-405 which is located down hydraulic gradient from the injection well. Laboratory analyses of groundwater samples collected from multiple monitoring wells during the pilot test indicate a decrease in total VOCs over time as shown on Figure 2. The most recent laboratory results for groundwater indicate the maximum observed decrease in total VOCs to be 99.96 % for MW-403 which is located 20 ft (6.1 m) from the injection well. In MW-403, acetone and isopropanol concentrations have reduced from 22,000 and 40,000 ppb respectively to below detection limits. Vinyl chloride, chloroethane, 1,1-dichloroethane and 1,1-dichloroethene concentrations were also reduced an estimated 90 to 98% in this well.

The minimum VOC reductions in the affected pilot test area were observed in MW-402 which is located 25 ft (7.6 m) up hydraulic gradient from the injection well. For MW-402, acetone and isopropanol concentrations were reduced by approximately 50%. The chlorinated VOC concentration reductions ranged from no change up to 62 %. This is attributed to a lack of air, nutrient, and methane delivery to this area created by the heterogeneous conditions of the subsurface. Figure 2 shows the total VOC changes over time for MW-1, MW-401, MW-403 and MW-405. The VOC changes for these monitoring wells are representative of VOC reductions across the pilot test area.

The parent chlorinated VOCs (TCE and 1,1,1 TCA) are sporadically detected in the site groundwater; therefore, it is difficult to ascertain the impact of the bioremediation system on these chemicals. This is in part due to the elevated detection levels created by high concentrations of acetone and isopropanol during baseline sampling events. Minor VOC reductions have been observed with the limited available data.

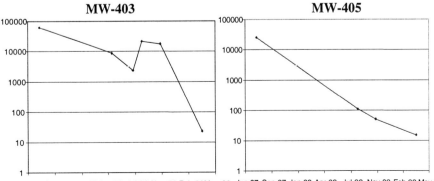

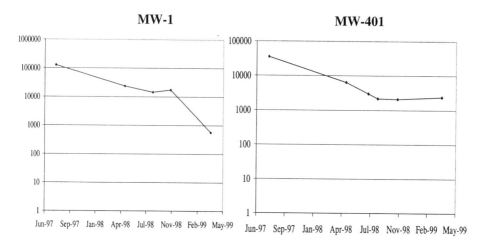

FIGURE 2 Total VOC concentrations in ug/l over time

Soil gas sampling and analysis was performed throughout the pilot testing to evaluate whether VOCs were collecting in the vadose zone as by-products of the air injection process. The VOC concentrations did not show a statistically significant increase of VOCs in the soil gas samples following the initiation of the injection system. Based on the increased microbial populations coupled with the decreases in VOCs, the injection of air, nutrients and methane appears to have enhanced bioremediation at this site.

CONCLUSIONS

Gaseous phase injection of air, nutrients, and methane in a clay and fractured rock system have significantly reduced the VOC concentrations in groundwater. Decreased VOC concentrations of up to 99.96% were observed in surrounding monitoring wells within one year of system operation. The treatment system affected monitoring wells 75 ft from the injection well with VOC reductions on the order of 99.9% in down hydraulic gradient monitoring wells.

Since completion of the pilot test, the enhanced bioremediation system has been expanded to full scale application to increase the delivery of the necessary amendments to complete the site restoration process in the source area.

ACKNOWLEDGEMENTS

This pilot test was performed in conjunction with the USEPA Superfund Innovative Technology Evaluation (SITE) program. For more information about the SITE program the reader is referred to www.epa.gov/ORD/SITE. We greatly appreciate their participation in the project and cooperative efforts regarding data and ideas. The majority of the microbial data and soil gas data were collected as part of the SITE Program and will be available through the USEPA SITE program at a later time.

REFERENCES

Hazen, T. C. 1995. *Preliminary Technology Report for the In Situ Bioremediation Demonstration (Methane Biostimulation) of the Savannah River Integrated Demonstration Project, DOE/OTD* $^{(u)}$. U. S. Department of Energy Report, WSRC-TR-93-670, Westinghouse Savannah River Company, Aiken, South Carolina.

A DESIGNING METHOD FOR IN-SITU GROUNDWATER BIOREMEDIATION THROUGH PRECISE EVALUATION OF TRANSPORT CHARACTERISTICS

Mitsutoshi Nakamura, Tatsushi Kawai and Jun'ichi Kawabata (Kajima Technical Research Institute, Tokyo, JAPAN)

INTRODUCTION

There are growing number of cases where groundwater solute transport phenomena are evaluated quantitatively so as to obtain information on environmental impact in surrounding area of contaminated ground and on remediation works. In reality, however, it is still difficult to give proper values of transport parameters in actual analyses such as dispersivity, retardation coefficient and so on. Quantitative evaluation of those transport parameters is important especially in designing of bioremediation system for contaminated groundwater, which affect significantly its system structure.

The authors proposed a designing method of bioremediation system for contaminated groundwater by trichloroethylene (TCE) with using solute transport calculation. Transport calculation is performed for the purpose of obtaining a proper mixing condition of additives (oxygen and nutrients), which allow effective cometabolic degradation of TCE of the in-situ biostimulation. Each solute of substance is artificially injected through wells respectively in different points or intermittently and alternately at a same point. In those calculations, transport parameters are determined by the results of in-situ tracer tests and laboratory column tests. The calculation results shows that proper quantitative evaluation of transport parameters is critically important to design an effective and efficient system. An effective remediation system is also proposed though this transport calculation study at an actual site in Japan.

OVERVIEW OF DEMONSTRATION TEST

Application to remediation system design. An aquifer to be treated was an unconfined aquifer (permeability coefficient is about $2.1*10^{-3}$cm/s) composed of diluvial fine sand (d_{50}=0.019cm). The remediation process considered here is biostimulation, which is one of in-situ bioremediation methods. Biostimulation is a technique for cleaning up a contaminated aquifer by making a condition in aquifers that should stimulate activities of microorganisms. The condition is created by injecting nutrients and oxygen into an aquifer and by mixing them with the groundwater at proper concentrations. The concept of this groundwater remediation system is illustrated in Figure 1. The remediation system is based on a groundwater circulation in which contaminated water is pumped up from well and mixed with the nutrients and oxygen, and then water is re-injected through injection wells. Biodegradation of TCE using methanotrophs has been demonstrated by many investigators. Some methanotrophs can degrade TCE aerobically with cometabolism, because their monooxygenases have a wide range of degradation activities. Since the remediation system uses methanotrophs, methane is actually used as a nutrient in this study.

Study procedure. The designing process for the remediation is shown in Figure 2. At first, an in-situ tracer test and a laboratory column test were conducted to investigate solute transport characteristics in the aquifer of an actual remediation site. Then, the tests were followed by modeling analysis. Finally, the remediation system was designed by a solute transport analysis.

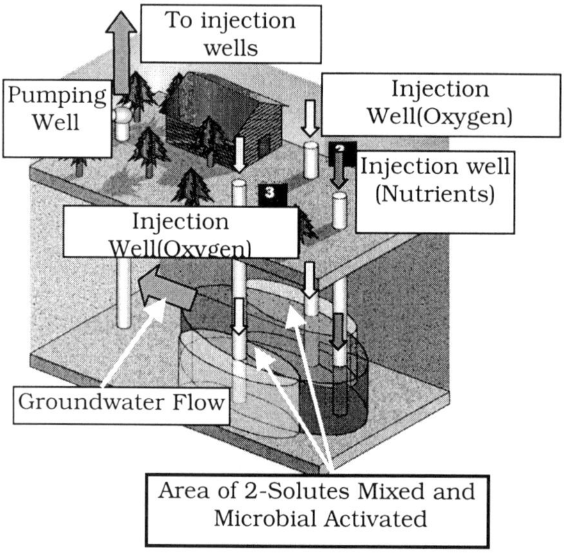

Fig.1. Example of the in-situ bioremediation 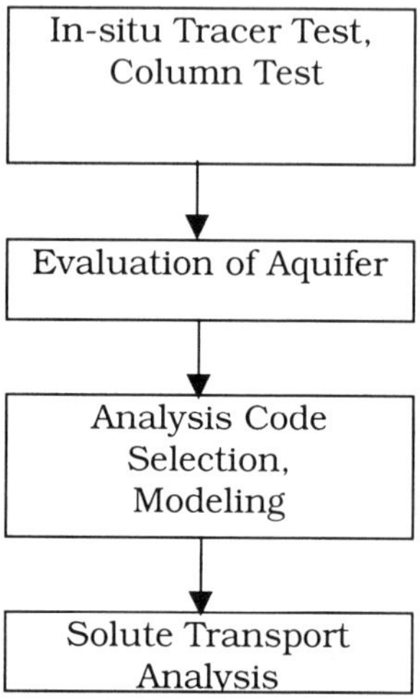   **Fig. 2. Designing process**

EVALUATION OF SOLUTE TRANSPORT CHARACTERISTUCS
Evaluation based on laboratory column test. The laboratory column test was conducted using solutes that should be actually injected in an aquifer (TCE, oxygen). The purpose of this test is to determine dispersivity and retardation coefficients that should be applied in the designing analysis. Dispersivity expresses dispersion characteristics.

A retardation coefficient is an indicator of adsorption and desorption characteristics of substance on soil particles in an aquifer. Both of them are major factors that should affect results of solute transport analysis.

In this column test, sand was taken from the site where in-situ tracer test was performed. Sand was packed in a column with the same density as the targeted aquifer. The outline of the test is shown in Figure 3. The test conditions are shown in Table 1. Chlorine ions (which were used in the in-situ tracer test), TCE and oxygen (which were used for in-situ bioremediation), were used as tracers. Solute transport characteristics were evaluated by examining breakthrough curves observed at the downstream end of the column.

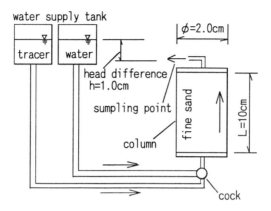

Fig. 3. Outline of the laboratory column test

Table 1. Conditions of column test

Case of Test	Case1-1	Case1-2
Initial Concentration of Each Tracer(ppm)	Cl : 1000 O_2 : 40	Cl : 100 TCE : 50
Method of Measurement	Cl : EC probe O_2 : DO probe	Cl:Ion-exchange Chromatography TCE : GC-MS

Figure 4 shows the test results. Breakthrough curves that indicate tracer arrivals is clearly observed at sampling point (Case1-1). The breakthrough curve of oxygen seems to be almost the same as the chlorine ion's curve. This result shows that the transport of oxygen was not retarded in this column which modelizes the targeted aquifer. The results of in-situ tracer test that was previously reported also indicate that oxygen has transport characteristics similar to those of conservative solutes (Robert.S. 1989). The dispersivities of oxygen and chlorine ions should be determined from the breakthrough curves of Figure 4.

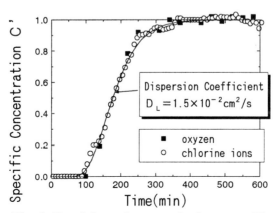

Fig. 4. Breakthrough curve of column test(Case1-1)

Figure 5 shows the test results. Breakthrough curves that indicate tracer arrivals are clearly observed in the sampling point (Case1-2). The TCE concentrations change with time is (retardation coefficient is about 1.1) more or less similar to the results of chlorine ions as a conservative tracer. This means low level of adsorption occurs in this column.

Retardation of organic chemical compound such as TCE are thought to be due largely to the organic content of the soil (William J.Dentsch . 1997). In these tests, the values of retardation factor close to 1.0 were obtained for TCE. This is mainly due to the low organic content of the sand in the aquifer. The dispersivities of TCE and chlorine ions obtained from the breakthrough curves shown in Figure 5 are very close.

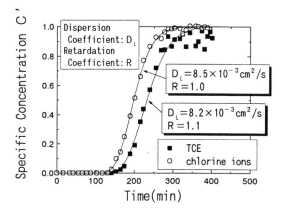

Fig.5 Breakthrough curve of column test(Case1-2)

Evaluation based on in-situ tracer test. An in-situ tracer test was conducted using the additives (methane solution, oxygen solution) to be used in an actual groundwater in-situ bioremediation system.

Advective-dispersive analysis was performed to estimate dispersivity from the test results. The numerical code used here was based on the upwind finite element method. The same analysis code was used in the remediation system study described in following chapter.

In the tracer test, automatic pumping and injecting system was used, which is able to keep a constant injecting tracer flow rate. Figure 6 illustrates the concept of the system. The test procedure was as follows. At first, groundwater was pumped up from pumping wells at a rate of Q=6.7 L/min (which is 80 % of the maximum pumping capacity for the well) to create an artificial groundwater flow field. Then, when the groundwater level stabilized, 100 ppm NaCl solution was injected continuously into the injection wells at a rate of 1.0 L/min. Finally, electrical conductivity is measured with time at the observation well located downstream of the injection well.

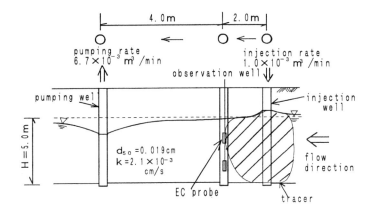

Fig. 6. Outline of the in-situ tracer test

Figure 7 shows the test results. The time required for the specific concentration (the ratio assuming the concentration at a time of injection as 1.0) of tracer to become 0.5 was used to calculate actual flow velocity and effective porosity. Two-dimensional advective-dispersive analysis was conducted to determine dispersivities that fit the breakthrough curves. Calculated values of effective porosity obtained from breakthrough curves were typical value of fine sand's aquifer (around 30%). Longitudinal dispersivity α_L (an indicator of dispersion in the direction of flow) were about 0.05 m.

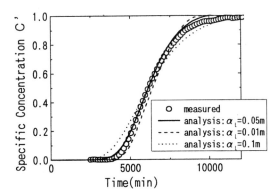

Fig. 7 Breakthrough curve of tracer test

Calculation of solute transport characteristics. Figure 8 shows a relationship between the dispersivities obtained from this in-situ tracer test, previously conducted in-situ tracer tests and laboratory column tests, and the observation scales L (the distance that the tracer travels). Nueman's equation was also shown in this figure (Xu and Eckstein. 1995). The plots relatively correspond well to Nueman's equation in this observation scale. It can be assumed from this result that the longitudinal dispersivity α_L to be used in solute transport analysis in designing should be about 0.05 m since the groundwater in-situ bioremediation system involves distances of several meters,

The dispersivity values(of the order of 10^{-4} m) obtained from the laboratory column test in this study are not only close to the values of previous column tests with similar scale, but also are reasonable if they are compared with the tracer test data from a point of scale effect view that Nueman showed. From these results, it is suggested that the retardation factors obtained from column tests can be applied for an in-situ transport modeling. Hence, it may be reasonably expected that in-situ injection of TCE and oxygen solutions will not result in retardation.

It is possible to sat that a methane solution will be transported in a manner that is similar to the way an oxygen transport in this aquifer, as it is already shown that characteristics of a methane solution and an oxygen solution on transport are almost the same as those of conservative solutes (Robert.S. 1989).

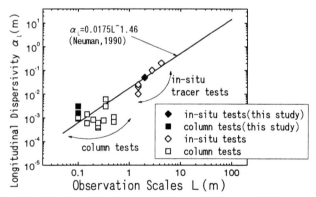

Fig. 8. Relationship between Dispersivities and Observation Scales

GROUNDWATER REMEDIATION SYSTEM

Analysis model. Solute transport analysis was conducted in order to investigate the conditions under which a methane solution and an oxygen solution are to be well mixed, with using characteristic values defined on the basis of the solute transport evaluation in preceding chapter. The parameters of the analysis are shown in Table 2.

Table 2. The parameters of the analysis

Permeability(cm/s)	$2.0*10^{-3}$
Static Head(m)	5.0
Radius of Influence (m)	10
Longitudinal Dispersivity (m)	0.05
Transverse Dispersivity (m)	0.005
Effective Porosity (%)	30
Pumping Rate(m^3/min)	$1.5*10^{-3}$

Mixing region. It has been reported that rapid multiplying of microorganisms would cause clogging and serious lowering of permeability at an injection well and its vicinity zone if a methane solution and an oxygen solution are injected simultaneously from a single well (John T.Cookson,JR . 1995). Continuous injection of each solute of substance from plural wells respectively or intermittent injection from a single well is methods of solution to prevent such phenomenon. It is suggested that mixing by using transverse dispersion along a streamline would not be enough for efficient mixing based on dispersivity values that we got from the investigation shown above (Kawabata et al. 1998). Intermittent injection from a single well is, therefore, applied as an injection method for the targeted area. Solute transport analysis was conducted in order to get a optimum intervals and flow rate, which should be effective for efficient mixing. The outline of the groundwater remediation system is shown in Figure 9. The condition of injection intervals applied for the analysis are shown in Table 3. One injection cycle consists of the injection of methane, oxygen, and water in this order.

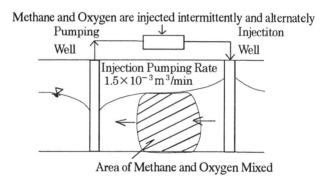

Fig.9. Outline of the remediation system

Table 3 Designing conditions

Case	Injection Time(min)			
	Methane Solution	Water	Oxygen Solution	Water
Case2-1	500	500	500	500
Case2-2	500	250	500	250
Case2-3	250	250	250	250

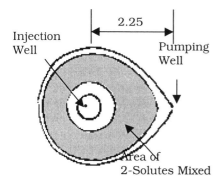

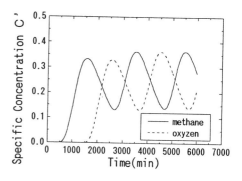

Fig. 10 Result of analysis(Case2-1) Fig. 11 Breakthrough curve (Case2-1)

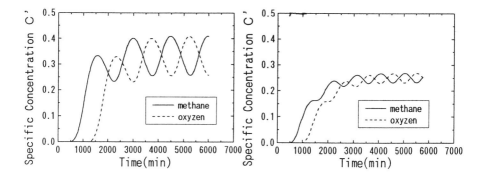

Fig. 12 Breakthrough curve(Case2-2) Fig. 13 Breakthrough curve (Case2-3)

Figure 10 shows a result of solute transport analysis. The Grey area shows a mixing zone of methane and oxygen when injected solution reaches the pumping well. Figure 11~13 shows changes with time of methane and oxygen concentrations at the observation well, which shows the two solutions in Case 2-2 tend to be mixed together with higher concentrations than in Case 2-1. It also shows that the concentration of both methane and oxygen in Case 2-3 is smaller than that in Case 2-1. These results indicate that the analysis model is capable of predicting a location of mixing zone with required values of each substance's concentration. Hence, it is possible to decide the required intervals among methane, oxygen and water injection and required concentration of each substance when injecting, from analysis shown here.

Conclusions

In this study, the in-situ tracer test indicated a longitudinal dispersivity of about 0.05 m, and the laboratory column test indicated that both oxygen and TCE are transported in aquifer with the same manner of typical conservative solutes. The mixing of additives were quantitatively evaluated by using a solute transport analysis on the basis of precise evaluation of transport characteristics. This analysis makes it possible to predict an optimum intermittent injection time and required concentration of each substance to be injected, which can create a proper condition to resolve TCE in groundwater efficiently. This method, therefore, should be useful in designing a biostimulation system for groundwater contaminated with TCE.

A part of test data and analytical results in this paper were collected and produced in the Industry's soil remediation technology development project implemented by the Research Institute of Innovative Technology for the Earth (RITE).This project is under a contract with the New Energy and Industrial Technology Development Organization (NEDO) as a part of the Ministry of International Trade of Japanese government.

REFERENCE

John T.Cookson,JR . 1995. Bioremediation Engineering. McGraw-Hill,Inc. 272-273.

Kawabata, Kawai and Nakamura . 1998. Designing Method using Transport Analysis for In-situ Bioremediation of Groundwater. Annual Report of Kajima Technical Research Institute. vol.46.

Kawabata, Kawai and Nakamura . 1997. Method of Evaluating Groundwater Flow Transport Characteristics. Annual Report of Kajima Technical Research Institute. vol.45.

Robert.S. 1989. In-situ Aquifer Restoration of Chlorinated Aliphatics by Methanotrophic Bacteria.. 172-196.

William J.Dentsch . 1997. Groundwater Geochemistry. Lewis Publishers. 54-55.
Xu and Eckstein . 1995. Use of least-Squares Method in Evaluation of Relationship between Dispersivity and Field Scale. Groundwater. vol.33. No.6. 905-908.

COMETABOLID DEGRADATION OF CHLORINATED SOLVENT CONTAMINANTS BY ACTINOMYCETES

Seung-Bong Lee, Stuart E. Strand and H. David Stensel
(University of Washington, Seattle, WA)

ABSTRACT: A phenol-fed actinomycetes bacterial enrichment has demonstrated sustained trichloroethylene (TCE) degradation ability in the gas treatment reactor. A pure culture was isolated from the enrichment and tested for the degradation cability of TCE, cis-dichloroethylene (*cis*-DCE) and vinyl chloride (VC). The culture could degrade TCE, *cis*-DCE and VC. The degradation kinetics in order of highest rates are *cis*-DCE>VC>TCE. The addition of phenol increased the degradation rates. In mixtures competitive inhibition has been observed. Compared to TCE and VC, c-DCE was degraded much faster and showed none or very little metabolite toxicity effects.

INTRODUCTION

TCE is one of the most common contaminants found at hazardous waste sites. Biological treatment of TCE has received much interest as a potential low cost remediation alternative. Two cometabolic systems have been extensively studied: microbes grown on gaseous substrates (methane and propane), and microbes grown on aromatic substrates (phenol and toluene). The pure cultures that have been extensively studied for TCE degradation include the methane-oxidizer *Methylosinus trichosporium* OB3b, and the phenol- and toluene-oxidizer *Burkholderia cepacia* G4. The OB3b shows the highest specific TCE degradation rate, but the use of OB3b is limited by metabolite toxicity. Hopkins et al. (1993) showed that in situ cometabolic degradation activity stimulated by phenol addition resulted in a higher percent TCE and *cis*-DCE removal than by methane addition. *Burkholderia cepacia* G4 was maintained in a gas treatment reactor for over 6 weeks at various TCE loading rates (Landa et al., 1994). The growth yield and the specific activity of the cells for toluene were not affected by TCE reactor loading rates of up to 145 µmol/liter-h. However, the toluene usage efficiency was very low with a transformation yield in the range of 0.02-0.07 mg TCE degraded/mg toluene used. In another pure culture study, the G4 strain was cultivated in a fed-batch reactor, together with three other toluene-degrading strains (Mars et al., 1998). After TCE addition, the only strain that remained in the culture could not degrade TCE. Though G4 has shown promise as a TCE-degrading organism there is no work demonstrating the ability to sustain TCE degradation with G4 in an open reactor.

In our work a phenol-degrading enrichment, dominated by actinomycetes bacteria, was able to outcompete another phenol-degrading culture and sustain TCE degradation for over 13 weeks in a gas treatment reactor (Lee et al., 1998). The culture exhibited an unusually high TCE transformation capacity and a stable operation with minimal phenol consumption. The phenol usage efficiency was very high with a transformation yield of 0.3 mg TCE degraded/mg phenol used.

Objective. The objective of this study was to isolate the actinomycete pure culture from the gas treatment reactor and to evaluate its degradation characteristics for TCE, *cis*-DCE and VC.

MATERIALS AND METHODS

A lab-scale reactor containing 2-liter liquid volume at 0.4-meter depth, with a diffuser stone located at the bottom of the column was used in continuous operation studies (Figure 1).

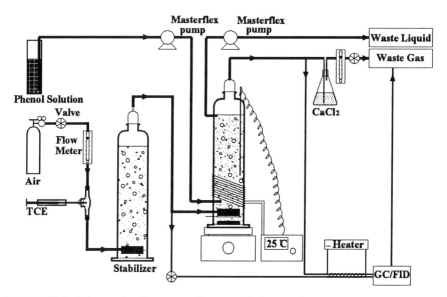

FIGURE 1. Schematic of a sparged suspended growth gas treatment reactor.

Pure TCE was introduced into the gas stream via a syringe pump. The reactor influent and effluent gases were automatically measured using SRI8610B gas chromatography (GC; Restek DB-624 megabore capillary column, oven 130°C, FID). Gas sampling lines were wrapped with heating tape. Gas standards were prepared by injecting weighted TCE pure liquids into 160-mL serum vials with 80mL deionized water. Headspace TCE concentrations were based on the liquid volume added and the Henry's constant. Reactor temperature was kept at 25±1°C. A stabilizer was installed in front of the gas reactor to stabilize influent gas concentration.

The reactor feed solution contained 1,000 mg/L phenol and the following nutrient concentrations: 700 mg/L KH_2PO_4, 1000 mg/L K_2HPO_4, 200 mg/L NH_4Cl, 30 mg/L $MgSO_4$, 66.5 mg/L $CaCl_2 \cdot 2H_2O$, 300 mg/L $NaHCO_3$, 55 µg/L $CuCl_2 \cdot 2H_2O$, 150 µg/L $ZnCl_2$, 20 µg/L $NiCl_2 \cdot 6H_2O$, 880 µg/L $FeSO_4 \cdot 7H_2O$, 135 µg/L Al$_2$(SO$_4$)$_3$·18H$_2$O, 280 µg/L MnCl$_2$·4H$_2$O, 55 µg/L CoCl$_2$·6H$_2$O, 30 µg/L NaMoO$_4$·2H$_2$O, and 50 µg/L H$_3$BO$_3$. The feed solution was fed every 2 hours with masterplux pump to minimize competitive inhibition of phenol with TCE.

Bacterial strain was isolated from the gas treatment bioreactor. Serial dilutions of mixed culture were smeared onto mineral medium agar plates containing 14 mg/L of phenol. From these plates a colony was isolated and plated onto another similarly prepared agar plate. Single colony was then transferred into a 160-mL serum bottle containing 1 % yeast extract. Yeast extract-grown culture was transferred onto an agar plate containing 14 mg/L of phenol. After three more transfers onto agar plates, an actinomycete pure culture was isolated and transferred to a chemostat fed phenol and operated at a 6-day SRT.

RESULTS AND DISCUSSION

Characterization of actinomycete pure culture. Microscopic examination of pure culture revealed the development of intensive mycelium. The culture developed a yellow color in the culture broth indicating *meta*-cleavage pathway of phenol degradation. The chloramphenicol sensitivity was investigated for further enzyme induction studies and a 10 mg/L of chloramphenicol concentration was found to inhibit cell synthesis.

The pure culture from the chemostat was used in carbon assimilation tests using Biolog MicroPlate™ to define the range of hydrocarbons that can be assimilated by the pure culture. The results shown in Table 1 may be used to investigate the ability to grow the actinomycete on a low cost substrate instead of phenol. However, such substrates can be degraded by many other bacteria, which would require developing a feeding strategy that favors actinomycetes dominance.

Degradation of TCE, *cis*-DCE and VC. The isolated pure culture could degrade TCE, *cis*-DCE, and VC (Figure 2). The degradation kinetics in order of highest rates are *cis*-DCE>VC>TCE. The addition of phenol significantly enhanced the degradation rates of chlorinated aliphatic compounds (CACs). Phenol might provide necessary reducing power for TCE oxidation and cellular energy for repairing cell damage by metabolite toxicity.

Competitive inhibition between TCE, *cis*-DCE and VC was observed in tests with chlorinated compounds mixture (Table 2). Compared to TCE and VC, *cis*-DCE is degraded much faster and shows none or very little metabolite toxicity effects. Degradation of TCE and VC showed more inhibition by the presence of other CACs.

Table 1. Carbon assimilation on Biolog GP2 MicroPlate™.

No Growth		Growth	
∀-Cyclodextrin,	∃-Cyclodextrin,	Glycogen	N-Acetyl-D-
Dextrin,	Inulin,	D-Arabitol	Glucosamine
Mannan,	Tween 40,	D-Galactose	D-Galacturonic
Tween 80,	N-Acetyl-D-	Gentiobiose	Acid
Amygdalin,	Mannosamine	D-Gluconic Acid	Lactulose
L-Arabinose,	Arbutin	Maltose	D-Mannose
D-Fructose	L-Fucose	∀-Methyl	3-Methyl Glucose
M-Inositol	∀-D-Lactose	D-Galactoside	∀-Methyl
D-Melezitose	∃-Methyl	∀-Methyl	D-Glucoside
D-Ribose	D-Glucoside	D-Mannoside	Palatinose
D-Tagatose	D-Xylose	D-Psicose	L-Rhamnose
∀-Hydroxybutyric	∀-Keto Valeric	Salicin	Sedohept u-losan
Acid	Acid	D-Sorbitol	Sucrose
Lactamide	D-Lactic Acid	D-Trehalose	Turanose
L-Lactic Acid	Methylester	Xylitol	Acetic Acid
D-Malic Acid	Succinic Acid	∃-Hydroxybutyric	(-Hydroxybutyric
Succinamic Acid	Alaninamide	Acid	Acid
N-Acetyl L-	L-Alanine	p-Hydroxyphenyl	∀-KetoGlutaric
Glutaminic Acid	Putrescine	Acetic Acid	Acid
2,3-Butanediol	Glycerol	L-Malic acid	Methyl Pyruvate
Adenosine	2—Deoxy	Mono-Methyl	Propionic Acid
Inosine	Adenosine	Succinate	D-Alanine
Uridine	Thymidine	L-Asparagine	L-Glutamic Acid
Thymidin	Adenosin	Glycyl-L-Glutamic	L-Pryoglutamic
e-5'-Mono-	e-5'-Mono-	Acid	Acid
phosphate	phosphate	L-Serine	∀-D-Glucose
Uridine-	Fructose-	Cellobiose	D-Mannitol
5'-Mono-	6-Phosphate	Maltotriose	∃-Methyl
phosphate	Glucose-	D-Melibiose	D-Galactoside
D-L-∀-	1-Phosphate	D-Raffinose	L-Alanyl-Glycine
Glycerol	Glucose-	Stachyose	
Phosphate	6-Phosphate	Pyruvic Acid	

Table 2. The specific degradation rate (k) of TCE, *cis*-DCE and VC in single compound and in mixture.

Compounds	k in single compound d^{-1}	k in mixture d^{-1}
TCE	0.13	0.02
cis-DCE	0.64	0.40
VC	0.34	0.03

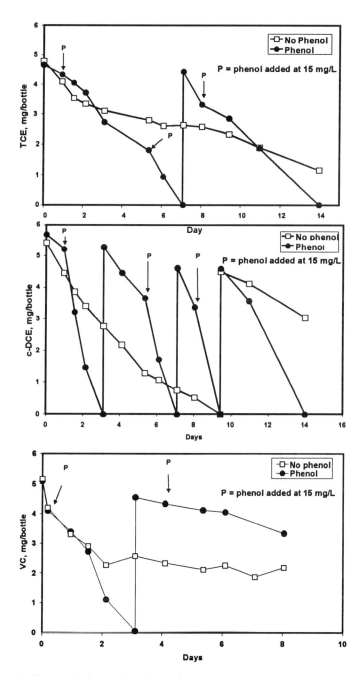

Figure 2. Degradations of TCE, *cis*-DCE and VC by actinomycete pure culture in single compound tests (VSS= 37 mg/L, Temperature = 23 °C).

CONCLUSIONS

The following conclusions can be drawn:
1. The actinomycete pure culture was successively isolated and showed degradation ability toward TCE, *cis*-DCE and VC.
2. The degradation kinetics in order of highest rates are *cis*-DCE>VC>TCE.
3. The addition of phenol increased the degradation rates of chlorinated aliphatic compounds.
4. Competitive inhibition was observed in mixtures of TCE, *cis*-DCE and VC.

ACKNOWLEDGMENTS

This research was funded by a grant from the National Institute of Environmental Health and Science (NIEHS), Superfund Basic Research Program (5P42 ESO4696-07) and through CRESP funded by Department of Energy (DS-FC-1-95-EW55084).

REFERENCES

Hopkins, G.D., L. Simprini and P. L. McCarty. 1993. "Microcosm and in situ field studies of enhanced biotransformation of trichloroethylene by phenol-utilizing microorganisms." *Appl. Environ. Microbiol.* 59: 2277-2285.

Landa, A.S., E. M. Sipkema, J. Weijma, A. A. Beenackers, J. Dolfing, and D. B. Janssen. 1994. "Cometabolic degradation of trichloroethylene by Pseudomonas cepacia G4 in a chemostat with toluene as the primary substrate." *Appl. Environ. Microbiol.* 60: 3368-3374.

Lee, S.B., J. P. Patton, S. E. Strand and H. D. Stensel. 1998. "Sustained biodegradation of trichloroethylene in a suspended growth gas treatment reactor." In : *First International Conference on Bioremediation of Chlorinated solvent and Recalcitrant Compounds*, Montray, CA. May. Battelle Press, Columbus, OH.

Mars, A. E., G. T. Prins, P. Wietzes, W. D. Koning and D. B. Janssen. 1998. "Effect of trichloroethylene on the competitive behavior of toluene-degrading bacteria." *Appl. Environ. Microbiol.* 64: 208-215.

PHYTOREMEDIATION OF ORGANIC SOLVENTS IN GROUNDWATER: PILOT STUDY AT A SUPERFUND SITE

Ari Ferro and Brandon Chard (Phytokinetics, Inc., Logan, Utah)
Michael Gefell (Blasland, Bouck, and Lee, Inc., Syracuse, New York)
Bruce Thompson (*de maximis*, inc., Simsbury, Connecticut)
Roger Kjelgren (Utah State University, Logan, Utah)

ABSTRACT: A phytoremediation system consisting of a dense stand of hybrid poplar, white willow, and six native tree species was installed at the SRSNE Superfund Site. The overall objective of the phytoremediation system is to biologically "pump and treat" contaminated groundwater, thereby reducing the need for mechanical pumps, at least on a seasonal basis. Initial greenhouse studies found that the concentration of total volatile organic compounds at the site did not limit tree growth. Water use by the stand of trees will initially be estimated by measuring sap velocity. This information will help us to assess when it will be feasible to reduce operation (and cost) of the mechanical pumps.

INTRODUCTION

Site Description. Groundwater at the Solvents Recovery Service of New England (SRSNE) Superfund Site in Southington, Connecticut is impacted with a complex mixture of volatile organic compounds (VOCs). The VOCs exist as dense non-aqueous phase liquid (DNAPL) in the saturated overburden and bedrock, as well as in a dissolved-phase plume. The majority of the DNAPL is localized in a containment area upgradient of a sheet pile wall installed as part of a Non-Time Critical Removal Action No. 1 (NTCRA 1). Groundwater from the 1.2 acre (0.49 ha) NTCRA 1 Containment Area is currently pumped from twelve extraction wells to a UV/oxidation treatment facility where approximately 10 million gallons (38 million liters) of water are treated per year. This process removes an average of 1875 lbs (850 kg) VOCs per year. An average pumping rate of 19 gallons (72 liters) per minute (gpm) is sufficient to maintain an inward hydraulic gradient across the sheet pile wall toward the Containment Area. Operation and maintenance of the treatment system currently costs approximately $500,000 per year.

Objectives. Phytoremediation is now being evaluated within the NTCRA 1 Containment Area. Approximately 950 trees, established on 0.8 acres (0.3 ha), constitute a biological "pump and treat" system. The objective of the phytoremediation system is to lower operation and maintenance costs at the site by reducing the amount of water treated by the UV/oxidation system. In addition, the phytoremediation system is expected to facilitate the biodegradation of a substantial mass of aqueous phase VOCs entering the Containment Area.

Phytoremediation Processes. Several processes act simultaneously in a phytoremediation system to degrade VOCs in groundwater. Dissolved organic groundwater contaminants entering the rhizosphere can be broken down by metabolically active bacteria and fungi that may use organic chemical waste as a source of energy and carbon (Rovira and Davey, 1971; Marschner, 1995; McFarlane, 1995; Crowley, Alvey, and Gilbert, 1997). Further, root exudates may contain surfactant molecules which can help solubilize hydrophobic organic compounds, thus making them more available for microbial attack (Soeder et al., 1996). Plant root exudates may also stimulate co-metabolic transformations: Certain organic contaminants cannot be used as a growth substrate by soil microbes, but in the presence of structurally related root exudates (co-metabolites), biodegradation is enhanced.

Many organic chemical contaminants are lipophilic and have a high affinity for the hydrophobic surfaces on organic matter. Such compounds may adsorb and bind to living or dead plant root tissue and become immobilized (Novak, et. al., 1995). This process has been termed "phytostabilization." Such bound residues cannot be removed from soil by conventional extraction techniques and therefore are thought to be less bioavailable, less mobile, and less toxic than the free species. Some organic chemical contaminants may be taken up into plants by a passive process related to the ability of the chemical to transverse the cell membrane. Once taken up, the chemical compound may be transformed (metabolized) or volatilized by the plant.

MATERIALS AND METHODS

Greenhouse studies. Prior to planting at the site, greenhouse studies were performed to evaluate the tolerance of hybrid poplar trees to the mixture of VOCs present in the dissolved phase plume (Ferro, et al., 1999). The mixture included chlorinated aliphatics, BTEX compounds, and alcohols (Table 1). The compounds listed in Table 1 are the aqueous phase VOCs detected in the shallow overburden groundwater in the containment area at concentrations >15µg/L. The concentrations listed for individual compounds are the highest observed at any shallow overburden well in the area.

Hybrid poplar cuttings (*Populus deltoides* x *P. nigra,* clone DN-34, 4-ft (1.2 m) long, 0.5 in. (1.3 cm) diameter) were grown in 50-gallon (189 L) polyurethane barrels packed with alternating layers of pea gravel and Kidman fine sandy loam soil with slow-release nutrients. In experiment A, three trees planted in each of four barrels were allowed to become established, and then were gradually treated for 88 days with a sub-irrigant containing either a low [42 mg/L total VOCs (TVOC)], medium (85 mg/L TVOC), or high (169 mg/L TVOC) dose of the VOC mixture, or water only (control). In Experiment B, trees were immediately irrigated with a low, medium, or high dose of the VOC cocktail, or with water only in the control barrel. The duration of this experiment was 55 days, with no period of plant establishment and no incremental dosing. Phytotoxicity was evaluated for both experiments by measuring the physiological parameters of stomatal conductance, shoot elongation, and biomass production. Greenhouse study results are described below.

TABLE 1. VOC cocktail; 1X (high dose) concentration.

Compound	Concentration (mg/L)	Compound	Concentration (mg/L)
1,1-Dichloroethane	25	m,p-Xylene	3.0
1,1-Dichloroethene	1.2	o-Xylene	1.0
Total 1,2-Dichloroethene	24	sec-Butanol	13
1,1,1-Trichloroethane	35	Ethanol	3.0
Ethylbenzene	21	Isopropanol	1.0
Toluene	40	Methanol	2.0

Field installation. The phytoremediation system was installed at the SRSNE Superfund Site beginning in late May 1998. Rows of hybrid poplar trees were densely planted in 3 ft (1 m)-deep trenches backfilled with sand and peat. In early April 1999, dead and marginal hybrid poplar trees were replaced with a mixed planting of rooted cuttings of white willows (*Salix alba*) and hybrid poplars. In addition, a small stand of native tree species was installed to determine the efficacy of natives in a phytoremediation system.

In the spring of 2000, a weather station and sap velocity measurement equipment (thermal dissipation probes) will be installed. The weather station will be fitted with sensors to record meteorological data including precipitation, humidity, wind speed, solar radiation, and soil and air temperature. Five trees with a minimum diameter of 2 inches (5.1 cm) will be selected and equipped with thermal dissipation probes. All of the sensors and probes will be connected to a data logger and multiplexer. Data will be retrieved from the data logger via a remote modem with attached cellular phone transceiver. This system will allow us to calculate water use by the stand.

RESULTS AND DISCUSSION

Greenhouse Study Results. Results of the greenhouse phytotoxicity study are detailed in Ferro, et al. (1999). Phytotoxic effects were not observed in either experiment. No VOC-dependent decreases in stomatal conductance, shoot elongation, or biomass production (leaf and root tissue) were observed. Thus, the installation of the poplar tree phytoremediation system at the SRSNE Superfund Site was considered to be feasible.

Full Scale Plantation Results. Most of the trees planted grew vigorously during the first growing season, with some trees growing more than 3 ft (1 m). However, approximately 20% mortality was observed in late fall of 1998. We believe this high mortality rate was due to high daytime temperatures during and after tree installation rather than phytotoxic effects, as evidenced by the results of the greenhouse experiment. Willows and poplar poles were planted throughout the containment area in early spring of 1999 to replace the dead trees. White willows were chosen to augment the phytoremediation system because they are host to a different suite of pathogens than poplars. Planting willows will increase species diversity, thereby improving the viability of the stand. White willows have similar growth characteristics to poplars. Both are large, fast growing trees, they use water at comparable rates, and, they are well adapted to the climate of the Southington area.

A small stand of native tree species was also planted in 1999 to determine the feasibility of using natives in a phytoremediation system. Seven individuals from each of six native tree species were planted to determine which native species could benefit a phytoremediation system. The six native tree species selected were pin oak (*Quercus palustris*), river birch (*Betula nigra*), sweet gum (*Liguidamabar styraciflua*), silver maple (*Acer saccharinum*), tulip tree (*Liriodendron tulipifera*), and eastern redbud (*Cercis canadensis*). These species were chosen because of their relatively rapid growth rates, tolerance of wet soils, height at maturity, water use, and disease resistance. If successful, mixed stands of native trees may offer additional advantages over hybrid poplar monocultures. Native species are often preferred by the public and may augment efforts to enhance naturally evolved ecosystems.

Currently, most of the tree species planted within the Containment Area are thriving, with a mortality rate of less than 5%. Tulip trees were decimated in the early spring by an insect infestation that was not controlled. Tulip trees were replaced in the fall of 1999 with eastern white pine (*Pinus strobus*). Conifers, such as pines, may offer additional benefits to phytoremediation systems in that they transpire water over a greater portion of the year.

Water Use and Groundwater Containment Projections. Evapotranspiration by the stand in its fifth growing season (Table 2) is predicted to produce a substantial hydraulic effect. During the summer months, it is likely that the phytoremediation system will reduce the rate and cost of pumping from the NTCRA 1 extraction wells.

One near-term objective for the project is to accurately estimate monthly water use during the growing season based on measured values for potential evapotranspiration (PET). The total volumetric water use by the trees (V_T) can be calculated using the following equation:

$$V_T = PET \bullet Kc \bullet LAI \bullet A$$

Where PET = potential evapotranspiration,
Kc = "crop coefficient" = rate of water use per leaf as a percentage of PET
LAI = leaf area index = the leaf area per unit area of soil
A = area of the stand of trees (0.8 acres) at the time of crown closure)

Values for PET used in the equation are site-specific (Table 2). Values for Kc and LAI are specific for a densely planted stand of hybrid poplar trees, and vary as a function of time (Figure 1). Note that V_T increases as the stand matures, and reaches a plateau value when the canopy closes during year 5. Values for LAI and Kc are conservative estimates, thus actual water use by the tree stand may be more than estimated.

TABLE 2. Volumetric water use estimations for a dense stand of poplars at the SRSNE Site, Southington, CT. Trees are planted on 6 ft (2 m) centers, total area = 0.8 acres (0.3 ha).

		V_T			
Month	PET (in)	Year 2 (gpm)	Year 3 (gpm)	Year 4 (gpm)	Year 5 (gpm)
April	4.14	0.47	1.88	5.11	7.50
May	5.62	0.62	2.46	6.70	9.85
June	6.35	0.72	2.87	7.83	11.50
July	7.08	0.78	3.10	8.44	12.41
August	6.08	0.67	2.66	7.25	10.65
September	4.24	0.48	1.92	5.22	7.68
October	2.68	0.29	1.17	3.19	4.69
Total:	36.20				
Avg:		0.57	2.30	6.25	9.18

Year	Kc	LAI
2	0.3	0.75
3	0.6	1.5
4	0.7	3.5
5	0.6	6

In the summer of 2000, meteorological data from an on-site weather station will be used to estimate PET. A portable leaf area meter will be used to measure LAI for certain trees. Estimation of V_T, based on measurements of sap velocity, will allow us to calculate Kc during several growing seasons. Thereafter, we will be able to accurately estimate V_T based on measured values for PET and LAI, and correlate V_T with groundwater elevations measured in a network of monitoring wells within the Containment Area.

As shown in Table 2, groundwater use by the plantation is expected to approach the historical dry-season pumping rates of the NTCRA 1 extraction wells (between 10 and 15 gpm). These data are consistent with the results of a preliminary groundwater flow (MODFLOW) model, which suggested the summertime evapotranspiration rate by the phytoremediation stand may be sufficient to hydraulically control groundwater within the sheetpile wall with the extraction wells inactive. Sap-flow measurements during the 2000 growing season will be used to refine the groundwater use estimates presented herein.

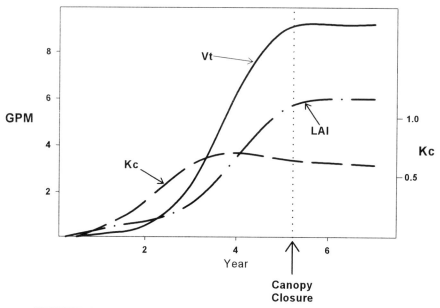

FIGURE 1. Estimated rate of water use by the 0.8 acre tree stand.

REFERENCES

Crowley, D.E., S. Alvey, and E.S. Gilbert. 1997. "Rhizosphere ecology of xenobiotic-degrading microorganisms." *In*: E.L. Kruger, T.A. Anderson, and J.R. Coats (eds.) *Phytoremediation of Soil and Water Contaminants*, pp. 20-36. American Chemical Society, Washington, D.C.

Ferro, A., J. Kennedy, R. Kjelgren, J. Rieder, and S. Perrin. 1999. "Toxicity assessment of volatile organic compounds in poplar trees." *Internat. J. Phytoremed.* 1:9-17.

Marschner, H. 1995. *Mineral nutrition of higher plants*. Academic Press, New York, NY.

McFarlane, J.C. 1995. "Anatomy and physiology of plant conductive systems. p. 13-36." *In* S. Trapp. and J.C. McFarlane (ed.) *Plant contamination: Modeling and simulation of organic chemical processes*. Lewis Publ., Boca Raton, FL.

Novak, J.M., K. Jayachandran, T.B. Moorman, and J.B. Weber. 1995. "Sorption and binding of organic compounds in soils and their relation to bioavailability." *In*: H.D. Skipper and R.F. Turco (eds.) *Bioremediation: Science and applications*. pp. 13-31. Soil Science Society of America, Madison, WI.

Rovira, A.D., and C.B. Davey. 1971. "Biology of the rhizosphere". p. 153-204. *In* E. W. Carson (ed.) *The plant root and its environment*. University Press of Virginia, Charlottesville, VA.

Soeder, C.J., A. Papaderos, M. Kleespies, H. Kneifel, F.H. Haegel, and L. Webb. 1996. "Influence of phytogenic surfactants (quillaya saponin and soya lecithin) on bio-elimination of phenanthrene and fluoranthene by three bacteria." *Appl. Microbial Biotechnol.* 44:654-659.

MONITORING SITE CONSTRAINTS AT NUWC KEYPORT'S HYBRID POPLAR PHYTOREMEDIATION PLANTATION

William L. Rohrer, URS Greiner Woodward Clyde-Dames & Moore-Radian International, Seattle, WA, USA
Lee Newman, Marietta Sharp, and Paul Heilman, College of Forest Resources, University of Washington, Seattle, WA, USA
B. Renee Wallis, Naval Facilities Engineering Field Activity Northwest, Poulsbo, WA, USA

ABSTRACT: Disposal of industrial and domestic waste at the Naval Underwater Weapons Center (NUWC) Keyport Landfill in Keyport, Washington resulted in contamination of groundwater by Volatile Organic Chemicals (VOCs). When the community advisory board balked at the projected high costs of conventional treatment and disposal methods evaluated under the CERCLA program to address the contamination, phytoremediation was proposed as a lower cost alternative. Part of the appeal of phytoremediation was that it was perceived as a "low-tech" solution compared to typical remediation systems, requiring a less rigorous and complex engineering and construction effort. Nonetheless, the implementation of this "green" technology was not as straightforward as originally envisioned. This paper discusses the site-specific constraints that were identified and resolved in the course of implementing the phytoremediation alternative at NUWC Keyport.

INTRODUCTION

NUWC Keyport Landfill Disposal History. The Naval Underwater Weapons Center Keyport (NUWC Keyport), 11 miles by water from Seattle, Washington is a Navy support facility for undersea weapons systems, countermeasures, and sonar systems. NUWC Keyport is located in Kitsap County in the central portion of Puget Sound (Figure 1). The facility occupies about 340 acres of a small peninsula in Liberty Bay. The former base landfill comprises about 9 acres in the western part of the base next to the tide flats and Dogfish Bay and is currently designated Operable Unit 1 (OU 1). Most of the landfill was formerly a marshland that extended from the tide flats toward the shallow lagoon.

Landfilling of the tidal marsh area with domestic and industrial wastes commenced from the 1930s until 1973. The result of the filling and disposal operations was documented contamination of shallow groundwater with volatile and other organic compounds. Remedial Investigation (RI) field studies of Operable Unit 1 landfill showed that VOCs migrate in a radial pattern from the landfill area toward adjacent tidal flats and Dogfish Bay on Puget Sound.

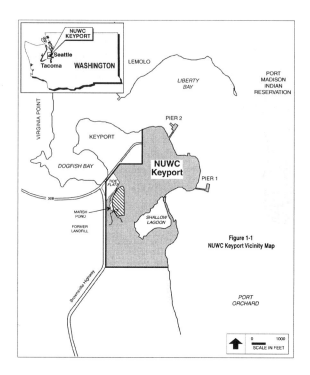

FIGURE 1. NUWC Keyport is located west of Seattle and borders Puget Sound.

The OU 1 landfill was the primary disposal area for both domestic and industrial wastes generated by the base from the 1930s until 1973 when the landfill was closed. Based upon interviews with base personnel, the following types of industrial wastes were likely disposed of at the landfill:

- Paints, lacquers, thinners, ketones, enamel, solvents, and other wastes from the paint and paint stripper shops,
- Residue from burning Otto fuel and solids contaminated with torpedo fuel,
- Cutting oils, acids, caustics, and lead slag from metal shops, and
- Pesticide rinsate from pest control shops.

The two tree plantations within OU 1 are located at the south end of the landfill and approximately two-thirds of the way from the south to the north end of the landfill (Figure 2).

Phytoremediation 469

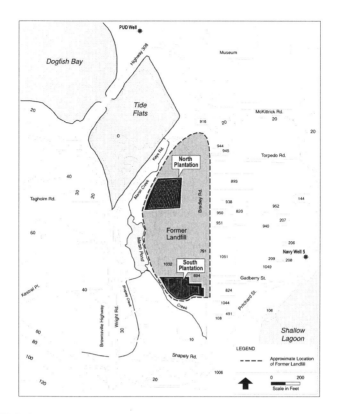

FIGURE 2. Local site conditions at north and south phytoremediation plantations.

Hydrogeologic Setting. Subsurface geologic units in the vicinity of the landfill are vertically and laterally variable. They include interbedded glacial deposits, non-glacial fluvial and floodplain deposits, and post-glacial estuarine deposits and 10 to 15 feet of manmade fill. The hydrostratigraphy under the landfill generally includes an unconfined upper aquifer, a middle aquitard, and an intermediate aquifer. The upper aquifer is composed of sands and silts; the aquitard is silt-rich in most places, but is locally sandy; and the intermediate aquifer is composed of sand, with local zones of gravel and silt. The aquitard is locally discontinuous, resulting in "windows" in the central, eastern, and northern portions of the landfill. Leakage between the two aquifers is likely enhanced in these areas.

In the unconfined aquifer, groundwater flow (at a depth of 15 to 20 feet below ground surface) is generally westward toward the marsh in the northern part of the landfill and to the southwest in the southern part.

PHYTOREMEDIATION PLANTATION DESIGN

Representatives of Naval Facilities Engineering Field Activity Northwest (EFA NW), NUWC Keyport, Washington Department of Ecology, EPA Region

10, the Suquamish Tribe, United States Geologic Survey (USGS), the Restoration Advisory Board (RAB), and the University of Washington collaborated over a period of five years to identify phytoremediation as the most cost-effective remedial alternative for cleaning up groundwater at the site. On a present-worth cost basis, phytoremediation was four times more effective than the most expensive alternative, and was selected as the remedy of choice in the CERCLA Record of Decision. Planting of approximately 900 hard cuttings of a specialized fast-growing cultivar of poplar tree was started on Earth Day 1999.

To specifically address site constraints that might affect the viability of the plantation and the effectiveness of the remedy, extensive site preparation construction activities were conducted, and a monitoring network of groundwater and vadose zone sampling locations was established. These constraints included: extensive industrial and construction debris in landfill soils, limited availability of nutrients, extremely compact subsoils, seasonal excess and deficit of soil moisture, and areas of plant toxicity.

Sampling has been carried out over the past year (1999-2000) to fulfill site work plan requirements and to: (1) calibrate the irrigation system to ensure proper soil moisture for tree growth and that VOC-contaminated irrigation water is not moving below the root zone, (2) assess the effect of the trees on groundwater levels and VOC concentrations in and near the plantations, (3) evaluate the potential for leaching of contaminants from the vadose zone during growing and dormant seasons, and (4) satisfy demonstration sampling requirements, as stipulated by the Record of Decision.

Specific sampling and monitoring media include: surface water, irrigation water, lysimeter water, soil, groundwater, tree tissue, air and transpiration vapor sampling; and groundwater elevations. Measurements taken on a weekly to semi-annual basis to assess the effectiveness of uptake of contaminated groundwater, transpiration of VOCs, and accumulation of degradation products and metabolites in root-zone soils and plant tissue, have not yet reflected the effects of phytoremediation on the shallow groundwater flow and chemistry regime.

SITE CONSTRAINTS

Site-specific constraints identified during the RI and pre-design field surveys included: extensive industrial and construction debris in landfill soils, limited availability of nutrients, extremely compact subsoils, seasonal excess and deficit of soil moisture, and areas of plant toxicity.

Because of the former presence of the landfill, the shallow subsurface soils contained abundant metallic, concrete, plastic, and other industrial and construction debris. Landfill debris that was exposed during soil ripping with ripping shanks on the back of a bulldozer was re-incorporated into the landfill where possible. Bulkier debris that could not be re-incorporated was either pressure washed and recycled (approximately 1.5 tons of metal) or sampled and transported for disposal (approximately 24 tons of concrete debris and 4 cubic yards of plastic and fabric sheeting). Based upon observations of the density of debris, degree of compaction of the landfill surface and widespread surface ponding, UW researchers recommended augmenting the surface with clean fill. The north plantation was covered with approximately 6 to12 inches of nearby borrow fill, while the south plantation was covered with approximately 12 to 18 inches of fill. Over 3,100 cubic yards of fill was spread on the two plantations.

Soil samples from 12 locations within the landfill (and from three depths) were analyzed by UW for agronomic suitability, including phosphorus, nitrogen, TOC, pH, and tree toxicity. The nutrient analyses indicated that soil amendment would be required, specifically to increase pH by adding lime and increasing nitrogen content by adding urea. After the initial soil preparation and drying of the new soil fill was accomplished, lime (2000 pounds per acre) and urea (300 pounds per acre) were applied with an orchard tractor and whirling spreader. A chisel plow was then used to incorporate the amendments to a depth of 12 to 18 inches. In subsequent months, granular fertilizer applications have been made to counteract a plantation-wide nitrogen deficiency.

Tree toxicity testing was conducted by monitoring the growth of poplar cuttings placed in samples of the landfill soils and grown in a greenhouse. Samples from 1 to 3 feet below ground surface at Location 1 in the south plantation caused tree stress and eventual plant mortality. Results of chemical analysis for full range of organic and inorganic compounds did not show significantly elevated concentrations, and no source of the tree mortality was identified.

The poplars were planted as 8-inch hardwood cuttings in imported borrow fill underlain by 10 to 15 feet of landfill soils. Consequently, it was necessary to irrigate the cuttings for the initial two growing seasons; the time required for the root systems to extend to the shallow aquifer. A drip irrigation system was designed consisting of subsurface driplines and drip emitters installed at 2-foot intervals along each row of cuttings. The irrigation system was designed to supply a minimum of 10 gallons per minute to the plantations, using shallow contaminated groundwater and tap water. Two shallow supply wells (one in the north plantation and one in the south) were constructed to provide the groundwater irrigation source. An elaborate system of water conditioning (for iron, sediment, and chlorine to control bacterial fouling), mixing (of groundwater and tap water), and telemetry was designed to control and monitor the application of water throughout both plantations in order to prevent under- and over-watering.

MONITORING REQUIREMENTS

Sampling has been carried out over the past year (1999-2000) to fulfill site work plan requirements and the following objectives:

- Calibrate the irrigation system to ensure proper soil moisture for tree growth and that VOC-contaminated irrigation water is not moving below the root zone
- Assess the effect of the trees on groundwater levels and VOC concentrations in and near the plantations

- Evaluate the potential for leaching of contaminants from the vadose zone during growing and dormant seasons
- Satisfy demonstration sampling requirements, as stipulated by the Record of Decision

Specific sampling and monitoring media include: surface water, irrigation water, soil pore water, soil, groundwater, tree tissue, air and transpiration vapor sampling; and groundwater elevations. Samples were taken from a fixed network of nine monitoring wells (to measure groundwater chemistry and groundwater levels), ten piezometers (to measure groundwater levels only), and six lysimeters (to monitor VOCs in vadose zone pore water), Figure 3.

PRELIMINARY RESULTS

To date, patterns of groundwater flow direction and gradient have reflected historical observations. In the north plantation, the general flow direction is toward the west-northwest, with average gradients ranging from 0.014 to 0.018 (Figure 4). The gradient locally steepens during low tide. In the south plantation, groundwater flow directions range from west to southwest in the eastern portion and west-northwest in the western portion, with gradients that vary from 0.004 to 0.006. Flow directions and gradients were observed to vary little on a seasonal basis, although slightly higher shallow water levels were observed in late spring relative to late fall. The shallow groundwater
elevation data do not as yet show any significant effect from the phytoremediation plantation. No such effect is anticipated until the trees reach maturity.

The automated irrigation system has supplied 211,175 gallons of water since its installation, of which 4,697 gallons, or approximately 4 percent, constituted contaminated well water. Based upon a water usage rate of 2,800 gallons per hour, the total irrigation volume represents approximately 31 irrigation events over a time period of five months.

Although both plantations required irrigation during the growing season, trees in the northeast quarter of the north plantation appear to be showing stress (red-colored leaves) as a result of oversaturated soil conditions during the rainy season. It may be even necessary to install some drainage relief in the future, if this condition continues. Those in the northwest corner showed stress resulting from decay of wood chips imported for a temporary roadway.

Replanting of cuttings was necessary at 34 locations (6.2 percent) in the north plantation and 36 locations (9.9 percent) in the south plantation. This initial mortality rate is within the normal range (0-10%) for new plantations, and there is no indication of soil toxicity effects based upon the spatial distribution of the replanted locations. Other agronomic indicators of tree health (at 25 locations) are summarized in Figure 5, specifically tree height and number of leaves per tree. Both tree height and number of leaves appear to be slightly higher in the north plantation, likely resulting from higher light intensities in the north plantation.

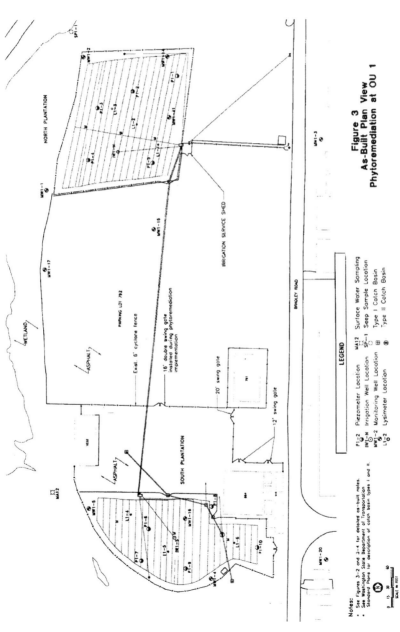

FIGURE 3. The NUWC Keyport phytoremediation monitoring network includes shallow monitoring wells, lysimeters, piezometers, and surface water sampling sites.

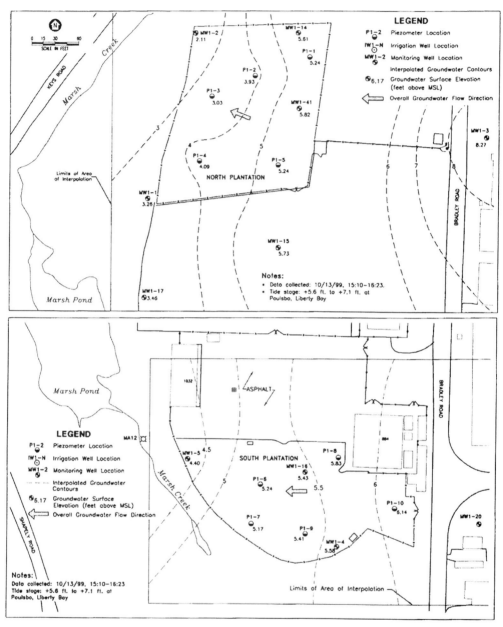

FIGURE 4. Shallow groundwater flow directions vary little from wet to dry seasons in the north and south plantations.

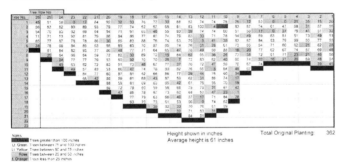

FIGURE 5. The spatial distribution of tree height shows vigorous growth after one season of nurturing. Maximum observed tree heights (115 inches) were recorded in the northern plantation.

Although only a limited set of groundwater and vadose zone samples have been collected, overall trends suggest that concentration trends for target VOCs (TCE, vinyl chloride, and cis-1,2-dichloroethene) are neither increasing nor decreasing and are within the range of concentrations historically detected within OU 1. As with the groundwater flow direction and gradient data, no significant effect on VOC concentrations is anticipated until the plantation trees mature.

As these effects do become more pronounced, the seasonal changes in saturated thickness will be compared to growth and transpiration rates of the cuttings. In a parallel manner, concentration trends in the vadose zone and the shallow aquifer will be compared to estimated chemical flux rates resulting from transpiration and metabolism of VOC within the plantation biomass.

REFERENCES

U.S. Navy. 2000. *Draft Phytoremediation Status Report, August-October 1999, Operable Unit Naval Undersea Warfare Center, Division Keyport, Washington.* Prepared by URS Greiner, Inc., for Engineering Field Activity Northwest, under CLEAN Contract No. N62474-89-D-9295. January 21, 2000.

———. 1999. *Phytoremediation Site Work Plan for Operable Unit 1. Naval Undersea Warfare Center, Division Keyport, Washington.* Prepared by URS Greiner, Inc., for Engineering Field Activity Northwest under CLEAN Contract No. N62474-89-D-9295. March 1999.

———. 1999. *Draft Phytoremediation Status Report, May-July 1999, Operable Unit Naval Undersea Warfare Center, Division Keyport, Washington.* Prepared by URS Greiner, Inc., for Engineering Field Activity Northwest, under CLEAN Contract No. N62474-89-D-9295. September 28, 1999.

———. 1998. *Final Record of Decision for Operable Unit 1, Naval Undersea Warfare Center Division, Keyport, Washington.* Prepared by URS Greiner, Inc., for Engineering Field Activity Northwest under CLEAN Contract No. N62474-89-D-9295. September 1998.

THE INFLUENCE OF AN INTEGRATED REMEDIAL SYSTEM ON GROUNDWATER HYDROLOGY

William H. Schneider (Roy F. Weston, Inc., West Chester, PA)
John G. Wrobel (U.S. Army Garrison, Aberdeen, MD)
Steven R. Hirsh (U.S. EPA, Philadelphia, PA)
Harry R. Compton (U.S. EPA, Edison, NJ)
Dale Haroski (Lockheed Martin, Inc., Edison, NJ)

ABSTRACT: This paper summarizes the development of a remedial system designed to hydraulically contain and ultimately reduce a plume consisting of primarily 1,1,2,2-tetrachloroethane (1,1,2,2-TeCA). The system consists of groundwater circulating wells or extraction wells located in the core of the 1,1,2,2-TeCA plume to provide active source control, combined with monitored natural attenuation (MNA) and phytoremediation instituted to further reduce dissolved-phase contaminants. Monitoring of natural attenuation parameters indicates that abiotic and biotic degradation is significantly reducing 1,1,2,2-TeCA concentrations. Phytoremediation is provided by a 4.5-year-old plantation of 172 hybrid poplars observed to be seasonally influencing groundwater hydrology. A 3D-geospatial model (earthVision®), which was constructed based on extensive geological, geophysical, and chemical data, defines both the hydrostratigraphic framework and the 1,1,2,2-TeCA distribution and is the basis for a 3D-groundwater flow (MODFLOW) and contaminant transport (RT3D) model. Model results indicate that the system may remove 85% of the total 1,1,2,2-TeCA mass after 30 years, with groundwater wells and MNA providing the bulk of mass removal. Phytoremediation emerges as a significant contributor by providing 7% of the total mass removal. Field evidence and modeling indicate that the integrated remedial system is capable of effectively reducing contaminant mass, thereby satisfying the remedial objective.

INTRODUCTION

Objective. The objective of this study was to determine the feasibility of deploying an integrated system to remediate a 1,1,2,2-TeCA plume. A 3D groundwater flow and contaminant transport model was developed to estimate the capacity of the proposed remedial technologies to hydraulically contain and ultimately reduce the 1,1,2,2-TeCA plume. The model was used to determine the well configurations that achieve optimal mass removal while minimizing impacts to a freshwater marsh. In addition, the model was used to assess mass removal generated by natural attenuation and phytoremediation and to estimate the contaminant loading to the freshwater marsh. Finally, the model was used to identify significant data gaps.

Site Description. The J-Field site is located in the Edgewood Area of Aberdeen Proving Ground, MD (Figure 1). Groundwater in a surficial aquifer is impacted by primarily 1,1,2,2-TeCA and trichloroethene (TCE) as a result of past disposal activities (Argonne, 1996). The 1,1,2,2-TeCA plume is bilobed and extends 82 meters toward the southwest and 110 meters to the east (Figure 2). The 1,1,2,2-TeCA source area resides within a local groundwater recharge area. Groundwater flow in the surficial aquifer is through a fine sand and clayey silt unit that exhibits a bulk hydraulic conductivity of approximately 0.3 to 1.5 m/day (Quinn et al., 1996).

MODELING APPROACH

A phased-modeling approach was used to simulate the fate and transport of the 1,1,2,2-TeCA plume and examine the effectiveness of the integrated remedial system in removing contaminant mass.

3D-Geospatial Model. A comprehensive 3D-geospatial model was constructed based on extensive geological, geophysical, and chemical data using earthVision® (Dynamic Graphics, 1999). The geospatial model defines both the hydrostratigraphic framework and the 1,1,2,2-TeCA distribution at the site and is the framework for a 3D-groundwater flow (MODFLOW) and contaminant transport (RT3D) models (McDonald and Harbaugh [1988], Clement [1998]). The 3D-geospatial model assisted the remedial design efforts by characterizing the primary hydrostratigraphic units and subsequently defined the model layers in MODFLOW. The geospatial model illustrates that the thickness of the surficial aquifer varies between 7 and 12 meters and the underlying first-confining unit extends continuously from the study area to beneath the Chesapeake Bay with no evidence of breaching by paleochannels (Figure 3).

The distribution of contaminants was modeled using 3D minimum tension interpolation techniques and integrated within the hydrostratigraphic framework (Schneider and Wrobel, 1998). The distribution of VOCs existing within the framework of the 3D-geospatial model was exported to the RT3D model and used to represent the initial conditions in RT3D. The geospatial model and existing hydrogeologic data were used to estimate both dissolved- and sorbed-phase contaminant mass (Table 1). 1,1,2,2-TeCA mass was also incorporated into the model to represent residual-phase materials that are suspected of feeding a local hot spot. Quantifying the contaminant mass enabled the project team to establish remedial goals by identifying the amount of VOC mass that requires removal.

TABLE 1. Estimated Contaminant Mass.

Contaminant	Dissolved Mass (lb)	Sorbed Mass (lb)	Estimated Residual Source Mass (lb)
1,1,2,2 - TeCA	1,320	1,820	1,400
TCE	475	1,310	-

Groundwater Flow Model. A groundwater flow model was developed by incorporating data from the geospatial model (e.g., hydrostratigraphic framework) into MODFLOW. The model grid was telescoped using 82 rows, 86 columns, and 6 layers with minimum lengths of 3 meters (Figure 4). The flow model was calibrated to mean groundwater elevations based on continuously measured data. Inverse modeling techniques were applied to achieve acceptable model calibration with the aid of UCODE (Poeter and Hill, 1998). Details of the MODFLOW model and the structured approach to model calibration are presented in Quinn et al. (1996) and WESTON (2000).

Contaminant Transport Model. The fate and transport of 1,1,2,2-TeCA, TCE, and their respective transformation products was simulated using a contaminant transport model (RT3D) that contains a multi-species, reactive transport module. This module uses first-order, rate-limited kinetics that characterize the sequential VOC degradation used to represent natural attenuation processes at J-Field. The degradation rates and pathways for VOCs were based on site-specific data and differed between hydrostratigraphic units. Details of natural attenuation are outlined in the following sections as well as in Yuen et al. (1998).

SIMULATED REMEDIAL PROGRAM

Each component of several potential integrated remedial systems: (1) groundwater circulating wells (GCWs), (2) extraction wells, (3) phytoremediation, and (4) monitored natural attenuation (MNA) was incorporated into the MODFLOW-RT3D model using the techniques presented in the following subsections.

Groundwater Circulating Wells. The GCWs were simulated using the well package in MODFLOW. The circulation process was replicated in the model by segmenting the surficial aquifer into multiple layers. The rates of extraction and reinjection applied to the model were 1.9 liters/min determined from aquifer tests. The withdrawal rates of the GCWs were held constant during the course of the 30-year simulations. The placement of the GCWs was focused in the area of highest 1,1,2,2-TeCA concentrations.

Groundwater Extraction Wells. Groundwater extraction wells were simulated using the well package in MODFLOW. The simulated flow rates were held steady at 1.9 liters/min based on aquifer tests. The placement of extraction wells coincided with the GCWs to compare their performance.

Phytoremediation. Phytoremediation provides a natural mechanism for providing additional VOC reduction. The plantation consists of 172 deep-rooted poplar trees planted in buffer groups adjacent to the marsh. The poplar trees were simulated using the MODFLOW well package based on the premise that the poplar trees act as solar-driven pumps that siphon from the groundwater table. The rate of groundwater withdrawal for each tree was calculated using a site-specific crop index (CI) generated from 4 years of sap flow, weather, leaf, stem, and land area data collected at J-Field.

Predicted withdrawal rates for each poplar tree were incorporated into the well package using 30 1-year stress periods. These stress periods are essentially an average rate of the groundwater withdrawal compiled for a given water-year cycle based on the respective seasonal uptake rates (spring, summer, fall, and winter). Based on the CI, these steady-state groundwater withdrawal rates were projected to increase as the plantation matures and trees increase in size. The CIs indicated the plantation withdrew approximately 4,162 liters/day upon reaching maturity at 10 years. The removal rates and methods used to determine the average flow rates for the poplar trees that were the basis of the modeling approach are outlined in a report detailing J-Field phytoremediation activities (Dynamax, 1999).

Estimates of the mass removal predicted for the trees required the use of a transpiration stream concentration factor (TSCF) to describe the degree of partitioning that occurs at the poplar tree root surfaces. TSCFs of 0.79 (1,1,2,2-TeCA) and 0.74 (TCE) were based on methods outlined by Schnoor (1996).

Natural Attenuation. A comprehensive natural attenuation study conducted at J-Field concluded that VOC degradation is actively occurring both in the surficial aquifer, where iron-reducing conditions exist and abiotic processes predominate, and in the marsh, where methanogenic conditions support biotic degradation (Yuen et al., 1998). Field data support these findings as evidenced by the widespread distribution of transformation products and the rapid decline in 1,1,2,2-TeCA and TCE concentrations in groundwater as the plume migrates to the marshes. A simplified representation of the primary degradation pathways and rates was simulated using the RT3D model (see Table 2).

TABLE 2. VOC Degradation Pathways and Half-Lives (Days).

Surficial Aquifer	1,1,2,2-TeCA	→ TCE	→ DCE	→ VC	→ Eth
(Upland Area)		1900	2320	2600	2930
Surficial Aquifer	1,1,2,2-TeCA	→ DCE	→ VC	→ Eth	
(Marsh Area)		90	460	1730	

MODEL RESULTS AND CONCLUSIONS

The combined models proved useful for evaluating the capacity for the integrated remedial system to effectively remove contaminant mass. Figure 5 illustrates an example of a 30-year simulation that indicates the integrated remedial system is capable of removing 78% of the total 1,1,2,2-TeCA mass after 30 years. Groundwater wells and natural attenuation processes contribute the bulk of predicted contaminant mass removal. The model results agree with existing data that J-Field contains an ideal hydrogeologic setting (highly reduced groundwater chemistry and low groundwater flow velocities) that promotes both biotic and abiotic degradation of 1,1,2,2-TeCA. In addition, existing low-permeability, organic-rich marsh sediments effectively retard contaminant migration. These conditions also appear ideal for successfully implementing phytoremediation based on the predicted 7.5% removal of total 1,1,2,2-TeCA mass generated by the poplar trees. Field data indicate that the plantation produces hydraulic containment of the southeastern edge of the 1,1,2,2-TeCA plume during the mid- to late-summer months.

While the hydrogeologic conditions encourage the successful application of natural attenuation and phytoremediation, the low-permeability sediments reduce source removal rates by limiting sustainable well yields to 1.9 liters/min.

Despite the low yields, the wells do provide an active mechanism for achieving 1,1,2,2-TeCA source reduction, and field data demonstrate that mass removal by wells is achievable. Field evidence and modeling indicate that the integrated remedial system is capable of effectively reducing contaminant mass at J-Field and satisfying the remedial objective.

REFERENCES

Argonne National Laboratory. 1996. *Remedial Investigation Report for J-Field.*

Clement, T.P. 1998. RT3D - *A Modular Computer Code for Simulating Reactive Multi-Species Transport in 3-Dimensional Groundwater Aquifers Version 1.0.*

Dynamax. 1999. *EPA Phytoremediation Study for Lockheed Martin.*

McDonald, M.G. and A.W. Harbaugh. 1988. *A Modular Three-Dimensional Finite-Difference Ground-Water Flow Model, Techniques of Water Resources Investigations, Book 6, Chapter A1*, U.S. Geological Survey, Reston, VA.

Poeter, E.P., and M.C. Hill, 1998. *Documentation of UCODE, A Computer Code for Universal Inverse Modeling.* USGS WRI Report 98-4080.

Quinn, J.J., R.L. Johnson, T.L. Patton, and L.E. Martino. 1996. *An Optimized Groundwater Extraction System for the Toxic Burning Pits Area of J-Field, Aberdeen Proving Ground, MD.* Argonne National Laboratory, Argonne, IL.

Schneider, W. and J. Wrobel. 1998. "Interfacing 3D Visualization Technology With MODFLOW-RT3D and MODPATH to Evaluate Remedial Alternatives." MODFLOW '98 Proceedings, Golden, CO.

Schnoor, J.L. 1996. *Environmental Modeling Fate and Transport of Pollutants in Water, Air, and Soil.* John Wiley & Sons, Inc., Wiley-Interscience Publication.

WESTON (Roy F. Weston, Inc.). 2000. *Draft: A Three-Dimensional Groundwater Flow and Contaminant Transport Model To Evaluate Remedial Alternatives at J-Field.* Prepared for DSHE, Aberdeen Proving Ground, MD.

Yuen, C.R., J. Quinn, L. Martino, R.P. Biang, and T. Patton. 1998. *Natural Attenuation Study of Groundwater at Toxic Burning Pits Area of Concern at J-Field, Aberdeen Proving Ground, Maryland.* ANL, Argonne, IL.

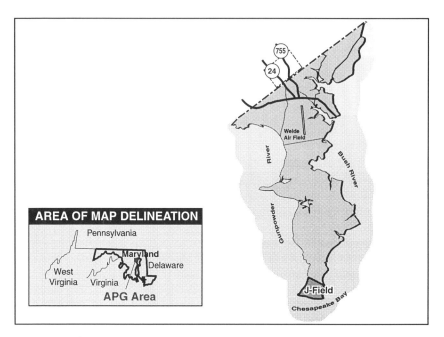

FIGURE 1. Location map for J-Field, Aberdeen Proving Ground, MD

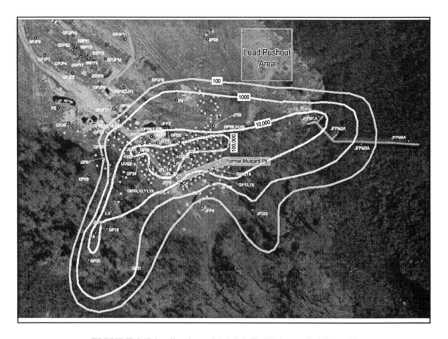

FIGURE 2. Distribution of 1,1,2,2-TeCA in surficial aquifer

Phytoremediation 483

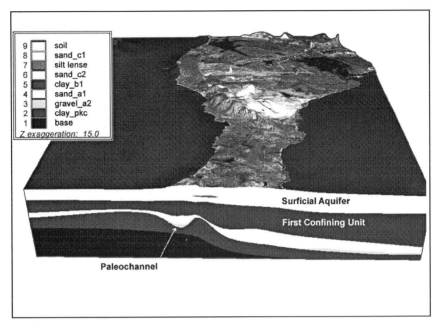

FIGURE 3. 3-D Geospatial model illustrating the hydrostratigraphic units

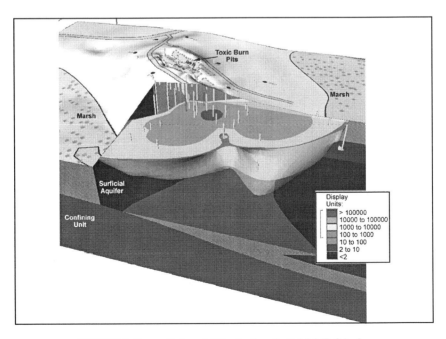

FIGURE 4. Geospatial model illustrating the 1,1,2,2-TeCA plume

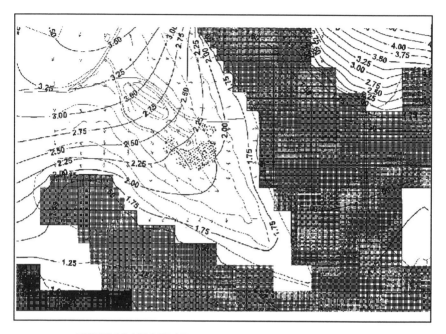

FIGURE 5. MODFLOW grid showing simulated hydraulic head

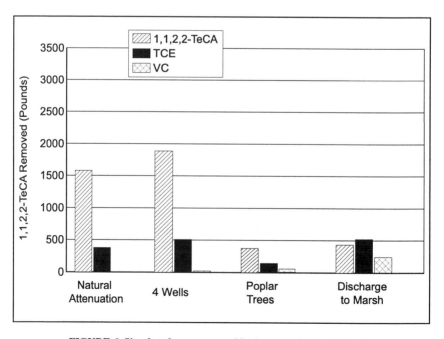

FIGURE 6. Simulated mass removal by integrated remedial system

UPTAKE OF ARSENIC BY TAMARISK AND EUCALYPTUS UNDER SALINE CONDITIONS

Robert W. Tossell (GeoSyntec Consultants, Guelph, Ontario, Canada)
Kinsley Binard (Geomatrix Consultants Inc., Oakland, California)
Michael T. Rafferty (S.S. Papadopulos & Associates, Inc., San Francisco, California)

ABSTRACT: The uptake of arsenic by Tamarisk (*Tamarix parviflora*) and Eucalyptus (*Eucalyptus camaldulensis*) was evaluated as part of a study to evaluate the use of trees to control flow of arsenic-bearing groundwater from a site in East Palo Alto, California. Results of a laboratory study, indicated that Tamarisk and Eucalyptus tolerated elevated concentrations of arsenic in soil (up to 1,000 mg/kg) and dissolved arsenic present in groundwater (up to 25 mg/L). Uptake of arsenic was considerably greater for Tamarisk compared to Eucalyptus under the same treatments. Average arsenic concentrations in roots, shoots and leaves of Tamarisk were 446, 40 and 35 mg/kg respectively across all treatments. Eucalyptus had a similar but lower arsenic concentration distribution averaging 269, 7 and 5 mg/kg in roots, shoots and leaves respectively. Both species tolerated sodium even at the concentrations up to 2,000 mg/L. However, Tamarisk had considerably greater concentrations of sodium in all tissues tested indicating it may have a greater tolerance to sodium. Both Tamarisk and Eucalyptus proved to be excellent candidate species for hydraulic control purposes given the elevated arsenic and sodium present at the Site. The results of the laboratory study were used to design a pilot-scale field test in which approximately 220 Tamarisk trees and 90 Eucalyptus trees were planted at the Site to control groundwater containing elevated concentrations of arsenic.

INTRODUCTION

Arsenic is an anthropogenic metal that has been found a number of sites throughout North America and Europe. Native soil arsenic concentrations generally range from 1 to 50 milligrams per kilogram (mg/kg), but a concentration of approximately 5 mg/kg is common (Bohn et al, 1979; Dragun, 1988). Arsenic (As) is present in soils primarily as As^{3-} (As III) and As^{5-} (As V). Both Arsenic III and V are weak acids and are strongly sorbed to soils under most common soil conditions. The mobility of arsenic as with other heavy metals depends on the metal concentration, nature of the absorbing surfaces (mineral and organic), the abundance of ligands, the alkalinity or acidity of the soil and oxidation/reduction state of the soil (Bohn et al, 1979).

A number of studies reported the affect of arsenic on plant toxicity and uptake, (Maharg et al., 1993; Salt et al., 1995) by herbaceous or aquatic species. Dushenko et al. (1995), reported concentrations in pond weed (*Potamogeton pectinatus*), grown in arsenic rich sediments, ranging from 190 to 4,990 mg/kg dry weight. This study indicated that pond weed, and possibly other plants, have a high tolerance to arsenic. Dushenko et al. further indicated that mycorrhiza and other microorganisms may play an important role in the phytotoxicity of arsenic at a given site. The primary response to arsenic by plants is reduced growth (Dushenko et al., 1995). Any significant reduction in growth or transpiration would be deleterious to the application of hydraulic control using trees. Sheppard (1992) reported significant growth reduction of herbaceous plants when exposed to arsenic. Studies reporting the effect of arsenic toxicity and uptake by perennial woody species have been limited.

This study evaluated the uptake of arsenic by Tamarisk and Eucalyptus as a means of controlling migration of arsenic-bearing groundwater. This study may be the first to report the uptake of arsenic by woody phreatophytes, Tamarisk and Eucalyptus and the significance of these species for phytoremediation of sites containing arsenic.

Objective. The objective of the laboratory study was to evaluate the effect of arsenic and sodium found in Site groundwater and soil on the uptake of arsenic and sodium by Tamarisk and Eucalyptus.

Site Conditions. Site is located adjacent to the San Francisco Bay next to a salt-water tidal wetland. Site geology consists of fine-grained (silts and clays) and coarse-grained (sands and gravel) alluvial and marine sediments. The shallow groundwater zone extends to approximately 18.5 meters below ground surface (m bgs) and has two water-yielding units. These units are referred to as the upper shallow groundwater zone, which is found between 1.5 and 4.6 m bgs, and the lower shallow groundwater zone, which is found at 7.7 to 10.8 m bgs. The flow in the shallow groundwater zones is generally in a south-easterly direction, toward the tidal wetland.

Arsenic is present in the groundwater at concentrations ranging from 0.01 to 0.2 milligrams per liter (mg/L) over most of the area, but can be as high as 200 mg/L in localized areas. Concentrations of arsenic in soil vary greatly, averaging 10 mg/kg and ranging from less than 1.0 mg/kg to greater than 1,000 mg/kg in localized areas. Sodium concentrations range from 150 to 2,000 mg/L and are greater near the tidal wetland.

MATERIALS AND METHODS

An eighteen-week greenhouse study evaluated the uptake and translocation of arsenic by Tamarisk and Eucalyptus under simulated field conditions. Treatments tested in the study included varying arsenic and sodium concentrations in water solution, varying phosphate concentration in soil, and the use of soils treated for arsenic stabilization and untreated soils. Arsenic and sodium treatments were applied to the trees in water solution at concentrations that represented an average and maximum concentration present at the site. Sodium Chloride was added at concentrations of 500 and 2,000 mg/L and sodium arsenate at concentrations of 2.5 and 25 mg/L. A control (water and nutrients only in untreated soil) was used to collect data on the baseline performance of trees without the affect of high concentrations of sodium and arsenic. Treated soil was composed of 20 percent (%) (by volume) cement treated soil (10% Portland cement), 60% native untreated soil, and 20% commercial grade peat. Untreated soil was composed of 80% native untreated soil, and 20% commercial grade peat. For high phosphate treatments, twenty grams of a water soluble solid phosphate fertilizer (0-20-0 commercial grade superphosphate) was added to the top of each pot at the time the trees were planted. This is approximately equivalent to a 120 kilograms/hectare [kg/ha] field application (~100 mg/L orthophosphate in soil solution).

Tamarisk saplings (approximately 0.9 to 1.2 m tall) and potted Eucalyptus seedlings (approximately 0.3 to 0.6 m tall) were planted in 13 L pots with drainage holes at the base. Each of these planting pots was placed in a larger (19 L) pot to contain the water and solutions, which were supplied to the trees (Tossell et al, 1998). Each treatment was replicated four (4) times to measure variability within treatments and for the purpose of conducting statistical analyses. A constant head reservoir (4 L) was used to supply treatments to the plant on-demand by maintaining 5 cm of standing water in each larger pot. At the termination of the greenhouse study on day 123, final size measurements were taken (Tossell et al., 1998). Soil and plant tissue samples (leaves, shoot tissue and root tissue) were collected from each pot/tree to analyze the concentrations of phosphorus, sodium and arsenic. Statistical analyses consisted of Analysis of Variance (ANOVA) with orthogonal contrast procedures and Multiple Range Tests. The Kolmogorov-Smirnov test was used to determine if data were normally distributed as required for the ANOVA analysis. The Kruskal-Wallace Test was used in the event that the data were not normally distributed.

RESULTS AND DISCUSSION

Soil Chemistry. Table 1 presents a summary of soil arsenic and sodium concentration data from samples collected at the termination of the study. Concentrations of arsenic in potted soils were quite variable, ranging from 596 to 2,408 mg/kg. Treatments containing Tamarisk had soil arsenic concentrations, which ranged from 596 to 2,226 mg/kg. Eucalyptus had similar soil arsenic concentrations, ranging from 614 to 2,408 mg/kg. Both the control treatments had similar and relatively low arsenic concentrations (Tamarisk = 687 mg/kg and Eucalyptus − 651 mg/kg). The highest arsenic concentrations were not always associated with the highest arsenic treatments and therefore no clear trend in soil arsenic was noted.

Sodium in soils was much higher than concentrations of arsenic, ranging from 626 mg/kg to 5,640 mg/kg across all treatments. The sodium concentration of the controls was significantly lower than those treatments receiving the highest doses of sodium in solution (Probability Value [PV] = 0.001). Tree type and the concentration of sodium applied in solution had the greatest affect on soil sodium concentrations. The soil in pots containing Tamarisks had higher sodium concentrations than the soils collected from pots bearing Eucalyptus trees. It is probable that this difference reflects the higher volume (i.e. greater mass use) of water used by Tamarisks compared to Eucalyptus. Higher rates of water use would necessitate that higher volumes of sodium bearing water be drawn into the potting media from the water supply vessels. The residual sodium concentrations in the soil would reflect the increased flux of sodium in solution over the course

TABLE 1: Summary of Study Treatments and Soil and Plant Tissue Chemistry

Treatment ID	Treatments			Final Soil Concentrations (mg/kg)		Final Plant Concentrations (mg/kg)						
						Roots		Shoots		Leaves		
	Arsenic (mg/L)	Sodium (mg/L)	Phosphate (mg/L$_{soil}$)	Soil (mg/kg)	Arsenic	Sodium	Arsenic	Sodium	Stem	Sodium	Arsenic	Sodium
Tamarisk												
ASp	25	2,000	NA	300	1,152	4,650	700	7,710	18.7	14,300	12.0	43,725
aSp	2.5	2,000	NA	50	596	5,640	75	6,680	3.0	8,320	4.0	47,100
ASP	25	2,000	100	50	724	5,000	453	7,970	67.2	15,100	143.9	51,050
aSP	2.5	2,000	100	300	2,147	5,050	879	8,860	10.4	6,260	3.0	55,825
Asp	25	500	NA	50	2,226	3,270	275	5,480	55.0	6,150	51.8	15,500
asp	2.5	500	NA	300	1,606	2,650	451	5,690	6.1	7,750	4.7	22,300
AsP	25	500	100	300	1,554	1,910	797	4,770	152.0	9,860	60.6	32,375
asP	2.5	500	100	50	669	2,560	314	5,050	46.6	9,420	26.5	26,567
C	NA	NA	100	50	687	1,150	72	2,300	4.0	4,480	5.1	4,967
Eucalyptus												
ASp	25	2,000	NA	300	1,101	2,750	370	4,170	10.8	2,570	4.5	3,658
aSp	2.5	2,000	NA	50	614	3,080	142	3,830	5.5	1,940	7.3	2,063
ASP	25	2,000	100	50	1,812	2,450	236	4,300	6.7	2,010	3.2	1,796
aSP	2.5	2,000	100	300	955	2,420	595	4,370	9.1	4,290	4.1	3,205
Asp	25	500	NA	50	943	1,940	213	3,230	2.5	782	6.8	1,758
asp	2.5	500	NA	300	1,587	1,780	136	3,780	13.3	2,400	5.9	3,518
AsP	25	500	100	300	2,408	1,700	498	3,750	5.2	2,220	5.7	1,318
asP	2.5	500	100	50	1,038	2,247	112	3,040	4.2	1,860	3.7	2,320
C	NA	NA	100	50	651	626	122	2,090	9.4	904	5.1	859

Notes:

Treatment ID: letter code corresponds to treatment (A = arsenic, S = sodium, P = Phosphate, C = Control); upper case corresponds to high treatment concentrations, lower case are low treatment concentrations

Arsenic and Sodium Treatments: applied in solution during watering

NA - no treatment applied (control)

of the study. In absence of the effect of tree type, the concentrations of sodium in soil correlated very strongly with the concentration of sodium applied in treatment solutions.

Plant Tissue Chemistry. Analytical characterization of arsenic and sodium in plant tissues is presented in Table 1. Sodium tissue concentrations were generally 1 to 3 orders of magnitude greater than arsenic, with the greatest difference occurring in the leaves. Arsenic concentrations were highest in the roots and the lowest in the leaves of both Tamarisk and Eucalyptus. Sodium concentrations in Tamarisk leaves were an order of magnitude greater than both shoot and root concentrations. Eucalyptus had similar concentrations of sodium in roots, shoots and leaves, but were generally lower than Tamarisk.

Roots. Concentrations of arsenic in root tissues were highly variable among the experimental treatments, ranging from 72 to 879 mg/kg (Figure 1a). Patterns of response between Tamarisk and Eucalyptus were virtually identical although concentrations of arsenic in the roots of Tamarisks were consistently higher than Eucalyptus for corresponding treatments. The average arsenic content of Tamarisk roots for all treatments was 446 mg/kg compared to 269 mg/kg in Eucalyptus roots. This likely reflects a difference in the capacities of the two species to uptake arsenic from soils and water. Overall, the experimental treatment that accounts for the greatest variability in observed concentrations of arsenic in the roots of both Tamarisk and Eucalyptus is soil arsenic concentration. Tamarisks and Eucalyptus exposed to soils with higher initial arsenic concentrations exhibited the highest concentrations of root tissue arsenic. Phosphorus treatments added to the soil at the start of the study did not effect concentrations of arsenic in roots as expected and reported in Meharg and MacNair (1994).

Sodium concentrations in root tissue were also variable. Tamarisk root sodium concentrations were consistently higher than those of Eucalyptus. The highest sodium concentrations were found in roots of those trees with the high sodium treatments. The control treatments of both Tamarisk and Eucalyptus had the lowest concentrations of sodium in root tissue. Phosphorus treatments did not exhibit any clear increase or decrease in sodium uptake in either species.

Shoots. Arsenic concentrations in shoots of both Eucalyptus and Tamarisks were much lower than root tissue concentrations (Table 1 and Figure 1b). Shoot arsenic concentrations ranged from 2.5 to 152 mg/kg, and averages were considerably higher in Tamarisk shoots (40.3 mg/kg), than in the shoots of Eucalyptus (7.4 mg/kg). The concentration of arsenic in Tamarisk shoots were influenced by the concentration of arsenic in soil and the applied treatment solutions. This is not surprising as Tamarisk had greater water use rates and would therefore uptake more arsenic in solution. The control treatments had the lowest arsenic shoot concentrations. No clear trends were noted regarding arsenic concentrations in Eucalyptus shoots.

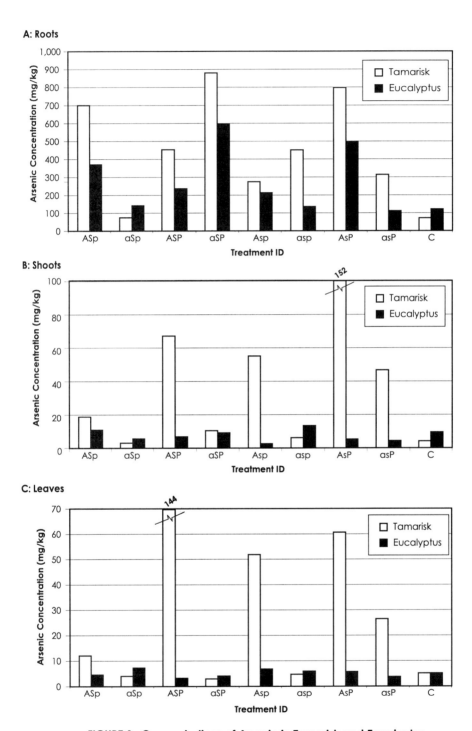

FIGURE 1. Concentrations of Arsenic in Tamarisk and Eucalyptus roots (a), shoots (b) and leaves (c).

Tamarisk sodium concentrations in shoots ranged from 4,480 to 15,100 mg/kg. Sodium concentrations in Tamarisk shoots were approximately 3 fold greater than that of Eucalyptus. Eucalyptus sodium concentrations ranged from 2,090 to 4,370 mg/kg.

Leaves. Concentrations of arsenic in the leaves, of both Tamarisks (3.0 mg/kg to 143.9 mg/kg) and Eucalyptus (3.2 mg/kg to 7.3 mg/kg), were similar in magnitude (within plant type) to the concentrations in the shoots and substantially lower than arsenic concentrations in root tissues of these trees (Table 1). Tamarisk leaf arsenic concentrations were approximately one order of magnitude greater than that of Eucalyptus at 35 mg/kg and 5 mg/kg respectively (Figure 1c). Tamarisk leaf arsenic concentrations from 3 of 4 high arsenic treatments, were found to be significantly higher than the control treatment (Probability Value [PV] = 0.001). However, no consistent trend was noted between high and low arsenic treatments. Eucalyptus treatments did not exhibit the same trend as no significant differences in treatments were noted.

High concentrations of arsenic in Tamarisk leaves corresponded to high phosphate levels, suggesting that phosphorus may increase the uptake of arsenic and translocation to the leaves. Phosphorus may therefore be used to aid with the extraction of arsenic from soils.

Concentrations of sodium in leaves were significantly greater in Tamarisk treatments than in Eucalyptus treatments. Sodium in leaves did not exhibit any significant differences among Eucalyptus treatments. However, all Tamarisk treatments containing any sodium addition were significantly greater than the Tamarisk control. No consistent trend was noted between high sodium treatments and low sodium treatments of either tree species.

CONCLUSIONS

This study evaluated the uptake of arsenic by Tamarisk and Eucalyptus as a means of controlling migration of arsenic-bearing groundwater. Tamarisk and Eucalyptus trees tolerated elevated concentrations of arsenic in soil (1,000 mg/kg) and dissolved arsenic present in groundwater (25 mg/kg). Uptake of arsenic was considerably greater for Tamarisk compared to Eucalyptus under the same treatments. It is believed that this study is the first to report the uptake of arsenic by woody phreatophytes, Tamarisk and Eucalyptus.

Both species tolerated sodium at the concentrations up to 2,000 mg/L but Tamarisk appears to have a greater tolerance to sodium. Both Tamarisk and Eucalyptus proved to be excellent candidate species for hydraulic control purposes given the elevated arsenic and sodium present at the Site. These species are well suited to the climate of the Southwestern United States and could prove useful for extraction of metals, or organic compounds from groundwater at depths as deep as 8 to 10 m bgs. As the next phase of this study, Tamarisk and Eucalyptus trees were planted and monitored as part of a pilot-scale remedy to control groundwater containing elevated concentrations of arsenic at the Site.

REFERENCES

Bohn, H.L., B.L. McNeal, and G.A. O'Conner. 1979. *Soil Chemistry*. John Wiley-Interscience Publications, New York.

Dragun J., 1988. *The Soil Chemistry of Hazardous Materials*. Hazardous Materials Control Research Institute, Silver Spring, Maryland.

Dushenko, W.T., and K.J. Reimer, 1995. "Arsenic Bioaccumulation and Toxicity in Aquatic Macrophytes Exposed to Gold-Mine Effluent: Relationships with Environmental Partitioning, Metal Uptake and Nutrients." Aquatic Botany 50 (1995): 141-158.

Meharg, A.A., Q.J. Cumbes, and M.R. MacNair, 1993. "Pre-Adaptation of Yorkshire Fog, Holcus lanatus L. (Poaceae) to Arsenate Tolerance." Evolution. 47(1): 313-316.

Meharg, A.A., and M.R. MacNair, 1994. "Relationship Between Plant Phosphorus Status and the Kinetics of Arsenic Influx in Clones of Deschampsia cespitosa (L.) Beauv. That Differ in Their Tolerance to Arsenate." Plant and Soil. 162: 99-106.

Salt, D.E., M.Blaylock, N.P.B.A. Kumar, V. Dusnhenkov, B.D. Ensley, I. Chet and I. Raskin, 1995. "Phytoremediation: A Novel Strategy for the Removal of Toxic Metals from the Environment Using Plants." Bio/Technology. 13: 468-474.

Sheppard, S.C., 1992. "Summary of Phytotoxic Concentrations of Soil Arsenic." Water, Air and Soil Pollution. 64: 539-550.

Tossell, R.W., K. Binard, L. Sangines-Uriarte, M. T. Rafferty and N. P Morris. 1998. "Evaluation of Tamarisk and Eucalyptus Transpiration for the Application of Pytoremediation." First International Conference on Remediation of Chlorinated and Recalcitrant Compounds, Monterey, CA. Bioremediation and Phytoremediation. 4: 257-262.

PHYTOREMEDIATION OF A CREOSOTE CONTAMINATED SITE – A FIELD STUDY

John T. Novak, Mark Widdowson, Mark Elliott and Sandra Robinson
Virginia Polytechnic Institute & State University
Blacksburg, VA 24061 USA

ABSTRACT: In 1997, a phytoremediation system consisting of approximately 1,000 hybrid poplar trees was undertaken at a 2.5-acre, creosote-contaminated site in Tennessee. A study was undertaken to characterize the extent of contamination (polycyclic aromatic hydrocarbons, PAHs) and to assess the performance of the trees for removal of contamination from the soil and groundwater and to control the movement of water across the site. The site is underlain with shale at a depth of approximately 3 meters. PAH-contaminated soil and groundwater was present throughout the site but creosote, present as a dense non-aqueous phase liquid (DNAPL), was perched on top of the shale at a depth of up to 25 cm. Multi-level samplers were used to characterize groundwater quality. Phytoremediation, groundwater extraction and natural processes, including aerobic and anaerobic biodegradation, are contributing to attenuation and a gradual reduction in soil and groundwater contamination over time but it is unclear the extent to which the phytoremediation system is contributing to the removal of PAHs. Monitoring will continue at this site for several more years to determine if phytoremediation is a suitable technology for this site.

INTRODUCTION

Phytoremediation is an attractive remediation method because of its low cost relative to other remediation technologies and its acceptance by the public. However, for this method to be used, the limitations of phytoremediation must be acceptable from a risk standpoint. The primary concerns with regard to phytoremediation are the time it takes for the plant system to become fully functional and the dormancy of the plants during winter months.

The site that was used for this study was heavily contaminated with creosote. The creosote was found during a stream rechanneling project and was discharging directly to the stream. To stop the creosote from moving off site, a trench was dug down to rock to intercept the groundwater flow (Figure 1) and the contaminated groundwater was pumped to a separation tank where the contaminants were removed and the water was discharged to a sewer line. Once the flow was controlled, phytoremediation became feasible because contamination could no longer move off site. In 1997, the state regulatory agency approved a demonstration study of phytoremediation. Approximately 1,000 hybrid poplar trees were planted in May 1997 and monitoring of the site began.

The site lies next to a railroad yard and not only was contaminated with creosote but also was underlain with discarded coal in layers up to two feet thick.

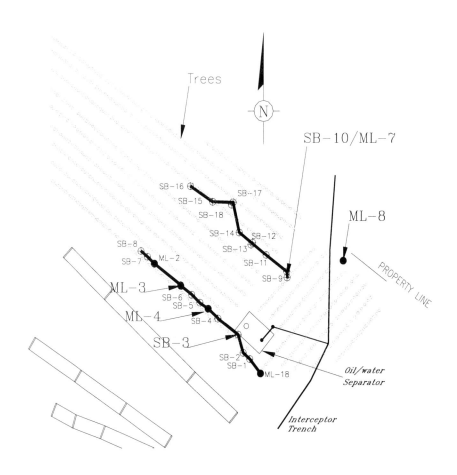

FIGURE 1. Location of site with soil transects.

The coal covered about 35% of the site. The soil which was excavated for the interceptor trench was piled up at the site and before the poplar trees were planted, this soil, which was contaminated with PAHs, was spread across the site at a depth of about two feet. The coal layers were not removed but were covered with the contaminated soil. The trees were planted in trenches (Figure 1) to which fertilizer was added.

As noted, the upper two feet contained a layer of material excavated from construction of the trench and this material was primarily sandy clay. At a depth of two to three feet, pieces of wood, coal, gravel and black plastic material might be found, depending on the specific location. Below three feet, the site was underlain by variably-thick layers of sandy clay and clay. At about eight feet, sand was present and beneath this, at around ten feet, was a confining layer of shale. Creosote was pooled on top of the shale over about 3/4 acre of the site and, in the low spots, was approximately one foot thick. Creosote was also found in the soil and there were distinct layers in some locations at a depth of five to six feet below ground surface.

METHODS AND MATERIALS

In order to evaluate the progress of remediation, 20 multilevel samplers were placed around the site. The sampling ports were arranged so that groundwater samples could be obtained at every foot. In addition, two soil transects (Figure 1) were used to characterize the soil contamination levels. The sampling transects were selected to characterize changes in the contaminant plume over time. Transect 1 cut across the contaminated area while Transect 2 started in the middle of the most contaminated area and moved to clean soil. Soil samples were obtained across both soil transects using a hand auger and samples were collected at every foot down to bedrock. Usually, the first three feet were not characterized because these soils were above the groundwater level and frequently were contaminated with coal. When coal was present in samples, no PAH analysis was made because the coal contributed non-creosote PAHs.

To characterize soil contamination, six PAHs were quantified: acenaphthene, fluorene, phenanthrene (three-ringed PAHs) and fluoranthene, pyrene and chrysene (four-ringed PAHs). These six compounds were almost always present when PAHs were found and made up the majority of PAHs. Soil samples were collected using a hand auger, placed in plastic bags, and stored on ice. They were then transported back to the lab where they were refrigerated at 4°C until analyzed. Extraction and analysis procedure is described in detail in Fetterolf et al. (1999).

In addition to the six PAHs mentioned above, four others were measured for groundwater analysis: naphthalene, acenaphthylene, anthracene, and benzo(b)fluoranthene. Groundwater samples were collected in 15ml amber bottles filled to the top. The vials were also stored on ice and transported to the lab for analysis. Samples were extracted with methylene chloride and quantified by gas chromatography. In addition to PAH analysis, groundwater was characterized in the field for dissolved oxygen (DO) using a Winkler titration method and iron(II) using a colorimetric technique. Laboratory analysis of sulfate, nitrate, and nitrite was by ion chromatography.

RESULTS AND DISCUSSION

Groundwater. Remediation of groundwater was evaluated by considering the changes in plume size and concentration over time. Three time snapshots of total PAH concentration distributions are shown in Figure 2, coinciding with sampling events scheduled between growing seasons (March 1998, January 1999 and December 1999). The plume was observed to be shrinking in size, especially in the Southwest area near the railroad tracks. The data for Figure 2 is for samples taken between three and six feet above bedrock. The December 1999 groundwater samples collected from the multilevel samplers along Transect 1 were generally clean from two feet above bedrock to the top of the groundwater table.

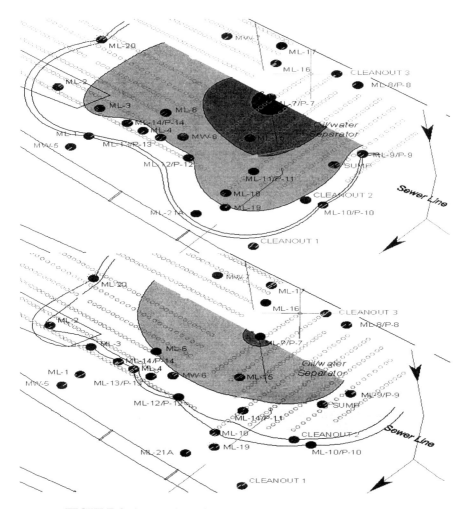

FIGURE 2. Reduction of total PAHs in groundwater plume (Top-March 1998, Bottom-December 1999).

Immediately above bedrock, PAHs were detected but PAH concentrations were also declining over time.

Naphthalene concentration distribution from Transect 1 for the three time snapshots are shown in Figure 3. It can be seen that in March 1998, high levels of contamination were present at ML-11 and ML-3. In September 1998, a major source of creosote was removed up gradient of ML-2 and ML-3 and this may have contributed to the reduction of naphthalene in this area. However, the decrease in naphthalene concentration at ML-11 is most likely due to a combination of natural attenuation and phytoremediation. Figure 3 shows significant decline in naphthalene in the water table region and also at the lower levels near bedrock. In Figure 4, the total PAHs are shown as a function of depth at ML-3. Groundwater PAH concentrations at ML-3 have declined to near zero at

Phytoremediation

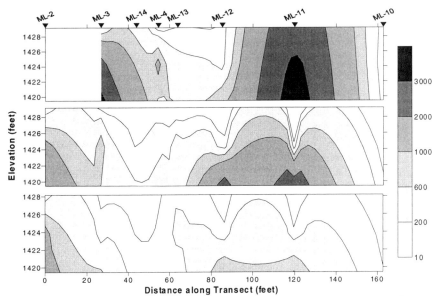

FIGURE 3. Changes in groundwater naphthalene concentration (µg/L) across Transect 1
(Top-March 1998, Middle-January 1999, Bottom-December 1999).

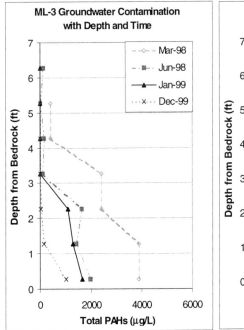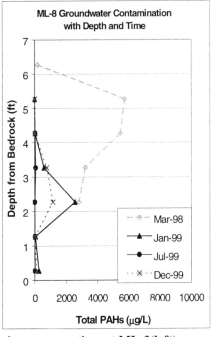

FIGURE 4. Groundwater PAH changes over time at ML-3 (left) and ML-8 (right).

four to six feet above bedrock and from approximately 4,000 to 1,000 µg/L at sampling ports located just above bedrock.

The primary reason this demonstration project was approved by the regulators was that the trench was in place to cut off the discharge of contaminants to the stream. Three multilevel samplers were placed down gradient of the interceptor trench to confirm contaminant capture by the interception and phytoremediation systems. Once the trench was operational, the discharge of contaminants was reduced to below detection in most samples. For ML-9 and ML-10, no PAHs were found. In ML-8 contaminant levels have been greatly reduced from March 1998 when the trench went on-line to the present. As shown in Figure 4, groundwater contamination in ML-8 has been reduced but not entirely eliminated. However, no contamination has been detected in the stream.

Soil Contamination. Soil samples were collected along the two transects over three years. For Transect 1, the PAH levels have declined in much the same manner as the groundwater (Figure 5). The concentration of chrysene has declined to near zero in the upper regions but has also declined significantly near bedrock.

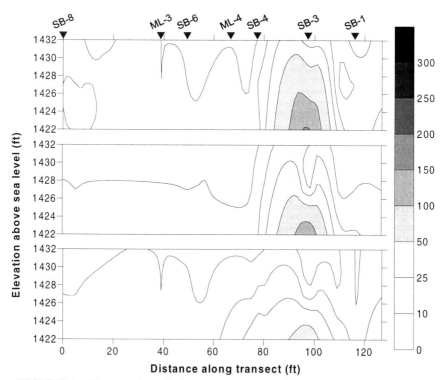

FIGURE 5. Changes in soil chrysene concentration (mg/kg) over time for Transect 1 (Top-July 1997, Middle-June 1998, Bottom-October 1999).

SB-3, where the contamination is greatest, is the same location as ML-11 so it can be seen that both the soil and groundwater are being remediated. The creosote contamination immediately above bedrock remains high but has undergone significant reduction across Transect 1.

Data for Transect 2 are not yet showning reduction in the soil or groundwater contamination levels. At the center of the plume, pure creosote can be pumped from the lowest port of ML-7 and this creosote appears to be lying in a low spot in the bedrock. Therefore, this is the location where contaminant reduction is least likely to be occurring. However, as the plume shrinks and the trees continue to grow, it is expected that the most severely contaminated areas will also improve.

The Role of Phytoremediation. The goal of this project was to evaluate phytoremediation for treatment of creosote contaminated soil and groundwater. There are formidable challenges at this site caused by the presence of layers of coal and the high concentration of creosote sitting on bedrock in nearly pure form. However, after 2 1/2 years following the planting of trees and removal of a major contaminant source, it is clear that the soil and groundwater has improved. There is also evidence that the creosote DNAPL has receded over part of the site.

There are three remediation methods being employed at the site: direct removal by interception by the trench, natural bioattenuation, and phytoremediation. Evidence for the first two is compelling. The contamination levels of the groundwater decline substantially after passing the trench so direct removal appears to be controlling the movement of contamination off-site. Groundwater data shows that when oxygen is present, PAHs are at or near zero, suggesting that aerobic degradation is occurring. Additional data from the MLS network show anaerobic biodegradation products are present in groundwater.

Evidence for phytoremediation is much less direct. Tree heights in some areas are at 30 feet, suggesting that the roots should penetrate to a depth of 8 to 10 feet below land surface. Additionally, soil contamination in the treed areas at a depth of four to five feet below ground surface has declined over time (Figure 4) and similarly, groundwater contamination in the upper region of the saturated zone has also declined (Figures 2 and 3). It has also been found (data not shown) that the trees have a significant impact on the groundwater table during the active growing season. Therefore, the roots are withdrawing groundwater, and the contaminants in the groundwater should be undergoing treatment by the trees. Examination of the tree roots in the contaminated area show no accumulation of PAHs or degradation products so it cannot be demonstrated that direct uptake is occurring. Tree growth has not been negatively impacted by the creosote and high PAH concentrations. In the most contaminated areas, tree growth is similar to locations where the soil is less contaminated.

It is most likely that phytoremediation will become a much more important factor now that the roots are penetrating to near bedrock. The test of phytoremediation will be to see if the free product sitting in low bedrock areas will be removed or if continued reliance on the interceptor trench will be needed.

Proof of phytoremediation for this site is likely to come over the next two years as the tree roots enter the most severely contaminated areas.

CONCLUSIONS

Although the site is being remediated through direct removal and natural attenuation, the value of poplar trees and phytoremediation has not yet been fully demonstrated. Continued monitoring over the next two years should provide the data to evaluate phytoremediation at this site. If the trees can eliminate the most mobile fractions of the creosote and allow discontinuation of the interceptor trench operation, the technology will be deemed a success. After 2 1/2 years of operation, the value of phytoremediation has yet to be determined.

REFERENCES

Fetterolf, G.J., Novak, J.T., Crosswell, S.B. and Widdowson, M.A. (1999) Phytoremediation of Creosote-Contaminated Surface Soil. *Phytoremediition and Innovative Strategies for Specialized Remedial Applications*, A. Leeson and B.C. Alleman, eds. Battelle Memorial Institute, Columbus, OH.

IN-SITU BIOREMEDIATION OF #2 FUEL OIL UTILIZING PHYTOREMEDIATION

Eric P. Carman (ARCADIS Geraghty & Miller, Milwaukee, Wisconsin)
Tom L. Crossman (ARCADIS Geraghty & Miller, Tampa, Florida)
Kevin L. Daleness (ARCADIS Geraghty & Miller, Milwaukee, Wisconsin)

ABSTRACT: A release of fuel oil occurred in the 1970's at an industrial facility in Wisconsin. Diesel range organics (DRO) that exceed Wisconsin cleanup levels are present in four generalized hotspots at the facility. Phytoremediation was implemented to enhance bioremediation in the hotspots, based on bioremediation feasibility testing, where DRO was shown to be reduced 40% to 90% within the laboratory and the results from an agronomic assessment. During 1996, 300 *Prairie cascade* willows were planted to enhance microbial growth within the rhizosphere of the trees. The growth of the trees has been encouraging and test pits show that the rhizosphere has extended into the DRO-impacted soil. Samples from within the rhizosphere were collected for analysis of DRO in October 1999 to aid in determining the effectiveness of the trees in enhancing bioremediation of fuel oil constituents. The results from the sampling show that the phytoremediation program is progressing as planned. Although there is variability with the soil sample results, concentrations of DRO in Hotspots 2 and 4 have decreased between 66 and 68 percent after three years, however there is no conclusive trend in the concentrations of DRO in Hotspot 1. Additional soil samples will be collected in 2002.

INTRODUCTION

Phytoremediation, the use of plants for remediation of soil, sediments, and water, is a technology that holds promise to cost effectively address sites contaminated with moderately hydrophobic compounds, such as petroleum hydrocarbons (PHCs), chlorinated solvents, munitions, and excess nutrients (Schnoor et al., 1995). In addition, phytoremediation holds promise toward the cleanup of sites contaminated with heavy metals (Azadpour and Matthews, 1996) and recalcitrant organics, such as those found at manufactured gas plants (MGP) and refinery sites (Schwab and Banks, 1994) and sites with soil impacted with petroleum hydrocarbons (Banks et al., 1999). Plants are reported to remediate contaminated environments by several mechanisms. These mechanisms include either direct uptake of contaminants by plants and the resulting accumulation, biodegradation, or volatilization of those contaminants; or enhancement of the biodegradation process in the rhizosphere. Enhanced biodegradation processes within the rhizosphere has recently been referred to as enhanced rhizosphere degradation (USEPA, 1998).

Certain trees, known as phyreatophytes, take up large volumes of water and can be used to hydraulically control and treat contaminated groundwater plumes. At a site in New Jersey, for example, nitrate concentration in groundwater was significantly reduced following implementation of a *TreeMediation*^{TM} program (Gatliff, 1994). For this project, enhanced rhizosphere degradation by a species of phyreatophyte has been implemented to stimulate the biodegradation of DRO in an aged fuel oil spill in Wisconsin.

Objective. The objective of this project is to remediate soil and fill materials contaminated with DRO within the four study areas at the facility to below 1,000 mg/kg DRO, as required by the Wisconsin Department of Natural Resources (WDNR). Excavation and treatment of the soil materials was not a preferred option, based on the potential costs associated with segregating construction debris from soil.

Site Description. Operations began at this site in the early 1900s. A heterogeneous mixture of fill material was used to extend the property boundary west to an adjacent river (Figure 1). During the late 1970s, a section of below-ground piping transferring No. 2 fuel oil from a larger aboveground storage tank (AST) to a smaller AST failed, resulting in a subsurface release. Approximately 15,000 gallons (56,800 liters) of fuel oil were recovered from shallow trenches installed at the site. Concentrations of hydrocarbon constituents in groundwater are below current WDNR-established groundwater quality standards.

Investigations at the site have determined that highly contaminated soil (concentrations greater than 1,000 mg/kg DRO) remain in four generalized hotspots at the site (Figure 1). The site is underlain by 3 to 15 feet (0.9 to 4.6 m) of heterogeneous fill material comprised of wood timbers, sawdust, construction debris, clay, sand, and gravel. The depth of the water table ranges from 3 to 9 feet (0.9 to 2.7 m), and the water table slopes west toward the adjacent river. Three of the hotspots (Hotspots 1, 2 and 4) are below a hard-packed gravel equipment storage area, and a fourth (Hotspot 3) is located below a vegetated area along the river.

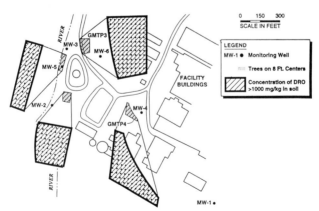

FIGURE 1. Site schematic and locations of the soil hotspots. A total of 300 *Prairie cascade* willow trees planted during May 1996 within the hotspots.

MATERIALS AND METHODS

Sample Collection and Soil Characterization. Soil samples were collected and analyzed during May 1994, October 1995, April 1996 and October 1999. Backhoe test pits were completed during May 1994 to collect soil samples that were used to determine initial concentrations of DRO in soil, microbial population densities and respiration rates, and to perform accelerated bioventing tests. The results from those samples were a major factor in recommending the phytoremediation program and are described in Carman et al. (1998). Six soil samples were collected from three test pits and were visually screened in the field with a flame ionization detector. Soil samples were collected from the capillary fringe at depths of between 3 feet and 5 feet (0.9 to 1.5 m) below land surface (ft bls). Soil samples were also collected during May 1996, October 1996, and October 1999 to establish the perimeter of each area to be planted, determine initial concentrations of DRO, and the concentration of DRO after three years of phytoremediation. Soil samples were collected during these events using hollow stem augers and standard splitspoon sampling techniques or geoprobe techniques.

During the October 1999 sampling, soil samples were collected from within 1 foot of previous boring locations to monitor the progress of biodegradation within the rhizosphere of the trees. The soil sampling locations were surveyed and samples were collected for analysis of DRO from the same depth intervals as samples collected in 1995, prior to implementing the phytoremediation program. Soil samples were collected both inside and outside of the planted areas so that a comparison could be made between the rate of decreased concentrations of DRO within the rhizosphere, and the rate of decreased concentrations of DRO in an unplanted area (to represent background).

A backhoe was also used to remove soil around the root area and expose the rhizosphere of a randomly selected tree in Hotspot 2 during October 1999. Soil samples were collected within densely cluster roots in the rhizosphere, outside the apparently rhizosphere, and midway between those two locations. The three soil samples were collected from within a distance of 6 inches (15.24 cm) from each other.

Soil samples for DRO analysis were submitted to a WDNR-certified laboratory for analysis. Chromatograms were requested from and provided by the laboratory for analysis.

Agronomic Assessment. In June 1995, soil samples were collected from Hotspots 2 and 3 to determine potential phyto-toxic effects on tree root development and concentrations of agronomic constituents of interest, including nitrate and ammonia-nitrogen, phosphorous, potassium, zinc, iron, manganese, copper, sulfur, calcium and sodium, pH, soluble salts (as specific conductance) and percent organic matter. The bench top root development study was performed by Applied Natural Sciences (ANS) of Fairfield, Ohio using soil collected from Hotspot 2 and Hotspot 3 and cuttings from willow trees indigenous to the site. Soil and cuttings were refrigerated by ANS until July 1995. Cuttings were removed from the refrigerator, transplanted into sand and rooted. The rooted cuttings were planted and were monitored for leaf and root development. After two months of development, willow and hybrid poplar plants were transplanted into soil collected from Hotspot 2 and Hotspot 3. Plant growth was observed for a period of 6 weeks, the plants were harvested and root systems were evaluated. Soil samples were submitted to Servi-Tech Inc. of Dodge City, Kansas for analysis.

RESULTS AND DISCUSSION

Agronomic Assessment. Both hybrid poplars and willows exhibited good aerial growth during the root development portion of the agronomic assessment. The willows, however, demonstrated a more pronounced tendency to establish rooting within the DRO-affected soil. With the exception of the relatively high concentration of soluble salts in subsurface soil, the analyzed constituents were within acceptable ranges for tree growth. Based on the results of the agronomic assessment, the decision was made to plant a species of willow in the four soil study areas.

Tree Planting. Hybrid willow trees (*Prairie cascade*) were planted in the four hot spots during the week of May 13, 1996. The trees were planted roughly at a spacing of 8 feet (2.4 m), with a total of 300 trees planted (Figure 1).

Trees were planted using *TreeMediation*^{TM}, which is a proprietary process developed by ANS that focuses rooting activity and rhizosphere development in the zone of contamination. The targeted root zone depths for Hotspots 1 and 2 is 4 to 6 feet (bls) (1.2 m to 1.8 m), Hotspot 3 is 8 to 10 feet bls (2.4 m to 3 m), and Hotspot 4 is 6 to 8 feet bls (1.8 m to 2.4 m). Site visits were made more frequently through the first summer in 1996 to monitor the growth of the trees. Precipitation was monitored, however unusually heavy precipitation during the summer and fall 1996 eliminated the need for supplemental watering during the initial year of the phytoremediation program.

Operation and Maintenance. Operation and maintenance (O&M) activities have been performed periodically to maintain the health of the trees and assure biodegradation of DRO within the rhizosphere. The periodic O&M activities have included tissue sampling, insecticide applications, watering, pruning, and observing the overall health and growth of the trees. During 1997 and 1999, test pits have been completed to visually monitor extent of the rhizosphere of the trees.

Leaf tissue samples have been collected during annual site visits beginning in 1996. The purpose of the sampling has been to determine concentrations of agronomic constituents in the trees and to determine the necessity, and adequacy of a fertilization program. In general, results from the leaf tissue sampling indicate the need for surface applications of high nitrogen fertilizer between the rows of trees. During the tissue sampling conducted in October 1996, significant insect damage was noted on the majority of the trees. A systemic insecticide was applied to the trees in the spring of 1997. The results of the insecticide injection were apparent in that there was a significantly reduced insect infestation. A commercial insecticide has been applied on an as-needed basis since 1997 to control periodic infestations. Rainfall in the area has been monitored and watering of the trees has been implemented during dry times by facility personnel. The trees have been pruned as needed, during annual site visits.

Growth of the Trees. In general the trees have exhibited excellent to fair growth since they were planted in May 1996. Trees have grown an average of 4 to 6 feet in height, and trunk diameters have increased from roughly ¾ of an inch at the time of planting up to 4 inches at the end of the third growing season.

The trees in Hotspots 1 and 2 have exhibited faster and more vibrant initial growth than those in Hotspots 3 and 4. The slower and less vibrant initial grow is attributed to the deeper targeted rhizosphere depths in Hotspots 3 and 4. The deeper rhizosphere depth was anticipated to slightly restrict the rate of initial root growth. However, trees in Hotspots 3 and 4 have exhibited apparent accelerated growth rates in the 1998 and 1999 seasons, compared to growth rates in 1996 and 1997.

Tree mortality has been significantly lower than expected. At the onset of the phytoremediation program, it was anticipated that 10% of the trees (total of 30 trees) would be die off during the first year. However, since the program was initiated, only seven trees, or approximately 2% of those that were initially planted have died. The majority of those trees died during the first or second year. One tree in Hotspot 2 was lost after inadvertently being sprayed with herbicide that was being applied for weed control.

During the 1997 and 1999 annual visits, tests pits were completed to monitor the rooting activity within the rhizosphere of the trees. In 1997, the roots had yet to penetrate the DRO-impacted soil. However during the 1999 visit, the test pits showed that the tree roots had begun to penetrate the DRO-impacted soils, which is critical to the success of the enhanced rhizosphere degradation.

Chromatograms and Soil DRO. The chromatograms from the samples analyzed during the previous bioremediation feasibility testing stage of this project exhibited a relative decrease in the proportion of the more water soluble, lower molecular weight DRO as compared to the higher molecular weight counterparts in the fuel oil. In addition to a decrease in concentration, that observed changed in the pattern of the chromatograms over time is consistent with biodegradation of DRO. Chromatograms from DRO analyses performed on samples collected during October 1999 are being analyzed to determine the presence of a pattern that would indicate enhanced rhizosphere degradation.

A statistical approach was used to assist in evaluating the results from the DRO analyses performed in 1995 and 1999. The approach involved calculating geometric means of the DRO results, as well as the median concentrations and standard deviations. These calculations were performed for the samples collected within each planted hotspot and samples collected outside of each hotspot. The statistical data is included in Table 1. For soil samples collected inside Hotspot 2 and 4, the geometric mean of the laboratory results has decreased 66%, and 68% following three growing seasons, respectively. The full data set from Hotspot 1 suggests there is not a conclusive trend in DRO concentrations. However, the geometric mean of samples collected inside Hotspot 1 in 1999 was 493 mg/kg, while the geometric mean for DRO concentrations in 1995 was 251 mg/kg. Only one soil sample was collected from Hotspot 3 due to inaccessibility of a wide profile drill rig, and inability for a narrower profile rig to drill to the requisite depth. The single sample taken from Hotspot 3 decreased from 1300 mg/kg in 1995 to non-detectable in 1999. The concentrations reported for DRO outside the hotspots (unplanted areas) also decreased, but generally to a lesser extent than the decrease observed in soil samples collected from within the planted hotspots (Table 1).

TABLE 1. Statistical summary of soil samples that were analyzed for DRO during 1995 (prior to planting) and during October 1999, after the third growing season.

	Hot Spot 1		Hot Spot 2	
	Inside DRO 1995/1999	Outside DRO 1995/1999	Inside DRO 1995/1999	Outside DRO 1995/1999
Geometric Mean	493/251	68/52	280/96	71/24
Standard Deviation	6862/6996	94/369	1631/576	97/9
Median	250/192	140/68	160/93	100/20

	Hot Spot 3*		Hot Spot 4	
	Inside DRO 1995/1999	Outside DRO 1995/1999	Inside DRO 1995/1999	Outside DRO 1995/1999
Geometric Mean	924/2.5	NS/NS	155/49	11/28
Standard Deviation	1871/NA	NA/NA	3292/663	25/64
Median	1020/2	NA/NA	110/60	5/66

DRO - Diesel Range Organics in milligrams/kilogram.
NS - Not sampled.
NA - Not applicable.
* - Represents only one soil sample.

Three soil samples were collected within close proximity to each other within the 1999 test pit to aid in determining if degradation of DRO was accelerated within the rhizosphere. These results are presented in Table 2. The sample collected within a densely rooted portion of the rhizosphere contained DRO at a concentration of 65 mg/kg. The soil sample that was collected from a portion of the rhizosphere with fewer roots contained DRO at a concentration of 250 mg/kg and the sample collected outside of the rhizosphere contained 280 mg/kg DRO. Although the changes in concentration may not be significant given the variability of DRO across the treatment unit, there was a measured trend toward lower concentrations of DRO closer to the roots.

TABLE 2. Results from soil samples that were collected from three locations in the rhizosphere during October 1999.

Constituent	Sample Location/ Observations	Within Rhizosphere/ Numerous Roots Present	Within Rhizosphere/ Some Roots Present	Outside Rhizosphere/ No Roots Present
DRO		65	250	280

DRO   Diesel Range Organic.
Concentrations presented in milligrams per kilogram (mg/kg)

Future Activities. The fertilization program will continue on a semi-annual basis and insecticide applications are expected to continue on an annual basis or as necessary. Direct observations of root growth in one or more study areas are scheduled for the fall of 2002. These direct observations will aid in determining the extent that roots are penetrating downward and laterally across the study areas. Another round of soil samples for DRO analysis will be collected from each phytoremediation area during Fall 2002 (6 years from planting) to continue to monitor the progress of enhanced rhizosphere degradation.

REFERENCES

Azadpour and E. Matthews. 1996. "Remediation of Metals Contaminated Sites Using Plants." Remediation. Summer 1996. pp 1-18.

Banks, M.K., Govindaraju, R.S. Schwab, A.P., Kulakow, P. and J. Finn, 1999, Phytoremediation of Hydrocarbon-Contaminated Soil. Edited by Stephanie Fiorenza, Carroll L. Oubre and C. Herb Ward. Lewis Publishers, Boca Raton FL.

Carman, Eric P., Crossman, Tom L. and Edward G. Gatliff, 1998, "Phytoremediation of No. 2 Fuel Oil-Contaminated Soil." Journal of Soil Contamination. 7(4):455-466.

Gatliff, G. 1994. "Vegetative Remediation Process Offers Advantages Over Traditional Pump and Treat Technologies." Remediation. Summer 1994. pp 343-352.

Schwab, A.P. and M. K. Banks. 1994. "Biologically Mediated Dissipation of Polyaromatic Hydrocarbons in the Root Zone." In T.A. Anderson and J.R. Coats (Eds). Bioremediation Through Rhizosphere Technology. pp. 132-141. American Chemical Society. ACS Symposium Services 563.

Schnoor, J.L., Licht, L.A., McCutcheon, S.C., Wolfe, N.L., and L.H. Carreira. 1995. "Phytoremediation of Organic and Nutrient Contaminants." Environmental Science & Technology. 29 (7) pp 318-323.

United States Environmental Protection Agency, 1998, A Citizen's Guide to Phytoremediation, EPA 542-F-98-011.

2000 AUTHOR INDEX

This index contains names, affiliations, and book/page citations for all authors who contributed to the seven books published in connection with the Second International Conference on Remediation of Chlorinated and Recalcitrant Compounds, held in Monterey, California, in May 2000. Ordering information is provided on the back cover of this book.

The citations reference the seven books as follows:

2(1): Wickramanayake, G.B., A.R. Gavaskar, M.E. Kelley, and K.W. Nehring (Eds.), *Risk, Regulatory, and Monitoring Considerations: Remediation of Chlorinated and Recalcitrant Compounds.* Battelle Press, Columbus, OH, 2000. 438 pp.

2(2): Wickramanayake, G.B., A.R. Gavaskar, and N. Gupta (Eds.), *Treating Dense Nonaqueous-Phase Liquids (DNAPLs): Remediation of Chlorinated and Recalcitrant Compounds.* Battelle Press, Columbus, OH, 2000. 256 pp.

2(3): Wickramanayake, G.B., A.R. Gavaskar, and M.E. Kelley (Eds.), *Natural Attenuation Considerations and Case Studies: Remediation of Chlorinated and Recalcitrant Compounds.* Battelle Press, Columbus, OH, 2000. 254 pp.

2(4): Wickramanayake, G.B., A.R. Gavaskar, B.C.Alleman, and V.S. Magar (Eds.) *Bioremediation and Phytoremediation of Chlorinated and Recalcitrant Compounds.* Battelle Press, Columbus, OH, 2000. 538 pp.

2(5): Wickramanayake, G.B. and A.R. Gavaskar (Eds.), *Physical and Thermal Technologies: Remediation of Chlorinated and Recalcitrant Compounds.* Battelle Press, Columbus, OH, 2000. 344 pp.

2(6): Wickramanayake, G.B., A.R. Gavaskar, and A.S.C. Chen (Eds.), *Chemical Oxidation and Reactive Barriers: Remediation of Chlorinated and Recalcitrant Compounds.* Battelle Press, Columbus, OH, 2000. 470 pp.

2(7): Wickramanayake, G.B., A.R. Gavaskar, J.T. Gibbs, and J.L. Means (Eds.), *Case Studies in the Remediation of Chlorinated and Recalcitrant Compounds.* Battelle Press, Columbus, OH, 2000. 430 pp.

Abbott, James (Battelle/USA) *2(7):*393
Abboud, Salim (Alberta Research Council/CANADA) *2(7):*153
Abrahams, Jennifer (HSI GeoTrans/USA) *2(1):*287
Abrams, Stewart H. (Envirogen/USA) *2(4):*157
Abrajano, Teofilo (Argonne National Laboratory/USA) *2(1):*149
Abriola, Linda M. (University of Michigan/USA) *2(2):*77

Adams, Andrew (Los Alamos National Laboratory/USA) *2(7):*319
Adams, Craig D. (University of Missouri-Rolla/USA) *2(7):*25
Adams, Timothy V. (ENSR Corporation/USA) *2(5):*167; *2(6):*217
Adesida, Adebola (University of Georgia/USA) *2(7):*101
Adriaens, Peter (University of Michigan/USA) *2(3):*161
Agrawal, Abinash (Wright State University/USA) *2(5):*253

Ahmed, Farahat M. (Environment Public Authority/KUWAIT) *2(7):*337

Albrecht, Iris D. (University of Michigan/USA) *2(3):*161

Al-Fayyomi, Ihsan A. (Metcalf & Eddy, Inc./USA) *2(4):*221

Allen, Harry L. (U.S. EPA/USA) *2(7):*161

Allen, Shelley A. (Colorado State University/USA) *2(5):*301

Allende, J.L. (Universidad Complutense de Madrid/SPAIN) *2(7):*237

Al-Meshan, Mishal A. (Public Authority for Applied Education & Training/KUWAIT) *2(7):*337

Alonso, R. (Universidad Politecnica/SPAIN) *2(7):*237

Alvarez, Pedro J. (University of Iowa/USA) *2(6):*339

Al-Yousfi, A. Basel (Union Carbide Corporation/USA) *2(3):*145

Andersen, Pete (HSI GeoTrans, Inc./USA) *2(4):*429

Anderson, David (State of Oregon/USA) *2(1):*23; *2(4):*213

Anderson, Grant A. (U.S Army Corp of Engineers/USA) *2(6):*273

Anderson, Jeff (Geomega/USA) *2(6):*257

Anderson, Sarah Elizabeth (University of Waterloo/CANADA) *2(2):*157

Antia, Jimmy (University of Cincinnati/USA) *2(7):*357

Appel, Lisa (South Carolina Dept of Health & Environmental Control/USA) *2(1):*23

Armstrong, Kevin G. (Montgomery Watson/USA) *2(6):*347

Arsenault, Marilyn (Arsenault Legg, Inc./USA) *2(1):*111

Atagana, Harrison I. (University of Natal/REP OF SOUTH AFRICA) *2(7):*131

Aubertin, Howie (U.S. Air Force/USA) *2(1):*1

Aziz, Carol E. (Groundwater Services, Inc./USA) *2(1):*117; *2(4):*71

Aziz, Julia J. (Groundwater Services, Inc./USA) *2(1):*319

Babel, Wolfgang (Umweltforschungszentrum/GERMANY) *2(4):*133; *2(7):*229

Baehr, John (U.S. Army Corps of Engineers/USA) *2(6):*153

Baker, Joseph L. (Honeywell Federal Manufacturing & Technologies/USA) *2(6):*417

Balba, M. Talaat (Conestoga-Rovers & Associates/USA) *2(6):*161; *2(7):*301

Balbierz, Bridget E. (Unocal/USA) *2(1):*191

Baral, Rishi (Bucknell University/USA) *2(4):*55

Barkovskii, Andrei L. (University of Michigan/USA) *2(3):*161

Barlaz, Morton A. (North Carolina State University/USA) *2(4):*47, 421

Barnes, Paul W. (Earth Tech/USA) *2(7):*81

Barton, Andrew C. (Battelle/USA) *2(3):*89

Basel, Michael D. (Montgomery Watson/USA) *2(2):*117

Batchelor, Bill (Texas A&M University/USA) *2(5):*221

Bauer, Nicholas (Alper Holdings USA, Inc) *2(6):*169

Baviello, Mary Ann (Environmental Liability Management, Inc./USA) *2(4):*337

Beadle, Deidra (SteamTech Environmental Services/USA) *2(5):*149

Beaty, J. Neal (Tait Environmental Management, Inc./USA) *2(7):*41

Becker, Petra Maria (TNO Institute of Environmental Sciences/GERMANY) *2(4):*63

Beckett, G.D. (AquiVer/USA) *2(1):*279

Beckman, Scott W. (Science Applications Intl Corp (SAIC)/USA) *2(7):*327

Becvar, Erica S.K. (Applied Research Associates/USA) *2(3):*175

Bensch, Jeffrey C. (HSI GeoTrans/USA) *2(4):*237

Author Index 511

Bergersen, Ove (SINTEF Oslo/NORWAY) *2(7):*385
Berini, Christopher M. (U.S. Army Corps of Engineers/USA) *2(6):*109
Beyke, Gregory (Current Environmental Solutions, LLC/USA) *2(5):*183, 191
Bienkowski, Lisa A. (IT Corporation/ USA) *2(4):*229
Binard, Kinsley (Geomatrix Consultants, Inc./USA) *2(4):*485
Bjerg, Poul L (Technical University of Denmark/DENMARK) *2(3):*9
Blanchet, Denis (Institut Francais Du Petrole/FRANCE) *2(7):*205
Blickle, Frederick W. (Conestoga Rovers & Associates/USA) *2(1):*133, 295; *2(2):*133
Blowes, David (University of Waterloo/CANADA) *2(6):*361
Blum, Brian A. (McLaren/Hart, Inc./ USA) *2(2):*25
Boettcher, Gary (ARCADIS Geraghty & Miller, Inc./USA) *2(1):*311
Boggs, Kevin G. (Wright State University/USA) *2(5):*253
Bokermann, Christian (Technical University of Berlin/GERMANY) *2(6):*433
Bollmann, Dennis D. (City and County of Denver/USA) *2(5):*113
Booth, Robert (XCG Consultants Ltd./ CANADA) *2(5):*135
Borch, Robert S. (USA) *2(7):*93
Borchert, Susanne (CH2M Hill/USA) *2(5):*19
Borden, Robert C. (North Carolina State University/USA) *2(4):*47, 421
Bosma, Tom N.P. (TNO Institute of Environmental Sciences/THE NETHERLANDS) *2(4):*63
Boulicault, Kent J. (Parsons Engineering Science/USA) *2(4):*1
Bow, William (CADDIS Inc./USA) *2(4):*15
Bowen, William B. (Advanced GeoServices Corp/USA) *2(1):*231
Boyd, Thomas J. (U.S. Navy/USA) *2(7):*189

Boyle, Susan L. (Haley & Aldrich, Inc./USA) *2(4):*255
Bradley, Paul M. (U.S. Geological Survey/USA) *2(3):*169; *2(7):*17
Brady, Warren D. (IT Corporation/ USA) *2(3):*201, 209
Brauning, Susan (Battelle/USA) *2(1):*245
Brenner, Richard C. (U.S. EPA/USA) *2(7):*393
Bricelj, Mihael (National Institute of Biology/SLOVENIA) *2(4):*123
Bridge, Jonathan R. (HSI GeoTrans, Inc./USA) *2(2):*149
Briseid, Tormod (SINTEF Oslo/NORWAY) *2(7):*385
Brooker, Daniel (Applied Power Concepts, Inc./USA) *2(4):*101
Brourman, Mitchell D. (Beazer East, Inc./USA) *2(2):*1
Brown, Richard A. (ERM/USA) *2(6):*125
Brown, Susan (Environment Canada/ CANADA) *2(5):*261
Bryant, J. Daniel (Geo-Cleanse International, Inc./USA) *2(5):*307
Buchanan, Ronald J. (DuPont Co./USA) *2(4):*77
Buggey, Thomas R. (Dames & Moore/ USA) *2(2):*141
Burdick, Jeffrey S. (ARCADIS Geraghty & Miller, Inc./USA) *2(4):*263
Burken, Joel G. (University of Missouri-Rolla/USA) *2(7):*25
Burnett, R. Donald (Morrow Environmental Consultants Inc./CANADA) *2(5):*35
Burwinkel, Stephen (University of Central Florida/USA) *2(6):*385
Butler, David (Applied Engineering & Sciences, Inc./USA) *2(5):*127
Caffoe, Todd M. (New York State-DEC/USA) *2(4):*255

Campbell, Ted R. (U.S. Geological Survey/USA) *2(1):*349
Cannata, Marc A. (Parsons Engineering Science/USA) *2(5):*9; *2(6):*385

Carl, Fred G. (Shell Technology Ventures/USA) *2(5):*197
Carlson, Rebecca A. (Leggette, Brashears & Graham, Inc./USA) *2(5):*101
Carman, Eric P. (ARCADIS Geraghty & Miller, Inc./USA) *2(4):*501
Carsley, Michael (U.S. Navy/USA) *2(1):*339
Carter, Gregory L. (Earth Tech/USA) *2(4):*437
Carver, Edward (United States Air Force/USA) *2(4):*1; *2(5):*9
Casaus, Benito (Duke Engineering & Services/USA) *2(2):*195
Casey, Cliff (U.S Navy/USA) *2(5):*307
Cha, Daniel K. (University of Delaware/USA) *2(6):*425
Chace, Peter A. (Battelle Memorial Institute/USA) *2(1):*141
Chambers, Jane Deni (Northgate Environmental Mgt, Inc./USA) *2(6):*49, 57
Chan, Nancy Lau (University of Central Florida/USA) *2(6):*385
Chang, Hung-Li (University of Southern California/USA) *2(6):*195
Chang, Sung-Woon (Lamar University/USA) *2(5):*293
Chang, Y.C. (Gifu University/JAPAN) *2(4):*197
Chard, Brandon (Phytokinetics, Inc./USA) *2(4):*461
Charrois, Jeffrey W.A. (Komex International, Ltd./CANADA) *2(1):*255
Chen, Bor-Yann (U.S. EPA/USA) *2(7):*307
Chen, Daniel H. (Lamar University/USA) *2(6):*241
Chen, Hui (Northwest Normal University/CHINA) *2(6):*233
Chen, Wei (Brown & Caldwell/USA) *2(1):*239
Cherepy, Nerine (Lawrence Livermore National Laboratory/USA) *2(5):*277
Cherry, John A. (University of Waterloo/CANADA) *2(6):*135
Chheda, Pradeep (University of Connecticut/USA) *2(6):*145

Chitwood, John (CADDIS Inc./USA) *2(4):*15
Chiu, Pei C. (University of Delaware/USA) *2(6):*425
Cho, Jong Soo (Environmental Strategies and Applications, Inc./USA) *2(7):*1
Choi, Heechul (Kwangju Institute of Science and Technology/KOREA) *2(6):*225
Chopra, Manoj (University of Central Florida/USA) *2(6):*323
Christ, John A. (U.S. Air Force/USA) *2(5):*253
Christensen, Anders Georg (NIRAS Consulting Engineers & Planners A/S/DENMARK) *2(3):*9; *2(5):*83
Christian, Barry (Earth Tech/USA) *2(5):*85
Chrostowski, Paul C. (CPF Associates, Inc./USA) *2(3):*41
Clark, Patrick (U.S. EPA/USA) *2(7):*327
Clarke, Bruce H. (Earth Tech/USA) *2(5):*85
Clarke, James N. (Law Engineering & Environmental Services/USA) *2(3):*105
Clausen, Christian A. (University of Central Florida/USA) *2(6):*323, 385
Clayton, Wilson S. (IT Corporation/USA) *2(6):*101, 125
Cline, Paul R. (AstraZeneca Corp/CANADA) *2(7):*251
Clough, Herbert F. (Hart Crowser, Inc./USA) *2(5):*197
Cochran, James R. (IT Corporation/USA) *2(1):*393
Cockrum, Dirk K. (Camp, Dresser & McKee, Inc.) *2(4):*175
Coffin, Richard B. (U.S. Navy/USA) *2(7):*189
Coffman, Richard (Tait Environmental Management, Inc./USA) *2(7):*41
Comeau, Yves (Ecole Polytechnique of Montreal/CANADA) *2(4):*353
Compton, Harry R. (U.S. EPA/USA) *2(4):*477
Conca, James L. (Los Alamos National Laboratory/USA) *2(7):*319

Author Index

Condee, Charles W. (Envirogen, Inc./USA) *2(4):*157, 165
Conley, Denis (Haley & Aldrich/USA) *2(5):*175, 197
Connor, John (Groundwater Services, Inc./USA) *2(1):*15
Conroy, Tina. (Rowan University/USA) *2(3):*65
Cook, Jim (Beazer East, Inc./USA) *2(2):*181; *2(4):*337; *2(7):*139
Cooley, Austin I. (Brown and Caldwell/USA) *2(1):*263
Coons, Darlene (Conestoga-Rovers & Associates/USA) *2(6):*161; *2(7):*301
Cooper, William J. (University of North Carolina-Wilmington/USA) *2(6):*209; *2(7):*57
Corcho, Diego (Sheffield University/UNITED KINGDOM) *2(4):*183
Cork, Philip (U.S. Air Force/USA) *2(4):*71
Cornuet, Thomas S. (Roy F. Weston, Inc./USA) *2(4):*373
Cox, Evan E. (GeoSyntec Consultants/CANADA) *2(3):*217
Creek, Daniel N. (Alpine Environmental Inc/USA) *2(7):*49
Cronan, Timothy J. (U.S. Army Corps of Engineers/USA) *2(6):*109
Crossman, Tom L. (ARCADIS Geraghty & Miller, Inc./USA) *2(4):*501
Cumming, Lydia (Battelle/USA) *2(6):*409

Dahlstrom, Kim (Danish EPA/USA) *2(3):*9
Dahmke, Andreas (Christian-Albrechts University of Kiel/GERMANY) *2(6):*433
Dai, Qunli (University of Windsor/CANADA) *2(6):*33
Daleness, Kevin L. (ARCADIS Geraghty & Miller, Inc./USA) *2(4):*501
Daly, Matthew H. (Environmental Resources Management/USA) *2(4):*405
Damera, Raveendra (General Physics Corp./USA) *2(1):*103

Davidson, James M. (Alpine Environmental Inc/USA) *2(7):*49
Davies, Neil (GeoSyntec Consultants/USA) *2(6):*299
Davis, Andy (Geomega/USA) *2(6):*257
Davis, James A. (U.S. Geological Survey/USA) *2(6):*281
Deardorff, Therese M. (U.S. Army Headquarters/USA) *2(7):*269
de Bont, Jan A M (Agricultural Univ of Wageningen/THE NETHERLANDS) *2(4):*63
DeFlaun, Mary F. (Envirogen, Inc./USA) *2(4):*93, 319
Deigaard, Lars (ScanRail Consult/DENMARK) *2(6):*377
Delfino, Thomas A. (Geomatrix Consultants, Inc./USA) *2(4):*361
DeLong, George (Lockheed Martin Energy Systems/USA) *2(5):*51, 261
Delshad, Mojdeh (University of Texas at Austin/USA) *2(2):*203
Demers, Gregg (Environmental Resources Management/USA) *2(4):*405
Demetriades-Shah, Tanvir (LI-COR, Inc./USA) *2(7):*189
DePercin, Paul (U.S. EPA/USA) *2(1):*61
DeReamer, Tom (General Physics Corp./USA) *2(4):*279
Desai, Naren (U.S. Army/USA) *2(1):*103
Desilva, Sumedha (Battelle/USA) *2(7):*65
Destaillats, Hugo (California Institute of Technology/USA) *2(6):*201
DeVane, Paul B. (U.S. Air Force/USA) *2(2):*49
Devlin, J.F. (Rick) (University of Waterloo/CANADA) *2(6):*393
DeZeeuw, Richard (OR Dept of Environ Quality/USA) *2(1):*9
Diaz, Art F. (San Jose State University/USA) *2(7):*33
Dick, Vincent B. (Haley & Aldrich/USA) *2(4):*255

Dieckmann, Paul (Honeywell Federal Manufacturing & Technologies/ USA) *2(6):*417

Dijk, John J.A. (Wageningen University/ THE NETHERLANDS) *2(4):*63

Dinicola, Richard S. (U.S. Geological Survey/USA) *2(3):*169

Di Palma, Luca (Universita degli Studi "La Sapienza"/ITALY) *2(7):*245

DiStefano, Thomas D. (Bucknell University/USA) *2(4):*55

Divine, Craig E. (ARCADIS Geraghty & Miller, Inc/USA) *2(2):*69

Dixon, Thomas N. (Tait Environmental Management, Inc./USA) *2(7):*41

Dobbs, Gregory M. (United Technologies Research Center/USA) *2(6):*145

Dolan, Mark E. (Oregon State University/USA) *2(5):*67

Dollar, Peter (GeoSyntec Consultants/ CANADA) *2(6):*299

Dooley, Maureen A. (Harding Lawson Associates/USA) *2(4):*287

Doppalapudi, Sudhakar R. (Great Lakes Soil & Environmental Consultants, Inc./USA) *2(6):*177

Dorrler, Richard (ARCADIS Geraghty & Miller, Inc./USA) *2(5):*141

Dosani, Majid (IT Corporation/USA) *2(7):*379

Downey, Douglas C. (Parsons Engineering Science, Inc./USA) *2(3):*175

Drescher, Eric (Battelle/USA) *2(6):*409

Drogos, Donna L. (Santa Clara Valley Water District/USA) *2(7):*33

Druelle, Vincent (CERSTA/FRANCE) *2(7):*205

Drummond, Chad D. (The University of Michigan/USA) *2(2):*77

Dudas, Marvin J. (University of Alberta/ CANADA) *2(7):*153

Dukes, Craig (South Carolina Dept of Health & Environmental Control/ USA) *2(1):*23

Dupras, John (Maxymillian Technologies/USA) *2(5):*157

Duran, Joe (Woodward-Clyde Pty Limited/AUSTRALIA) *2(6):*401

Durant, Neal D. (HSI GeoTrans, Inc./USA) *2(4):*429

Dwarakanath, Varadarajan (Duke Engineering & Services/USA) *2(2):*17

Easley, Diane (U.S. EPA/USA) *2(4):*245

Eddy-Dilek, Carol A. (Westinghouse Savannah River Co/USA) *2(2):*33, 41

Edelman, Michael J. (Dames & Moore/USA) *2(2):*141

Edwards, Elizabeth A. (University of Toronto/CANADA) *2(3):*17, 217

Eick, Matthew (Virginia Polytechnic Institute & State University/USA) *2(3):*201, 209

Eidsa, Gudmunn (SINTEF/NORWAY) *2(7):*385

Ek, Scott (Tait Environmental Management, Inc./USA) *2(7):*41

Eliason, W. Mark (Roy F. Weston, Inc./USA) *2(4):*373

Elliott, Mark. (Virginia Polytechnic Inst & State Univ/USA) *2(4):*493

Ellis, David E. (Dupont Company/USA) *2(4):*77

Ellis, Jamelle H. (Clemson University/ USA) *2(1):*175

Elsholz, Allen (Lawerence Livermore National Laboratory/USA) *2(5):*277

Enfield, Carl G. (U.S. EPA/USA) *2(2):*101; *2(5):*61

Erickson, James R. (HSI GeoTrans/ USA) *2(7):*139

Esler, Charles T. (AGRA Earth & Environmental, Inc./USA) *2(1):*79; *2(3):*129

Estuesta, Paul (Minnesota Pollution Control Agency/USA) *2(3):*49

Etter, Terry (Unisys Corporation/USA) *2(1):*387

Evans, John C. (Pacific Northwest National Laboratories/USA) *2(6):*369

Evans, Patrick J. (Camp Dresser & McKee/USA) *2(1):*167, 387

Everett, Jess W. (Rowan University/ USA) *2(3):*65

Everett, Lorne G. (University of California–Santa Barbara/USA) *2(2):*59
Ewing, John E. (Duke Engineering & Services/USA) *2(2):*9

Fagervold, Sonja K. (University of Oslo/NORWAY) *2(7):*385
Fain, Stephen (Radian International/USA) *2(1):*1
Falta, Ronald W. (Clemson University/USA) *2(5):*1
Fam, Sami A. (Innovative Engineering Solutions, Inc./USA) *2(4):*23
Fang, Yanjun (Northwest Normal University/CHINA) *2(6):*233
Farone, William (Applied Power Concepts, Inc./USA) *2(4):*39, 101
Farrell, James (University of Arizona/USA) *2(6):*353
Farrell, Kristi (University of Central Florida/USA) *2(6):*385
Fayolle, Francoise (Institut Francais du Petrole/FRANCE) *2(4):*141
Ferland, Derek R. (Air Force Institute of Technology/USA) *2(5):*253
Fernandes, Leta F. (University of Ottawa/CANADA) *2(4):*381
Ferrer, E. (Universidad Complutense de Madrid/SPAIN) *2(7):*237
Ferrey, Mark L. (Minnesota Pollution Control Agency/USA) *2(3):*49, 57
Ferro, Ari M. (Phytokinetics, Inc./USA) *2(4):*461
Fiacco, R. Joseph (Environmental Resources Management/USA) *2(4):*405
Figura, Michael (U.S. Navy/USA) *2(3):*113
Finder, Kevin B. (IT Corporation/USA) *2(6):*265
Findlay, Margaret (Bioremediation Consulting, Inc./USA) *2(4):*23
Findley, Joseph (US Army Corps of Eng/USA) *2(6):*153
Fischer, Erling V. (NIRAS/DENMARK) *2(5):*83
Fisher, Arthur (U.S. Navy/USA) *2(3):*89, 137

Fisher, Gary M. (Lucent Technologies/USA) *2(2):*25
Fishman, Michael (Dynamac Corp/USA) *2(2):*101
Fitzgerald, James D. (Environmental Resources Management/USA) *2(4):*405
Flanders, Jonathan (Envirogen, Inc./USA) *2(4):*157
Fleming, David (Current Environmental Solutions/USA) *2(5):*167
Focht, Robert M. (EnviroMetal Technologies, Inc./CANADA) *2(6):*265, 417
Fogel, Samuel (Bioremediation Consulting, Inc./USA) *2(4):*23
Folkes, David J. (EnviroGroup, Ltd./USA) *2(5):*245
Forman, Sarah (URS - Dames & Moore/USA) *2(5):*51, 261
Fortman, Tim (Battelle Marine Sciences Laboratory/USA) *2(7):*181
Fortun, A. (Universidad Complutense de Madrid/SPAIN) *2(7):*237
Fosbrook, Cristal (U.S. Army AK/USA) *2(7):*269
Fossati, Frank R. (Shell Oil Company/USA) *2(5):*197
Foster, Sarah A. (CPF Associates, Inc./USA) *2(3):*41
Fox, Tad C. (SAIC/USA) *2(7):*9
Franz, Thomas J. (Franz Environmental Inc./CANADA) *2(3):*113
Freethey, Geoffrey W. (U.S. Geological Survey/USA) *2(6):*281
Frigon, Jean-Claude (National Research Council of Canada/CANADA) *2(4):*353
Frisch, Samuel (Envirogen/USA) *2(7):*115
Fruchter, John S. (Pacific Northwest National Laboratories/USA) *2(6):*369
Fu, Shang (University of Michigan/USA) *2(3):*161
Fuglsang, Inger (Danish Environmental Protection Agency/DENMARK) *2(5):*83
Fujita, Masanori (Osaka University/JAPAN) *2(1):*183

Fuller, Christopher C. (U.S. Geological Survey/USA) *2(6):*281

Gannon, David John (AstraZeneca Corp./CANADA) *2(7):*251
Gao, Jinzhang (Northwest Normal University/CHINA) *2(6):*233
Garbi, C. (Universidad Complutense de Madrid/SPAIN) *2(7):*237
Garcia, Luis (Rowan University/USA) *2(3):*65
Gascon, J.A. (Gaiker/SPAIN) *2(7):*293
Gates, Kimberly (U.S. Army Garrison/USA) *2(5):*51, 261
Gaule, Chris (Malcolm Pirnie, Inc./USA) *2(6):*273
Gavaskar, Arun R. (Battelle/USA) *2(1):*47, 141; *2(2):*49; *2(6):*409
Gefell, Michael J. (Blasland, Bouck & Lee, Inc./USA) *2(4):*461
Geiger, Cherie L. (University of Central Florida/USA) *2(6):*323, 385
Germon, Matt (CH2M Hill/USA) *2(4):*303
Gerritse, Jan (TNO Institute of Environmental Sciences/THE NETHERLANDS) *2(4):*63
Getman, Gerry D. (Commodore Advanced Sciences, Inc./USA) *2(7):*349
Giattino, Robert (RRM, Inc./USA) *2(7):*65
Gibbons, Robert D. (The University of Illinois at Chicago/USA) *2(1):*133
Gibbs, James T. (Battelle/USA) *2(1):*215; *2(7):*9, 65
Gibson, Rhonda (Earth Tech/USA) *2(5):*85
Ginn, Jon S. (U.S. Air Force/USA) *2(1):*61; *2(2):*17
Glaser, John A. (U.S. EPA/USA) *2(7):*379
Glass, David J. (D Glass Associates Inc/USA) *2(1):*33
Glass, Scott (U.S. Navy/USA) *2(5):*75
Godard, Steve (Radian Internation/URS Corp./USA) *2(1):*1, 61
Goering, Catherine E.B. (Chemical Waste Management/USA) *2(4):*361

Goldstein, Kenneth J. (Malcolm Pirnie, Inc./USA) *2(6):*273; *2(7):*213
Goltz, Mark N. (Air Force Institute of Technology/USA) *2(5):*253
Gonzales, James R. (U.S. Air Force/USA) *2(1):*117, 319
Gordon, E. Kinzie (Parsons Engineering Science, Inc./USA) *2(3):*81
Gossett, James M. (Cornell University/USA) *2(3):*175
Gotpagar, J.K. (Union Carbide Corporation/USA) *2(3):*145
Govind, Rakesh (University of Cincinnati/USA) *2(7):*307
Govoni, John W. (U.S. Army Corps of Engineers/USA) *2(6):*109
Granade, Steve (U.S. Navy/USA) *2(4):*229
Granley, Brad A. (Leggette, Brashears & Graham, Inc/USA) *2(5):*101
Graves, Duane (IT Corporation/USA) *2(1):*393; *2(3):*121, 209; *2(4):*205
Gray, Murray R. (University of Alberta/CANADA) *2(7):*153
Gray, Neil C.C. (AstraZeneca Canada Inc./CANADA) *2(7):*251
Green, Donald J. (U.S. Army/USA) *2(4):*279; *2(5):*51, 261
Green, Mark R. (Envirogen/USA) *2(7):*115
Gross, John (Weyerhaeuser Company/USA) *2(4):*303
Grosse, Douglas W. (U.S. EPA/USA) *2(7):*169
Grossl, Paul R. (Utah State University/USA) *2(3):*201
Guarini, William J. (Envirogen, Inc./USA) *2(7):*115
Guarnaccia, Joseph F. (Ciba Specialty Chemicals Corp./USA) *2(2):*101
Guest, Peter R. (Parsons Engineering Science, Inc./USA) *2(1):*369
Guiot, Serge R. (Biotechnology Research Institute/CANADA) *2(4):*353
Gupta, K.G. (Panjab University/INDIA) *2(7):*277
Gupta, Neeraj (Battelle/USA) *2(1):*47; *2(2):*49; *2(6):*409

Guswa, John H. (HSI GeoTrans, Inc./USA) *2(2):*149

Haas, Patrick E. (U.S. Air Force/USA) *2(1):*357; *2(4):*1, 31, 71
Hadlock, Gregg L. (Radian International/USA) *2(1):*1
Haeseler, Frank (Institut Francais du Petrole/FRANCE) *2(7):*205
Haff, James (Meritor Heavy Vehicle Systems, LLC/USA) *2(4):*221
Hahn, Ernest J. (The University of Michigan/USA) *2(2):*77
Hajali, Paris A. (Haley & Aldrich Inc./USA) *2(6):*209
Hampton, Mark (Groundwater Services, Inc.) *2(4):*71
Hansen, AnneMette (EPA Copenhagen/DENMARK) *2(3):*9
Hansen, Kevin (Advanced GeoServices Corp/USA) *2(1):*231
Hansen, Kirk S. (Ground Technology, Inc./USA) *2(5):*197
Hansen, Lance D. (U.S. Army Corps of Engineers/USA) *2(7):*145
Hansen, Michael A. (ARCADIS Geraghty & Miller Inc./USA) *2(4):*263
Harbage, Todd A. (Clemson University/USA) *2(7):*221
Hardison, Wayne C. (Haley & Aldrich Inc./USA) *2(6):*209
Hare, Paul W. (General Electric Co./USA) *2(1):*377
Harkness, Mark R. (GE Corporate R&D Center/USA) *2(3):*17, 113; *2(4):*9
Harms, Willard D. (URS Greiner Woodward Clyde/USA) *2(4):*295
Haroski, Dale M. (Lockheed Martin/USA) *2(4):*477
Harris, James C. (U.S. EPA/USA) *2(1):*303; *2(7):*123
Harris, Kenneth A. (Stone Container Corp/USA) *2(7):*161
Hart, Michael A. (Carpenter Technology Corp./USA) *2(2):*141
Hartsfield, Brent (Florida Dept of Environmental Protection/USA) *2(1):*23

Harwell, Jeffrey (University of Oklahoma/USA) *2(2):*219
Hasegawa, Mark (Surbec-ART Environmental, LLC/USA) *2(2):*219
Hatsu, M. (Gifu University/JAPAN) *2(4):*197
Hatzinger, Paul B. (Envirogen Inc./USA) *2(4):*165; *2(7):*115, 213
Hauck, Regine (Technical University Berlin/GERMANY) *2(4):*107
Hausmann, Tom (Battelle Marine Sciences Laboratory/USA) *2(7):*181
Hawkins, Andrew C. (The University of Iowa/USA) *2(6):*339
Haynes, R.J. (University of Natal/REP OF SOUTH AFRICA) *2(7):*131
Heaston, Mark S. (Earth Tech/USA) *2(7):*81
Heckelman, Curtis A. (U.S. Army Corps of Engineers/USA) *2(7):*213
Heffron, Mike (Foster Wheeler Environmental/USA) *2(5):*237
Hegemann, Werner (Technische Universitat Berlin/GERMANY) *2(4):*107
Heilman, Paul (University of Washington/USA) *2(4):*467
Heller, Paula (UFA Ventures, Inc./USA) *2(7):*319
Henning, Leo G. (Kansas Dept of Health & Environment/USA) *2(1):*9
Heraty, Linnea J. (Argonne National Laboratory/ER-203/USA) *2(1):*149
Herrmann, Ronald (U.S. EPA/USA) *2(7):*379
Heron, Gorm (SteamTech Environmental Services/USA) *2(5):*61, 149
Heron, Tom (NIRAS/DENMARK) *2(5):*61
Herrington, R. Todd (Parsons Engineering Science, Inc./USA) *2(3):*175
Herrmann, Ronald (U.S. EPA/USA) *2(7):*379
Hewitt, Alan D. (U.S. Army/USA) *2(1):*87, 327, 333; *2(6):*109
Hicken, Steve (U.S. Air Force/USA) *2(1):*1

Hickman, Gary (CH2M Hill/USA) *2(4):*303
Hicks, James E. (Battelle/USA) *2(1):*47; *2(6):*409; *2(7):*9
Hicks, John (Parsons Engineering Science, Inc/USA) *2(3):*175
Hiebert, Franz K. (RMT, Inc./USA) *2(3):*73
Hinchee, Robert E. (Parsons Engineering Science, Inc./USA) *2(4):*1
Hindi, Nizar (Earth Tech/USA) *2(5):*85
Hines, Robert D. (U.S. Filter/Envirex, Inc./USA) *2(4):*175
Hirsh, Steven R. (U.S. EPA/USA) *2(4):*477
Hoag, George E. (University of Connecticut/USA) *2(6):*145
Hocking, Grant (Golder Sierra LLC/USA) *2(6):*307
Hoenke, Karl A. (Chevron Chemical Company/USA) *2(6):*257
Hoffmann, Michael R. (California Inst of Technology/USA) *2(6):*201
Holcomb, David H. (Focus Environmental, Inc./USA) *2(1):*207
Holdsworth, Thomas (U.S. EPA/USA) *2(2):*49
Holish, Lawrence L. (Patrick Engineering, Inc./USA) *2(7):*259
Holt, Ben D. (Argonne National Laboratory/ER-203/USA) *2(1):*149
Holzmer, Frederick J. (Duke Engineering & Services/USA) *2(2):*187, 195, 203
Honniball, James H. (Geomatrix Consultants, Inc./USA) *2(4):*361
Hood, Chris (CH2M Hill/USA) *2(5):*19
Hood, Eric D. (GeoSyntec Consultants/CANADA) *2(6):*9, 83
Hopkins, W. Alan (Brown & Caldwell/USA) *2(1):*263
Hough, Benjamin (Tetra Tech EM, Inc./USA) *2(5):*43
Hoxworth, Scott W. (Parsons Engineering Science, Inc./USA) *2(4):*1, 327; *2(5):*9
Huang, Kun-Chang (University of Connecticut/USA) *2(6):*145

Huang, Lin (Argonne National Laboratory/USA) *2(1):*149
Huesemann, Michael H. (Battelle/USA) *2(1):*271; *2(7):*181
Hughes, Joseph B. (Rice University/USA) *2(4):*31
Hull, John H. (Hull & Associates Inc./USA) *2(7):*369
Huscher, Theodore (Nebraska Department of Environmental Quality/USA) *2(4):*245
Hwang, Inseong (Texas A&M University/USA) *2(5):*221
Hyman, Michael R. (North Carolina State University/USA) *2(4):*149

Ibaraki, Motomu (Ohio State University/USA) *2(2):*125
Ickes, Jennifer (Battelle/USA) *2(7):*393
Ike, Michihiko (Osaka University/JAPAN) *2(1):*183
Ingram, Sherry (IT Corporation/USA) *2(3):*121
Isacoff, Eric G. (Rohm and Haas Company/USA) *2(7):*57
Islam, Jahangir (The University of Auckland/NEW ZEALAND) *2(1):*223; *2(4):*115
Isosaari, Pirjo (National Public Health Institute/FINLAND) *2(7):*343
Istok, Jonathan D. (Oregon State University/USA) *2(4):*303

Jackson, Dennis (Westinghouse Savannah River Co./USA) *2(2):*33, 41; *2(4):*429; *2(7):*221
Jackson, John A. (Parsons Engineering Science, Inc./USA) *2(3):*81
Jackson, Richard E. (Duke Engineering & Services/USA) *2(2):*9, 17, 85, 195
Jackson, W. Andrew (Texas Tech University/USA) *2(4):*279
Jakobsen, Rasmus (Technical University of Denmark/DENMARK) *2(3):*9
Janosy, Robert J. (Battelle/USA) *2(6):*409
Jenkins, Thomas F. (U.S. Army Corps of Engineers/USA) *2(1):*87

Jersak, Joseph M. (Hull & Associates, Inc./USA) *2(7):*369
Jiang, Zhenhua (Argonne National Laboratory/USA) *2(5):*121
Jin, Minquan (Duke Engineering & Services/USA) *2(2):*9, 17, 85
Jin, Xinglong (Northwest Normal University/CHINA) *2(6):*233
Johansen, Paul (Boeing Corp./USA) *2(3):*97
Johnson, Christian D. (Battelle/USA) *2(4):*229
Johnson, Mark (Hydro-Environmental Technologies/USA) *2(1):*125
Johnson, Steven E. (Roy F. Weston, Inc./USA) *2(4):*373
Jones, Alison (Geomatrix Consultants, Inc./USA) *2(2):*1
Jones-Meehan, JoAnne (U.S. Naval Research Laboratory/USA) *2(7):*145
Joseph, Jiju M. (California Institute of Technology/USA) *2(6):*201
Journell, Scot (Southwest Ground-Water Consultants, Inc./USA) *2(7):*161
Jurgens, Robert Dean (Kansas Dept of Health & Environment/USA) *2(1):*23
Jurka, Valdis (Lucent Technologies/USA) *2(5):*167, 183

Kabala, Zbigniew (Duke University/USA) *2(5):*43
Kampbell, Donald H. (U.S. EPA/USA) *2(3):*57
Kan, Amy T. (Rice University/USA) *2(1):*239
Kawabata, Junichi (Kajima Technical Research Institute/JAPAN) *2(4):*445
Kawai, Tatsushi (Kajima Corporation/JAPAN) *2(4):*445
Kean, Judie (Florida Dept of Environmental Protection/USA) *2(4):*205
Kellar, Edward M. (Environmental Science & Engineering, Inc./USA) *2(6):*153
Keller, Arturo A. (University of California–Santa Barbara/USA) *2(2):*59; *2(7):*73
Keller, Carl (Flexible Liner Underground Technologies, Ltd./USA) *2(2):*33
Kelley, Mark E. (Battelle/USA) *2(1):*215, 245; *2(3):*89, 137
Kennedy, Lonnie G. (Earth Sciences Consulting/USA) *2(3):*65
Kerfoot, William B. (K-V Associates, Inc./USA) *2(1):*53; *2(5):*27; *2(6):*187
Khan, Tariq A. (Groundwater Services, Inc./USA) *2(4):*31
Khodadoust, Amid P. (University of Cincinnati/USA) *2(7):*357
Kiilerich, Ole (HOH Water Technology/DENMARK) *2(6):*377
Kilkenny, Scott (Kinder Morgan Energy Partners/USA) *2(4):*175
Kim, Jeongkon (Hydrogeologic, Inc./USA) *2(6):*225
King, Mark (Groundwater Insight/CANADA) *2(2):*181
Kinsall, Barry L. (Oak Ridge National Laboratory/USA) *2(4):*395
Kjelgren, Roger (Utah State University/USA) *2(4):*461
Klecka, Gary. M. (The Dow Chemical Co,/USA) *2(3):*17
Klens, Julia L. (IT Corporation/USA) *2(1):*393; *2(3):*209
Knowlton, Robert G. (Decision FX, Inc./USA) *2(3):*33
Knox, Robert C. (University of Oklahoma/USA) *2(2):*219
Koenigsberg, Stephen S. (Regenesis/USA) *2(4):*39, 101, 213, 287; *2(7):*87
Kohler, Robert (Centre for Environmental Research/GERMANY) *2(6):*331
Komoski, R. E. (Johnson & Johnson/USA) *2(5):*141
Kopania, Andrew A. (EMKO Environmental/USA) *2(1):*287
Kopinke, Frank-Dieter (Centre for Environmental Research/GERMANY) *2(6):*331
Koprivanac, Natalija (University of Zagreb/USA) *2(5):*215

Korte, Nic (Oak Ridge National Laboratory/USA) *2(6):*417
Kram, Mark (University of California–Santa Barbara/USA) *2(2):*59; *2(7):*73
Kramer, Julie (Battelle/USA) *2(1):*215, 339; *2(7):*9
Kranz, Scott (AGRA Earth and Environmental/USA) *2(1):*79
Krishnamoorthy, Rajagopal (U.S. Navy/USA) *2(3):*137
Krug, Thomas (GeoSyntec Consultants/CANADA) *2(6):*299
Kryczkowski, Rosann (ITT Night Vision/USA) *2(4):*437
Kueper, Bernard H. (Queen's University/CANADA) *2(2):*165
Kukor, Jerome J. (Rutgers University/Cook College/USA) *2(7):*197
Kurz, David W. (EnviroGroup Limited/USA) *2(5):*245

Laase, Alan D. (Oak Ridge National Laboratory/USA) *2(6):*417
LaBrecque, Douglas (Multi-Phase Technologies, LLC/USA) *2(5):*149
Lago, Rafael (Tetra Tech Environmental/USA) *2(2):*219
Landmeyer, James E. (U.S. Geological Survey/USA) *2(3):*169; *2(7):*17
LaPoint, Edward (GE Corporate Environmental Programs/USA) *2(2):*149
Larsen, Jan Wodschow (HOH Water Technology/DENMARK) *2(6):*377
Larsen, Thomas H. (Hedeselskabet/DENMARK) *2(3):*9; *2(5):*83
Lawrence, Kathryn (U.S. EPA, Region IX/USA) *2(7):*161
Lazorchak, James W. (U.S. EPA, ORD/USA) *2(7):*357
Leavitt, Alan (Northgate Environmental Mgt, Inc./USA) *2(6):*49, 57
Lee, Cindy (Clemson University/EE&S/USA) *2(7):*221
Lee, Jennifer H. (Parsons Engineering Science, Inc./USA) *2(3):*81
Lee, K.C. (Union Carbide Corporation/USA) *2(3):*145

Lee, Michael D. (Terra Systems, Inc./USA) *2(4):*47, 77, 405
Lee, Minho (University of Delaware/USA) *2(6):*425
Lee, Seung-Bong (University of Washington/USA) *2(4):*455
Leeson, Andrea (Battelle/USA) *2(5):*67
Legg, Michael (Arsenault Legg, Inc./USA) *2(1):*111
Lehmicke, Leo G. (Exponent Environmental Group/USA) *2(3):*97
Leigh, Daniel P. (IT Corporation/USA) *2(4):*229
Lemke, Lawrence D. (The University of Michigan/USA) *2(2):*77
Lemos, Barbara B. (Earth Tech/USA) *2(4):*437
Lenschow, Soren Rygaard (Ribe Amt/DENMARK) *2(5):*83
Lenzo, Frank (ARCADIS Geraghty & Miller, Inc./USA) *2(4):*263
Leonard, Wendy (IT Corporation/USA) *2(6):*125
Le Penru, Yann (Institut Francais du Petrole/FRANCE) *2(4):*141
Lerner, David N. (University of Sheffield/UNITED KINGDOM) *2(4):*183
Lesage, Suzanne (Environment Canada/CANADA) *2(5):*261
Leskovsek, Hermina (University of Ljubljana/SLOVENIA) *2(4):*123
Leslie, Greg (Orange County Water District/USA) *2(6):*209
Levin, Richard S. (Environmental Science & Engineering, Inc./USA) *2(6):*153
Lewis, Mark (General Physics Corporation/USA) *2(1):*103
Lewis, Richard W. (IT Corp/USA) *2(6):*125
Lewis, Ronald F. (U.S. EPA/USA) *2(7):*327
Li, Dong X. (Unocal Corporation/USA) *2(1):*191
Li, Kuyen (Lamar University/USA) *2(5):*293; *2(6):*241
Li, Tie (University of Arizona/USA) *2(6):*353

Li, Tong (Tetra Tech EM Inc./USA) *2(5)*:43
Li, X. David (The Ohio State University/USA) *2(6)*:41
Lieberman, M. Tony (Solutions Industrial & Environmental Services/USA) *2(4)*:47
Lightner, Alison (U.S. Air Force/USA) *2(5)*:67; *2(6)*:409
Lim, Hyung-Nam (Kwangju Institute of Science and Technology/KOREA) *2(6)*:225
Lin, Cynthia (Conestoga-Rovers & Associates/USA) *2(6)*:161; *2(7)*:301
Lindhult, Eric C. (Dames & Moore Inc/USA) *2(2)*:141
Lindsey, Michele E. (University of New Orleans/USA) *2(6)*:181
Ling, Meng. (University of Houston/USA) *2(1)*:319
Lingens, Rolf (RRM, Inc./USA) *2(7)*:65
Linn, William (Florida Dept. of Env. Protection/USA) *2(1)*:23
Linnemeyer, Harold (Duke Engineering & Services/USA) *2(2)*:195
Liu, Steven Y.C. (Lamar University/USA) *2(6)*:241
Lizotte, Craig (Envirogen Inc./USA) *2(4)*:319
Llewellyn, Tim (Dames & Moore/USA) *2(5)*:51, 261
Lobo, C. (Inst. Madrileno de Investigacion Agraria y Alimentaria/SPAIN) *2(7)*:237
Lodato, Michael (IT Corporation/USA) *2(4)*:205
Logan, Bruce E. (Penn State University/USA) *2(7)*:87
Loncaric Bozic, Ana (University of Zagreb/CROATIA) *2(5)*:215
Londergan, John T. (Duke Engineering & Services/USA) *2(2)*:195
Lonie, Christopher M. (U.S. Navy/USA) *2(5)*:175
Looney, Brian B. (Westinghouse Savannah River Co./WSRC/USA) *2(2)*:41
Lorbeer, Helmut (UFZ-Centre for Environmental Research/GERMANY) *2(4)*:133

Lowry, Gregory V. (Stanford University/USA) *2(5)*:229
Lu, Jia (University of New Orleans/USA) *2(6)*:181
Lu, Ningping (Los Alamos National Laboratory/USA) *2(7)*:319
Lu, Xiaoxia (Peking University/CHINA) *2(4)*:63
Ludwig, Ralph D. (U.S. EPA/USA) *2(6)*:361
Lundegard, Paul D. (Unocal Corporation/USA) *2(1)*:157, 279
Lundy, James (Minnesota Pollution Control Agency/USA) *2(3)*:49
Lundy, William (BioManagement Services, Inc./USA) *2(6)*:177; *2(7)*:259
Luperi, Kenneth J. (Environmental Liability Management, Inc./USA) *2(2)*:181

MacDonald, Jacqueline A. (RAND Corporation/USA) *2(3)*:1
Mackenzie, Katrin (UFZ-Center for Environmental Research/GERMANY) *2(6)*:331
MacKinnon, Leah (GeoSyntec Consultants/CANADA) *2(6)*:9
Madill, Scott (Radian International, LLC/USA) *2(3)*:73
Magar, Victor S. (Battelle/USA) *2(5)*:67; *2(7)*:393
Mahaffey, William R. (Pelorus Environmental & Biotechnology/USA) *2(7)*:139
Mahoney, Laura A. (Brown and Caldwell/USA) *2(1)*:263
Major, David W. (GeoSyntec Consultants/CANADA) *2(3)*:217
Makdisi, Richard S. (Stellar Environmental Solutions/USA) *2(3)*:185
Maki, Eric (Current Environmental Solutions, LL/USA) *2(5)*:191
Man, Malcolm K. (Morrow Environmental Consultants Inc./CANADA) *2(5)*:35
Manning, Joseph W. (Envirogen Inc./USA) *2(7)*:115

Marley, Michael C. (XDD, LLC/USA) *2(4):*319
Marrin, Donn L. (San Diego State University/USA) *2(5):*285
Marsh, Russel (U.S Army Corp of Engineers/USA) *2(6):*273
Marshall, Melanie (University of Windsor/CANADA) *2(6):*25
Martin, Hank (Environmental Liability Management, Inc./USA) *2(2):*181
Martin, Margarita (Universidad Complutense de Madrid/SPAIN) *2(7):*237
Martin-Montalvo, D. (Universidad Complutense de Madrid/SPAIN) *2(7):*237
Martinez-Inigo, M.J. (Inst. Madrileno de Investigacion Agraria y Alimentaria/ SPAIN) *2(7):*237
Marvin, Bruce K. (IT Corporation/USA) *2(6):*101
Mason, Anna (Marshall Macklin Monaghan/CANADA) *2(3):*113
Maughon, Mike (U.S. Navy/USA) *2(1):*339; *2(5):*307
May, Ira (Army Environmental Center/USA) *2(5):*121
Mayes, John (Ministry of the Environment/CANADA) *2(2):*173
McAlary, Todd A. (GeoSyntec Consultants/CANADA) *2(6):*299
McCall, Sarah (Battelle/USA) *2(7):*9
McCall, Wesley (Geoprobe Systems/USA) *2(1):*71
McCauley, Paul T. (U.S. EPA/USA) *2(7):*379
McClay, Kevin (Envirogen, Inc./USA) *2(4):*165
McCue, Terrence (University of Central Florida/USA) *2(4):*327
McDonald, Shane (Malcolm Pirnie, Inc,/USA) *2(6):*273
McGee, Bruce C.W. (McMillan-McGee Corporation/CANADA) *2(5):*207
McGill, William B. (University of Alberta/CANADA) *2(1):*255
McGregor, Rick (Conor Pacific Environmental/CANADA) *2(6):*361

McHugh, Thomas E. (Groundwater Services, Inc./USA) *2(1):*15
McInnis, Dean L. (CFR Technical Services, Inc./USA) *2(4):*279
McKay, Daniel J. (U.S. Army Corps of Engineers/USA) *2(6):*109
McKinley, Scott (CH2M Hill, Inc./USA) *2(4):*303
Medved, Joe (General Motors/USA) *2(2):*133
Meiggs, Ted (Foremost Solutions) *2(6):*291
Meinardus, Hans W. (Duke Engineering & Services/USA) *2(1):*61; *2(2):*9, 17, 195
Meixner, Jarolim (DIOKI d.d./ CROATIA) *2(5):*215
Melby, Jeff (LFR Levine - Fricke/USA) *2(6):*49, 57
Mengs, G. (Universidad Complutense de Madrid/SPAIN) *2(7):*237
Merli, Carlo (Universita degli Studi "La Sapienza"/ITALY) *2(7):*245
Metzger, Keith (XCG Consultants Ltd./CANADA) *2(5):*135
Metzler, Donald R. (U.S. Dept of Energy/USA) *2(3):*33
Meyer, Chris (Unocal Corporation/USA) *2(1):*157
Miles, Rick A. (ENSR Corporation/USA) *2(7):*269
Millar, Kelly (Environment Canada /CANADA) *2(5):*261
Miller, David R. (Sandia National Laboratories/USA) *2(3):*193
Miller, John A. (Schlumberger Oilfield Services/USA) *2(3):*73
Miller, Michael D. (Conestoga-Rovers & Associates/USA) *2(2):*133
Miller, T. Ferrell (Lockheed Martin/ REAC/USA) *2(7):*161
Moes, Michael (Erler & Kalnowski, Inc./USA) *2(6):*117
Monot, Frederic (Institut Francais du Petrole/FRANCE) *2(4):*141
Montgomery, Michael T. (U.S. Navy/ USA) *2(7):*189
Moore, Beverly (Los Alamos National Laboratory/USA) *2(7):*319

Morgan, Scott (URS - Dames & Moore/USA) *2(5)*:51
Mori, Kazuhiro (Yamanashi University/JAPAN) *2(1)*:183
Moring, B.K. (U.S. Navy/USA) *2(1)*:339
Morkin, Mary (GeoSyntec Consultants/USA) *2(6)*:393
Morrison, Stan J. (Environmental Sciences Laboratory/USA) *2(6)*:281
Moser, G.P. (AstraZeneca Corp./CANADA) *2(7)*:251
Mott-Smith, Ernest (IT Corporation/USA) *2(6)*:101, 125
Mountjoy, Keith (Conor Pacific Environmental/CANADA) *2(6)*:361
Mowder, Carol S. (Dames & Moore/USA) *2(5)*:261
Mravik, Susan C. (U.S. EPA/USA) *2(5)*:269
Mueller, James G. (URS Corporation/Dames & Moore/USA) *2(7)*:139, 189
Muller, Roland H. (UFZ Centre for Environmental Research/GERMANY) *2(7)*:229
Munakata, Naoko (Stanford University/USA) *2(5)*:229
Murali, Dev M. (General Physics Corporation/USA) *2(1)*:103
Murdoch, Lawrence C. (Clemson University/USA) *2(5)*:127; *2(6)*:291
Murphy, Mark (Parsons Engineering Science, Inc./USA) *2(5)*:113
Murphy, Sean (Komex International Ltd./CANADA) *2(7)*:153
Murray, Willard A. (Harding Lawson Associates/USA) *2(4)*:287
Murt, Victoria (Nebraska Dept of Environmental Quality/USA) *2(4)*:245
Murthy, D.V.S. (Indian Institute of Technology Madras/INDIA) *2(4)*:413

Naber, Steve (Battelle/USA) *2(1)*:141; *2(2)*:49
Naftz, Dave (U.S. Geological Survey/USA) *2(6)*:281
Nakamura, Mitsutoshi (Kajima Technical Research Institute/JAPAN) *2(4)*:445
Nam, Kyoungphile (Rutgers University/USA) *2(7)*:197
Napolitan, Mark R. (General Motors Remediation Team/USA) *2(1)*:133
Narjoux, Adeline (National Research Council of Canada/CANADA) *2(4)*:353
Nelson, Mark D. (Leggette, Brashears & Graham, Inc./USA) *2(5)*:101
Nelson, Christopher H. (Isotec/USA) *2(2)*:117
Nelson, Matthew D. (University of Waterloo/CANADA) *2(6)*:135
Nelson, Nicole T. (Daniel B. Stephens & Associates, Inc./USA) *2(3)*:153
Nelson, Timothy Michael (Hill Air Force Base/USA) *2(1)*:1
Nemeth, Gary (General Physics Corp./USA) *2(4)*:279
Nestler, Cathy (U.S. Army/USA) *2(7)*:145
Neville, Scott L. (Aerojet General Corp./USA) *2(7)*:93
Nevokshonoff, Barry (Sequoia Environmental Remediation, Inc./CANADA) *2(5)*:207
Newell, Charles J. (Groundwater Services, Inc./USA) *2(1)*:117, 319; *2(4)*:31
Newman, Lee A. (University of Washington/USA) *2(4)*:467
Niekamp, Scott (Wright State University/USA) *2(5)*:253
Nielsen, Charlotte (HOH Water Technology/DENMARK) *2(6)*:377
Neilsen, H.H. (NIRAS/DENMARK) *2(5)*:83
Nishino, Shirley (U.S. Air Force/USA) *2(3)*:175
Noftsker, Christina (Exponent/USA) *2(3)*:97
Norris, Robert D. (Brown and Caldwell/USA) *2(1)*:263
Novak, John T. (Virginia Polytechnic Inst & State Univ/USA) *2(4)*:493

Nuttall, H. Eric (University of New Mexico/USA) *2(6)*:177; *2(7)*:259
Nyer, Evan (ARCADIS Geraghty & Miller, Inc./USA) *2(1)*:311; *2(2)*:109
Nygaard, Tove (NIRAS/DENMARK) *2(5)*:83
Nzengung, Valentine A. (University of Georgia/USA) *2(4)*:347; *2(7)*:101

Oberle, Daniel W. (SECOR International Inc./USA) *2(6)*:91
Ochsner, Mark (Ecology & Environment/USA) *2(4)*:213
Oh, Byung-Taek (The University of Iowa/USA) *2(6)*:339
O'Hannesin, Stephanie (EnviroMetal Technologies, Inc./CANADA) *2(6)*:273
Okeke, Benedict C. (Gifu University/JAPAN) *2(4)*:197
O'Neill, J. Edward (Smithville Phase IV Bedrock Remed Prgm/CANADA) *2(2)*:173
O'Niell, Walter L. (University of Georgia/USA) *2(4)*:347; *2(7)*:101
O'Reilly, Kirk T. (Chevron Research & Technology Co/USA) *2(4)*:149
Ospina, Rafael I. (Golder Sierra LLC/USA) *2(6)*:307
Ostergaard, Henrik (Frederiksborg Amt/DENMARK) *2(5)*:83
Ottesen, Patricia (Lawrence Livermore National Laboratory/USA) *2(1)*:111
Ozaki, Hiroaki (Kyoto University/JAPAN) *2(4)*:85

Pabbi, Sunil (National Centre for Conservation and Utilisation of Blue-Green Algae/INDIA) *2(7)*:277
Pac, Timothy (IT Corporation/USA) *2(6)*:101
Palmer, Tracy (Applied Power Concepts, Inc./USA) *2(4)*:101
Palumbo, Anthony V. (Oak Ridge National Laboratory/USA) *2(4)*:389, 395
Panka, Brian (University of Missouri-Rolla/USA) *2(7)*:25

Papic, Sanja (University of Zagreb/CROATIA) *2(5)*:215
Pardue, John H. (Louisiana State University/USA) *2(4)*:279
Parker, Beth L. (University of Waterloo/CANADA) *2(6)*:135
Parker, Gary (Los Alamos National Laboratory/USA) *2(7)*:319
Parkin, Gene F. (The University of Iowa/USA) *2(7)*:107
Peabody, Carey E. (Erler & Kalinowski, Inc./USA) *2(6)*:117
Pearsall, Lorraine J. (CPF Associates, Inc./USA) *2(3)*:41
Peeples, James A. (Metcalf & Eddy, Inc./USA) *2(4)*:221
Pehlivan, Mehmet (Tait Environmental Management, Inc./USA) *2(7)*:41
Pennell, Kurt D. (Georgia Institute of Technology/USA) *2(2)*:211
Peramaki, Matthew P. (Leggette, Brashears & Graham, Inc./USA) *2(5)*:101
Peterson, David M. (Duke Engineering & Services, Inc./USA) *2(3)*:33, 193
Peterson, Michelle L. (AGRA Earth & Environmental, Inc./USA) *2(1)*:79; *2(3)*:129
Petrucci, Elisabetta (Universita degli Studi "La Sapienza"/ITALY) *2(7)*:245
Peven-McCarthy, Carole (Battelle /USA) *2(7)*:393
Pfiffner, Susan M. (University of Tennessee/USA) *2(4)*:389
Phelps, Tommy J. (Oak Ridge National Laboratory/USA) *2(4)*:389
Phillips, Theresa (GRACE Bioremediation Technologies/CANADA) *2(7)*:285
Piana, Michael J. (U.S. Geological Survey /USA) *2(6)*:281
Piazza, Mark (Elf Atochem North America/USA) *2(7)*:285
Pirelli, Tony (Van Water & Rogers, Inc./USA) *2(4)*:23
Piveteau, Pascal (Institut Francais du Petrole/FRANCE) *2(4)*:141

Plaehn, William A. (Parsons Engineering Science, Inc./USA) *2(5):*113
Plaza, E. (Universidad Complutense de Madrid/SPAIN) *2(7):*237
Pohlman, John W. (Geo-Centers, Inc./USA) *2(7):*189
Pohlmann, Dirk (IT Corporation/USA) *2(3):*121
Poland, John (Queens University/CANADA) *2(1):*41
Pope, Gary A. (University of Texas at Austin/USA) *2(2):*85, 187, 203
Pound, Michael J. (U.S. Navy/USA) *2(3):*81
Powers, Susan E. (Clarkson University/USA) *2(7):*57
Pratt, Randy (CH2M Hill, Inc./USA) *2(4):*303
Preziosi, Damian V. (CPF Associates, Inc./USA) *2(3):*41
Pritchard, P. H. (Hap) (U.S. Navy/USA) *2(7):*145
Ptak, Thomas (University of Tuebingen/GERMANY) *2(1):*95

Quinn, Jacqueline W. (National Aeronautics & Space Admin/USA) *2(6):*385
Quinnan, Joseph (Envirogen/USA) *2(4):*157, 319

Radel, Steve (Beazer East, Inc./USA) *2(4):*337
Radosevich, Mark (University of Delaware/USA) *2(6):*425
Raes, Eric (Engineering & Land Planning Assoc., Inc./USA) *2(2):*181; *2(4):*337
Rafferty, Michael T. (S.S. Papadopulos & Associates, Inc./USA) *2(4):*485
Ramirez, Jorge (IT Corporation/USA) *2(6):*125
Ramsburg, C. Andrew (Georgia Institute of Technology/USA) *2(2):*211
Ramsdell, Mark (Haley & Aldrich Inc./USA) *2(4):*255

Randall, Andrew A. (University of Central Florida/USA) *2(4):*327; *2(6):*323
Ranney, Thomas (Science & Technology Corporation/USA) *2(1):*87
Rao, Venkat (BioManagement Services, Inc./USA) *2(6):*177
Rathfelder, Klaus M. (University of Michigan/USA) *2(2):*77
Rawson, James R.Y. (General Electric Corporate R&D/USA) *2(2):*149
Reardon, Ken (Colorado State University/USA) *2(5):*301
Reifenberger, Trish (Brown and Caldwell/USA) *2(5):*191
Reinhard, Martin (Stanford University/USA) *2(5):*229
Reinhart, Debra R. (University of Central Florida/USA) *2(6):*323, 385
Reitman, Christopher T. (Advanced GeoServices Corp/USA) *2(1):*231
Reitsma, Stanley (University of Windsor/CANADA) *2(6):*25, 33
Repta, Cory (University of Waterloo/CANADA) *2(6):*393
Reynolds, David A. (Queen's University/CANADA) *2(2):*165
Rhine, E. Danielle (University of Delaware/USA) *2(6):*425
Rice, Barry N. (Roy F. Weston, Inc./USA) *2(5):*93
Richards, Sarah A. (Lehigh University/USA) *2(6):*249
Richardson, William K. (Advanced GeoServices Corp/USA) *2(1):*231
Richter, Amy P. (University of Cincinnati/USA) *2(4):*191
Ridley, Maureen (Lawrence Livermore National Lab./USA) *2(1):*111
Rifai, Hanadi S. (University of Houston/USA) *2(1):*319
Riha, Brian D. (Westinghouse Savannah River Company/USA) *2(1):*175; *2(2):*33, 41
Riis, Charlotte (NIRAS Consulting Engineers & Planners A/S/DENMARK) *2(3):*9

Rijnaarts, Huub H.M. (TNO Institute of Environmental Science/THE NETHERLANDS) *2(4):*63
Ringelberg, David (U.S. Army/USA) *2(7):*145
Rittmann, Bruce E. (Northwestern University/USA) *2(3):*1, 25
Roberts, Eric (Excalibur Group/USA) *2(6):*169
Robinson, Sandra (Virginia Polytechnic Inst & State Univ/USA) *2(4):*493
Rohrer, William L. (URS Greiner/USA) *2(4):*467
Rosansky, Stephen H. (Battelle/USA) *2(1):*339; *2(2):*49
Rose, Thomas M. (CES/USA) *2(1):*103
Rossabi, Joseph (Westinghouse Savannah River Company/USA) *2(1):*175; *2(2):*33, 41, 59
Rowland, Ryan C. (U.S. Geological Survey /USA) *2(6):*281
Rudolf, Elizabeth (City College of New York/USA) *2(7):*327
Ruiz, Nancy (GeoSyntec Consultants/USA) *2(6):*385
Rush, Richard (XCG Environmental Services/CANADA) *2(5):*135
Rutter, Allison (Queen's University/CANADA) *2(1):*41

Sabatini, David A. (University of Oklahoma/USA) *2(2):*219
Salas, O. (Gaiker/SPAIN) *2(7):*293
Sanchez, M. (Universidad Complutense de Madrid/SPAIN) *2(7):*237
Sandefur, Craig A. (Regenesis/USA) *2(4):*39, 213, 373
Sanford, William E. (Colorado State University/USA) *2(2):*69
Santillan, Javier (U.S. Air Force/USA) *2(1):*369
Sass, Bruce M. (Battelle/USA) *2(2):*49; *2(6):*409
Sayles, Gregory D. (U.S. EPA/USA) *2(1):*61
Schad, Hermann (I.M.E.S. GmbH/GERMANY) *2(6):*315

Schirmer, Mario (UFZ Centre for Environmental Research Leipzig-Halle/GERMANY) *2(1):*95
Schmidt, Robin (Wisconsin Department of Natural Resources/USA) *2(1):*9
Schneider, William H. (Roy F. Weston Inc./USA) *2(4):*477
Schreier, Cindy G. (PRIMA Environmental/USA) *2(6):*49, 57
Schroder, David L. (SECOR International, Inc/USA) *2(6):*91
Schuhmacher, Thea (LFR Levine Fricke/USA) *2(4):*15
Schulze, Bertram (ARCADIS Trischler & Partner/GERMANY) *2(6):*315
Schwartz, Franklin (The Ohio State University/USA) *2(2):*125; *2(6):*1, 17, 41
Scrocchi, Susan (Conestoga-Rovers & Associates/USA) *2(6):*161
Seagren, Eric A. (University of Maryland/USA) *2(1):*125
Seech, Alan G. (GRACE Bioremediation Technologies/CANADA) *2(7):*285
Sei, Kazunari (Yamanashi University/JAPAN) *2(1):*183
Semprini, Lewis (Oregon State University/USA) *2(4):*303; *2(5):*67
Senick, Maira (Watervliet Arsenal) *2(6):*273; *2(7):*213
Seol, Yongkoo (The Ohio State University/USA) *2(6):*17
Sepic, Ester (University of Ljubljana/SLOVENIA) *2(4):*123
Serna, Carlos (Roy F. Weston, Inc./USA) *2(4):*373
Sewell, Guy W. (U.S. EPA/USA) *2(5):*269
Sfeir, Hala (University of Central Florida/USA) *2(6):*323
Shah, Sunil (Union Carbide Corporation/USA) *2(3):*145
Shangraw, Timothy C. (Parsons Engineering Science, Inc./USA) *2(5):*113
Sharma, Kusumakar (Asian Institute of Technology/THAILAND) *2(4):*85
Sharp, Marietta (University of Washington/USA) *2(4):*467

Sheldon, Jack K. (Montgomery Watson/USA) *2(6):*347
Sherwood-Lollar, Barbara (University of Toronto/CANADA) *2(3):*17
Shiau, Bor-Jier (Ben) (Surbec-ART Environmental, LLC/USA) *2(2):*219
Shoemaker, Christine (Cornell University/USA) *2(4):*311
Shrout, Joshua D. (University of Iowa/USA) *2(7):*107
Siegrist, Robert (Colorado School of Mines/USA) *2(6):*67, 75, 117
Silva, Jeff A.K. (Duke Engineering & Services/USA) *2(2):*195
Simon, Michelle A. (U.S. EPA/USA) *2(5):*43
Singhal, Naresh (University of Auckland/NEW ZEALAND) *2(1):*223; *2(4):*115
Singletary, Michael A. (Duke Engineering & Services, Inc./USA) *2(1):*61; *2(3):*193
Sirivithayapakorn, Sanya (University of California/USA) *2(7):*73
Skeen, Rodney S. (Battelle PNWD/USA) *2(4):*229
Sklarew, Deborah S. (Pacific Northwest National Laboratories/USA) *2(6):*369
Slack, William W. (FRx Inc./USA) *2(5):*127; *2(6):*291
Slater, Greg F. (University of Toronto/CANADA) *2(3):*17
Sleep, Brent (University of Toronto/CANADA) *2(3):*17
Slenska, Michael (Beazer East, Inc./USA) *2(2):*1; *2(7):*189
Sminchak, Joel R. (Battelle/USA) *2(1):*47; *2(2):*49
Smith, Ann P. (Radian International/USA) *2(1):*117
Smith, Gregory J. (Radian International/USA) *2(1):*149; *2(5):*167, 183; *2(6):*217
So, Juho (Drycleaner Environmental Response Trust Fund/USA) *2(1):*23
Sorensen, H. (Storstroems County/DENMARK) *2(5):*83

Sowers, Hank (Steamtech Environmental Services/USA) *2(5):*149
Spadaro, Jack T. (AGRA Earth & Environmental Inc/USA) *2(3):*129
Spain, Jim (U.S. Air Force/USA) *2(3):*175
Spargo, Barry J. (U.S. Navy/USA) *2(7):*189
Spivack, Jay L. (GE Corporate R&D Center/USA) *2(3):*17
Srinivasan, S.V. (Central Pollution Control Board/INDIA) *2(4):*413
Sriwatanapongse, Watanee (Stanford University/USA) *2(5):*229
Stainsby, Ross R. (Camp Dresser & McKee/USA) *2(1):*387
Stark, Jeffrey A. (U.S. Army Corps of Engineers/USA) *2(6):*109
Steager, Claire (Rowan University/USA) *2(3):*65
Steele, Julia K. (GEO-CENTERS, Inc./USA) *2(7):*189
Steffan, Robert J. (Envirogen, Inc./USA) *2(4):*93, 157, 165
Stegemeier, George L. (GLS Engineering Inc./USA) *2(5):*197
Steiner, Michael F. (Parsons Engineering Science, Inc./USA) *2(5):*113
Steiof, Martin (Technical University of Berlin/GERMANY) *2(6):*433
Stening, James (SHE Pacific Pty Limited/AUSTRALIA) *2(6):*401
Stensel, H. David (University of Washington/USA) *2(4):*455
Stephens, Daniel B. (Daniel B. Stephens & Associates, Inc./USA) *2(3):*153
Stotler, Chris (URS Greiner Woodward Clyde/USA) *2(2):*17
Stout, Scott (Battelle/USA) *2(7):*393
Strand, Stuart E. (University of Washington/USA) *2(4):*455
Stringfellow, William T. (Lawrence Berkeley National Laboratory/USA) *2(4):*175
Struse, Amanda M. (The IT Group/USA) *2(6):*67
Studer, James E. (Duke Engineering & Services/USA) *2(1):*61; *2(3):*193

Sturchio, Neil C. (Argonne National Laboratory/USA) *2(1):*149
Suidan, Makram T. (University of Cincinnati/USA) *2(4):*191; *2(7):*357
Suk, Heejun (Penn State University/USA) *2(2):*93
Sullivan, Mark (RRM, Inc./USA) *2(7):*65
Sullivan, Tom (Bascor Environmental, Inc./USA) *2(4):*23
Susaeta, Inaki (Gaiker/SPAIN) *2(7):*293
Sutherland, Justin (University of Missouri-Rolla/USA) *2(7):*25
Suthersan, Suthan (ARCADIS Geraghty & Miller, Inc./USA) *2(4):*263; *2(5):*141
Sweeney, Robert (Unocal/USA) *2(1):*191
Swingle, Todd (Parsons Engineering Science/USA) *2(4):*1; *2(5):*9
Szecsody, James E. (Pacific Northwest National Laboratories/USA) *2(6):*369
Szerdy, Frank S. (Geomatrix Consultants, Inc./USA) *2(2):*1

Tabak, Henry H. (U.S. EPA/USA) *2(7):*307, 357
Taer, Andrew (Fugro GeoSciences, Inc./USA) *2(2):*1
Tagawa, Lori T. (Waste Management, Inc./USA) *2(5):*113
Takamizawa, Kazuhiro (Gifu University/JAPAN) *2(4):*197
Talley, Jeffrey W. (U.S. Army Corps of Engineers/USA) *2(7):*213
Tarr, Matthew A. (University of New Orleans/USA) *2(6):*181
Tate, Tim (Conestoga-Rovers & Associates/USA) *2(7):*301
Taylor, Christine (North Carolina State University/USA) *2(4):*149, 295
Taylor, Kristin A. (URS Greiner Woodward Clyde/USA) *2(4):*295
Taylor, Brian S. (URS Greiner Woodward Clyde/USA) *2(4):*295
Teutsch, Georg (University of Tuebingen/GERMANY) *2(1):*95

Thompson, Bruce R. (de maximis, inc./USA) *2(4):*461
Thompson, Leslie C. (Pintail Systems, Inc./USA) *2(7):*327
Thompson, Paul D. (Boeing/USA) *2(3):*97
Thomson, Neil R. (University of Waterloo/CANADA) *2(2):*157; *2(6):*9, 83
Thorpe, Steve (Radian International/USA) *2(1):*61
Tischuk, Michael D. (Hanson North America/USA) *2(1):*287; *2(7):*139
Togna, A. Paul (Envirogen Inc/USA) *2(7):*115
Tomasko, David (Argonne National Laboratory/USA) *2(5):*121
Tomson, Mason B. (Rice University/USA) *2(1):*239
Tornatore, Paul M. (Haley & Aldrich, Inc./USA) *2(6):*209; *2(7):*57
Tossell, Robert W. (GeoSyntec Consultants/CANADA) *2(4):*485
Tovanabootr, Adisorn (Oregon State University/USA) *2(5):*67
Toy, Patrick C. (University of Central Florida/USA) *2(6):*385
Travis, Bryan (Los Alamos National Laboratory/USA) *2(4):*429
Trasarti, Fausto (Universita degli Studi "La Sapienza"/ITALY) *2(7):*245
Trevors, Jack T. (University of Guelph/CANADA) *2(7):*285
Trippler, Dale (Minnesota Pollution Control Agency/USA) *2(1):*9
Trone, Paul M. (AGRA Earth & Environmental/USA) *2(1):*79; *2(3):*129
Tuhkanen, Tuula (Tampere University of Technology/FINLAND) *2(7):*343
Tunks, John. (Parsons Engineering Science, Inc./USA) *2(1):*369
Turpie, Andrea (Envirogen/USA) *2(4):*319
Tuta, Zane H. (IT Corporation/USA) *2(6):*265

Unz, Richard (Pennsylvania State University/USA) *2(7):*87

Urynowicz, Michael A. (Envirox LLC/USA) *2(6):*75, 117
Utgikar, Vivek P. (U.S.EPA/USA) *2(7):*307

Vail, Christopher H. (Focus Environmental, Inc./USA) *2(1):*207
Vainberg, Simon (Envirogen, Inc./USA) *2(4):*165
VanBriesen, Jeanne Marie (Carnegie Mellon University/USA) *2(3):*25
Vancheeswaran, Sanjay (CH2M Hill/USA) *2(4):*303
Vandecasteele, Jean-Paul (Institut Francais du Petrole/FRANCE) *2(7):*205
Vartiainen, Terttu (National Public Health Institute & University of Kuopio/FINLAND) *2(7):*343
Venosa, Albert (U.S. EPA/USA) *2(4):*191
Vermeul, Vince R. (Pacific Northwest National Laboratories/USA) *2(6):*369
Vierkant, Gregory P. (Lucent Technologies, Inc./USA) *2(6):*217
Vilardi, Christine L. (STV Incorporated/USA) *2(1):*199
Vincent, Jennifer C. (Earth Tech/USA) *2(4):*437
Vinegar, Harold J. (Shell E&P Technology/USA) *2(5):*197
Vogan, John L. (EnviroMetal Technologies Inc/CANADA) *2(6):*401, 417
Vogt, Carsten (UFZ-Centre for Environmental Research/ GERMANY) *2(4):*133
Vroblesky, Don A. (U.S. Geological Survey/USA) *2(1):*349; *2(7):*17

Waisner, Scott A. (TA Environmental, Inc./USA) *2(7):*213
Wallace, Mark N. (U.S. Army Corps of Engineers/USA) *2(7):*269
Wallis, B. Renee Pahl (U.S. Navy/USA) *2(4):*467
Wallis, F.M. (University of Natal/REP OF SOUTH AFRICA) *2(7):*131

Walsh, Matthew (Envirogen, Inc./USA) *2(4):*157
Walti, Caryl (Northgate Environmental Mgt, Inc./USA) *2(6):*49, 57
Wanty, Duane (The Gillette Company/ USA) *2(4):*405
Warburton, Joseph M. (Metcalf & Eddy, Inc./USA) *2(4):*221
Ware, Leslie (Anniston Army Depot/USA) *2(6):*153
Warith, Mustafa A. (University of Ottawa/CANADA) *2(4):*381
Warner, Scott D. (Geomatrix Consultants, Inc./USA) *2(4):*361
Warren, Randall J. (Shell Canada Products, Ltd./USA) *2(5):*207
Watkinson, Robert J. (Sheffield University/UNITED KINGDOM) *2(4):*183
Weeber, Phil (HSI Geotrans/USA) *2(4):*429
Weiss, Holger (Centre for Environmental Research/GERMANY) *2(6):*331
Wellendorf, William G. (Southwest Ground-water Consultants, Inc./ USA) *2(7):*161
Wells, Samuel L. (Golder Sierra LLC/USA) *2(6):*307
Werner, Peter (Technische Universitat Dresden/GERMANY) *2(7):*205
West, Brian (U.S. Army Corps of Engineers/USA) *2(7):*269
Westerheim, Michael (Unisys Corporation/USA) *2(1):*387
Weston, Alan (Conestoga-Rovers & Associates/USA) *2(6):*161; *2(7):*301
Wharry, Stan (U.S. Army/USA) *2(7):*81
Whiter, Terri M. (Focus Environmental, Inc./USA) *2(1):*207
Wickramanayake, Godage B. (Battelle/USA) *2(1):*339
Widdowson, Mark A. (Virginia Polytechnic Inst & State Univ/USA) *2(4):*493
Wiedemeier, Todd H. (Parsons Engineering Science, Inc./USA) *2(1):*357; *2(3):*81; *2(4):*1
Wildenschild, Dorthe (Lawerence Livermore National Laboratory/ USA) *2(5):*277

Wildman, Mathew J. (The University of Iowa/USA) *2(6):*339
Williams, Gustavious P. (Argonne National Laboratory/USA) *2(5):*121
Williams, Jerry (ENSR Corporation/USA) *2(7):*269
Williams, Mark D. (Pacific Northwest National Lab./USA) *2(6):*369
Williamson, Dean F. (CH2M Hill/USA) *2(6):*257
Williamson, Travis (Battelle/USA) *2(3):*137
Willis, Matthew B. (Cornell University/USA) *2(4):*311
Wilson, Glenn (Desert Research Institute/USA) *2(4):*395
Wilson, Gregory W. (University of Cincinnati/USA) *2(4):*191
Wilson, James T. (Geo-Cleanse International, Inc./USA) *2(5):*307; *2(6):*153
Wilson, John T. (U.S. EPA/USA) *2(3):*57; *2(7):*1
Witt, Michael E. (The Dow Chemical Co,/USA) *2(3):*17
Wolf, Michael (Earth Tech/USA) *2(5):*85
Wood, A. Lynn (U.S. EPA/USA) *2(2):*101; *2(5):*269
Woodward, David S. (Earth Tech/USA) *2(1):*401
Woody, Bernard A. (United Technologies Research Center/USA) *2(6):*145
Wright, Judith (UFA Ventures, Inc./USA) *2(7):*319
Wrobel, John G. (U.S. Army/USA) *2(4):*477
Wu, Jun (The Pennsylvania State University/USA) *2(7):*87

Wuensche, Lothar (UFZ-Centre for Environmental Research/GERMANY) *2(4):*133
Wyatt, Jeff (Chevron Chemical Company/USA) *2(6):*257

Xu, Guoxiang (University of New Orleans/USA) *2(6):*181

Yang, Ruiqiang (Northwest Normal University/CHINA) *2(6):*233
Yeates, Robert M. (IT Corporation/USA) *2(6):*265
Yeh, G.T. (Penn State University/USA) *2(2):*93
Yeh, S. Laura (U.S. Navy/USA) *2(2):*187, 203, 219
Yen, Teh Fu (University of Southern California/USA) *2(6):*195
Yoon, Woong-Sang (Sam) (Battelle/USA) *2(6):*409
Young, Byron L. (U.S. Army Corps of Engineers/USA) *2(6):*109
Yu, Chung-Ching (Lamar University/USA) *2(5):*293

Zafiro, Alan (IT Corporation/USA) *2(7):*379
Zahiraleslamzadeh, Zahra (FMC Corporation/USA) *2(4):*237
Zenker, Matthew J. (North Carolina State University/USA) *2(4):*47, 421
Zhang, Hubao (Duke Engineering & Services/USA) *2(3):*33; *2(6):*1
Zhang, Husen (The Pennsylvania State University/USA) *2(7):*87
Zhang, Weixian (Lehigh University/USA) *2(6):*249

2000 KEYWORD INDEX

This index contains keyword terms assigned to the articles in the seven books published in connection with the Second International Conference on Remediation of Chlorinated and Recalcitrant Compounds, held in Monterey, California, in May 2000. Ordering information is provided on the back cover of this book.

In assigning the terms that appear in this index, no attempt was made to reference all subjects addressed. Instead, terms were assigned to each article to reflect the primary topics covered by that article. Authors' suggestions were taken into consideration and expanded or revised as necessary to produce a cohesive topic listing. The citations reference the seven books as follows:

2(1): Wickramanayake, G.B., A.R. Gavaskar, M.E. Kelley, and K.W. Nehring (Eds.), *Risk, Regulatory, and Monitoring Considerations: Remediation of Chlorinated and Recalcitrant Compounds.* Battelle Press, Columbus, OH, 2000. 438 pp.

2(2): Wickramanayake, G.B., A.R. Gavaskar, and N. Gupta (Eds.), *Treating Dense Nonaqueous-Phase Liquids (DNAPLs): Remediation of Chlorinated and Recalcitrant Compounds.* Battelle Press, Columbus, OH, 2000. 256 pp.

2(3): Wickramanayake, G.B., A.R. Gavaskar, and M.E. Kelley (Eds.), *Natural Attenuation Considerations and Case Studies: Remediation of Chlorinated and Recalcitrant Compounds.* Battelle Press, Columbus, OH, 2000. 254 pp.

2(4): Wickramanayake, G.B., A.R. Gavaskar, B.C.Alleman, and V.S. Magar (Eds.) *Bioremediation and Phytoremediation of Chlorinated and Recalcitrant Compounds.* Battelle Press, Columbus, OH, 2000. 538 pp.

2(5): Wickramanayake, G.B. and A.R. Gavaskar (Eds.), *Physical and Thermal Technologies: Remediation of Chlorinated and Recalcitrant Compounds.* Battelle Press, Columbus, OH, 2000. 344 pp.

2(6): Wickramanayake, G.B., A.R. Gavaskar, and A.S.C. Chen (Eds.), *Chemical Oxidation and Reactive Barriers: Remediation of Chlorinated and Recalcitrant Compounds.* Battelle Press, Columbus, OH, 2000. 470 pp.

2(7): Wickramanayake, G.B., A.R. Gavaskar, J.T. Gibbs, and J.L. Means (Eds.), *Case Studies in the Remediation of Chlorinated and Recalcitrant Compounds.* Battelle Press, Columbus, OH, 2000. 430 pp.

A

α-ketoglutarate-dependent cleavage **2(7):**229
abiotic release date **2(7):**181
acetate **2(4):**23, 107, 389, 437
acid mine drainage, *see* mine waste
acid-enhanced degradation **2(7):**33
acridine orange (AO) **2(6):**233
actinomycetes **2(4):**455

activated carbon **2(6):**257, 315
advanced oxidation technology (AOT) **2(6):**201, 209, 217, 225, 233, 241, 249; **2(7):**25
aeration **2(5):**237
air monitoring **2(1):**207
air sparging, *see* sparging
air stripping **2(5):**293

algae **2(7)**:101
alluvium **2(2)**:17
amendment recipes **2(7)**:81
AOT, *see* advanced oxidation technology
AOX method **2(5)**:215
AquaBlock™ **2(7)**:369
aromatic compounds (*see also specific chemicals*) **2(1)**:149, 183; **2(6)**:181
Arrenhius equation **2(7)**:245
arsenic **2(4)**:485
attenuation, *see* natural attenuation
azo dyes **2(4)**:85; **2(6)**:201

B

barometric pumping **2(5)**:83
barriers, *see* permeable reactive barriers
bedrock **2(6)**:217
    fractured **2(2)**:25, 149, 173; **2(4)**:405
benzene (*see also* benzene, toluene, ethylbenzene, and xylenes) **2(1)**:215; **2(4)**:183
benzene, toluene, ethylbenzene, and xylenes (BTEX) (*see also specific chemicals*) **2(1)**:167, 191; **2(4)**:165, 337; **2(6)**:187, 291; **2(7)**:17, 41
bioattenuation **2(6)**:323, 347
bioaugmentation **2(4)**:77, 389, 395, 405
bioavailability **2(1)**:167, 239, 245; **2(3)**:209; **2(7)**:181, 357
biobarrier **2(4)**:71, 115; **2(6)**:347
bioclogging **2(4)**:115
biodegradation **2(3)**:17, 25, 65, 89, 97, 113, 121, 129, 145, 153, 175; **2(4)**:15, 31, 123, 141, 213, 287, 311, 501; **2(5)**:113, 167, 261, 269, 301; **2(6)**:339; **2(7)**:107, 115, 123, 139, 145, 161, 181, 189, 197, 205, 213, 229, 277, 285, 357
    aerobic biodegradation **2(3)**:121; **2(4)**:107, 373; **2(7)**:139
    anaerobic biodegradation **2(3)**:121, 145, 169; **2(4)**:1, 55, 63, 77, 107, 139, 373; **2(7)**:1, 139, 385
    anaerobic/aerobic biodegradation **2(3)**:81; **2(4)**:85, 319, 327, 353

    enhanced anaerobic biodegradation **2(4)**:9, 93, 221, 205, 229, 237, 245, 255, 279, 287, 295, 303, 319, 405
biomineralization **2(7)**:327
biopolymer slurry **2(6)**:265
bioreactor **2(4)**:107, 165, 175, 191, 347, 353, 421, 455; **2(7)**:101, 115
bioslurping **2(1)**:339; **2(5)**:7, 85
bioslurry **2(7)**:379
biosorption of metals **2(7)**:307
biosparging, *see* sparging
bioventing **2(1)**:61
biowall, *see* biobarrier
BiOx® process **2(6)**:177
brownfields **2(1)**:33
BTEX, *see* benzene, toluene, ethylbenzene, and xylenes
Burkholdera cepacia **2(4)**:141

C

capping **2(7)**:369
carbaryl **2(7)**:277
carbon tetrachloride **2(4)**:93, 389; **2(6)**:353
catalytic reaction **2(5)**:229; **2(6)**:225, 331
cathodic hydrogen **2(6)**:339
cDCE, *see* dichloroethene (DCE)
characterization **2(1)**:79; **2(2)**:17, 41, 49
chemical flushing **2(2)**:125
chemical oxidation, *see* oxidation
chemical stabilization **2(5)**:221; **2(7)**:301, 319
chlorinated benzene **2(3)**:175
chlorinated solvents (*see also specific chemicals*) **2(1)**:103, 117, 263; **2(3)**:97, 113; **2(4)**:31, 237, 287; **2(6)**:49, 57, 153, 323
chlorobenzene **2(3)**:175; **2(4)**:133, 337
chloroethanes (*see also specific chemicals*) **2(3)**:169; **2(5)**:27
chloroethenes (*see also specific chemicals*) **2(3)**:89; **2(5)**:27
chloroform **2(1)**:199; **2(4)**:93
chromium (*see also* metals) **2(6)**:49
circulation wells, *see* groundwater circulation wells
cis-1,2-dichloroethene, *see* dichloroethene

Clostridium bifermentans **2(4)**:197
coal tar **2(1)**:157
cold regions **2(1)**:41; **2(7)**:269
column testing **2(4)**:381
cometabolism **2(1)**:61; **2(4)**:141, 149, 157, 183, 421, 429, 437, 445, 455; **2(5)**:67
community participation **2(3)**:1
composting **2(7)**:81, 251
cone penetrometer testing (CPT) **2(2)**:1, 33, 41, 59
containment **2(4)**:477
contaminated sediments, *see* sediments
copper (*see also* metals) **2(7)**:307
corn oil **2(4)**:47
corn syrup **2(4)**:93
cosolvent **2(5)**:269
cost **2(1)**:103; **2(2)**:211; **2(4)**:55; **2(7)**:49
CPT, *see* cone-penetrometer testing
creosote **2(1)**:255, 287, 303; **2(2)**:1; **2(4)**:493; **2(7)**:131, 145
crosshole flow **2(5)**:141
cyanobacteria **2(4)**:347
cyclohexane **2(4)**:183
cytochrome P450 **2(4)**:183

D

DARAMEND® **2(7)**:285
data management system (DMS) **2(1)**:111
data quality objectives **2(1)**:141
DCA, *see* dichloroethane
DCE, *see* dichloroethene
decay rate **2(7)**:9
deep reactive barrier, *see* permeable reactive barrier
degradation, *see* biodegradation
dehalogenase **2(4)**:197
dehydrogenation **2(6)**:249
dense, nonaqueous-phase liquids (DNAPLs), *see subheading under* nonaqueous-phase liquids
dichlorobenzene **2(3)**:175
2,2-dichlorodiisopropylether **2(4)**:107
dichloroethane (DCA) **2(3)**:217; **2(5)**:67, 75
dichloroethene (DCE) **2(3)**:153; **2(4)**:47, 197, 221, 221, 229, 237, 255, 287, 295, 303, 319, 327, 373, 405, 455; **2(5)**:19, 67, 75, 93, 113, 245, 307; **2(6)**:49, 409
dichloromethane **2(4)**:197
dichloropropane **2(4)**:107, 197
diesel **2(5)**:85
diffusion samplers **2(1)**:349, 369, 377
diffusion transport cell **2(6)**:67
1,4-dioxane **2(4)**:421
dioxins **2(1)**:207; **2(3)**:161; **2(7)**:169, 349
DIPE **2(4)**:183
dipole-flow test **2(5)**:43
DNAPLs, *see subheading under* nonaqueous-phase liquid
dry cleaning **2(1)**:9, 23; **2(4)**:205, 213, 245
dual-phase extraction (*see also* multiphase extraction) **2(4)**:23; **2(5)**:93; **2(7)**:41

E

EDBE **2(4)**:183
edible oil **2(4)**:77
electrical heating, *see* soil heating technologies
electrokinetics **2(5)**:277
electron acceptors **2(4)**:63, 133
electron beam **2(6)**:209
electron donors **2(4)**:1, 15, 23, 39, 47, 55, 71, 93, 107, 263, 279, 319
electroosmosis **2(5)**:277
enhanced anaerobic biodegradation, *see* biodegradation
enhanced reductive dechlorination **2(4)**:205, 213, 263, 271, 361
environmental forensics **2(1)**:157, 191
error analysis **2(2)**:85
ethane **2(4)**:221
ethanol **2(4)**:93, 141, 353
ethene **2(3)**:217; **2(4)**:221, 229, 237, 255, 287, 295, 405
explosives **2(1)**:33, 87; **2(6)**:339; **2(7)**:81, 87

F

Fenton's oxidation **2(5)**:301, 307; **2(6)**:91, 153, 161, 169, 177, 181, 187, 195; **2(7)**:197

fingerprinting, *see* environmental forensics
fluidized-bed reactor (FBR) **2(7)**:115
fluoranthene **2(4)**:123
flushing (*see also* surfactants) **2(2)**:195, 219; **2(3)**:33
forensic geochemistry **2(1)**:157
formate **2(4)**:141
fractured bedrock, *see* bedrock
fractured media **2(2)**:157, 165
free-product recovery **2(5)**:127
fuel oxygenates **2(4)**:175, 191
fungal treatment **2(4)**:413
funnel-and-gate **2(6)**:257, 315
furans **2(7)**:43

G

GAC, *see* granular activated carbon
gas chromatography (GC) **2(1)**:87; **2(4)**:123
gasoline **2(4)**:149; **2(5)**:207
GC, *see* gas chromatography
GCW, *see* groundwater circulation wells
gene probes **2(1)**:183
geographical information system (GIS) **2(1)**:111, 117, 125, 133, 141
geostatistics **2(2)**:77
GIS, *see* geographical information system
Gordonia termae **2(4)**:141
granular activated carbon (GAC) **2(5)**:135; **2(7)**:49
graphical tools **2(1)**:125
groundwater aeration, *see* sparging
groundwater circulation wells (GCW) **2(5)**:19, 51, 253
groundwater extraction **2(3)**:97
groundwater modeling **2(1)**:199; **2(7)**:9

H

heating, *see* soil heating technologies
heavy metals, *see* metals
heptachlorodibenzofuran **2(7)**:343
herbicides, *see* pesticides/herbicides
hexachlorocyclohexane (HCH) **2(7)**:285, 293
hexavalent chromium **2(3)**:185; **2(6)**:57
hot air injection **2(5)**:141

HRC™, *see* hydrogen release compound
hydration **2(5)**:285
hydraulic control **2(4)**:485
hydraulic fracturing **2(5)**:127; **2(6)**:291
hydrogen **2(1)**:393; **2(3)**:9, 57; **2(4)**:31, 255; **2(6)**:347, 433
hydrogen peroxide **2(4)**:327; **2(5)**:307; **2(6)**:153; **2(7)**:25
hydrogen release compound **2(4)**:15, 39, 101, 205, 213, 237, 255, 287, 295, 373
hydrolysis **2(6)**:249; **2(7)**:245
hydrous pyrolysis **2(5)**:149

I

immobilization **2(2)**:133; **2(7)**:327
immunoassay **2(7)**:221
ISCO (in situ chemical oxidation), *see* oxidation
indoor air **2(5)**:245
internal combustion engine **2(7)**:73
intrinsic bioremediation, *see* natural attenuation
in-well heater **2(5)**:141
in-well stripping **2(5)**:19; **2(7)**:41
iron, used as a permeable reactive barrier medium (*see also* permeable reactive barrier) **2(6)**:273, 291, 307, 339, 353, 369, 385, 393, 401, 409, 417, 425, 433; **2(7)**:107
iron-reduction **2(1)**:167; **2(3)**:65
ISCO (in situ chemical oxidation), *see* oxidation
isoalkanes **2(4)**:149
isobutane **2(4)**:149
isopropanol **2(4)**:437
isotope, carbon **2(3)**:17

K

kinetics **2(3)**:41, 201; **2(4)**:149; **2(5)**:1; **2(6)**:75, 245, 353, 393

L

lactate **2(4)**:15, 23, 39, 93, 101, 229, 245
land treatment **2(7)**:123
landfarming **2(7)**:131, 145, 213

landfill **2(1)**:223; **2(2)**:133; **2(4)**:115, 319, 361, 429
laser-induced fluorescence (LIF) **2(1)**:175; **2(2)**:59
leach model **2(5)**:221
leachate/leaching **2(1)**:223; **2(4)**:115; **2(7)**:93, 205
LIF, *see* laser-induced fluorescence
life-cycle design **2(1)**:339
light, nonaqueous-phase liquids (LNAPLs) *see subheading under* nonaqueous-phase liquids)
lindane, *see* pesticides/herbicides
LNAPL, *see* light, nonaqueous-phase liquids
long-term monitoring **2(1)**:1, 263, 311, 319, 327, 333, 339, 357
low-flow purging/sampling techniques **2(1)**:401
low-permeability **2(2)**:187

M

Mann-Kendall Regression **2(1)**:319
manufactured gas plant (MGP) **2(1)**:157; **2(7)**:205
market trends **2(1)**:33
mass spectrometry **2(4)**:123
mass transfer **2(2)**:157; **2(5)**:1; **2(6)**:75
MC, *see* methylene chloride
membrane **2(6)**:331; **2(7)**:73
membrane-induced probe (MIP) **2(2)**:59
mercury (*see also* metals) **2(7)**:301
metalloporphyrins **2(6)**:233
metals **2(3)**:209; **2(6)**:307, 361; 327; **2(7)**:293, 307, 319 , 327
methanol **2(4)**:107, 175
methanotroph **2(4)**:429, 437, 445
methylene chloride (MC) **2(1)**:401; **2(4)**:23
methyl tertiary butyl ether (MTBE) **2(1)**:33, 215; **2(4)**:141, 149, 157, 165, 175, 183, 191; **2(6)**:187, 195, 209, 249; **2(7)**:1, 9, 17, 25, 33, 41, 49, 57, 65, 73
MGP, *see* manufactured gas plant
microbial filter **2(4)**:381
microbial mats **2(4)**:347; **2(7)**:101
microbial transport **2(4)**:395
microcosm **2(4)**:133, 327

microsparging, *see* sparging
mine waste **2(7)**:307, 319
MIP, *see* membrane-induced probe
MNA (monitored natural attenuation), *see* natural attenuation
modeling (*see* also numerical modeling) **2(1)**:223; **2(3)**:25, 89, 113; **2(4)**:311, 429, 477; **2(5)**:1, 51; **2(6)**:33
molasses **2(4)**:221, 263
monitored natural attenuation (MNA), *see* natural attenuation
monitoring, *see* long-term monitoring
MPE **2(5)**:135
MTBE, *see* methyl tertiary butyl ether
multilevel sampling **2(1)**:95
multiphase extraction (*see also* dual-phase extraction) **2(5)**:101, 135
multiphase flow **2(2)**:93, 101
munitions, *see* explosives

N

naphthalene **2(4)**:381
NAPLs, *see* nonaqueous-phase liquids
National Research Council **2(3)**:1
natural attenuation **2(1)**:125, 149, 167, 263, 357, 393, 401; **2(3)**:1, 9, 17, 25, 33, 41, 49, 57, 65, 73, 81, 89, 97, 105, 113, 121, 129, 137, 145, 153, 161, 169, 175, 185, 193, 201, 209, 217; **2(4)**:1; 271, 337, 477; **2(5)**:307; **2(5)**:35; **2(6)**:217; **2(7)**:1, 393
natural organic matter (NOM) **2(6)**:181; **2(7)**:49
nitrate **2(4)**:389
nitrogen **2(4)**:337
nitrous oxide **2(4)**:437
NOM, *see* natural organic matter
nonaqueous-phase liquids (NAPLs) **2(2)**:9; **2(6)**:1, 33
  dense, nonaqueous-phase liquids (DNAPLs) **2(1)**:79, 175, 287, 387; **2(2)**:1, 9, 17, 25, 33, 41, 49, 59, 69, 77, 85, 93, 101, 109, 117, 125, 133, 141, 149, 157, 165, 173, 181, 187, 195, 203, 211, 219; **2(5)**:61, 141, 149, 167,

183; **2(6):**9, 17, 25, 41, 67, 83, 117, 125, 135, 153, 161
light, nonaqueous-phase liquids (LNAPLs) **2(1):**175, 279
numerical modeling **2(2):**77, 93, 125, 203; **2(5):**51, 121; **2(6):**1, 83
nutrients **2(4):**85, 337

O

off-gas treatment **2(1):**339
on-site analysis **2(1):**87
optimization **2(1):**339; **2(4):**311
ORC™, *see* oxygen release compound
oxidation (*see also* advanced oxidation technology *and* Fenton's oxidation) **2(2):**117; **2(5):**27, 149, 301, 307; **2(6):**1, 9, 25, 49, 57, 67, 75, 83, 91, 101, 109, 117, 125, 135, 145, 153, 161, 169, 177, 187; **2(7):**197, 237, 259, 349
oxygenase **2(4):**149
oxygenates **2(4):**183; **2(7):**33
oxygen release compound **2(4):**373
ozone **2(5):**27; **2(6):**187, 225

P

PAHs, *see* polycyclic aromatic hydrocarbons
palladium **2(5):**229, 253; **2(6):**331, 353
passive remediation **2(5):**83; **2(6):**135, 273
Pasteurella sp IFA **2(4):**123
PCA, *see* tetrachlorothane
PCBs, *see* polychlorinated biphenyls
PCE, *see* tetrachlorothene
PCP, *see* pentachlorophenol
pentachlorophenol (PCP) **2(1):**303; **2(5):**35, 301; **2(7):**123, 161, 169
perchlorate **2(6):**209; **2(7):**87, 93, 101, 107, 115
perchloroethane, *see* tetrachloroethane
perchloroethene, *see* tetrachloroethene
perchloroethylene, *see* tetrachloroethene
performance assessment **2(2):**9, 187
permanganate oxidation **2(6):**1, 9, 17, 25, 33, 41, 49, 57, 67, 75, 83, 91, 101, 109, 117, 125, 135, 145, 161

permeability reduction **2(4):**115
permeable barrier, *see* permeable reactive barrier
permeable reactive barrier (PRB) **2(4):**47; **2(6):**257, 265, 273, 281, 291, 299, 307, 315, 323, 361, 369, 377, 385, 393, 401, 409, 417, 425, 433; **2(7):**301
peroxide **2(6):**91
pesticides/herbicides **2(1):**33; **2(6):**257; **2(7):**229, 237, 245, 251, 259, 269, 277, 285
petroleum hydrocarbons **2(1):**157, 183, 191; **2(6):**225
phase transfer catalyst (PTC) **2(6):**17
phenanthrene (*see also* pesticides/herbicides) **2(6):**225
phosphorus **2(4):**337
photocatalytic oxidation **2(5):**135; **2(6):**233, 241
photodegradation **2(7):**343
photolytic oxidation, *see* photocatalytic oxidation
phytoremediation **2(4):**461, 467, 477, 485, 493, 501; **2(7):**17, 101
Pinellas culture **2(4):**405
partitioning interwell tracer test (PITT) **2(2):**9, 17, 69, 85, 195
PITT, *see* partitioning interwell tracer test
plume containment **2(1):**357
polychlorinated biphenyls (PCBs) **2(1):**41; **2(2):**173; **2(5):**175; **2(6):**161; **2(7):**337, 349, 385, 393
polycyclic aromatic hydrocarbons (PAHs) **2(1):**255; **2(4):**493; **2(5):**197, 301; **2(6):**225, 315; **2(7):**123, 145, 153, 181, 189, 197, 205, 213, 221, 357, 379, 393
polylactate **2(7):**87
poplar **2(4):**461, 467, 493
potassium permanganate, *see* permanganate oxidation
PRB, *see* permeable reactive barrier
propachlor (*see also* pesticides/herbicides) **2(7):**237
propane **2(4):**157; **2(5):**67
Pseudomonas cepacia **2(7):**277
Pseudomonas putida **2(4):**395
Pseudomonas stutzeri **2(4):**389
p-xylene **2(4):**175
pyrovate **2(4):**107

R

radionuclides **2(6)**:281
Raman spectroscopy **2(1)**:175
RBCA (Risk-Based Corrective Action), *see* risk
RDX, *see* explosives
reactive barrier, *see* permeable reactive barrier
reactive walls, *see* permeable reactive barrier
recirculation wells, *see* groundwater circulation wells
reductive dechlorination **2(3)**:121; **2(4)**:1, 15, 39, 47, 71, 77, 205, 245, 303, 311; **2(5)**:221, 269
reductive dehalogenation **2(4)**:353
refrigerant **2(1)**:199
remedial endpoint **2(1)**:339
rhizosphere **2(4)**:501
ribbon sampler **2(2)**:33
risk assessment/management **2(1)**:15, 199, 207, 215, 223, 231, 239, 245, 255, 263, 279

S

sequencing batch reactor (SBR) **2(4)**:85
SEAR, *see* surfactant-enhanced aquifer remediation
sediments **2(3)**:161, 369; **2(7)**:189, 357, 369, 379, 385, 393
SEM/AVS, *see* simultaneously extracted metal and acid volatile sulfide analysis
simultaneously extracted metal and acid volatile sulfide analysis (SEM/AVS) **2(3)**:209
site characterization **2(1)**:47, 61, 191; **2(2)**:33, 69; **2(3)**:65
site classification **2(1)**:15, 23
site closure **2(1)**:1, 231
Six-Phase Heating™ (SPH) (*see also* soil heating technologies) **2(5)**:167, 183, 191
slurry treatment **2(7)**:379
sodium permanganate, *see* permanganate oxidation
soil gas **2(1)**:215, 327

soil heating technologies (*see also* steam injection) **2(5)**:149, 157, 167, 175, 183, 191, 197, 207
soil vapor extraction (SVE) **2(1)**:387; **2(4)**:23; **2(5)**:75, 83, 113, 121, 207; **2(7)**:65, 73
solar-powered **2(5)**:237
solidification/stabilization, *see* chemical stabilization
solvent extraction **2(7)**:293
sonication **2(6)**:195, 201, 385
source treatment **2(2)**:125; **2(3)**:49; **2(4)**:295; **2(6)**:9, 217
source zone mapping **2(2)**:25
soybean oil **2(4)**:47
sparging **2(3)**:49; **2(4)**:31, 429; **2(5)**:1, 9, 27, 35, 61, 67, 75; **2(6)**:187; **2(7)**:65
SPH, *see* Six-Phase Heating™
stable isotope analysis **2(1)**:149
steam injection **2(5)**:141, 149, 167
sucrose **2(4)**:361
sulfate/sulfate reduction **2(3)**:65; **2(4)**:229; **2(6)**:361; **2(7)**:307, 385
surfactants **2(2)**:187, 195, 203, 211, 219; **2(7)**:153
surfactant-enhanced aquifer remediation (SEAR) **2(2)**:77, 195, 203
SVE, *see* soil vapor extraction
system optimization, *see* optimization

T

TAME **2(4)**:183
TBE **2(4)**:141, 149, 165, 183
TBF **2(4)**:149
TCA, *see* trichloroethane
TCE, *see* trichloroethene
TCFE **2(4)**:303
technical impracticability **2(1)**:279, 287, 295, 303
tetrachloroethane (PCA) **2(2)**:181; **2(4)**:279, 477
tetrachloroethene (PCE) **2(1)**:9, 23, 149, 349, 401; **2(2)**:101, 187, 211; **2(3)**:49; **2(4)**:15, 23, 55, 197, 205, 213, 245, 287, 303, 327, 347; **2(5)**:19, 61, 83, 113, 135, 141, 177, 221, 241, 269, 307; **2(6)**:409

tetrachloroethylene, *see* tetrachloroethene
tetrahydrofuran **2(4)**:421
thermal remediation technologies, *see* soil heating technologies
three-dimensional (3-D) groundwater flow measurement **2(1)**:53
TNT, *see* explosives
TOC, *see* total organic carbon
toluene (*see also* benzene, toluene, ethylbenzene, and xylenes) **2(4)**:175, 183
total organic carbon (TOC) **2(7)**:49
toxaphene **2(7)**:251
tracers (*see also* partitioning interwell tracer test) **2(4)**:303
TRAMPP **2(4)**:429
trichloroacetyl chloride (TCAC) **2(6)**:241
trichloroethane (TCA) **2(1)**:401; **2(3)**:57, 153; **2(4)**:23, 197, 437; **2(5)**:75, 167, 183; **2(6)**:49, 67, 75, 101; 109, 117, 145, 169
trichloroethene (TCE) **2(1)**:1, 149, 191, 231, 327, 349, 377; **2(2)**:25, 109, 141; **2(3)**:17, 57, 129, 153; **2(4)**:1, 44, 77, 101, 175, 197, 229, 237, 255, 287, 295, 303, 327, 347, 353, 405, 429, 437, 445, 455; **2(5)**:1, 19, 67, 75, 93, 113, 149, 167, 183, 229, 307; **2(6)**:17, 109, 125, 145, 217, 291, 299, 347, 353, 369, 385, 409, 417, 425
trichloroethylene, *see* trichloroethene
trichlorofluoromethane **2(1)**:199
triethyl phosphate **2(4)**:437
trihalomethanes (THMs) **2(6)**:117
trinitrotoluene (TNT), *see* explosives

U–Z

ultrasound, *see* sonication **2(6)**:195, 201, 385
ultraviolet (UV) **2(7)**:25, 343
uncertainty analysis **2(1)**:223; **2(3)**:41
underground storage tanks (USTs) **2(1)**:215
uranium (*see also* metals) **2(6)**:281
USTs, *see* underground storage tanks
UV, *see* ultraviolet
vacuum extraction **2(2)**:141; **2(5)**:135
vapor extraction, *see* soil vapor extraction
VC, *see* vinyl chloride
vegetable oil **2(4)**:1
vinyl chloride (VC) **2(3)**:121; **2(4)**:47, 221, 229, 237, 255, 287, 295, 303, 319, 373, 405, 429, 455; **2(5)**:9, 75, 93, 307
visualization tools **2(1)**:125; **2(2)**:157
vitamin **2(5)**:261
wastewater **2(4)**:85, 413; **2(5)**:215
water treatment **2(1)**:339; **2(7)**:49
wetland **2(4)**:279
willow **2(4)**:461; **2(7)**:101
wood treatment **2(7)**:123, 131, 139, 145, 153, 161, 169
zero-valent iron, *see* iron, used as a permeable reactive barrier medium
zinc (*see also* metals) **2(7)**:307